NEUROCHEMICAL ASPECTS OF ALZHEIMER'S DISEASE

NEUROCHEMICAL ASPECTS OF ALZHEIMER'S DISEASE

RISK FACTORS, PATHOGENESIS, BIOMARKERS, AND POTENTIAL TREATMENT STRATEGIES

AKHLAQ A. FAROOQUI

Department of Molecular and Cellular Biochemistry, The Ohio State University, Columbus, Ohio, United States

ACADEMIC PRESS

An imprint of Elsevier

Academic Press is an imprint of Elsevier
125 London Wall, London EC2Y 5AS, United Kingdom
525 B Street, Suite 1800, San Diego, CA 92101-4495, United States
50 Hampshire Street, 5th Floor, Cambridge, MA 02139, United States
The Boulevard, Langford Lane, Kidlington, Oxford OX5 1GB, United Kingdom

Notices
Knowledge and best practice in this field are constantly changing. As new research
and experience broaden our understanding, changes in research methods, professional
practices, or medical treatment may become necessary.

Practitioners and researchers must always rely on their own experience and knowledge
in evaluating and using any information, methods, compounds, or experiments described
herein. In using such information or methods they should be mindful of their own safety
and the safety of others, including parties for whom they have a professional responsibility.

To the fullest extent of the law, neither the Publisher nor the authors, contributors, or
editors, assume any liability for any injury and/or damage to persons or property as a
matter of products liability, negligence or otherwise, or from any use or operation of any
methods, products, instructions, or ideas contained in the material herein.

Library of Congress Cataloging-in-Publication Data
A catalog record for this book is available from the Library of Congress

British Library Cataloguing-in-Publication Data
A catalogue record for this book is available from the British Library

ISBN: 978-0-12-809937-7

For information on all Academic Press publications visit our website at
https://www.elsevier.com/books-and-journals

Working together
to grow libraries in
developing countries

www.elsevier.com • www.bookaid.org

Publisher: Nikki Levy
Acquisitions Editor: Melanie Tucker
Editorial Project Manager: Kristi Anderson
Production Project Manager: Chris Wortley
Designer: Matthew Limbert

Typeset by TNQ Books and Journals

This monograph is dedicated to my beloved father "late Sharafyab Ahmed Sahab" whose guidance and influence continue to inspire and support me.

Akhlaq A. Farooqui

Contents

3. Contribution of Neural Membrane Phospholipids, Sphingolipids, and Cholesterol in the Pathogenesis of Alzheimer's Disease

4. Contribution of Nucleic Acids in the Pathogenesis of Alzheimer's Disease

5. Type II Diabetes and Metabolic Syndrome as Risk Factors for Alzheimer's Disease

6. Contribution of Neuroinflammation in the Pathogenesis of Alzheimer's Disease

10. Perspective, Summary, and Directions for Future Research on Alzheimer's Disease

About the Author

Dr. Akhlaq A. Farooqui is a leader in the field of signal transduction; brain phospholipases A_2; bioactive ether lipid metabolism; polyunsaturated fatty acid metabolism; glycerophospholipid-, sphingolipid-, and cholesterol-derived lipid mediators; glutamate-induced neurotoxicity; and modulation of signal transduction by phytochemicals. Dr. Farooqui has discovered the stimulation of plasmalogen-selective phospholipase A_2 (PlsEtn-PLA$_2$) and diacyl- and monoacylglycerol lipases in brains from patients with Alzheimer's disease. Stimulation of PlsEtn-PLA$_2$ produces plasmalogen deficiency and increases levels of eicosanoids, which may be related to the loss of synapses in brains of patients with Alzheimer's disease. Dr. Farooqui has published cutting-edge research on the generation and identification of glycerophospholipid-, sphingolipid-, and cholesterol-derived lipid mediators in kainic acid–mediated neurotoxicity by lipidomics. Dr. Farooqui has authored 11 monographs: *Glycerophospholipids in Brain: Phospholipase A₂ in Neurological Disorders* (2007); *Neurochemical Aspects of Excitotoxicity* (2008); *Metabolism and Functions of Bioactive Ether Lipids in Brain* (2008); *Hot Topics in Neural Membrane Lipidology* (2009); *Beneficial Effects of Fish Oil in Human Brain* (2009); *Neurochemical Aspects of Neurotraumatic and Neurodegenerative Diseases* (2010); *Lipid Mediators and their Metabolism in the Brain* (2011); *Phytochemicals, Signal Transduction, and Neurological Disorders* (2012); *Metabolic Syndrome: An Important Risk Factor for Stroke, Alzheimer Disease, and Depression* (2013), *Inflammation and Oxidative Stress in Neurological Disorders* (2014), *High Calorie Diet and the Human Brain* (2015), and *Therapeutic Potentials of Curcumin for Alzheimer Disease* (2016). All monographs are published by Springer, New York and Springer International Publishing AG, Basel, Switzerland.

In addition, Dr. Akhlaq A. Farooqui has edited 10 books [*Biogenic Amines: Pharmacological, Neurochemical and Molecular Aspects in the CNS* (2010) Nova Science Publisher, Hauppauge, NY; *Molecular Aspects of Neurodegeneration and Neuroprotection*, Bentham Science Publishers Ltd. (2011); *Phytochemicals and Human Health: Molecular and Pharmacological Aspects* (2011), Nova Science Publisher, Hauppauge, NY; *Molecular Aspects of Oxidative Stress on Cell Signaling in Vertebrates and Invertebrates* (2012), Wiley Blackwell Publishing Company, NY; *Beneficial Effects of Propolis on Human Health in Chronic Diseases* (2012) Vol 1, Nova Science Publishers, Hauppaauge, NY; *Beneficial Effects of Propolis on Human Health in Chronic Diseases* (2012) Vol 2, Nova Science Publishers, Hauppaauge, NY; *Metabolic Syndrome and*

Neurological Disorders (2013), Wiley Blackwell Publishing Company, NY; *Diet and Exercise in Cognitive Function and Neurological Diseases* (2015), Wiley Blackwell Publishing Company, NY; *Trace Amines and Neurological Disorders: Potential Mechanisms and Risk Factors* (2016), Elsevier, NY; and *Neuroprotective Effects of Phytochemicals in Neurological Disorders* (2017), Wiley-Blackwell, John Wiley and Sons, Inc., Hoboken, NJ].

Preface

Alzheimer's disease (AD) is a progressive neurodegenerative disease that primarily affects the regions of the brain that are associated with mood changes, memory, and cognition. AD is the most common cause of dementia among the elderly, accounting for two-thirds of all dementia cases; AD is characterized by the accumulation of senile plaques [enriched in β-amyloid (Aβ) peptide] and neurofibrillary tangles (NFTs; enriched in hyperphosphorylated Tau protein). The accumulation of these hallmarks leads to massive synaptic and neuronal loss in the brain. Aβ peptide is derived by the action of β- and γ-secretases on amyloid precursor protein. Accumulation of Aβ results in aggregation and deposition in the form of extracellular amyloid plaques [senile plaques (SPs)]. NFTs are composed of hyperphosphorylated Tau protein, which is involved in axonal transport. In addition to SPs and NFTs, oxidative stress, neuroinflammation, aberrant cell cycle reentry, mitochondrial dysfunction, alterations in the intracellular calcium homeostasis, and activation of microglial cells and astrocytes are among the other pathological characteristics observed in AD brains. The cause of AD is not known, and controversy persists over which abnormalities initiate the pathogenesis of AD, which contribute to neurodegenerative process, and even whether some of these abnormalities represent neuroprotective mechanisms in the brain. Majority of AD cases (95%) are sporadic (late-onset form). These patients are older than 65 years. Only 5% cases are primarily genetic (early-onset familial form) involving apolipoprotein E (APOE), amyloid precursor protein (*APP*), presenilin 1 (*PSEN 1*), and presenilin 2 (*PSEN 2*) genes. The causes of sporadic AD are not known.

The involvement of Aβ in the pathogenesis of AD is accepted by many researchers. However, recent studies have indicated that there are important limitations to Aβ cascade hypothesis. First, a direct correlation between Aβ aggregates [Aβ-derived diffusible ligands (ADDLs)] and dementia severity has not been clearly established, because some patients with AD without amyloid deposition display severe memory deficits, whereas other patients with cortical Aβ deposits have no dementia symptoms. These observations along with the failure of anti-Aβ therapies to preserve or rescue cognitive function suggest that Aβ aggregation and accumulation may not be universally neurotoxic, but other mechanisms directly or indirectly related to ADDL may contribute to the pathogenesis of AD. Another important point in the pathogenesis of AD is that neither SPs nor NFTs are specific for AD. Other neuropathological conditions such as Parkinson's disease, stroke, hereditary cerebral hemorrhage with amyloidosis of Dutch origin, and sporadic cerebral amyloid angiopathy also show amyloid pathology similar to AD without any dementia, suggesting that Aβ alone is insufficient to cause neurodegeneration and cognitive symptoms observed in AD. Based on the aforementioned information, there has been a passionate debate about the acceptance of Aβ cascade hypothesis in recent years. Recent thinking is that AD is a complex multifactorial disease, which cannot be due to one factor (Aβ), but may involve changes in a number of signal transduction processes associated with oxidative stress, neuroinflammation, and neurodegeneration in the aging brain. In my judgment, more studies are required on neuronal membrane–related signal transduction processes that contribute not only to the generation and accumulation of Aβ and hyperphosphorylation of Tau but also to other processes that aid chronic neurodegeneration in AD.

In the light of this information, I have decided to provide readers with a comprehensive and cutting-edge description of risk factors, pathogenesis, biomarkers, and potential treatment strategies for AD. This monograph has 10 chapters. The first chapter describes information on validity evaluation of Aβ cascade hypothesis of AD. Chapter 2 describes information on risk factors for the pathogenesis of AD. Chapter 3 describes information on the contribution of glycerophospholipid-, sphingolipid-, and cholesterol-derived lipid mediators in the pathogenesis of AD. Chapter 4 describes cutting-edge information on the contribution of nucleic acids in the pathogenesis of AD. Chapter 5 describes information on the contribution of type 2 diabetes and metabolic syndrome in the pathogenesis of AD. Chapter 6 describes information on the contribution of neuroinflammation in the pathogenesis of AD. Chapter 7 describes information on biomarkers of AD. Chapter 8 is devoted to cutting-edge information on potential strategies for the treatment of AD. Chapter 9 summarizes studies on immunotherapy of AD in human subjects. Chapter 10 provides readers with a perspective that will be important for future research work on AD. My presentation and demonstrated ability to present complicated information on signal transduction processes in AD makes this book particularly accessible to neuroscience graduate students, teachers, and fellow researchers. It can be used as supplement text for a range of neuroscience courses. Clinicians, neuroscientists, neurologists, and pharmacologists will find this book useful for understanding the molecular aspects of AD and its treatment. To the best of my knowledge, no one has written a monograph on risk factors, pathogenesis, biomarkers, and potential treatment strategies for AD. This monograph is the first to provide a comprehensive description of signal transduction processes associated with the pathogenesis and treatment of AD.

The choice of topics presented in this monograph are personal. They are based not only on my interest in the pathogenesis of AD and effects of diet on the brain but also on areas in which major progress has been made. The key objective of this monograph is to critically evaluate the information on risk factors, pathogenesis, biomarkers, and potential treatment strategies for AD in the brain. Each chapter of this monograph contains an extensive list of references, which are arranged alphabetically, to works that are cited in the text. I have tried to ensure uniformity and mode of presentation as well as a logical progression of subjects from one topic to another and have provided an extensive bibliography. For the sake of simplicity and uniformity, a large number of figures with chemical structures of dietary components along with line diagrams of colored signal transduction pathways are also included. I hope that my attempt to integrate and consolidate the knowledge on risk factors, pathogenesis, biomarkers, and potential treatment strategies for AD in the brain will initiate more studies on molecular mechanisms and treatment of AD in the human brain. This knowledge will be useful for the optimal health of young, boomer, and preboomer American generations.

Akhlaq A. Farooqui
Columbus, Ohio, USA

Acknowledgments

I thank my wife, Tahira, for critical reading of this monograph, offering valuable advice, useful discussion, and evaluation of the subject matter. Without her help and participation, this monograph neither could nor would have been completed. I would also like to express my gratitude to Melanie Tucker (Senior Acquisitions Editor) and Kristi Anderson (Senior Editorial Project Manager) and Chris Wortley (Book Production Project Manager) at Elsevier/Academic Press for their full cooperation, quick responses to my queries and professional manuscript handling.

Akhlaq A. Farooqui

List of Abbreviations

Phosphatidylcholine	PtdCho
Phosphatidylethanolamine	PtdEtn
Phosphatidylinositol	PtdIns
Phoshatidylinositol 4-phosphate	PtdIns4P
Phosphatidylinositol 4,5-bisphosphate	PtdIns(4,5)P_2
Inositol-1,4,5-trisphosphate	Ins-1,4,5-P_3
Arachidonic acid	ARA
Docosahexaenoic acid	DHA
Eicosapentaenoic acid	EPA
Phospholipase A_2	PLA_2
Cyclooxygenase	COX
Lipoxygenase	LOX
Epoxygenase	EPOX
Protein kinase C	PKC
Alzheimer's disease	AD
β-Amyloid	Aβ
Neurofibrillary tangles	NFTs
Aβ-derived diffusible ligands	ADDLs
Cerebrospinal fluid	CSF
Glycogen synthase-3	GSK-3
Reactive oxygen species	ROS
Reactive nitrogen species	RNS
Metabolic syndrome	MetS
Advanced glycation endproducts	AGE
Amyloid precursor protein	APP
Insulin growth factor	IGF
Metabolic syndrome	MetS

Interleukin	IL
Receptor for advanced glycation endproducts	RAGE
Tumor necrosis factor-alpha	TNF-α
Long-term potentiation	LTP
Superoxide dismutase	SOD
Catalase	CAT
Glutathione peroxidase	GPx
Nitric oxide	NO
Blood–brain barrier	BBB
Positron emission tomography	PET
Single photon emission computed tomography	SPECT

Neurochemical Aspects of β-Amyloid Cascade Hypothesis for Alzheimer's Disease: A Critical Evaluation

INTRODUCTION

Alzheimer's disease (AD), the most common form of dementia, is a complex disease characterized by the accumulation of extracellular β-amyloid (Aβ) plaques (senile plaques) and intracellular neurofibrillary tangles (NFTs) composed of Tau amyloid fibrils (Hardy, 2006, 2009) leading to the loss of synapses and degeneration of neurons in multiple brain regions (cortical and subcortical areas and hippocampus), causing a loss of cognitive brain functions, along with progressive impairment of activities of daily living and often behavioral and physiological changes like apathy and depression. How these factors ultimately contribute to memory impairments and cognitive deficits that clinically characterize the disease remains elusive. According to the 2010 World Alzheimer Report, there are an estimated 36 million people worldwide living with dementia at a total cost of more than US$600 billion in 2010, and the incidence of AD throughout the world is expected to double in the next 20 years (AD International, 2013). It is sixth leading cause of death in the United States victimizing about 5.5 million in the year 2012 (Alzheimer's Association, 2012, 2013). By 2050, this number is expected to jump to 16 million, and in the next 20 years it is projected that AD will affect one in four Americans, rivaling the current prevalence of obesity and diabetes (Brookmeyer et al., 1998). At the neuropathological level the hallmarks of AD are the presence of senile plaques and tangles in brain. Major components of the Aβ deposits are hydrophobic Aβ peptides, which are 38- to 43-amino-acid-long fragments derived from proteolytic processing of the amyloid precursor protein (APP) (Aguzzi and Haass, 2003). Aβ forms aggregates, which accumulate in different subcellular organelles of neurons in patients with AD. Aβ plaques (senile plaques) first appear in the frontal cortex, and then spread over the entire cortical region, whereas hyperphosphorylated Tau and insoluble tangles initially appear in the limbic system (entorhinal cortex, hippocampus, and dentate gyrus) and then progress to the cortical region. Neurofibrillary tangles appear before the deposition of plaque in brains of patients with AD, and that tangle pathology is more closely associated with disease severity than the plaque load (Braak and Braak, 1998; Josephs et al., 2008). Other changes in AD include cerebral amyloid angiopathy (CAA), age-related brain atrophy, white matter rarefaction, and granulovacuolar degeneration. CAA is the most prevalent disturbance, appearing in about 70% of patients with AD along with agitation, which appears in about 50% of patients (Frisoni et al., 1999).

The majority of AD cases (>90%–95%) are sporadic [late-onset AD (LOAD)]. These patients are older than 65 years. Only 5%–7% cases are primarily genetic (early-onset familial form) involving apolipoprotein E (APOE), *APP*, presenilin 1 (*PSEN 1*), and presenilin 2 (*PSEN 2*) genes (Goate et al., 1991; van der Flier and Scheltens, 2005; Kowalska et al.,

2004). All the aforementioned AD genes have been reported to upregulate the cerebral Aβ levels, with the majority of early-onset familial AD mutations increasing the ratio of Aβ42 to Aβ40, which enhances the oligomerization of Aβ into neurotoxic assemblies (Hardy and Selkoe, 2002). Among APOE genes, the *APOE4* gene is the strongest and only confirmed genetic risk factor for the development of LOAD, which enhances the risk level by three times in heterozygous individuals and by 12 times in homozygous individuals (Bertram, 2009) compared to *APOE3*, whereas *APOE2* decreases AD risk by approximately twofold per allele. Mechanism(s) by which *APOE4* increases AD risk include both Aβ-dependent effects, i.e., modulation of Aβ levels, aggregation, neurotoxicity, and neuroinflammation, and Aβ-independent effects, i.e., neuronal development, glucose metabolism, brain activity, and lipid metabolism (Liu et al., 2013). APOE4 protein not only regulates Aβ aggregation and clearance but also is an essential regulator of brain cholesterol metabolism. APOE4 plays an important role in cerebral energy metabolism, modulation of chronic inflammation, neurovascular function, neurogenesis, and synaptic plasticity (Kim et al., 2009, 2014). Sporadic AD is not only accompanied by an accumulation of Aβ plaques and neurofibrillary tangles (Hardy, 2006, 2009), but also by many metabolic, pathological, and neurochemical alterations, including hypometabolism (Mosconi et al., 2008), disruption of blood–brain barrier (BBB) (Zlokovic, 2011), alterations in lipid and glucose metabolism, onset of diabetes and metabolic syndrome, activation of microglial and astroglial cells, and onset of oxidative stress (Mrak and Griffin, 2005; Prokop et al., 2013; Farooqui, 2013, 2014). Among these factors, oxidative stress occurs at early stages of AD before the appearance of amyloid plaques and neurofibrillary tangles and acts to exacerbate the disease progression. Many hypotheses have been proposed to explain the neurodegeneration in AD including: (1) selective vulnerability of cholinergic neurons in the basal forebrain, (2) aluminum deposit hypothesis, (3) Aβ cascade hypothesis, (4) reduction in neurotrophic factors, (5) protein misfolding and aggregation hypothesis, (6) amyloid cascade inflammatory hypothesis, (7) neurovascular hypothesis, (8) insulin insensitivity hypothesis, and (9) dendritic hypothesis (Katzman and Saitoh, 1991; Castellani et al., 2009; Karran et al., 2011; McGeer and McGeer, 2013; Farooqui, 2013; Zlokovic, 2011; de la Monte and Tong, 2014; Cochran et al., 2014). Among the aforementioned hypotheses, two major hypotheses are the cholinergic hypothesis, which ascribes the clinical features of dementia to the deficit cholinergic neurotransmission, and the amyloid cascade hypothesis, which emphasizes the deposition of insoluble peptides formed due to the faulty cleavage of the APP. Although Aβ cascade hypothesis does not explain all features of AD, it has dominated research studies on AD from the past 20 years. This hypothesis was proposed in the late 1980s

(Allsop et al., 1988; Selkoe, 1989) and was formalized in 1992 in a review by Hardy and Higgins (Hardy and Higgins, 1992). According to the Aβ cascade hypothesis an imbalance between production and clearance of Aβ and its accumulation and aggregation in the brain is linked to the development of AD (Hardy, 2006, 2009). Aβ cascade hypothesis is supported by genetics, biochemistry, and molecular biology studies (Hardy, 2006, 2009). Thus in early-onset type of AD (familial forms), genetic alterations increase the production of Aβ due to mutation involving APP and presenilin genes (PSEN1, PSEN2) (Scheuner et al., 1996). Aβ dysregulation in the far more common late-onset "sporadic" AD is less well understood. The deposition of Aβ in the cerebral vasculature causes significant damage to the brain endothelial cells, contributing to a range of characteristic CAA-associated neurovascular anomalies including lobar hemorrhage, cerebral microbleeds, ischemic stroke, and chronic vascular inflammation (Auriel and Greenberg, 2012). Accumulation of Aβ and its aggregation contribute to a variety of cytotoxic effects. For example, Aβ not only affects the mitochondrial redox activity, increases the production of reactive oxygen species (ROS), damages the intracellular calcium homeostasis, and induces the formation of selective calcium channels, but also promotes the release of cytokines through the increase in the phospholipases A_2 (PLA_2), C, and D activities along with alterations in the organization and dynamics of the actin cytoskeleton initiated by filament-dynamizing proteins in the ADF/cofilin family, whereas Tau hyperphosphorylation and NFT formation contribute to the increased rate of protein misfolding, generation of amyloidogenic oligomers, underactivity of repair systems such as chaperones and ubiquitin–proteasome system, or a failure of energy supply and antioxidant defense mechanisms (Hardy and Selkoe, 2002; Zhang and Saunders, 2007). These processes result in abnormal and unbalanced functional activities leading to neuronal dysfunction and ultimately causing neural cell death (Bertram and Tanzi, 2008; Maloney and Bamburg, 2007; Hardy, 2009; Farooqui, 2010).

The involvement of Aβ in the pathogenesis of AD is accepted by many researchers; however, studies by Zheng and Koo, 2011 have unveiled a more complicated picture of APP-derived fragments suggesting that some APP-derived peptides produce neurotoxic effects, whereas others harbor neuroprotective effects. The proteolytic degradation of APP by multiple proteases results in the generation of soluble APP peptides (sAPPα, sAPPβ), various C-terminal fragment (CTF) and N-terminal fragment, p3, and APP intracellular domain (AICD) fragments (Zheng and Koo, 2011). Caspase-mediated cleavage of APP in the cytosolic region releases a cytotoxic peptide, C31, which plays a role in synapse loss and neuronal death. Moreover, other fragments such as Jcasp and YENPTY (motif in the cytoplasmic domain of APP) induce cytotoxic as well as neuroprotective effects (Zheng and Koo, 2011).

PROPERTIES AND ROLES OF AMYLOID PRECURSOR AND TAU PROTEINS IN ALZHEIMER'S DISEASE

Aβ peptides are produced from the proteolytic cleavage of APP, a larger type I transmembrane-spanning glycoprotein. Its gene is located on chromosome 21 in humans. The APP promoter sequence indicates that the *APP* gene is the housekeeping gene. The APP promoter lacks typical TATA and CAAT boxes, but contains consensus sequences for the binding of a number of transcription factors including SP-1, AP-1, and AP-4 sites; a heat shock control element; and two Alu-type repetitive sequences (Izumi et al., 1992; Quitschke and Goldgaber, 1992). APP consists of multiple structural and function domains such as E1 and E2 domain, TMD domain, C-terminal domain, and Kunitz domain (Dawkins and Small, 2014). E1 domain consists of growth factor–like domain and copper-binding domain (CuBD). E1 and E2 domains are connected by a potentially flexible, less well conserved linker region of unknown function, primarily composed of acidic amino acids. A second linker of undefined structure, containing the cleavage sites of α- and β-secretases, anchors the whole extracellular part to the single transmembrane helix. The E1 domain of APP also contains a heparin-binding loop, which is involved in the heparin-induced dimerization (Gralle et al., 2006). The contribution of the E2 domain to the oligomerization of APP is controversial (Wang and Ha, 2004; Kaden et al., 2009). Therefore the role of APP domains in neural metabolism is not known. During transcription, differential splicing of APP messenger RNA (mRNA) produces a number of APP splice variants. The major expressed isoforms of APP have 770, 751, or 695 amino acid residues. APP751 and APP770 are expressed in most tissues and contain a 56-amino-acid Kunitz protease inhibitor (KPI) domain within their extracellular regions. APP695 is predominantly expressed in neurons and lacks the KPI domain (Rohan de Silva et al., 1997; Kang and Muller-Hill, 1990). APP exhibits both neurotoxic and neurotrophic protective effects in the brain (Zheng and Koo, 2011). It is difficult to analyze and study the properties of full-length APP because overexpression or downregulation of APP in various cell systems may generate many proteolytic products, thus rendering it virtually impossible to isolate the precise physiological role of uncleaved APP. Furthermore, the presence of APLP1 and APLP2 in the system may complicate the results on studies on downregulation of APP (Zheng and Koo, 2011).

APP plays an important role in neuroprotection, ion transport, synapse formation, and transcriptional signaling (Fig. 1.1). APP functions as a molecular switch, which controls both neuroplasticity-related processes and pathogenesis of AD. After its synthesis in the endoplasmic reticulum (ER) in the neuronal soma, APP enters the intracellular transport along the secretory, endocytic, and recycling routes. Along these routes, APP undergoes cleavage into defined sets of fragments, which themselves are

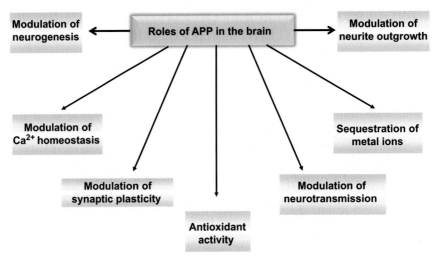

FIGURE 1.1 **Roles of amyloid precursor protein (APP) in the brain.** *5-LOX*, 5-lipoxy-genase; *AD*, Alzheimer's disease; *ARA*, arachidonic acid; *Bcl-2*, B-cell lymphoma 2; *COX-2*, cyclooxygenase-2; *cPLA₂*, cytosolic phospholipase A₂; *cyto-c*, cytochrome; *Glu*, glutamate; *I-κB*, inhibitory subunit of NF-κB; *IL-1β*, interleukin-1β; *IL-6*, interleukin-6; *iNOS*, inducible nitric oxide synthase; *lyso-PtdCho*, lysophosphatidylcholine; *MCP-1*, monocyte chemoat-tractant protein-1; *MMP-9*, matrix metalloproteinase-9; *NF-κB*, nuclear factor-κB; *NF-κB-RE*, nuclear factor-κB-response element; *NFT*, neurofibrillary tangles; *NMDA*, N-methyl-D-aspartate; *NMDAR*, NMDA receptor; *PtdCho*, phosphatidylcholine; *ROS*, reactive oxygen species; *sPLA₂*, secretory phospholipase A₂; *TNFα*, tumor necrosis factor α.

transported—mostly independently—to distinct sites in neurons, where they act as trophic factors and promote neurite outgrowth and synaptogenesis, as well as growth, cell proliferation, and neuronal migration (Muresan et al., 2009, 2013; Dawkins and Small, 2014; Hughes et al., 2014). The molecular mechanisms contributing to APP-mediated cell proliferation, neuronal migration, neurite outgrowth, and synaptogenesis are not fully understood. However, it is suggested that the structure of APP resembles that of a cell surface receptor (Kang et al., 1987), but a receptor function for APP has not been fully established. APP has been reported to interact with adaptor proteins through its conserved NPXY domain; extracellularly, APP interacts with a component of the extracellular matrix, F-spondin (Ho and Sudhof, 2004). APP also contains a CuBD (Barnham et al., 2003) and possesses ferroxidase activity (Duce et al., 2010). In addition, the cytoplasmic tail of APP, AICD, has been reported to participate in the transcriptional regulation (Cao and Sudhof, 2001). To study other physiological roles of APP, mice lacking APP have been generated. APP knockouts show enhanced excitatory synaptic activity and neurite growth (Priller et al., 2006), which is consistent with the finding that APP-deficient mice are more susceptible to glutamate-induced toxicity (Steinbach et al., 1998). Overexpression of APP not only has been reported to modulate $Ca_v1.2$ L-type calcium channel levels and significant reduce the expression of two proteasome subunits, and proteasome subunit α type-5 and proteasome subunit β, leading to the inhibition of regulator of calcineurin, but also influences GABAergic short-term plasticity (Yang et al., 2009; Wu et al., 2015). Furthermore, APP contributes to postsynaptic mechanisms via the regulation of the surface trafficking of excitatory N-methyl-D-aspartate (NMDA) receptors (Innocent et al., 2012).

The other hallmark of AD, neurofibrillary tangles, are made up of Tau protein, which is a member of the family of microtubule-associated proteins (MAPs) (Binder et al., 1985; Maccioni et al., 2001). Tau gene is located on chromosome 17q21.1 (Neve et al., 1986). It is primarily found in axonal region of neurons, where Tau protein plays a fundamental role in the assembly and stabilization of microtubules, promotion of axonal transport, and induction of neurite outgrowth (Fig. 1.2) (Binder et al., 1985; Maccioni et al., 2001). In addition, Tau protein also contributes to the maintenance of neuronal polarity and in the stabilization of the morphology of differentiated neurons. Tau is a primarily cytosolic protein. Tau is also found in the nucleolus and is associated with the nucleolar organizer regions, where it plays an important role in the nucleolar organization and/or heterochromatization of ribosomal RNA genes (Sjöberg et al., 2006). Furthermore, it is reported that Tau also causes damage much before the development of filamentous aggregates. Tau is a scaffolding protein. The increase in levels of Tau and alterations in its subcellular localization (due to increased insolubility and impaired clearance) result in the interaction of Tau with neural cell proteins leading to the

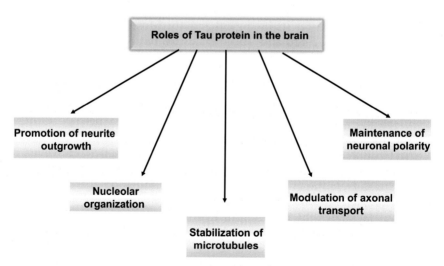

FIGURE 1.2   Roles of Tau protein in the brain.

impairments in their physiological functions. Thus interactions of Tau with membrane is a highly dynamic process, which depends on the process of phosphorylation, such that inhibition of casein kinase 1 (CK1) or glycogen synthase kinase 3β (GSK3β) significantly increases Tau translocation to the membrane, and Tau N-terminal phosphorylation prevents the Tau-membrane localization (Pooler et al., 2012). Tau has also been identified in the lipid rafts of the Tg2576 mouse brain, the AD brain (Kawarabayashi et al., 2004), and lipid rafts of primary neurons, where it is regulated by Aβ oligomers (Williamson et al., 2008). In neurons, it localizes in good quantity within the synapses (Sahara et al., 2014). Collective evidence suggests that different localization of Tau provides evidence for its role in non–microtubule-associated functions, such as signal transduction (Lee, 2005).

Tau is encoded by a single gene located on chromosome 17 (17q21). Tau gene possesses 16 exons in its primary transcript. Mature protein length consists of about 352 up to 441 amino acid residues, and a molecular mass of 45–65 kDa depending on the Tau isoforms (Goedert et al., 1989; Farías et al., 2011). Tau consists of four regions: an N-terminal projection region, a proline-rich domain, a microtubule-binding domain, and a C-terminal region (Mandelkow et al., 1996). The C-terminal region of Tau contains a domain containing the microtubule-binding repeats, which is critical for microtubule assembly (Maccioni et al., 2001). The affinity of Tau for microtubules is regulated by the phosphorylation. Hyperphosphorylation of Tau is catalyzed by a number of protein kinases including cyclin-dependent kinase-5 (Cdk-5), GSK3, CaM kinase II, casein kinase II, stress-activated kinase, c-Jun N-terminal kinase (JNK), kinase p38, and Fyn kinase (Gong and Iqbal, 2008; Avila et al., 2010). Phosphorylation of Tau changes its shape and regulates its biological activity. Most of the phosphorylation sites are on Ser–Pro and Thr–Pro motifs, but a number of sites on other residues have also been reported (Bretteville et al., 2009). Among the Tau phosphorylating enzymes, GSK-3 is the key kinase that mediates Tau hyperphosphorylation. The molecular mechanisms by which accumulation, hyperphosphorylation, and aggregation of Tau contribute to the pathogenesis of AD are unclear. Few in vivo studies have been performed on the roles of the aforementioned Tau kinases or phosphatases in mediating Tau toxicity and inducing AD-related memory deficit. However, it is proposed that Aβ triggers the phosphorylation of Tau leading to the dissociation from the microtubules and its accumulation at the dendritic compartments. Phosphorylated Tau shows stronger interaction with Fyn and thus facilitates the targeting of Fyn to dendritic spines. The targeting of Fyn to postsynaptic density sensitizes the NMDA receptors and renders neurons more vulnerable to the Aβ toxicity in the postsynaptic compartment (Ittner and Götz, 2011). It is not known whether hyperphosphorylation of Tau occurs in situ in the dendritic spines due to altered kinase/phosphatase activities there or occurs elsewhere in the neuron and is then transported to the dendritic spines. On the basis of mathematical modeling experiments, it is also suggested that the bulk accumulation of Tau aggregates in cell bodies may depress neuronal energy metabolism through molecular crowding leading to long-term alterations in neuronal physiology (Vazquez, 2013). Hyperphosphorylation of Tau makes Tau resistant to calcium-activated proteases, calpains, and the ubiquitin-proteasome pathway. Hyperphosphorylation of Tau may also worsen the accumulation of insoluble fibrillar Tau (fibrillar Tau), which exerts its neurotoxic effects by increasing oxidative stress, neuronal apoptosis, mitochondrial dysfunction, collapse of the microtubule-based cytoskeleton, inducing neuritic dystrophy, and subsequent neuronal demise (Mandelkow et al., 2003; Arnaud et al., 2006; Oddo et al., 2008). In AD, Tau exhibits several abnormal characteristics including aggregation, abnormal posttranslational modifications, somatodendritic mislocalization, and a putative role as a cell-to-cell transmissible protein (Cochran et al., 2014). Based on immunochemical studies, it is proposed that Tau is absent from dendrites. In AD, it is mislocalized into dendrites and becomes easily visible by immunostaining (Cochran et al., 2014). Furthermore, dendritic Tau mislocalization can be induced by exogenous application of Aβ onto primary neurons (Zempel et al., 2010). Tau is also found in the nucleus under normal physiological conditions. Its role in the nucleus remains unknown. However, it is proposed that Tau protects against DNA damage in its phosphorylated state (Sultan et al., 2011). Interestingly, Tau has been shown to induce chromatin relaxation, which may subsequently lead to DNA damage and global changes in the transcription (Frost et al., 2014). Collective evidence suggests that both hallmarks of AD (senile plaques and neurofibrillary tangles) induce abnormal signal transduction processes related to excitotoxicity, oxidative stress, and neuroinflammation resulting in neurodegeneration in AD.

AMYLOID PRECURSOR PROTEIN PROCESSING

APP processing occurs through two pathways, namely, (1) nonamyloidogenic or (2) amyloidogenic pathways, which are initiated by two endopeptidases called α- and β-secretases (Schmitz et al., 2002) (Fig. 1.3). Cleavage by α-secretase or β-secretase (BACE-1) results in the shedding of nearly the entire ectodomain yielding large soluble APP derivatives (called APPsα (C83) and APPsβ (C99)) along with the generation of membrane-tethered α- or β-CTFs. CTF processing by γ-secretase generates the harmless P3 peptide (nonamyloidogenic pathway) or Aβ peptides ranging in size from 35 to 42 amino acids (amyloidogenic pathway), plus the AICD fragment (Vassar et al., 1999; Takami and Funamoto, 2012). This CTF is further

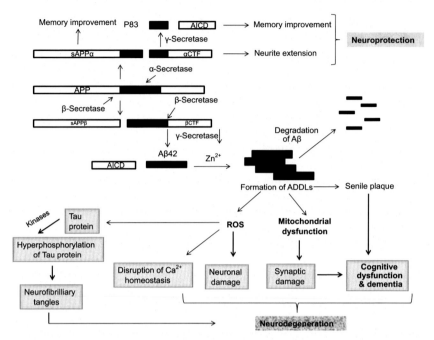

FIGURE 1.3   **Amyloid precursor protein processing, formation of β-amyloid (Aβ) and effect of ADDLs on neurodegeneration in the Alzheimer's disease.** *ADDL*, Aβ-derived diffusible ligands; *AICD*, APP intracellular domain; *APP*, amyloid precursor protein; *CTF*, C-terminal fragment; *ROS*, reactive oxygen species; *sAPP*, soluble APP.

hydrolyzed by γ-secretase to generate Aβ. Several zinc metalloproteinases, such as TACE/ADAM17, ADAM9, ADAM10, and MDC-9, and an aspartyl protease BACE2 can also hydrolyze APP at the α-secretase site located within the Aβ domain (Allinson et al., 2003), essentially precluding the generation of intact Aβ. γ-Secretase is a protein complex consisting of presenilin 1 (PSEN1)/presenilin 2 (PSEN2), nicastrin, anterior pharynx-defective 1, and presenilin enhancer 2 (Kimberly et al., 2003; Li et al., 2003; Yu et al., 2000). It not only hydrolyzes APP but also acts on Notch (a protein that resides on the surface for receiving signal as a heterodimeric receptor). In the membrane α-secretase is located in phospholipid-rich domains, whereas both β- and γ-secretases reside in cholesterol-rich lipid rafts of plasma membrane (Cordy et al., 2003). Based on detailed investigations, it is proposed that alterations in levels of cholesterol or/and ratio of cholesterol to phospholipids in cellular membrane modulate APP-processing pathways through secretases (Wolozin, 2004; Kaether and Haass, 2004). The cleavage of APP by α-secretase precludes formation of Aβ, whereas APP cleavage by β- and γ-secretases releases Aβ peptide. Thus α-secretase competes with β-secretase for the APP binding and APP hydrolysis (Skovronsky et al., 2000; Postina et al., 2004). Most of intracellular Aβ normally occurs in the neuronal cytosol, but it is also colocalized with different organelles depending on where APP as well as β- and γ-secretase reside. In particular, it has been reported to be produced in the secretory pathway–related organelles including ER, medial Golgi saccules, as well as trans-Golgi network (Greenfield et al., 1999).

NONAMYLOIDOGENIC SIGNALING IN THE NORMAL BRAIN

The majority of APP enters the nonamyloidogenic pathway through the involvement of α-secretase. Factors such as mutations, environmental stimuli, and aging modulate APP processing and influence this pattern of APP processing, but the mechanism remains unclear (Querfurth and LaFerla, 2010). Sirtuin1, a NAD-dependent deacetylase activates the transcription of α-secretase. Upregulation of α-secretase activity through the 5-hydroxytryptamine 4 (5-HT4) receptor has been reported to reduce the production of Aβ, decrease Aβ plaque load, and improve cognitive impairment in transgenic mouse models of AD (Pimenova et al., 2014). The molecular mechanism associated with 5-HT4-receptor-stimulated proteolysis of APP is not fully understood. However, it is reported that G protein and Src-dependent activation of phospholipase C and casein kinase 2 along with inositol trisphosphate phosphorylation are closely associated with the upregulation of α-secretase activity (Pimenova et al., 2014). Upregulation of α-secretase activity initiates the activation of notch pathway by cleaving the membrane-bound notch receptor thus liberating an intracellular domain that activates nuclear genes for neurogenesis (Costa et al., 2005). In addition to notch pathway, physiological neurogenesis in the adult brain is regulated by numerous cell extrinsic and intrinsic factors, including local cytokine/chemokine signals, CDK5 (Lagace et al., 2008; Johnson et al., 2009), and Wnt/bone morphogenetic protein (Lie et al., 2005) cascades. The products of α-sectretase-catalyzed reaction produce several neurochemical effects. Thus sAPPα enhances memory when injected into the cerebral ventricles of mice (Meziane et al., 1998) and AICD improves memory in transgenic mice (Laird et al., 2005; Ma et al., 2007; Konietzko, 2012). The cleavage of sAPP also plays an important role in axon pruning during development, and trophic factor deprivation stimulates β-secretase-dependent cleavage of an N-terminal fragment of APP, which binds to death receptor-6 and triggers axon degeneration (Nikolaev et al., 2009). CTF also facilitates neurite extension. Treatment of primary hippocampal neurons and SH-SY5Y neuroblastoma cells with recombinant sAPPα protects these cells from trophic factor deprivation–mediated cell death (Milosch et al., 2014). The neuroprotective effect of APLP can be abrogated in neurons from APP knockout animals and APP-depleted SH-SY5Y cells, but not in cells depleted of APP-like proteins 1 and 2 (APLP1 and APLP2), indicating that expression of membrane-bound holo-APP is required for sAPPα-dependent neuroprotection (Milosch et al., 2014). The molecular mechanism associated with neuroprotective effects is not fully understood. However, based on detailed investigations, it is proposed that the neuroprotective mechanism may involve G-protein-coupled activation of the Akt pathway (Milosch et al., 2014). Collectively, these studies suggest that the soluble N-terminal sAPPα and CTF have neuroprotective properties.

AMYLOIDOGENIC SIGNALING IN ALZHEIMER'S DISEASE

In amyloidogenic pathway, the rate-limiting step is the cleavage of APP by the β-secretase (BACE-1), which is located in the physical proximity of APP (Thinakaran and Koo, 2008). The remaining membrane-anchored stubs are degraded within the membrane plane by the γ-secretase complex, releasing the AICD into the cytosol. This leads to the generation of the multiple forms of Aβ (ranging from 37 to 43 amino acids), which form aggregates and accumulate in different subcellular organelles of neurons in patients with AD (Fig. 1.4). As stated earlier, amyloidogenic processing of APP results in the production of a CTF corresponding to AICD. This has been reported to translocate into the nucleus and activate gene transcription (Pardossi-Piquard and Checler, 2012).

APP processing generates Aβ as the predominant species. A trace amount of Aβ42 is produced in a ratio of approximately 99:1. Chemical cross-linking studies have indicated that although both Aβ40 and Aβ42 are capable of forming fibrils, they maintain distinct oligomer distributions. For example, Aβ40 and Aβ42 monomers form dimers, trimers, and tetramers in solution. Aβ42 also form pentamers and hexamers, called paranuclei, which self-associate to form dodecamers, protofibrils, and fibrils (Bitan et al., 2003). Aβ*56 (a 12-mer) is an endogenous product found in APP transgenic mice. Secreted Aβ42 are dimers and trimers, which are produced by cultured cells and are resistant to proteolytic degradation. It should be noted that the heterogeneity of Aβ42 reflects not only biological variation but also technical variation in the methods to produce synthetic oligomers or to isolate them from biological tissues (Benilova et al., 2012). In vivo studies in mice and humans indicate that dodecamers of Aβ42 may be the proximate neurotoxins in AD (Lesne et al., 2006). Thus Aβ40 is benign, whereas Aβ42 due to the presence of two hydrophobic amino acids mediates amyloidogenic effects producing neurodegeneration. Aβ42 peptides are not only highly immunogenic and proinflammatory but also may self-organize into "annular ring" structures, which allow hydrophobic side chains to face and interact with the plasma

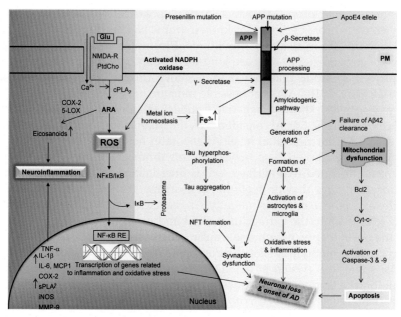

FIGURE 1.4 **Processes contributing to the pathogenesis in the brain in patients with AD.** *5-LOX*, 5-lipoxygenase; *AD*, Alzheimer's disease; *APP*, amyloid precursor protein; *ARA*, arachidonic acid; *Bcl-2*, B-cell lymphoma 2; *COX-2*, cyclooxygenase-2; *cPLA₂*, cytosolic phospholipase A_2; *cyto-c*, cytochrome; *Glu*, glutamate; *I-κB*, inhibitory subunit of NF-κB; *IL-1β*, interleukin-1β; *IL-6*, interleukin-6; *iNOS*, inducible nitric oxide synthase; *lyso-PtdCho*, lysophosphatidylcholine; *MCP-1*, monocyte chemoattractant protein-1; *MMP-9*, matrix metalloproteinase-9; *NF-κB*, nuclear factor-κB; *NF-κB-RE*, nuclear factor-κB-response element; *NFT*, neurofibrillary tangles; *NMDA*, N-methyl-D-aspartate; *NMDAR*, NMDA receptor; *PtdCho*, phosphatidylcholine; *ROS*, reactive oxygen species; *sPLA₂*, secretory phospholipase A_2; *TNFα*, tumor necrosis factor α.

membrane proteins forming solvated channel pores. This process may result in disturbance in cellular ion homeostasis through uncontrolled leakage of ions into and/or out of the cell (Connelly et al., 2012; Jang et al., 2014). High levels of Aβ42 peptide due to excessive production may not only cause cellular toxicity directly through altered Aβ42 peptide–plasma membrane interactions and channel-mediated destabilization of ionic homeostasis and also through direct binding with cell adhesion molecules such as neuroligins and neurexins located in the postsynaptic cleft (Connelly et al., 2012; Jang et al., 2014; Sindi and Dodd, 2015) leading to alterations in neurotransmission and memory formation before plaques build up. Collectively, these studies support the view that Aβ42 peptide may be directly related to the pathogenesis of AD (Haass and Selkoe, 2007). Aβ42 is the predominant species of Aβ in senile plaques (Verdile et al., 2004). For the sake of simplicity, I will refer to Aβ42 as Aβ, a generic term, which describes the binding of Congo red with β-pleated sheet of fibrillar proteins and displays optical polarization properties. The physiological role of Aβ is not known, but it has been reported to occur throughout life in healthy individuals. Picomolar concentrations of Aβ have been reported to increase long-term potentiation (LTP) resulting in improved synaptic plasticity and memory (Puzzo and Arancio, 2013; Morley et al., 2010). Another possible function of Aβ is to inhibit the activity of γ-secretase through negative feedback (Kamenetz et al., 2003). High concentrations of Aβ are considered as the main culprit in the pathogenesis of AD. At the molecular level, Aβ mediates neurotoxicity by triggering signaling cascades on the postsynaptic membrane. Neuronal culture studies have indicated that surface APP is internalized by clathrin-mediated endocytosis and that proteolytic Aβ cleavage occurs in the endosomal pathway in the axon. Subsequently, Aβ recycles to the cell surface, where it is released into the extracellular fluid and mediates signaling cascades on the postsynaptic membrane, causing increased synaptic alpha-amino-3-hydroxy-5-methylisoxazole-4-propionate receptor endocytosis and dendritic spine loss (Hsieh et al., 2006). It has also been reported that the inhibition of endocytosis reduces APP internalization and immediately lowers Aβ levels in vivo (Cirrito et al., 2008). Thus it is tempting to speculate that in its early stages, AD is a disorder of synaptic receptor trafficking and synapse dysfunction. At high concentration, Aβ and its aggregates impair synaptic function and initiate neuronal degeneration (Gong et al., 2003). Based on in vitro studies, it is proposed that aggregation of Aβ is a multistep process that is modulated by the conformation, environment (pH, electrical charge, hydrophilicity, or hydrophobicity), phosphorylation, presence of metal ions, membrane composition, and biophysical properties of the membrane (Zhu et al., 2015). The phosphorylation of extracellular Aβ by protein kinase A on serine residue 8 promotes aggregation by stabilizing the β-sheet conformation of Aβ and promotes the formation of oligomeric Aβ aggregates (dimers or trimers or as large as dodecamers) that initiate fibrillization (Kumar and Walter, 2011; Larson and Lesné, 2012). Phosphorylated Aβ has been reported to occur not only in the brains of transgenic mice but also in the brains of human patients with AD supporting the view that phosphorylation-mediated aggregation of Aβ may be relevant in the pathogenesis of LOAD (Kumar and Walter, 2011). The aggregated form of Aβ also termed as Aβ-derived diffusible ligands (ADDLs). The size of ADDLs ranges from tetramers to dodecamers. Thus multiple forms of ADDL appear to exist and coexist in brain tissue. The exact identity of the nature of the relationships between these various forms is unknown. However, it is becoming increasingly evident that ADDLs act as an initiators of AD not only by inducing synaptic loss and progressive cognitive decline but also by mediating the development of Tau pathology and synaptic dysfunction (Tu et al., 2014; Viola and Klein, 2015; Selkoe, 2008; Bloom, 2014). In addition, at very low concentrations (nanomolar) ADDLs not only inhibit LTP in brain slices but also produce dendritic spine retraction from pyramidal cells and impair rodent spatial memory (Lacor et al., 2004; Walsh et al., 2002; Selkoe, 2008). Furthermore, increased exposure of ADDLs accelerates neural cell death, through a signaling-dependent mechanism requiring the protein tyrosine kinase Fyn. Based on these observations, it is proposed that early memory loss in AD and the progressively catastrophic dementia are caused by neural signaling dysfunction triggered by soluble ADDLs (Klein et al., 2001). ADDLs interact with a number of specialized synaptic proteins, including scaffolding proteins and ion channels, which are clustered at dendritic spines. These spines are specialized protrusions from a dendrite's shaft, where neurons form synapses to receive and integrate information (Fu and Zuo, 2011). The spines that contain glutamate receptors and postsynaptic density components are the primary locations of excitatory synapses. The loss of dendritic spines directly correlates with the loss of synaptic function. ADDL-mediated synaptic protein depletion and synaptic loss (Reddy et al., 2005; Scheff et al., 2007) occurs in the hippocampal and cortical regions of human AD as well as in animal models of AD. The level of synaptic protein reduction closely parallels the degree of premortem cognitive impairment suggesting that the elevation and synaptic loss of ADDLs, rather than Aβ plaque load, may represent the best indicators of the severity of dementia or cognitive impairment in AD (Klein et al., 2001; Reddy et al., 2005; Scheff et al., 2007). Converging evidence suggests that the generation of ADDLs and loss of dendritic spines are closely associated with the pathogenesis of AD, along with hyperphosphorylation of Tau, which acts downstream to Aβ as a modulator of the disease progression. ADDL mediates its neurotoxic effects by enhancing the generation of ROS not only directly but also through indirect mechanisms. Thus ADDLs directly produce H_2O_2 not only through the activation of a copper-dependent superoxide dismutase-like activity (Fang et al., 2010) and activation of NADPH-oxidase in astrocytes, but also through the modulation of mitochondrial ROS generation via regulation of activity of enzymes such as Aβ-binding alcohol dehydrogenase and α-ketoglutarate dehydrogenase (Borger et al., 2013). Mitochondria are the major intracellular targets of ADDLs. High levels of ADDLs produce mitochondrial dysfunction, leading not only to the overproduction of ROS and inhibition of cellular respiration but also to the reduction of ATP production and damage to the mitochondrial structure (Starkov and Beal, 2008; Reddy and Beal, 2008). In cultures of cortical neurons, ADDLs not only induce apoptosis through the JNK–c-Jun–FasL–caspase-dependent extrinsic apoptotic pathway but also modulate redox factor-1(Tan et al., 2009). It is proposed that ADDL promotes neurodegeneration through the induction of oxidative stress and neuroinflammation. Many of aforementioned effects of ADDLs can be neutralized by a nonamidated 8-residue peptide (RGTFEGKF). This peptide is called as ADDL or Aβ-oligomer interacting peptide (AIP). AIP not only suppresses self-aggregation of ADDLs into protofibrillar structures but also promotes the disaggregation of amyloid oligomers from fibrils. ADDLs are stable in the presence of excess AIP (Barucker et al., 2015). In this regard, AIP may act as a chaperone, not only causing an apparent dissociation of Aβ complexes into monomers but also competing with the addition of further monomers or oligomers to effectively prevent the growth of ADDLs into larger assemblies. The net results of these processes can be attributed to the potential immunogenicity of amyloid protein misfolding and aggregation, which most probably occurs 10–15 years before the clinical manifestation of AD (Querfurth and LaFerla, 2010). Two features are necessary for the misfolding of Aβ peptide. One change involves the structural change from α-helix configuration of native status to form a cross-β structure. Second change is the presence of one or two hydrophobic core fragments. During the formation of misfolded Aβ protein, the hydrophobic interaction is the driving force and the cross-β structure provides the platform for assembling. Compared to the native status of the protein, the misfolded assembly provides a changed microenvironment, which is normally more hydrophobic and is a very important feature for developing sensitive fluorescent probes (Leandro and Gomes, 2008).

INVOLVEMENT OF Aβ AND TAU IN NEURODEGENERATION IN ALZHEIMER'S DISEASE

Neurodegeneration in AD is a complex, progressive, and multifactorial process. It is initiated by alterations in signal transduction processes due to the accumulation and aggregation of Aβ and hyperphosphorylated Tau protein. The molecular mechanism of neurodegeneration in AD is not fully understood. However, mounting evidence from genetic, pathological, and functional studies indicate that an imbalance between the production and clearance of Aβ peptides in the brain results in accumulation and aggregation of Aβ (Figs. 1.3 and 1.4). The toxic Aβ aggregates in the form of ADDLs, intraneuronal Aβ, and amyloid plaques produce multiple effects on neurons and glial cells leading to oxidative damage to synapses, mitochondrial dysfunction, calcium dysregulation, inflammation, and ER stress, which may induce neurodegeneration through apoptotic cell death of neurons (Farooqui, 2010). Continuous ADDL formation or sustained elevation of Aβ may also produce a chronic effect on the innate immune system through the involvement of Toll-like receptor 2 (TLR2), TLR4, and TLR6, and their coreceptors CD14, CD36, and CD47 in activated microglial cells. This process can destroy functional neurons by direct phagocytosis (Neniskyte et al., 2011;

Liu et al., 2012). Besides this, the activation of microglial cells may result in inflammatory response through the release of inflammatory mediators (complement factors, eicosanoids, chemokines, and proinflammatory cytokines), which can impair microglial clearance of Aβ and the neuronal debris and increase in microglial cell–mediated neurodegeneration and loss of neuronal synapses, contributing greatly to AD pathogenesis. These observations support the "β-amyloid cascade hypothesis" (Hardy and Selkoe, 2002; Blennow et al., 2006). Synaptic loss is one of the major features of AD, which correlates with the degree of dementia. NMDA receptors (NMDARs) have been shown to mediate downstream effects of the Aβ in AD models. Other receptor and proteins that bind to Aβ include prion proteins (Lauren et al., 2009) and A7 nicotinic acetylcholine receptor (Wang et al., 2000). This binding contributes to the pathogenesis of neuropathy. Furthermore, in silico assay revealed that several neurotransmitters including acetylcholine can bind Aβ more favorably than their corresponding receptors. These observations support the view that the binding of neurotransmitters with Aβ may be associated with the AD pathogenesis through the disturbance of the normal signaling of neurotransmitters. Binding of Aβ with prion protein indicates that AD is a prionlike disease with misfolded Aβ molecules acting as nucleation seeds and initiating aggregate formation by recruiting additional unfolded or oligomeric Aβ peptides and thereby accelerating amyloid growth (Yin et al., 2014). In addition, AD is accompanied by mitochondrial dysfunction and increase in levels of transition metal ions such as iron (Fe^{3+}), copper (Cu^{2+}), zinc (Zn^{2+}), which promote ROS formation and increase.

Tau (tubulin-associated unit) is a low-molecular-weight protein, which has ability to promote microtubule assembly in vitro. It is expressed in both neuronal and nonneuronal cells, but predominantly in neurons (Martin et al., 2011). In developing neurons, Tau activity is crucial not only for the morphogenesis of the growth cones but also for promoting axonal growth (Martin et al., 2011). Tau is encoded by a single gene located on chromosome 17 (17q21), possessing 16 exons in its primary transcript. Mature Tau protein contains 352–441 amino acid residues with a molecular mass of 45–65 kDa depending on the Tau isoforms (Martin et al., 2011). The C-terminal region of Tau contains the microtubule-binding repeats, which are critical for microtubule assembly. The affinity of Tau for microtubules is regulated by phosphorylations. This process leads to the dissociation of Tau protein from microtubule, Tau and Aβ aggregation, and NFT formation. The crossing of human APP transgenic mice with human Tau transgenic mice leads to marked increase in the aggregation of Tau with concomitant dendritic spine loss and acceleration in cognitive impairment (Chabrier et al., 2014). On the basis of extensive investigation, it is proposed that the cross-talk between Aβ and Tau not only contributes to induction and enhancement of Tau phosphorylation but also to the Aβ-mediated proteasomal impairment of Tau degradation and dysregulation of axonal transport with possible bidirectional effects, leading to increase in Aβ as well as Tau (Blurton-Jones and Laferla, 2006). Accumulation of Aβ and phosphorylation of Tau has been reported to downregulate the density and function of synapses leading to abnormalities in signal transduction processes in neuronal network within the brain (Selkoe, 2008; Iqbal et al., 2009). The accumulation of Aβ and the spatial distribution and severity of NFT formation closely correlate with cognitive impairment and brain atrophy observed in AD (Nelson et al., 2012). ADDL toxicity can be prevented by Tau reduction or deficiency as evidenced by transgenic mouse studies (Roberson et al., 2007).

Some investigators propose that in the pathogenesis of AD, tauopathy develops first, but mainly remains associated with subcortical and medial temporal limbicareas. Tauopathy is a characteristic feature of aging (Price and Morris, 1999). The accumulation of Aβ in the neocortex takes place later and independent from medial temporal tauopathy, which is Tau-mediated neuronal loss. This temporal tauopathy is ultimately responsible for the clinical symptoms of AD (Rabinovici et al., 2008; Wolk et al., 2012). This proposal does not conflict with the Aβ cascade hypothesis in that it acknowledges the central role of Aβ in the pathogenesis of AD. In addition, many investigators have reported that the degree of Tau-related pathology is better correlated with the degree of dementia when compared to the amyloid plaque burden; hence, making Tau as a desirable target in symptomatic patients with AD (Bancher et al., 1993; Terry, 1996). Further support for this idea comes from the results from the human immunization trials, where the reduction in amyloid plaque load does not produce cognitive benefits in symptomatic subjects with AD.

Activation of microglial cells by Aβ and ROS, stimulation of PLA_2, cyclooxygenase-2, and 5-lipoxygenase (5-LOX) in AD contribute to high levels of proinflammatory cytokines and increased levels of prostaglandins (Farooqui, 2010). These mediators intensify oxidative stress and neuroinflammation. Oxidative stress and neuroinflammation are interrelated processes. In AD, neuroinflammatory changes are caused by the activation of microglia, astrocytes, and macrophages particularly in the amyloid deposits (Heneka and O'Banion, 2007). The degeneration of neurons in the brain leads to the release of large amounts of proinflammatory mediators, including cytokines, ROS, and prostaglandins, all of which increase the generation of Aβ (Velez-Pardo et al., 2002). Aβ-mediated respiratory burst in microglia also produces tumor necrosis factor α (TNFα), which aggravates Aβ deposition and further neuronal dysfunction eventually leading to neural cell death (Liu et al., 2002; Agostinho et al., 2010; Farooqui, 2010). With time, the buildup of plaques and tangles contribute to the neurodegeneration in AD (Meyer-Luehmann et al., 2008). In AD, it is not clear when the disease sets in and how long does it take for neuropathological changes to appear. However, it is generally believed that in AD neurodegeneration starts 6–10 years before the symptoms. In addition, epigenetic factors that modulate genes may also contribute to neurodegeneration through alterations in the methylation or oxidation of CpG dinucleotides in DNA (Zawia et al., 2009). Converging evidence suggests that increased cross-talk among ADDL and Tau-mediated oxidative stress, neuroinflammation, abnormal protein dynamics with defective protein degradation, autoimmunity, and abnormal mitochondrial function leading to increase in Ca^{2+} levels, impairment in energy metabolism along with synaptic and dendritic injury and disturbances in the process of adult neurogenesis, circuitry dysfunction, and aberrant innervation may contribute to neurodegeneration in AD (Farooqui, 2010; Jellinger, 2009, 2010; Crews et al., 2010).

DEGRADATION OF Aβ IN THE BRAIN AND FACTORS THAT MODULATE Aβ CLEARANCE

As stated, Aβ is generated from APP by sequential cleavages by β- and γ-secretases. In normal adult brains, Aβ synthesis and clearance rates have been determined in the cerebrospinal fluid (CSF) and estimated to be 7.6% and 8.3%, respectively (Bateman et al., 2006). Thus accumulation of Aβ is unlikely in the normal brain. However, extracellular and intracellular accumulation of Aβ peptides along with small defects in Aβ clearance can be sufficient to cause Aβ accumulation leading to alterations in the balance between production, aggregation, and degradation. Many studies have indicated that in the brain decrease in Aβ clearance is responsible for the accumulation of Aβ and development of AD (Weller et al., 2008). This suggestion is also supported by the accumulation of Aβ in the blood vessel walls in addition to the brain, leading in CAA, which is present in approximately 90% of patients with AD (Love, 2004) and is the most common cause of lobar intracerebral hemorrhage in the elderly (Viswanathan and Greenberg, 2011).

ADDLs are removed from the brain by various clearance systems, such as enzymatic degradation and cellular uptake, transport across the BBB and blood–CSF barrier, interstitial fluid bulk flow, and CSF absorption into the circulatory and lymphatic systems. However, there are three major mechanisms, namely, global, local, and microglial-based receptor–sensor phagocytosis mechanisms for Aβ removal. The former mechanism requires vascular transport across the BBB by binding/transport proteins. In the global mechanism low-density lipoprotein receptor–related protein1 (LRP1) binds the Aβ, at the abluminal side of the BBB, and initiates Aβ clearance from brain to blood via transcytosis across the BBB (Bell et al., 2007) and is the primary receptor mediating transport of Aβ across the BBB into circulation, thereby clearing it from the brain (Fig. 1.5). LRP1 is abundantly expressed in various brain cell types including neurons and glial cells in brain parenchyma and smooth muscle cells and pericytes in cerebral vasculature to mediate cellular uptake of an array of ligands including APP, apolipoprotein E, α2-macroglobulin, ABC transporter, and receptor-associated protein, all of which are involved in either Aβ production or clearance (Bu, 2009). Furthermore, LRP1 is also coupled with other cell surface signaling receptors to regulate signal transduction (Herz and Strickland, 2001). In the liver, LRP1 mediates systemic clearance of Aβ (Tamaki et al., 2006). β-secretase cleaves the N-terminus extracellular domain of LRP (von Arnim et al., 2005), which releases soluble LRP1 (sLRP1). sLRP1 normally circulates in plasma (Quinn et al., 1997) and contributes to the removal of Aβ from the luminal side. Elimination of Aβ by LRP1 is a two-step process. Step 1 takes place at the site of the BBB on the surface of endothelial membrane facing the brain (abluminal side). In this step,

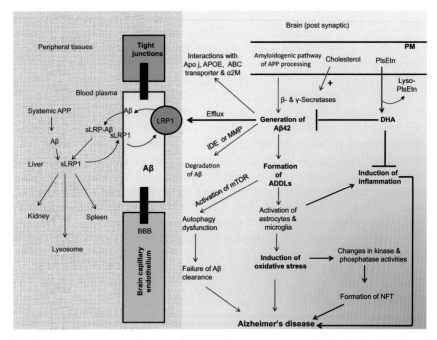

FIGURE 1.5 Schematic diagram showing the involvement of LRP1, IDE or MMP, and mTOR in the clearance of Aβ from the brain into the blood. *ABC*, ATP-binding cassette transporter; *apoJ*, apolipoprotein J; *APP*, amyloid precursor protein; *BBB*, blood–brain barrier; *DHA*, docosahexaenoic acid; *LRP1*, low-density lipoprotein receptor–related protein1; *lyso-PlsEtn*, lysoethanolamine plasmalogen; *mTOR*, mammalian target of rapamycin; *PlsEtn*, ethanolamine plasmalogen; *sLRP1–Aβ*, sLRP1–Aβ complexes; *TG*, tight junction; *α2M*, α2-macroglobulin.

the membrane-bound LRP1 binds cerebral Aβ and rapidly initiates its clearance by transcytosis into the blood (luminal) side (Deane et al., 2004; Bell et al., 2007; Sagare et al., 2007). Step 2 takes place in the plasma, where sLRP1 in the periphery contributes to Aβ clearance by sequestering free Aβ from the brain into the circulation (Fig. 1.5) (Sagare et al., 2007, 2013; Ramanathan et al., 2015). Coimmunoprecipitation of sLRP1-bound Aβ in neurologically normal humans has indicated that circulating sLRP1 can sequester 70%–90% of plasma Aβ (Sagare et al., 2007, 2012), thereby driving the Aβ gradient in favor of efflux across the BBB. sLRP1 also contributes to the removal of Aβ from peripheral tissues such as liver, kidney, and spleen (Sagare et al., 2007). Apolipoprotein J (apoJ), APOE2, APOE3, and APOE4 and α2-macroglobulin also influence differentially LRP1-mediated Aβ clearance at the BBB. Statins, which are potent inhibitor of cholesterol synthesis, upregulate the expression of LPR1 leading to Aβ clearance in the vasculature (Shinohara et al., 2010) supporting the view that levels of Aβ in the brain are modulated by lipid metabolism. Free Aβ can also be transported from the circulation into the interstitium by receptor for advanced glycation end products (RAGE) (Pascale et al., 2011; Deane et al., 2012). Soluble transporters such as the soluble form of RAGE (sRAGE), anti-Aβ IgG, serum amyloid P component (Zlokovic and Frangione, 2003), and sLRP, which binds 70%–90% of plasma Aβ, interact with soluble Aβ and inhibit its binding to RAGE, thereby retarding Aβ from entering the interstitium (Zlokovic et al., 2010). It is also reported that insulin injection increases Aβ clearance from the brain (Vandal et al., 2014). Thus in 3xTg-AD mice fed with the high fat diet for 9 months, insulin injection restores cortical ADDL levels back to the level found in mice fed with the control diet. The AD-lowering effect of insulin is accompanied with improvement in memory function in HFD-fed 3xTg-AD mice (Vandal et al., 2014). These processes also lead to changes in molecular markers associated with Aβ production, such as increased in α-APP, increased X11α, decreased BACE, and decreased autophagy-related proteins (Kondo et al., 2010; Son et al., 2012a).

The local mechanisms involve the degradation of Aβ through several endopeptidases. Thus neprilysin (NEP or EC 3.4.24.11) is a zinc metalloendopeptidase (molecular mass 97 kD) that degrades Aβ. This enzyme preferentially cleaves both Aβ40 and Aβ42 in the brain. In neurons, NEP is subcellularly localized along axons and synapses (Fukami et al., 2002) where NEP-mediated Aβ degradation mainly occurs (Iwata et al., 2004). This subcellular distribution underlines the role of NEP in the degradation of several neuropeptides, for example, enkephalins, substance P, neuropeptide Y, tachykinins, bradykinin, and somatostatin (Barnes et al., 1995). The in vivo functions of NEP have been analyzed by utilizing NEP gene–disrupted mice. These animals show enhanced lethality to endotoxin treatment, probably due to the role of NEP in the metabolism of proinflammatory peptides, lower blood pressure, and higher microvascular permeability (Lu et al., 1995). NEP has been identified as a critical Aβ-degrading enzyme (ADE) in the brain (Iwata et al., 2001). A distinct matrix metalloproteinase (MMP) called insulin-degrading enzyme (IDE) also cleaves and facilitates the removal of Aβ in the brain. Many studies have indicated that NEP mRNA and protein expression levels are reduced in association with age or in subjects with AD (Caccamo et al., 2005; Maruyama et al., 2005; Wang et al., 2010; Marr and Hafez, 2014). However, these studies have been seriously challenged. Using highly specific enzyme immunocapture/activity assay, it is shown that NEP activity levels increase with age and during the progression of AD (Miners et al., 2009, 2010, 2011). This is similar to the consensus on most endopeptidase expression levels in association with AD (Miners et al., 2011) and may reflect a homeostatic response to the abundance of Aβ substrate and/or to the inflammatory environment occurring in AD. Another mechanism that may contribute to the increase in brain Aβ involves the influx or reentry of Aβ into the brain mainly through the receptor for RAGE. Since the steady-state levels of brain Aβ represent a dynamic equilibrium between synthesis, reuptake and clearance, any factor that results in reduced rate of Aβ removal is likely to cause Aβ accumulation. Interactions of Aβ and other ligands with RAGE activate multiple intracellular signaling pathways involving MAPKs such as ERK1/2, p38, JNK, PI3K, Src kinase, JAK/STAT, TGFβ/Smad, and members of the Rho GTPase signaling pathway. Moreover, Aβ–RAGE interactions also cause the generation of ROS, which influences cellular homeostasis and inflammatory response leading to chronic metabolic and neurological diseases such as cancer, diabetes, and AD (Farooqui, 2013).

Microglial-based receptor–sensor phagocytosis mechanism is mediated by a receptor called TREM2 (Jiang et al., 2013). This receptor is a glycosylated 230-amino-acid microglial membrane-spanning stimulatory and signaling sensor protein, which is encoded in mice on chromosome 17 and in humans on chromosome 6p21.1. In the brain, TREM2 is associated with sensing, phagocytosis, and removal of Aβ40 and Aβ42 peptides. Alterations in TREM2 function due to mutations (Hickman and El Khoury, 2014) may contribute not only to progressive accumulation and aggregation of Aβ42 as observed in sporadic AD but also to abnormalities in phagocytosis by microglial cells (Jiang et al., 2013). This suggestion is supported by studies in hAPP tg mice. It is reported that alterations in TREM2 in a subset of microglia results in compromised phagocytosis during plaque development (Guerreiro et al., 2013). Deletion of one TREM2 allele in hAPP tg mice significantly reduces the number of microglial cells associated with Aβ deposits (Ulrich et al., 2014). Conversely, TREM2 overexpression in hAPP tg mice not only decreases amyloid plaque burden but also reduces neuroinflammation, synapse loss, and spatial memory deficits (Jiang et al., 2014). TREM2 mutations may also alter its transport to the cell surface and shedding, associated with impaired phagocytic function (Kleinberger et al., 2014). These studies have led to the suggestion that levels of the shed ectodomain in extracellular fluid and CSF are lower in AD cases associated with TREM2 mutations. As stated earlier, in AD, activated microglia cells participate in the phagocytosis of Aβ and thus prevent the deposition of Aβ and the formation of amyloid plaques. However, long-term abnormalities in TREM2 function may induce deposition of Aβ along with microglial cell dysfunction. This may not only lead to the accumulation of Aβ but also to overproduction of ROS and increased expression of proinflammatory cytokines, subsequently damaging neurons and synapses due to oxidative stress and neuroinflammation (Farooqui, 2014).

Mammalian target of rapamycin (mTOR), a 289-kD serine/threonine protein kinase plays an important role in the regulation of cell growth, proliferation, and survival, and also in regulating protein synthesis. Activation of mTOR not only enhances Aβ generation and deposition (Cai et al., 2012), but also inhibits autophagy, decreases the Aβ clearance of the autophagy/lysosome system, and modulates the metabolism of APP by regulating β- and γ-secretase (Cai et al., 2012; Pei and Hugon, 2008; Spilman et al., 2010). Furthermore, mTOR also interacts with several key signaling pathways and regulates Aβ generation or Aβ clearance, including PtdIns 3K/Akt (Bhaskar et al., 2009), GSK-3, AMP kinase (Cai et al., 2012), and insulin/insulinlike growth factor 1 (Pei and Hugon, 2008). Several studies have also indicated that the activation of mTOR leads to the failure of Aβ removal from the brain since the dysfunction of autophagy triggered by mTOR facilitates the process of Aβ generation and decreases its clearance (Son et al., 2012b; Li et al., 2013; Chen et al., 2009). Thus converging evidence suggests that abnormalities in Aβ42 removal from brain by aforementioned mechanisms may put a tremendous amyloid burden on neural cells in the brain leading to loss of metabolic homeostatic and function.

MODULATION OF LEVELS OF Aβ-DERIVED DIFFUSIBLE LIGANDS BY LIPIDS AND CARBOHYDRATE METABOLISM

Activities of α-, β-, and γ-secretases and APP processing are modulated by lipids and fatty acids [docosahexaenoic acid (DHA) ethanolamine plasmalogen, and trans–fatty acids]. Thus trans–fatty acids, cholesterol, and GM1 ganglioside increase the generation of Aβ by upregulating the activity of β- and γ-secretases (Zha et al., 2004; Wolozin, 2004), whereas DHA and sphingomyelin decrease amyloidogenic processing of APP by downregulating the activity of β- and γ-secretases (Fig. 1.5) (Grimm et al., 2007, 2011a,b, 2012, 2013). Alternatively, APP can also by cleaved by α-secretase leading to the generation of sAPPα along with the liberation of peptide p3 (Hooper and Turner, 2002). The production of Aβ also depends on the bioavailability of cholesterol in neurons, since the activity balance of the α- and β-secretases is related to the lipid composition of neurons. High concentrations of cholesterol in the lipid raft domains of neuronal membranes result in an increase in amyloidogenic APP process catalyzed by β-secretase, whereas at lower levels, cholesterol metabolism stimulates the α-secretase-mediated APP processing. The hypothesis that statins (the cholesterol lowering drugs) can be used for the treatment of AD is based on the above mentioned information (Schmitt et al., 2004).

Long-term consumption of high-calorie diet (western diet), which is enriched in processed macronutrients (fats, cholesterol, proteins, and sugars), high in salt (sodium chloride), but low in fiber results in accumulation of Aβ in normal rodent brain (Fig. 1.6) (Sparks et al., 1994; Farooqui, 2015)

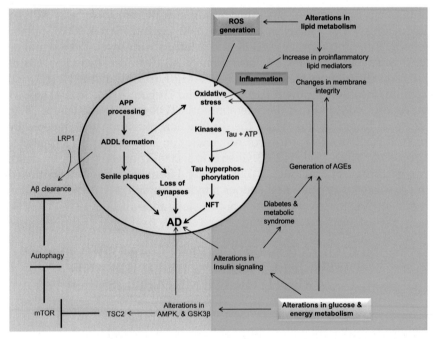

FIGURE 1.6 **Modulation of Aβ production and Tau phosphorylation by glucose and lipid metabolism in the brain.** *ADDL*, Aβ-derived diffusible ligands; *AGE*, advanced glycation end products; *AMPK*, AMP kinase; *APP*, amyloid precursor protein; *GSK3β*, glycogen synthase kinase3β; *LRP1*, low-density lipoprotein receptor–related protein1; *mTOR*, mammalian target of rapamycin; *NFT*, neurofibrillary tangle; *ROS*, reactive oxygen species; *TSC2*, tuberous sclerosis 2.

as well as in APP Tg mice (Ho et al., 2004). In a murine model of AD, consumption of high-calorie diet not only leads to acute hyperglycemia but also increases the production of Aβ (Macauley et al., 2015). Furthermore, alterations in insulin and insulin-mediated signaling pathway may produce changes in Aβ levels in the brain through proteolysis by IDEs (Vekrellis et al., 2000) and/or Aβ clearance from the brain (Shiiki et al., 2004). An alternative mechanism, which promotes the accumulation of autophagosomes and enhances amyloidogenic APP processing (Son et al., 2012b), is the upregulation of the activity of BACE1 (Guglielmotto et al., 2012). Antidiabetic drug, pioglitazone, has been reported to stimulate Aβ degradation by both microglia and astrocytes in an APOE-dependent manner (Mandrekar-Colucci et al., 2012). Thus glucose metabolism may contribute to the modulation of Aβ levels in the brain. Beside this, high levels of glucose due to a high-calorie-diet consumption not only increases the levels of Tau in the brain but also increases Tau phosphorylation through stimulation of c-Jun N-terminal kinase, GSK3β, and AMP kinase, contributing to neurodegeneration in type 2 diabetes, metabolic syndrome, and AD (Farooqui, 2013). Antidiabetic drug, metformin induces protein phosphatase 2A activity and reduces Tau phosphorylation in vitro and in animal models (Kickstein et al., 2010). These studies support the view that glucose energy metabolism is closely related to modulation of Tau and its hyperphosphorylation.

INTERACTIONS OF Aβ-DERIVED DIFFUSIBLE LIGANDS WITH VARIOUS RECEPTORS AND OTHER PROTEINS

ADDL treatment results in the induction of AD-like cellular pathology. The molecular mechanisms involved in this process are not fully understood. However, in the brain, ADDLs may interact with a number of proteins targets including glutamate receptors (both ionotropic and metabotropic), insulin receptors, acetylcholine receptors (both muscarinic and nicotinic), insulin receptor, advanced glycation end products (RAGE), Wnt receptor, neuroligin, cofilin, cellular prion protein (PrP^C), and EphB2 (Fig. 1.7) (De Felice et al., 2009; Lauren et al., 2009; Sturchler et al., 2008; Kessels et al., 2010; Renner et al., 2010; Cisse et al., 2011). Although the precise binding site remains controversial, ADDLs presumably act via multiple receptors at synapse, thereby contributing to the range of issues that characterize AD. ADDL not only impairs synaptic plasticity but also modulates NMDAR-mediated LTP and long term depression (LTD). The induction of LTP contributes to dendritic spine enlargement and increases in synapse density, whereas induction of LTD leads to dendritic spine shrinkage and synapse collapse (Matsuzaki et al., 2004; Zhou et al., 2004).

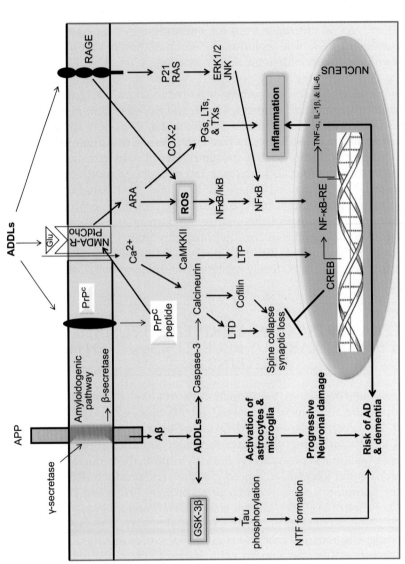

FIGURE 1.7   **Contribution of NMDA, PrP^C, and RAGE in ADDL-mediated synaptic dysfunction in Alzheimer's disease.** *ADDL,* Aβ-derived diffusible ligand; *APP,* amyloid precursor protein; *ARA,* arachidonic acid; *Aβ,* beta-amyloid; *CAMKKII,* Ca^{2+}/calmodulin-dependent protein kinase II; *COX-2,* cyclooxygenase-2; *cPLA$_2$,* cytosolic phospholipase A$_2$; *ERK1/2,* extracellular signal-regulated kinase1/2; *Glu,* glutamate; *I-κB,* inhibitory subunit of NF-κB; *IL-1β,* interleukin-1β; *IL-6,* interleukin-6; *JNK,* c-jun-N-terminal kinase; *NF-κB,* nuclear factor-κB; *NF-κB-RE,* nuclear factor-κB-response element; *NMDA,* N-methyl-ᴅ-aspartate; *NMDAR,* NMDA receptor; *p21,* cyclin-dependent kinase inhibitor 1; *PtdCho,* phosphatidylcholine; *RAGE,* receptor for advanced glycation end products; *ROS,* reactive oxygen species; *TNFα,* tumor necrosis factor-α. ADDLs interact with PrP^C, NMDA, and RAGE initiating ROS and cytokine-mediated signaling.

Signal transduction processes associated with dendritic spine enlargement and shrinkage are supported by protein kinases (p38 mitogen-activated protein kinase, calcium calmodulin–dependent protein kinase II, GSK3β, and ephrin receptor B2). These kinases modulate LTP (Tackenberg and Brandt, 2009). In contrast, protein phosphatases and proteases (calcineurin and caspases) contribute to the induction of LTD (Citri and Malenka, 2008). Increase in expression of cyclic AMP response element–binding protein contributes to the continuous induction of LTP by increasing the expression of several genes expressing and modulating brain-derived neurotrophic factor and nitric oxide synthase (Mayr and Montminy, 2001). These processes modulate memory formation. ADDLs contribute to synaptotoxicity by interacting and activating NMDA receptor. These processes result in the accumulation of the glutamate and D-serine in the synaptic cleft. Overactivation of NMDA receptor triggers abnormal calcium influx, which may have noxious impacts on neurons (Paula-Lima et al., 2013). In addition, the binding of ADDL with NMDA receptor abolishes the induction of LTP, a phenomenon relevant to memory formation. Similarly, the binding of ADDL with cellular prion protein (PrP^C) results in the phosphorylation of NMDA type of glutamate receptor through the activation of Fyn tyrosine kinase, a process that is closely associated with ADDL-mediated synaptotoxicity (Chin et al., 2004). Interactions between ADDL and PrP106-126, a peptide-derived from PrP^C, result in the activation of PLA$_2$ and 5-LOX. These enzymes release arachidonic acid, which is transformed into prostaglandins, leukotrienes, thromboxanes, and ROS (Stewart et al., 2001). The overexpression of PrP^C has been reported to activate membrane NADPH oxidase leading to the generation of more ROS locally initiating the activation of cofilin and formation of cofilin-saturated actin bundles (rods) (Walsh et al., 2014). Rods sequester cofilin from synaptic regions where it is needed for maintaining synaptic plasticity, a process associated with learning and memory. Rods can also interrupt vesicular transport by occluding the neurite within which they form. Through either or both mechanisms, rods may directly mediate the synaptic dysfunction associated with the pathogenesis of AD (Walsh et al., 2014). ADDLs are internalized by the astrocytes, and their internalization induces calcium influx and generation of ROS (Narayan et al., 2014). ADDLs also target microglial cells and their interactions promote TNFα and interleukin-1β release (Ledo et al., 2013). TNFα, in turn, stimulates stress kinase activity in neurons, resulting in impairment in insulin signaling and cognition in mice (Bomfim et al., 2012). Signal transduction mechanisms contributing to ADDL-mediated synaptic dysfunction involve calcium dyshomeostasis, activation of caspases and calcineurin, along with modulation of synaptic excitatory receptors and receptor tyrosine kinases, instigating a cascade of molecular events that culminate in the inhibition of LTP, facilitation of LTD, and synapse loss (Fig. 1.7).

The exposure of rodent brain slices to ADDL also reduces surface NMDAR expression (Snyder et al., 2005; Wang et al., 2000) leading to the internalization of NMDAR. ADDLs also downregulate plasma membrane insulin receptors, via a mechanism involving calcium calmodulin–dependent kinase II and casein kinase II inhibition (De Felice et al., 2009). Endogenous insulin signaling is an obligatory component of normal hippocampal function, and peripheral glucose metabolism. Disruption of insulin signaling by ADDL may lead to rapid impairment of spatial working memory, whereas delivery of exogenous insulin to the hippocampus enhances both memory and metabolism. A number of studies have indicated that insulin resistance, type 2 diabetes, and metabolic syndrome, which are characterized by failed glucose utilization, confers an approximate two- to threefold relative risk for the development of AD (Schrijvers et al., 2010; Farooqui, 2013). Collective evidence suggests that the interactions of ADDL with various receptors on neural cell membranes may trigger multiple effects modulating diverse signal transduction pathways leading to (1) dysregulation of Ca^{2+} homeostasis, (2) mitochondrial dysfunction, (3) reduction in neurogenesis, (4) inhibition of axonal transport, (5) induction of insulin resistance, (6) induction of oxidative stress, (7) activation of cofilin and formation of rods, and (8) modulation of cell cycle reentry (Viola and Klein, 2015). These processes gradually result in neurodegeneration as well as cognitive manifestations of AD (Mucke and Selkoe, 2012). Studies on behavioral consequences of intracerebral injections of ADDLs have been contradictory and depend on the site of injection and the nature of the peptide (Chambon et al., 2011). However, according to many studies intrahippocampal infusion of ADDLs affects working memory when memory performance is evaluated only 10 min after the injection, supporting the view that such deficits can be caused by the acute neurotoxic effect of the ADDLs (Pearson-Leary and McNay, 2012). In contrast, the long-term effects of ADDLs on working memory, which is strongly affected in early stages of AD, remain poorly characterized. Collectively, these studies suggest that binding of ADDLs with a number of receptors, which are embedded in neuronal plasma membrane, may potentially contribute to receptor-mediated ADDL synaptotoxic pathways (Benilova and De Strooper, 2011).

LIMITATIONS OF Aβ CASCADE HYPOTHESIS

It is clear from the earlier discussion that a large number of studies support the amyloid hypothesis. However, there are important limitations, which need to be mentioned. First, a direct correlation between ADDL and dementia severity has not been clearly established, since some patients without amyloid deposition show severe memory deficits, whereas other patients with cortical Aβ deposits have no dementia symptoms (Aizenstein et al., 2008). These observations along with the failure of some anti-Aβ therapies to preserve or rescue cognitive function suggest that Aβ may not be universally neurotoxic (Extance, 2010; Robakis, 2011; Tayeb et al., 2013), but that other mechanisms directly or indirectly related to ADDL may contribute to the pathogenesis of AD. In fact, several investigators have indicated that in animal models, ADDLs may not be toxic (Lee et al., 2007; Robakis, 2011), but activate kinases (Tabaton et al., 2010), acting as antioxidant (Zou et al., 2002), containing antimicrobial activity (Soscia et al., 2010), and modulating cholesterol transport (Yao and Papadopoulos, 2002). It is also reported that ADDLs are normally present in the brain and their concentration is higher in younger than in older individuals in the absence of dementia (van Helmond et al., 2010). Second, it is very difficult to study the pathophysiological role of each ADDLs and their aggregation status, because the composition of ADDL solutions and their molarities used in in vitro and in vivo conditions are hard to establish precisely (Puzzo and Arancio, 2013). This is due to the conformational changes that occur in ADDLs when they are dissolved in water or buffer. Another limitation is that ADDLs stick to the tubes, altering the final concentration of the solution. Finally, it must be stated that the results obtained in experimental models using a single ADDL species are not easily transposable in vivo, mainly because the brain normally produces a variety of ADDL peptides, which are either transported by transport proteins (LRP1, APOE, and ABC transporter) or hydrolyzed by several enzymes such as IDE and MMP (Puzzo and Arancio, 2013). Furthermore, behavioral abnormalities observed in animal models overexpressing APP on which Aβ cascade hypothesis is based cannot unambiguously be assigned to ADDLs because APP is metabolized to a large number of APP-derived fragments some of which are neurotoxic and others are not (Gauthier et al., 1997; Nalbantoglu et al., 1997). For example, an APP fragment derived from sAPPβ interacts with DR6 to trigger axon pruning and neuron death (Nikolaev et al., 2009). The short AID/AICD is a biologically active intracellular peptide, which modulates cell death, gene transcription, and Ca^{2+} homeostasis (Hamid et al., 2007; Passer et al., 2000; Kim et al., 2003). Caspase-derived APP fragments C31 (Lu et al., 2000) and Jcasp (Madeira et al., 2005; Bertrand et al., 2001) mediate in vitro toxic activities. Because of these evidence, various APP-derived fragments AID/AICD (Passer et al., 2000; Ghosal et al., 2009; Galvan et al., 2006), C31, and JCasp (Madeira et al., 2005), and sAPPβ (Nikolaev et al., 2009) have been reported to cause cytotoxicity and implicated in neurodegenerative processes. Thus at present the absence of information supporting disease-associated increases in ADDLs is a serious weakness of the theory that such ADDLs are the causative agents of the neurodegeneration of AD (Giliberto et al., 2010). In addition, the overexpression of APP in animal brain may cause neurotoxicity due to trafficking abnormalities. Based on the aforementioned studies, it is still unclear whether the behavioral abnormalities detected in Tg APP models are due to specific ADDLs, to toxicities derived from other toxic metabolites of APP, or to interference with cellular pathways due to the high levels of exogenous APP. Importantly, soluble Aβ peptides are normal components of human serum and cerebrovascular fluid and reports from several groups suggest they may have useful biological functions (Paris et al., 2004; Chen and Dong, 2009; Giuffrida et al., 2009).

Another important point is that neither Aβ plaques nor phospho-Tau neurofibrillary tangles are specific for AD. About 30% of normal-aged people have as many Aβ plaques in their brains as in typical cases of AD (Tomlinson et al., 1970; Dickson et al., 1992; van Duinen et al., 1987). Furthermore, other neuropathological conditions such as Parkinson disease (Petrou et al., 2015), which is characterized by monoaminergic dysfunction, Lewy body pathology, and cerebrovascular disease along with cognitive impairment, and hereditary cerebral hemorrhage with amyloidosis of Dutch origin and sporadic CAA, which show amyloid pathology similar to AD without any dementia, suggest that amyloid alone is insufficient to cause neuronal loss and cognitive symptoms observed in AD (Coria et al., 1987). Furthermore, neurotraumatic diseases such as stroke, head injury, and chronic traumatic encephalopathy are also accompanied by the accumulation of Aβ in the brain (Farooqui, 2010). Formation of neurofibrillary tangles with hyperphosphory-lated Tau, the other characteristic feature of AD, is a hallmark of several neurodegenerative diseases called tauopathies, which include frontotemporal dementia with parkinsonism linked to chromosome-17 Tau, Pick disease, corticobasal degeneration, postsupranuclear palsy, dementia pugilistica/traumatic brain injury/chronic traumatic encephalopathy, and Guam parkinsonism–dementia complex. Thus several different mechanisms may contribute to the etiopathogenesis of both plaques and tangles in the aforementioned neurological conditions. AD is a multifactorial disease, and its cause is complex. Aβ cascade hypothesis alone may not be able to account for all aspects of AD. The fact that vast overproduction of ADDLs in the brain of transgenic mouse models has failed to cause neurodegeneration raises the question as to whether accumulation of ADDLs is indeed the culprit for neurodegeneration in AD. Many studies have indicated that ADDL/ADDL-independent factors, including the actions of AD-related genes (MAP Tau, polymorphisms of apolipoprotein E4), inflammation, and oxidative stress, may also contribute to AD pathogenesis. Converging evidence suggests that the main mechanisms responsible for the neuronal cell death in AD are still poorly understood. It is unclear, for example, why certain neuronal populations such as cholinergic neurons are more vulnerable to AD than other neurons and how factors like the APOE4 allele, the process of aging, and genetic familial Alzheimer's disease (FAD) mutations contribute to degeneration of specific neuronal populations. Other studies have indicated that environmental factors such as oxidative stress and inflammatory processes may also contribute to the neuronal cell death of AD (Farooqui, 2010), although neither antioxidants nor antiinflammatory agents have been reported to have a significant effect on the course of AD-type dementia. This may be due to the fact that by the time

AD is detected after a significant number of neurons have been degenerated and these are difficult to replace. However, it is possible that in sporadic AD, the final outcome of neurodegenerative process is determined by both genetic and environmental factors. At present, the Aβ cascade hypothesis based on FAD mutations leading to ADDL-mediated neurodegeneration is the best available model for studying the cellular and molecular mechanisms of pathogenesis of sporadic disease. Since the clinical manifestations and neuropathology of FAD and sporadic AD are similar, lessons learned from studying the mechanisms of FAD can also be applicable to the pathogenesis of sporadic AD. Based on the aforementioned information, there has been a passionate debate about the acceptance of the Aβ cascade hypothesis (Herrup, 2015; Musiek and Holtzman, 2015). According to investigators, who oppose Aβ cascade hypothesis (Herrup, 2015), AD is a complex multifactorial disease, which cannot be due to Aβ. The pathogenesis of AD may involve changes in a number of signal transduction processes associated with neurodegeneration in the brain aging process (Herrup, 2015) and it is also reported that in AD, the accumulation of Aβ aggregates acts as a natural antibiotic against bacterial infection. Aβ aggregates provide neuroprotection by trapping and imprisoning the bacterial pathogen (*Salmonella typhimurium*) (Kumar et al., 2016). It is suggested that this information can be used for developing new potential therapies for the treatment of AD. In my judgment, more studies are required on neuronal membrane–related signal transduction processes and neuronal network that contribute to the generation of not only Aβ but also other processes that aid chronic neurodegeneration in AD.

CONCLUSION

It is well established that APP processing occurs via two pathways: β-secretase-mediated amyloidogenic pathway and α-secretase-catalyzed nonamyloidogenic pathway. APP processing through amyloidogenic and nonamyloidogenic processing depends on the cellular levels of α- and β-secretases and the traffic of APP to subcellular organelles expressing these proteases. In nonamyloidogenic pathway, cleavage occurs by α-secretase within the Aβ domain forming a large soluble N-terminal fragment (sAPPα) and a nonamyloidogenic CTF of 83 amino acid residues (C83). Further cleavage of this CTF by γ-secretase produces the nonamyloidogenic peptide (P3) and AICD. These APP-derived products contribute not only to memory improvement and axonal pruning but also to neurite extension in mutant mice. In amyloidogenic pathway, APP processing takes place by β-secretase at the beginning of the Aβ domain forming a soluble N-terminal fragment (sAPPβ) and an amyloidogenic CTF of 99 residues (C99). This CTF is further processed by γ-secretase generating Aβ, which undergoes oligomerization forming ADDLs. This species modulates the mitochondrial redox activity, increases the production of ROS, damages the intracellular calcium homeostasis, and promotes the formation of selective calcium channels, and also promotes the expression and the release of cytokines leading to neuroinflammation. Very little is known about the molecular mechanism(s) of endogenous ADDL assemblies. So exclusive attempts must be made to study the molecular mechanism(s) of ADDL isolation and characterization by various procedures. In addition, it is important to standardize the relative concentrations of ADDLs needed for neurotoxicity. Only then can investigators reduce the results on contradictory observations that plague the progress in identifying specific signaling cascades responsible for the pathogenesis of AD along with other factors (hyperphosphorylation of Tau protein) and subsequent reasonable diagnostic and therapeutic procedures for the treatment of AD. Although considerable progress has been made on the cause of AD, the pathogenesis of AD is still unclear. Accumulated data show that senile plaques and NTFs are associated with the neurodegeneration in AD. However, these two major pathological hallmarks may be just a consequence of the disease process rather than an initial event that causes AD. The failed result of several major clinical trials targeting Aβ is one of the strongest supports for this viewpoint. It goes with saying that more studies on the pathophysiological mechanisms associated with the formation of these two major pathological hallmarks of AD are fundamental and of critical importance not only for elucidating the cause of AD but also for the development of a disease-modifying treatment for AD.

References

AD International, 2013. World Alzheimer Report. Alzheimer's Disease International Consortium, Illinois, USA.

Agostinho, P., Cunha, R.A., Oliveira, C., 2010. Neuroinflammation, oxidative stress and the pathogenesis of Alzheimer's disease. Curr. Pharm. Des. 16, 2766–2778.

Aguzzi, A., Haass, C., 2003. Games played by rogue proteins in prion disorders and Alzheimer's disease. Science 302, 814–818.

Aizenstein, H.J., Nebes, R.D., Saxton, J.A., Price, J.C., Mathis, C.A., Tsopelas, N.D., et al., 2008. Frequent amyloid deposition without significant cognitive impairment among the elderly. Arch. Neurol. 65, 1509–1517.

Allinson, T.M., Parkin, E.T., Turner, A.J., Hooper, N.M., 2003. ADAMs family members as amyloid precursor protein α-secretases. J. Neurosci. Res. 74, 342–352.

Allsop, D., Wong, C.W., Ikeda, S., Landon, M., Kidd, M., Glenner, G.G., 1988. Immunohistochemical evidence for the derivation of a peptide ligand from the amyloid β-protein precursor of Alzheimer disease. Proc. Natl. Acad. Sci. U. S. A. 85, 2790–2794.

Alzheimer's Association, 2012. 2011 Alzheimer's diseases facts and figures: prevalence. Alzheimers Dement. 7, 12–13.

Alzheimer's Association, 2013. Alzheimer's disease facts and figures. Alzheimer Dement. 2013 (9), 208–245.

Arnaud, L., Robakis, N.K., Figueiredo-Pereira, M.E., 2006. It may take inflammation, phosphorylation and ubiquitination to 'tangle' in Alzheimer's disease. Neurodegener. Dis. 3, 313–319.

Auriel, E., Greenberg, S.M., 2012. The pathophysiology and clinical presentation of cerebral amyloid angiopathy. Curr. Atheroscler. Rep. 14, 343–350.

Avila, J., Gomez de Barreda, E., Engel, T., Lucas, J.J., Hernandez, F., 2010. Tau phosphorylation in hippocampus results in toxic gain-of-function. Biochem. Soc. Trans. 38, 977–980.

Bancher, C., Braak, H., Fischer, P., Jellinger, K.A., 1993. Neuropathological staging of Alzheimer lesions and intellectual status in Alzheimer's and Parkinson's disease patients. Neurosci. Lett. 162, 179–182.

Barnes, K., Doherty, S., Turner, A.J., 1995. Endopeptidase-24.11 is the integral membrane peptidase initiating degradation of somatostatin in the hippocampus. J. Neurochem. 64, 1826–1832.

Barnham, K.J., McKinstry, W.J., Multhaup, G., Galatis, D., Morton, C.J., Curtain, C.C., et al., 2003. Structure of the Alzheimer's disease amyloid precursor protein copper binding domain. A regulator of neuronal copper homeostasis. J. Biol. Chem. 278, 17401–17407.

Barucker, C., Bittner, H.J., Chang, P.K., Cameron, S., Hancock, M.A., Liebsch, F., et al., 2015. Aβ42-oligomer Interacting Peptide (AIP) neutralizes toxic amyloid-β42 species and protects synaptic structure and function. Sci. Rep. 5, 15410.

Bateman, R.J., Munsell, L.Y., Morris, J.C., Swarm, R., Yarasheski, K.E., Holtzman, D.M., 2006. Human amyloid-β synthesis and clearance rates as measured in cerebrospinal fluid in vivo. Nat. Med. 12, 856–861.

Bell, R.D., Sagare, A.P., Friedman, A.E., Bedi, G.S., Holtzman, D.M., Deane, R., 2007. Transport pathways for clearance of human Alzheimer's amyloid β-peptide and apolipoproteins E and J in the mouse central nervous system. J. Cereb. Blood Flow Metab. 27, 909–918.

Benilova, I., De Strooper, B., 2011. An overlooked neurotoxic species in Alzheimer's disease. Nat. Neurosci. 14, 949–950.

Benilova, I., Karran, E., De Strooper, B., 2012. The toxic Aβ oligomer and Alzheimer's disease: an emperor in need of clothes. Nat. Neurosci. 15, 349–357.

Bertram, L., 2009. Alzheimer's disease genetics current status and future perspectives. Int. Rev. Neurobiol. 84, 167–184.

Bertram, L., Tanzi, R.E., 2008. Thirty years of Alzheimer's disease genetics: the implications of systematic meta-analyses. Nat. Rev. Neurosci. 9, 768–778.

Bertrand, E., Brouillet, E., Caille, I., Bouillot, C., Cole, G.M., et al., 2001. A short cytoplasmic domain of the amyloid precursor protein induces apoptosis in vitro and in vivo. Mol. Cell. Neurosci. 18, 503–511.

Bhaskar, K., Miller, M., Chludzinski, A., Herrup, K., Zagorski, M., Lamb, B.T., 2009. The PI3K-Akt-mTOR pathway regulates Aβ oligomer induced neuronal cell cycle events. Mol. Neurodegener. 4, 14.

Binder, L.I., Frankfurter, A., Rebhun, L.I., 1985. The distribution of tau in the mammalian central nervous system. J. Cell Biol. 101, 1371–1810.

Bitan, G., Kirkitadze, M.D., Lomakin, A., Vollers, S.S., Benedek, G.B., Teplow, D.B., 2003. Amyloid β-protein (Aβ) assembly: Aβ40 and Aβ42 oligomerize through distinct pathways. Proc. Natl. Acad. Sci. U. S. A. 100, 330–335.

Blennow, K., de Leon, M.J., Zetterberg, H., 2006. Alzheimer's disease. Lancet 368, 387–403.

Bloom, G.S., 2014. Amyloid-β and Tau: the trigger and bullet in Alzheimer disease pathogenesis. JAMA Neurol. 71, 505–508.

Blurton-Jones, M., Laferla, F.M., 2006. Pathways by which Aβ facilitates tau pathology. Curr. Alzheimer Res. 3, 437–448.

Bomfim, T.R., Forny-Germano, L., Sathler, L.B., Brito-Moreira, J., Houzel, J.C., Decker, H., et al., 2012. An anti-diabetes agent protects the mouse brain from defective insulin signaling caused by Alzheimer's disease–associated Aβ oligomers. J. Clin. Invest. 122, 1339–1353.

Borger, E., Aitken, L., Du, H., Zhang, W., Gunn-Moore, F.J., Yan, S.S., 2013. Is amyloid binding alcohol dehydrogenase a drug target for treating Alzheimer's disease? Curr. Alzheimer Res. 10, 21–29.

Braak, H., Braak, E., 1998. Evolution of neuronal changes in the course of Alzheimer's disease. J. Neural Transm. Suppl. 53, 127–140.

Bretteville, A., Ando, K., Ghestem, A., Loyens, A., Bégard, S., et al., 2009. Two-dimensional electrophoresis of tau mutants reveals specific phosphorylation pattern likely linked to early tau conformational changes. PLoS One 4, 6.

Brookmeyer, R., Gray, S., Kawas, C., 1998. Projections of Alzheimer's disease in the United States and the public health impact of delaying disease onset. Am. J. Public Health 88, 1337–1342.

Bu, G., 2009. Apolipoprotein E and its receptors in Alzheimer's disease: pathways, pathogenesis and therapy. Nat. Rev. Neurosci. 10, 333–344.

Caccamo, A., Oddo, S., Sugarman, M.C., Akbari, Y., LaFerla, F.M., 2005. Age- and region-dependent alterations in Aβ-degrading enzymes: implications for Aβ-induced disorders. Neurobiol. Aging 26, 645–654.

Cai, Z., Zhao, B., Li, K., Quazi, S.H., Tan, Y., 2012. Mammalian target of rapamycin: a valid therapeutic target through the autophagy pathway for Alzheimer's disease? J. Neurosci. Res. 90, 1105–1118.

Cao, X., Sudhof, T.C., 2001. A transcriptively active complex of APP with Fe65 and histone acetyltransferase Tip60. Science 293 (5527), 115–120.

Castellani, R.J., Zhu, X., Lee, H.-G., Smith, M.A., Perry, G., 2009. Molecular pathogenesis of Alzheimer's disease: reductionist versus expansionist approaches. Int. J. Mol. Sci. 10, 1386–1406.

Chabrier, M.A., Cheng, D., Castello, N.A., Green, K.N., LaFerla, F.M., 2014. Synergistic effects of amyloid-β and wild-type human tau on dendritic spine loss in a floxed double transgenic model of Alzheimer's disease. Neurobiol. Dis. 64, 107–117.

Chambon, C., Wegener, N., Gravius, A., Danysz, W., 2011. Behavioural and cellular effects of exogenous amyloid-β peptides in rodents. Behav. Brain Res. 225, 623–641.

Chen, Y., Dong, C., 2009. Aβ40 promotes neuronal cell fate in neural progenitor cells. Cell Death Differ. 16, 386–394.

Chen, T.J., Wang, D.C., Chen, S.S., 2009. Amyloid-β interrupts the PI3K-Akt-mTOR signaling pathway that could be involved in brain-derived neurotrophic factor-induced Arc expression in rat cortical neurons. J. Neurosci. Res. 87, 2297–2307.

Chin, J., Palop, J.J., Yu, G.Q., Kojima, N., Masliah, E., Mucke, L., 2004. Fyn kinase modulates synaptotoxicity, but not aberrant sprouting, in human amyloid precursor protein transgenic mice. J. Neurosci. 24, 4692–4697.

Cirrito, J.R., Kang, J.E., Lee, J., Stewart, F.R., Verges, D.K., Silverio, L.M., et al., 2008. Endocytosis is required for synaptic activity-dependent release of amyloid-β in vivo. Neuron 58, 42–51.

Cisse, M., Halabisky, B., Harris, J., Devidze, N., Dubal, D.B., Sun, B., et al., 2011. Reversing EphB2 depletion rescues cognitive functions in Alzheimer model. Nature 469, 47–52.

Citri, A., Malenka, R.C., 2008. Synaptic plasticity: multiple forms, functions, and mechanisms. Neuropsychopharmacology 33, 18–41.

Cochran, J.N., Hall, A.M., Roberson, E.D., 2014. The dendritic hypothesis for Alzheimer's disease pathophysiology. Brain Res. Bull. 103, 18–28.

Connelly, L., Jang, H., Arce, F.T., Capone, R., Kotler, S.A., Ramachandran, S., et al., 2012. Atomic force microscopy and MD simulations reveal pore-like structures of all-D-enantiomer of Alzheimer's β-amyloid peptide: relevance to the ion channel mechanism of AD pathology. J. Phys. Chem. B 116, 1728–1735.

Cordy, J.M., Hooper, N.M., Turner, A.J., 2003. Exclusively targeting β-secretase to lipid rafts by GPI-anchor addition up-regulates β-site processing of the amyloid precursor protein. Proc. Natl. Acad. Sci. U. S. A. 100, 11735–11740.

Coria, F., Castaño, E.M., Frangione, B., 1987. Brain amyloid in normal aging and cerebral amyloid angiopathy is antigenically related to Alzheimer's disease β-protein. Am. J. Pathol. 129, 422–428.

Costa, R.M., Drew, J., Silva, A.J., 2005. Notch to remember. Trends Neurosci. 28, 429–435.

Crews, L., Rockenstein, E., Masliah, E., 2010. APP transgenic modeling of Alzheimer's disease: mechanisms of neurodegeneration and aberrant neurogenesis. Brain Struct. Funct. 214, 111–126.

Dawkins, E., Small, D.H., 2014. Insights into the physiological function of the β-amyloid precursor protein: beyond Alzheimer's disease. J. Neurochem. 129, 756–769.

De Felice, F.G., Wu, D., Lambert, M.P., Fernandez, S.J., Velasco, P.T., Lacor, P.N., 2009. Alzheimer's disease-type neuronal tau hyperphosphorylation induced by Aβ oligomers. Neurobiol. Aging 29, 1334–1347.

de la Monte, S.M., Tong, M., 2014. Brain metabolic dysfunction at the core of Alzheimer's disease. Biochem. Pharmacol. 88, 548–559.

Deane, R., Wu, Z., Sagare, A., Davis, J., Du Yan, S., Hamm, K., et al., 2004. LRP/amyloid β-peptide interaction mediates differential brain efflux of Aβ isoforms. Neuron 43, 333–344.

Deane, R., Singh, I., Sagare, A.P., Bell, R.D., Ross, N.T., LaRue, B., et al., 2012. A multimodal RAGE-specific inhibitor reduces amyloid β-mediated brain disorder in a mouse model of Alzheimer disease. J. Clin. Invest. 122, 1377.

Dickson, D.W., Crystal, H.A., Mattiace, L.A., Masur, D.M., Blau, A.D., Davies, P., et al., 1992. Identification of normal and pathological aging in prospectively studied nondemented elderly humans. Neurobiol. Aging 13, 179–189.

Duce, J.A., Tsatsanis, A., Cater, M.A., James, S.A., Robb, E., Wikhe, K., 2010. Iron-export ferroxidase activity of β-amyloid precursor protein is inhibited by zinc in Alzheimer's disease. Cell 142, 857–867.

Extance, A., 2010. Alzheimer's failure raises questions about disease-modifying strategies. Nat. Rev. Drug Discov. 9, 749–751.

Fang, C.L., Wu, W.H., Liu, Q., Sun, X., Ma, Y., Zhao, Y.F., Li, Y.M., 2010. Dual functions of β-amyloid oligomer and fibril in Cu(II)-induced H_2O_2 production. Regul. Pept. 163, 1–6.

Farías, G., Cornejo, A., Jiménez, J., Guzmán, L., Maccioni, R.B., 2011. Mechanisms of tau self-aggregation and neurotoxicity. Curr. Alzheimer Res. 8, 608–614.

Farooqui, A.A., 2010. Neurochemical Aspects of Neurotraumatic and Neurodegenerative Diseases. Springer, New York.

Farooqui, A.A., 2013. Metabolic Syndrome: An Important Risk Factor for Stroke, Alzheimer, and Depression. Springer Science-Business, New York.

Farooqui, A.A., 2014. Inflammation and Oxidative Stress in Neurological Disorders. Springer International Publishing, Switzerland.

Farooqui, A.A., 2015. High Calorie Diet and the Human Brain. Springer International Publishing, Switzerland.

Frisoni, G.B., Rozzini, L., Gozzetti, A., Binetti, G., Zanetti, O., Bianchetti, A., Trabucchi, M., Cummings, J.L., 1999. Behavioral syndromes in Alzheimer's disease: description and correlates. Dement. Geriatr. Cogn. Disord. 10, 130–138.

Frost, B., Hemberg, M., Lewis, J., Feany, M.B., 2014. Tau promotes neurodegeneration through global chromatin relaxation. Nat. Neurosci. 17, 357–366.

Fu, M., Zuo, Y., 2011. Experience-dependent structural plasticity in the cortex. Trends Neurosci. 34, 177–187.

Fukami, S., Watanabe, K., Iwata, N., Haraoka, J., Lu, B., et al., 2002. Aβ-degrading endopeptidase, neprilysin, in mouse brain: synaptic and axonal localization inversely correlating with Aβ pathology. Neurosci. Res. 43, 39–56.

Galvan, V., Gorostiza, O.F., Banwait, S., Ataie, M., Logvinova, A.V., et al., 2006. Reversal of Alzheimer's-like pathology and behavior in human APP transgenic mice by mutation of Asp664. Proc. Natl. Acad. Sci. U. S. A. 103, 7130–7135.

Gauthier, S., Panisset, M., Nalbantoglu, J., Poirier, J., 1997. Alzheimer's disease: current knowledge, management and research. CMAJ 157, 1047–1052.

Ghosal, K., Vogt, D.L., Liang, M., Shen, Y., Lamb, B.T., et al., 2009. Alzheimer's disease-like pathological features in transgenic mice expressing the APP intracellular domain. Proc. Natl. Acad. Sci. U. S. A. 106, 18367–18372.

Giliberto, L., d'Abramo, C., Acker, C.M., Davies, P., D'Adamio, L., 2010. Transgenic expression of the amyloid-β precursor protein-intracellular domain does not induce Alzheimer's disease–like traits in vivo. PLoS One 5, e11609.

Giuffrida, M.L., Caraci, F., Pignataro, B., Cataldo, S., De Bona, P., Bruno, V., 2009. β-Amyloid monomers are neuroprotective. J. Neurosci. 29, 10582–10587.

Goate, A., Chartier-Harlin, M.C., Mullan, M., Brown, J., Crawford, F., Fidani, L., Giuffra, L., Haynes, A., Irving, N., James, L., 1991. Segregation of a missense mutation in the amyloid precursor protein gene with familial Alzheimer's disease. Nature 349, 704–706.

Goedert, M., Spillantini, M.G., Jakes, R., Rutherford, D., Crowther, R.A., 1989. Multiple isoforms of human microtubule-associated protein tau: sequences and localization in neurofibrillary tangles of Alzheimer's disease. Neuron 3, 519–526.

Gong, C.X., Iqbal, K., 2008. Hyperphosphorylation of microtubule-associated protein tau: a promising therapeutic target for Alzheimer disease. Curr. Med. Chem. 15, 2321–2328.

Gong, Y., Chang, L., Viola, K.L., Lacor, P.N., Lambert, M.P., Finch, C.E., Krafft, G.A., Klein, W.L., 2003. Alzheimer's disease-affected brain: presence of oligomeric Aβ ligands (ADDLs) suggests a molecular basis for reversible memory loss. Proc. Natl. Acad. Sci. U. S. A. 100, 10417–10422.

Gralle, M., Oliveira, C.L., Guerreiro, L.H., McKinstry, W.J., Galatis, D., Masters, C.L., Cappai, R., Parker, M.W., Ramos, C.H., Torriani, I., Ferreira, S.T., 2006. Solution conformation and heparin-induced dimerization of the full-length extracellular domain of the human amyloid precursor protein. J. Mol. Biol. 357, 493–508.

Greenfield, J.P., Tsai, J., Gouras, G.K., Hai, B., Thinakaran, G., Checler, F., et al., 1999. Endoplasmic reticulum and trans-Golgi network generate distinct populations of Alzheimer β-amyloid peptides. Proc. Natl. Acad. Sci. U. S. A. 96, 742–747.

Grimm, M.O., Grimm, H.S., Hartmann, T., 2007. Amyloid β as a regulator of lipid homeostasis. Trends Mol. Med. 13, 337–344.

Grimm, M.O., Rothhaar, T.L., Grösgen, S., Burg, V.K., Hundsdörfer, B., Haupenthal, V.J., Friess, P., Kins, S., Grimm, H.S., Hartmann, T., 2011a. Trans fatty acids enhance amyloidogenic processing of the Alzheimer amyloid precursor protein (APP). J. Nutr. Biochem. 23, 1214–1223.

Grimm, M.O., Kuchenbecker, J., Grösgen, S., Burg, V.K., Hundsdörfer, B., Rothhaar, T.L., et al., 2011b. Docosahexaenoic acid reduces amyloid β production via multiple pleiotropic mechanisms. J. Biol. Chem. 286, 14028–14039.

Grimm, M.O., Zinser, E.G., Grösgen, S., Hundsdörfer, B., Rothhaar, T.L., Burg, V.K., et al., 2012. Amyloid precursor protein (APP) mediated regulation of ganglioside homeostasis linking Alzheimer's disease pathology with ganglioside metabolism. PLoS One 7, e34095.

Grimm, M.O., Zimmer, V.C., Lehmann, J., Grimm, H.S., Hartmann, T., 2013. The impact of cholesterol, DHA, and sphingolipids on Alzheimer's disease. Biomed. Res. Int. 814390.

Guerreiro, R., Wojtas, A., Bras, J., Carrasquillo, M., Rogaeva, E., et al., 2013. TREM2 variants in Alzheimer's disease. N. Engl. J. Med. 368, 117–127.

Guglielmotto, M., Aragno, M., Tamagno, E., Vercellinatto, I., Visentin, S., Medana, C., et al., 2012. AGEs/RAGE complex upregulates BACE1 via NF-κB pathway activation. Neurobiol. Aging 33, 196.e13–196.e27.

Haass, C., Selkoe, D.J., 2007. Soluble protein oligomers in neurodegeneration: lessons from the Alzheimer's amyloid β-peptide. Nat. Rev. Mol. Cell Biol. 8, 101–112.

Hamid, R., Kilger, E., Willem, M., Vassallo, N., Kostka, M., et al., 2007. Amyloid precursor protein intracellular domain modulates cellular calcium homeostasis and ATP content. J. Neurochem. 102, 1264–1275.

Hardy, J., 2006. Alzheimer's disease: the amyloid cascade hypothesis: an update and reappraisal. J. Alzheimers Dis. 9 (Suppl. 3), 151–153.

Hardy, J., 2009. The amyloid hypothesis for Alzheimer's disease: a critical reappraisal. J. Neurochem. 110, 1129–1134.

Hardy, J.A., Higgins, G.A., 1992. Alzheimer's disease: the amyloid cascade hypothesis. Science 256, 184–185.

Hardy, J., Selkoe, D.J., 2002. The amyloid hypothesis of Alzheimer's disease: progress and problems on the road to therapeutics. Science 297, 353–356.

Heneka, M.T., O'Banion, M.K., 2007. Inflammatory processes in Alzheimer's disease. J. Neuroimmunol. 184, 69–91.

Herrup, K., 2015. The case for rejecting the amyloid cascade hypothesis. Nat. Neurosci. 18, 794–799.

Herz, J., Strickland, D.K., 2001. LRP: a multifunctional scavenger and signaling receptor. J. Clin. Invest. 108, 779–784.

Hickman, S.E., El Khoury, J., 2014. TREM2 and the neuroimmunology of Alzheimer's disease. Biochem. Pharmacol. 88, 495–498.

Ho, A., Sudhof, T.C., 2004. Binding of F-spondin to amyloid-β precursor protein: a candidate amyloid-β precursor protein ligand that modulates amyloid-β precursor protein cleavage. Proc. Natl. Acad. Sci. U. S. A. 101, 2548–2553.

Ho, L., Qin, W., Pompl, P.N., Xiang, Z., Wang, J., Zhao, Z., et al., 2004. Diet-induced insulin resistance promotes amyloidosis in a transgenic mouse model of Alzheimer's disease. FASEB J. 18, 902–904.

Hooper, N.M., Turner, A.J., 2002. The search for α-secretase and its potential as a therapeutic approach to Alzheimer s disease. Curr. Med. Chem. 9, 1107–1119.

Hsieh, H., Boehm, J., Sato, C., Iwatsubo, T., Tomita, T., Sisodia, S., et al., 2006. AMPAR removal underlies Aβ-induced synaptic depression and dendritic spine loss. Neuron 52, 831–843.

Hughes, T.M., Lopez, O.L., Evans, R.W., Kamboh, M.I., Williamson, J.D., Klunk, W.E., et al., 2014. Markers of cholesterol transport are associated with amyloid deposition in the brain. Neurobiol. Aging 35, 802–807.

Innocent, N., Cousins, S.L., Stephenson, F.A., 2012. NMDA receptor/amyloid precursor protein interactions: a comparison between wild-type and amyloid precursor protein mutations associated with familial Alzheimer's disease. Neurosci. Lett. 515, 131–136.

Iqbal, K., Liu, F., Gong, C.X., Alonso, A.C., Grundke-Iqbal, I., 2009. Mechanisms of tau-induced neurodegeneration. Acta Neuropathol. 118, 53–69.

Ittner, L.M., Götz, J., 2011. Amyloid-β and tau—a toxic pas de deux in Alzheimer's disease. Nat. Rev. Neurosci. 12, 65–72.

Iwata, N., Tsubuki, S., Takaki, Y., Shirotani, K., Lu, B., Gerard, N.P., et al., 2001. Metabolic regulation of brain Aβ by neprilysin. Science 292, 1550–1552.

Iwata, N., Mizukami, H., Shirotani, K., Takaki, Y., Muramatsu, S., et al., 2004. Presynaptic localization of neprilysin contributes to efficient clearance of amyloid-β peptide in mouse brain. J. Neurosci. 24, 991–998.

Izumi, R., Yamada, T., Yoshikai, S., Sasaki, H., Hattori, M., Sakaki, Y., 1992. Positive and negative regulatory elements for the expression of the Alzheimer's disease amyloid precursor-encoding gene in mouse. Gene 112, 189–195.

Jang, H., Arce, F.T., Ramachandran, S., Kagan, B.L., Lal, R., Nussinov, R., 2014. Disordered amyloidogenic peptides may insert into the membrane and assemble into common cyclic structural motifs. Chem. Soc. Rev. 43, 6750–6764.

Jellinger, K.A., 2009. Recent advances in our understanding of neurodegeneration. J. Neural Transm. 116, 1111–1162.

Jellinger, K.A., 2010. Basic mechanisms of neurodegeneration: a critical update. J. Cell. Mol. Med. 14, 457–487.

Jiang, T., Yu, J.T., Zhu, X.C., Tan, L., 2013. TREM2 in Alzheimer's disease. Mol. Neurobiol. 48, 180–185.

Jiang, T., Tan, L., Zhu, X.C., Zhang, Q.Q., Cao, L., et al., 2014. Upregulation of TREM2 ameliorates neuropathology and rescues spatial cognitive impairment in a transgenic mouse model of Alzheimer's disease. Neuropsychopharmacology 39, 2949–2962.

Johnson, M.A., Ables, J.L., Eisch, A.J., 2009. Cell-intrinsic signals that regulate adult neurogenesis in vivo: insights from inducible approaches. BMB Rep. 42, 245–259.

Josephs, K.A., Whitwell, J.L., Ahmed, Z., Shiung, M.M., Weigand, S.D., Knopman, D.S., et al., 2008. β-Amyloid burden is not associated with rates of brain atrophy. Ann. Neurol. 63, 204–212.

Kaden, D., Voigt, P., Munter, L.M., Bobowski, K.D., Schaefer, M., Multhaup, G., 2009. Subcellular localization and dimerization of APLP1 are strikingly different from APP and APLP2. J. Cell Sci. 122 (Pt 3), 368–377.

Kaether, C., Haass, C., 2004. A lipid boundary separates APP and secretases and limits amyloid β-peptide generation. J. Cell Biol. 167, 809–812.

Kamenetz, F., Tomita, T., Hsieh, H., Seabrook, G., Borchelt, D., Iwatsubo, T., et al., 2003. APP processing and synaptic function. Neuron 37925–37937.

Kang, J., Muller-Hill, B., 1990. Differential splicing of Alzheimer's disease amyloid A4 precursor RNA in rat tissues: PreA4$_{695}$ mRNA is predominantly produced in rat and human brain. Biochem. Biophys. Res. Commun. 166, 1192–1200.

Kang, J., Lemaire, H.G., Unterbeck, A., Salbaum, J.M., Masters, C.L., Grzeschik, K.H., et al., 1987. The precursor of Alzheimer's disease amyloid A4 protein resembles a cell-surface receptor. Nature 325, 733–736.

Karran, E., Mercken, M., Strooper, B.D., 2011. The amyloid cascade hypothesis for Alzheimer's disease: an appraisal for the development of therapeutics. Nat. Rev. Drug Discov. 10, 698–712.

Katzman, R., Saitoh, T., 1991. Advances in Alzheimer's disease. FASEB J. 5, 278–286.

Kawarabayashi, T., Shoji, M., Younkin, L.H., Wen-Lang, L., Dickson, D.W., et al., 2004. Dimeric amyloid β protein rapidly accumulates in lipid rafts followed by apolipoprotein E and phosphorylated tau accumulation in the Tg2576 mouse model of Alzheimer's disease. J. Neurosci. 24, 3801–3809.

Kessels, H.W., Nguyen, L.N., Nabavi, S., Malinow, R., 2010. The prion protein as a receptor for amyloid-β. Nature 466, E3–E4.

Kickstein, E., Krauss, S., Thornhill, P., Rutschow, D., Zeller, R., Sharkey, J., et al., 2010. Biguanide metformin acts on tau phosphorylation via mTOR/protein phosphatase 2A (PP2A) signaling. Proc. Natl. Acad. Sci. U. S. A. 107, 21830–21835.

Kim, H.S., Kim, E.M., Lee, J.P., Park, C.H., Kim, S., et al., 2003. C-terminal fragments of amyloid precursor protein exert neurotoxicity by inducing glycogen synthase kinase-3β expression. FASEB J. 17, 1951–1953.

Kim, J., Basak, J.M., Holtzman, D.M., 2009. The role of apolipoprotein E in Alzheimer's disease. Neuron. 63, 287–303.

Kim, J., Yoon, H., Basak, J., Kim, J., 2014. Apolipoprotein E in synaptic plasticity and Alzheimer's disease: potential cellular and molecular mechanisms. Mol. Cells 37, 767–776.

Kimberly, W.T., LaVoie, M.J., Ostaszewski, B.L., Ye, W., Wolfe, M.S., Selkoe, D.J., 2003. γ-Secretase is a membrane protein complex comprised of presenilin, nicastrin, aph-1, and pen-2. Proc. Nat. Acad. Sci. U. S. A. 100, 6382–6387.

Klein, W.L., Krafft, G.A., Finch, C.E., 2001. Targeting small Aβ oligomers: the solution to an Alzheimer's disease conundrum? Trends Neurosci. 24, 219–224.

Kleinberger, G., Yamanishi, Y., Suarez-Calvet, M., Czirr, E., Lohmann, E., et al., 2014. TREM2 mutations implicated in neurodegeneration impair cell surface transport and phagocytosis. Sci. Transl. Med. 6, 243ra286.

Kondo, M., Shiono, M., Itoh, G., Takei, N., Matsushima, T., Maeda, M., et al., 2010. Increased amyloidogenic processing of transgenic human APP in X11-like deficient mouse brain. Mol. Neurodegener. 5, 35.

Konietzko, U., 2012. AICD nuclear signaling and its possible contribution to Alzheimer's disease. Curr. Alzheimer Res. 9, 200–216.

Kowalska, A., Pruchnik-Wolińska, D., Florczak, J., Modestowicz, R., Szczech, J., et al., 2004. Genetic study of familial cases of Alzheimer's disease. Acta Biochim. Pol. 51, 245–252.

Kumar, S., Walter, J., 2011. Phosphorylation of amyloid beta (Aβ) peptides – a trigger for formation of toxic aggregates in Alzheimer's disease. Aging (Albany NY) 3 PMC: 3184981.

Kumar, D.K., Choi, S.H., Washicosky, K.J., Eimer, W.A., Tucker, S., et al., 2016. Amyloid-β peptide protects against microbial infection in mouse and worm models of Alzheimer's disease. Sci. Transl. Med. 8(340), 340ra72.

Lacor, P.N., Buniel, M.C., Chang, L., Fernandez, S.J., Gong, Y., Viola, K.L., Lambert, M.P., et al., 2004. Synaptic targeting by Alzheimer's-related amyloid β oligomers. J. Neurosci. 24, 10191–10200.

Lagace, D.C., Benavides, D.R., Kansy, J.W., Mapelli, M., Greengard, P., Bibb, J.A., Eisch, A.J., 2008. Cdk5 is essential for adult hippocampal neurogenesis. Proc. Natl. Acad. Sci. U. S. A. 105, 18567–18571.

Laird, F.M., Cai, H., Savonenko, A.V., Farah, M.H., He, K., Melnikova, T., et al., 2005. BACE1, a major determinant of selective vulnerability of the brain to amyloid-β amyloidogenesis, is essential for cognitive, emotional, and synaptic functions. J. Neurosci. 25, 11693–11709.

Larson, M.E., Lesné, S.E., 2012. Soluble Aβ oligomer production and toxicity. J. Neurochem. 120 (Suppl. 1), 125–139.

Lauren, J., Gimbel, D., Nygaard, H., Gilbert, J., Strittmatter, S., 2009. Cellular prion protein mediates impairment of synaptic plasticity by amyloid-β oligomers. Nature 457, 1128–1132.

Leandro, P., Gomes, C.M., 2008. Protein misfolding in conformational disorders: rescue of folding defects and chemical chaperoning. Mini Rev. Med. Chem. 8, 901–911.

Ledo, J.H., Azevedo, E.P., Clarke, J.R., Ribeiro, F.C., Figueiredo, C.P., Foguel, D., et al., 2013. Amyloid-β oligomers link depressive-like behavior and cognitive deficits in mice. Mol. Psychiatry 2013 (18), 1053–1054.

Lee, G., 2005. Tau and src family tyrosine kinases. Biochim. Biophys. Acta 1739, 323–330.

Lee, H.-G., Zhu, X., Castellani, R.J., Nunomura, A., Perry, G., Smith, M.A., 2007. Amyloid-β in Alzheimer disease: the null versus the alternate hypotheses. J. Pharmacol. Exp. Ther. 321, 823–829.

Lesne, S., Koh, M.T., Kotilinek, L., Kayed, R., Glabe, C.G., Yang, A., Gallagher, M., Ashe, K.H., 2006. Nature 440, 352–357.

Li, T., Ma, G., Cai, H., Price, D.L., Wong, P.C., 2003. Nicastrin is required for assembly of presenilin/γ-secretase complexes to mediate notch signaling and for processing and trafficking of β-amyloid precursor protein in mammals. J. Neurosci. 23, 3272–3277.

Li, L., Zhang, S., Zhang, X., Zhang, X., Li, T., Tang, Y., Liu, H., Yang, W., Le, W., 2013. Autophagy enhancer carbamazepine alleviates memory deficits and cerebral amyloid-β pathology in a mouse model of Alzheimer's disease. Curr. Alzheimer Res. 10, 433–441.

Lie, D.C., Colamarino, S.A., Song, H.J., Desire, L., Mira, H., Consiglio, A., et al., 2005. Wnt signalling regulates adult hippocampal neurogenesis. Nature 437, 1370–1375.

Liu, Y., Qin, L., Wilson, B.C., An, L., Hong, J.-S., Liu, B., 2002. Inhibition by naloxone stereoisomers of β-amyloid peptide (1–42)-induced superoxide production in microglia and degeneration of cortical and mesencephalic neurons. J. Pharmacol. Exp. Ther. 302, 1212–1219.

Liu, S., Liu, Y., Hao, W., Wolf, L., Kiliaan, A.J., Penke, B., et al., 2012. TLR2 is a primary receptor for Alzheimer's amyloid β peptide to trigger neuroinflammatory activation. J. Immunol. 188, 1098–1107.

Liu, C.C., Kanekiyo, T., Xu, H., Bu, G., 2013. Apolipoprotein E and Alzheimer disease: risk, mechanisms and therapy. Nat. Rev. Neurol. 9, 106–118.

Love, S., 2004. Contribution of cerebral amyloid angiopathy to Alzheimer's disease. J. Neurol. Neurosurg. Psychiatry 75, 1–4.

Lu, B., Gerard, N.P., Kolakowski Jr., L.F., Bozza, M., Zurakowski, D., et al., 1995. Neutral endopeptidase modulation of septic shock. J. Exp. Med. 181, 2271–2275.

Lu, D.C., Rabizadeh, S., Chandra, S., Shayya, R.F., Ellerby, L.M., et al., 2000. A second cytotoxic proteolytic peptide derived from amyloid β-protein precursor. Nat. Med. 6, 397–404.

Ma, H., Lesne, S., Kotilinek, L., Steidl-Nichols, J.V., Sherman, M., Younkin, L., et al., 2007. Involvement of β-site APP cleaving enzyme 1 (BACE1) in amyloid precursor protein-mediated enhancement of memory and activity-dependent synaptic plasticity. Proc. Natl. Acad. Sci. U. S. A. 104, 8167–8172.

Macauley, S.L., Stanley, M., Caesar, E.E., Yamada, S.A., Raichle, M.E., Perez, R., et al., 2015. Hyperglycemia modulates extracellular amyloid-β concentrations and neuronal activity in vivo. J. Clin. Invest. 125, 2463–2467.

Maccioni, R.B., Munoz, J.P., Barbeito, L., 2001. The molecular bases of Alzheimer's disease and other neurodegenerative disorders. Arch. Med. Res. 32, 367–381.

Madeira, A., Pommet, J.M., Prochiantz, A., Allinquant, B., 2005. SET protein (TAF1β, I2PP2A) is involved in neuronal apoptosis induced by an amyloid precursor protein cytoplasmic subdomain. FASEB J. 19, 1905–1907.

Maloney, M.T., Bamburg, J.R., 2007. Cofilin-mediated neurodegeneration in Alzheimer's disease and other amyloidopathies. Mol. Neurobiol. 35, 21–44.

Mandelkow, E.M., Schweers, O., Drewes, G., Biernat, J., Gustke, N., Trinczek, B., Mandelkow, E., 1996. Structure, microtubule interactions, and phosphorylation of tau protein. Ann. N. Y. Acad. Sci. 777, 96–106.

Mandelkow, E.M., Stamer, K., Vogel, R., Thies, E., Mandelkow, E., 2003. Clogging of axons by tau, inhibition of axonal traffic and starvation of synapses. Neurobiol. Aging 24, 1079–1085.

Mandrekar-Colucci, S., Karlo, J.C., Landreth, G.E., 2012. Mechanisms underlying the rapid peroxisome proliferator-activated receptor-γ-mediated amyloid clearance and reversal of cognitive deficits in a murine model of Alzheimer's disease. J. Neurosci. 32, 10117–10128.

Marr, R.A., Hafez, D.M., 2014. Amyloid-β and Alzheimer's disease: the role of neprilysin-2 in amyloid-β clearance. Front. Aging Neurosci. 6, 187.

Martin, L., Latypova, X., Terro, F., 2011. Post-translational modifications of tau protein: implications for Alzheimer's disease. Neurochem. Int. 58, 458–471.

Maruyama, M., Higuchi, M., Takaki, Y., Matsuba, Y., Tanji, H., Nemoto, M., et al., 2005. Cerebrospinal fluid neprilysin is reduced in prodromal Alzheimer's disease. Ann. Neurol. 57, 832–842.

Matsuzaki, M., Honkura, N., Ellis-Davies, G.C., Kasai, H., 2004. Structural basis of long-term potentiation in single dendritic spines. Nature 429, 761–766.

Mayr, B., Montminy, M., 2001. Transcriptional regulation by the phosphorylation-dependent factor CREB. Nat. Rev. Mol. Cell Biol. 2, 599–609.

McGeer, P.L., McGeer, E.G., 2013. The amyloid cascade-inflammatory hypothesis of Alzheimer disease: implications for therapy. Acta Neuropathol. 126, 479–497.

Meyer-Luehmann, M., Spires-Jones, T.L., Prada, C., Garcia-Alloza, M., de Calignon, A., Rozkalne, A., Koenigsknecht-Talboo, J., Holtzman, D.M., Bacskai, B.J., Hyman, B.T., 2008. Rapid appearance and local toxicity of amyloid-β plaques in a mouse model of Alzheimer's disease. Nature 451, 720–724.

Meziane, H., Dodart, J.C., Mathis, C., Little, S., Clemens, J., Paul, S.M., Ungerer, A., 1998. Memory-enhancing effects of secreted forms of the β-amyloid precursor protein in normal and amnestic mice. Proc. Natl. Acad. Sci. U. S. A. 95, 12683–12688.

Milosch, N., Tanriöver, G., Kundu, A., Rami, A., François, J.C., Baumkötter, F., 2014. Holo-APP and G-protein-mediated signaling are required for sAPPα-induced activation of the Akt survival pathway. Cell Death Dis. 5, e1391.

Miners, J.S., Baig, S., Tayler, H., Kehoe, P.G., Love, S., 2009. Neprilysin and insulin-degrading enzyme levels are increased in Alzheimer disease in relation to disease severity. J. Neuropathol. Exp. Neurol. 68, 902–914.

Miners, J.S., van Helmond, Z., Kehoe, P.G., Love, S., 2010. Changes with age in the activities of β-secretase and the Aβ-degrading enzymes neprilysin, insulin-degrading enzyme and angiotensin-converting enzyme. Brain Pathol. 20, 794–802.

Miners, J.S., Morris, S., Love, S., Kehoe, P.G., 2011. Accumulation of insoluble amyloid-β in down's syndrome is associated with increased BACE-1 and neprilysin activities. J. Alzheimers Dis. 23, 101–108.

Morley, J.E., Farr, S.A., Banks, W.A., Johnson, S.N., Yamada, K.A., Xu, L., 2010. A physiological role for amyloid-β protein:enhancement of learning and memory. J. Alzheimers Dis. 19, 441–449.

Mosconi, L., Pupi, A., De Leon, M.J., 2008. Brain glucose hypometabolism and oxidative stress in preclinical Alzheimer's disease. Ann. N. Y. Acad. Sci. 1147, 180–195.

Mrak, R.E., Griffin, W.S., 2005. Potential inflammatory biomarkers in Alzheimer's disease. J. Alzheimers Dis. 8, 369–375.

Mucke, L., Selkoe, D.J., 2012. Neurotoxicity of amyloid β-protein: synaptic and network dysfunction. Cold Spring Harb. Perspect. Med. 2, a006338.

Muresan, V., Varvel, N.H., Lamb, B.T., Muresan, Z., 2009. The cleavage products of amyloid-β precursor protein are sorted to distinct carrier vesicles that are independently transported within neurites. J. Neurosci. 29, 3565–3578.

Muresan, V., Villegas, C., Ladescu Muresan, Z., 2013. Functional interaction between amyloid-β precursor protein and peripherin neurofilaments: a shared pathway leading to Alzheimer's disease and amyotrophic lateral sclerosis. Neurodegener. Dis. 13, 122–125.

Musiek, E.S., Holtzman, D.M., 2015. Three dimensions of the amyloid hypothesis: time, space and 'wingmen'. Nat. Neurosci. 18, 800–806.

Nalbantoglu, J., Tirado-Santiago, G., Lahsaini, A., Poirier, J., Goncalves, O., Verge, G., Momoli, F., Welner, S.A., Massicotte, G., Julien, J.P., Shapiro, M.L., 1997. Impaired learning and LTP in mice expressing the carboxy terminus of the Alzheimer amyloid precursor protein. Nature 387, 500–505.

Narayan, P., Holmstrom, K.M., Kim, D.H., Whitcomb, D.J., Wilson, M.R., St George-Hyslop, P., et al., 2014. Rare individual amyloid-β oligomers act on astrocytes to initiate neuronal damage. Biochemistry 53, 2442–2453.

Nelson, P.T., Alafuzoff, I., Bigio, E.H., Bouras, C., Braak, H., Cairns, N.J., et al., 2012. Correlation of Alzheimer disease neuropathologic changes with cognitive status: a review of the literature. J. Neuropathol. Exp. Neurol. 71, 362–381.

Neniskyte, U., Neher, J.J., Brown, G.C., 2011. Neuronal death induced by nanomolar amyloid β is mediated by primary phagocytosis of neurons by microglia. J. Biol. Chem. 286, 39904–39913.

Neve, R.L., Harris, P., Kosik, K.S., Kurnit, D.M., Donlon, T.A., 1986. Identification of cDNA clones for the human microtubule-associated protein tau and chromosomal localization of the genes for tau and microtubule-associated protein 2. Mol. Brain Res. 1, 271–280.

Nikolaev, A., McLaughlin, T., O'Leary, D.D., Tessier-Lavigne, M., 2009. APP binds DR6 to trigger axon pruning and neuron death via distinct caspases. Nature 457, 981–989.

Oddo, S., Caccamo, A., Tseng, B., Cheng, D., Vasilevko, V., Cribbs, D.H., LaFerla, F.M., 2008. Blocking Aβ$_{42}$ accumulation delays the onset and progression of tau pathology via the C terminus of heat shock protein70-interacting protein: a mechanistic link between Aβ and tau pathology. J. Neurosci. 28, 12163–12175.

Pardossi-Piquard, R., Checler, F., 2012. The physiology of the β-amyloid precursor protein intracellular domain AICD. J. Neurochem. 2012 (120), 109–124.

Paris, D., Townsend, K., Quadros, A., Humphrey, J., Sun, J., Brem, S., 2004. Inhibition of angiogenesis by Aβ peptides. Angiogenesis 7, 75–85.

Pascale, C.L., Miller, M.C., Chiu, C., Boylan, M., Caralopoulos, I.N., et al., 2011. Amyloid-β transporter expression at the blood–CSF barrier is age-dependent. Fluids Barriers CNS 8, 21.

Passer, B., Pellegrini, L., Russo, C., Siegel, R.M., Lenardo, M.J., et al., 2000. Generation of an apoptotic intracellular peptide by γ-secretase cleavage of Alzheimer's amyloid β protein precursor. J. Alzheimers Dis. 2, 289–301.

Paula-Lima, A.C., Brito-Moreira, J., Ferreira, S.T., 2013. Deregulation of excitatory neuro-transmission underlying synapse failure in Alzheimer's disease. J. Neurochem. 2013 (126), 191–202.

Pearson-Leary, J., McNay, E.C., 2012. Intrahippocampal administration of amyloid-β(1–42) oligomers acutely impairs spatial working memory, insulin signaling and hippocampal metabolism. J. Alzheimers Dis. 30, 413–422.

Pei, J.J., Hugon, J., 2008. mTOR-dependent signalling in Alzheimer's disease. J. Cell Mol. Med. 12 (6B), 2525–2532.

Petrou, M., Dwamena, B.A., Foerster, B.R., MacEachern, M.P., Bohnen, N.I., Müller, M.L., Albin, R.L., Frey, K.A., 2015. Amyloid deposition in Parkinson's disease and cognitive impairment: a systematic review. Mov. Disord. 30, 928–935.

Pimenova, A.A., Thathiah, A., De Strooper, B., Tesseur, I., 2014. Regulation of amyloid pre-cursor protein processing by serotonin signaling. PLoS One 9, e87014.

Pooler, A.M., Usardi, A., Evans, C.J., Philpott, K.L., Noble, W., et al., 2012. Dynamic associa-tion of tau with neuronal membranes is regulated by phosphorylation. Neurobiol. Aging 33, e427–e438.

Postina, R., Schroeder, A., Dewachter, I., Bohl, J., Schmitt, U., Kojro, E., et al., 2004. A disinte-grin-metalloproteinase prevents amyloid plaque formation and hippocampal defects in an Alzheimer disease mouse model. J. Clin. Invest. 113, 1456–1464.

Price, J.L., Morris, J.C., 1999. Tangles and plaques in nondemented aging and "preclinical" Alzheimer's disease. Ann. Neurol. 45, 358–368.

Priller, C., Bauer, T., Mitteregger, G., Krebs, B., Kretzschmar, H.A., Herms, J., 2006. Synapse formation and function is modulated by the amyloid precursor protein. J. Neurosci. 26, 7212–7221.

Prokop, S., Miller, K.R., Heppner, F.L., 2013. Microglia actions in Alzheimer's disease. Acta Neuropathol. 126, 461–477.

Puzzo, D., Arancio, O., 2013. Amyloid-β peptide: Dr. Jekyll or Mr. Hyde? J. Alzheimers Dis. 33 (Suppl. 1), S111–S120.

Querfurth, H.W., LaFerla, F.M., 2010. Alzheimer's disease. N. Engl. J. Med. 362, 329–344.

Quinn, K.A., Grimsley, P.G., Dai, Y.P., Tapner, M., Chesterman, C.N., Owensby, D.A., 1997. Soluble low density lipoprotein receptor-related protein (LRP) circulates in human plasma. J. Biol. Chem. 272, 23946–23951.

Quitschke, W.W., Goldgaber, D., 1992. The amyloid β-protein precursor promoter. A region essential for transcriptional activity contains a nuclear factor binding domain. J. Biol. Chem. 267, 17362–17368.

Rabinovici, G.D., Jagust, W.J., Furst, A.J., Ogar, J.M., Racine, C.A., et al., 2008. Aβ amyloid and glucose metabolism in three variants of primary progressive aphasia. Ann. Neurol. 64, 388–401.

Ramanathan, A., Nelson, A.R., Sagare, A.P., Zlokovic, B.V., 2015. Impaired vascular-mediated clearance of brain amyloid β in Alzheimer's disease: the role, regulation and restoration of LRP1. Front. Aging Neurosci. 7, 136.

Reddy, P.H., Beal, M.F., 2008. Amyloid β, mitochondrial dysfunction and synaptic damage: implications for cognitive decline in aging and Alzheimer's disease. Trends Mol. Med. 14, 45–53.

Reddy, P.H., Geethalakshmi, M., Byung, S.P., Joline, J., Geoffrey, M., William Jr., W., Jeffrey, K., Maria, M., 2005. Differential loss of synaptic proteins in Alzheimer's disease: implications for synaptic dysfunction. J. Alzheimers Dis. 7, 103–117.

Renner, M., Lacor, P.N., Velasco, P.T., Xu, J., Contractor, A., Klein, W.L., Triller, A., 2010. Deleterious effects of amyloid β oligomers acting as an extracellular scaffold for mGluR5. Neuron 66, 739–754.

Robakis, N.K., 2011. Mechanisms of AD neurodegeneration may be independent of Aβ and its derivatives. Neurobiol. Aging 32, 372–379.

Roberson, E.D., Scearce-Levie, K., Palop, J.J., Yan, F., Cheng, I.H., Wu, T., Gerstein, H., Yu, G.Q., Mucke, L., 2007. Reducing endogenous tau ameliorates amyloid β-induced deficits in an Alzheimer's disease mouse model. Science 316, 750–754.

Rohan de Silva, H.A., Jen, A., Wickenden, C., Jen, L.S., Wilkinson, S.L., Patel, A.J., 1997. Cell-specific expression of β-amyloid precursor protein isoform mRNAs and proteins in neurons and astrocytes. Brain Res. Mol. Brain Res. 47, 147–156.

Sagare, A., Deane, R., Bell, R.D., Johnson, B., Hamm, K., Pendu, R., et al., 2007. Clearance of amyloid-β by circulating lipoprotein receptors. Nat. Med. 13, 1029–1031.

Sagare, A.P., Deane, R., Zlokovic, B.V., 2012. Low-density lipoprotein receptor-related protein 1: a physiological Aβ homeostatic mechanism with multiple therapeutic opportunities. Pharmacol. Ther. 136, 94–105.

Sagare, A.P., Bell, R.D., Srivastava, A., Sengillo, J.D., Singh, I., Nishida, Y., et al., 2013. A lipoprotein receptor cluster IV mutant preferentially binds amyloid-β and regulates its clearance from the mouse brain. J. Biol. Chem. 288, 15154–15166.

Sahara, N., Murayama, M., Higuchi, M., Suhara, T., Takashima, A., 2014. Biochemical distribution of tau protein in synaptosomal fraction of transgenic mice expressing human p301l tau. Front. Neurol. 5, 26.

Scheff, S.W., Price, D.A., Schmitt, F.A., DeKosky, S.T., Mufson, E.J., 2007. Synaptic alterations in CA1 in mild Alzheimer disease and mild cognitive impairment. Neurology 68, 1501–1508.

Scheuner, D., Eckman, C., Jensen, M., Song, X., Citron, M., Suzuki, N., et al., 1996. Secreted amyloid β-protein similar to that in the senile plaques of Alzheimer's disease is increased in vivo by the presenilin 1 and 2 and APP mutations linked to familial Alzheimer's disease. Nat. Med. 2, 864–870.

Schmitt, B., Bernhardt, T., Moeller, H.-J., Heuser, I., Frölich, L., 2004. Combination therapy in Alzheimer's disease. CNS Drugs 18, 827–844.

Schmitz, A., Tikkanen, R., Kirfel, G., Herzog, V., 2002. The biological role of the Alzheimer amyloid precursor protein in epithelial cells. Histochem. Cell Biol. 117, 171–180.

Schrijvers, E.M., Witteman, J.C., Sijbrands, E.J., Hofman, A., Koudstaal, P.J., Breteler, M.M., 2010. Insulin metabolism and the risk of Alzheimer disease: the Rotterdam study. Neurology 75, 1982–1987.

Selkoe, D.J., 1989. Amyloid β protein precursor and the pathogenesis of Alzheimer's disease. Cell 58, 611–612.

Selkoe, D.J., 2008. Soluble oligomers of the amyloid β-protein impair synaptic plasticity and behavior. Behav. Brain Res. 192, 106–113.

Shiiki, T., Ohtsuki, S., Kurihara, A., Naganuma, H., Nishimura, K., Tachikawa, M., et al., 2004. Brain insulin impairs amyloid-β(1-40) clearance from the brain. J. Neurosci. 24, 9632–9637.

Shinohara, M., Sato, N., Kurinami, H., Takeuchi, D., Takeda, S., Shimamura, M., et al., 2010. Reduction of brain β-amyloid (Aβ) by fluvastatin, a hydroxymethylglutaryl-CoA reductase inhibitor, through increase in degradation of amyloid precursor protein C-terminal fragments (APP-CTFs) and Aβ clearance. J. Biol. Chem. 285, 22091–22102.

Sindi, I.A., Dodd, P.R., 2015. New insights into Alzheimer's disease pathogenesis: the involvement of neuroligins in synaptic malfunction. Neurodegener. Dis. Manag. 5, 137–145.

Sjöberg, M.K., Shestakova, E., Mansuroglu, Z., Maccioni, R.B., Bonnefoy, E., 2006. Tau protein binds to pericentromeric DNA: a putative role for nuclear tau in nucleolar organization. J. Cell Sci. 119, 2025–2034.

Skovronsky, D.M., Moore, D.B., Milla, M.E., Doms, R.W., Lee, V.M., 2000. Protein kinase C-dependent α-secretase competes with β-secretase for cleavage of amyloid-β precursor protein in the trans-golgi network. J. Biol. Chem. 275, 2568–2575.

Snyder, E.M., Nong, Y., Almeida, C.G., Paul, S., Moran, T., et al., 2005. Regulation of NMDA receptor trafficking by amyloid-β. Nat. Neurosci. 8, 1051–1058.

Son, S.M., Song, H., Byun, J., Park, K.S., Jang, H.C., Park, Y.J., Mook-Jung, I., 2012a. Altered APP processing in insulin-resistant conditions is mediated by autophagosome accumulation via the inhibition of mammalian target of rapamycin pathway. Diabetes 61, 3126–3138.

Son, S.M., Song, H., Byun, J., Park, K.S., Jang, H.C., Park, Y.J., et al., 2012b. Accumulation of autophagosomes contributes to enhanced amyloidogenic APP processing under insulin-resistant conditions. Autophagy 8, 1842–1844.

Soscia, S.J., Kirby, J.E., Washicosky, K.J., Tucker, S.M., Ingelsson, M., Hyman, B., 2010. The Alzheimer's disease-associated amyloid β-protein is an antimicrobial peptide. PLoS One 5, e9505.

Sparks, D.L., Scheff, S.W., Hunsaker III, J.C., Liu, H., Landers, T., Gross, D.R., 1994. Induction of Alzheimer-like β-amyloid immunoreactivity in the brains of rabbits with dietary cholesterol. Exp. Neurol. 126, 88–94.

Spilman, P., Podlutskaya, N., Hart, M.J., et al., 2010. Inhibition of mTOR by rapamycin abolishes cognitive deficits and reduces amyloid-β levels in a mouse model of Alzheimer's disease. PLoS One 5, e9979.

Starkov, A.A., Beal, F.M., 2008. Portal to Alzheimer's disease. Nat. Med. 14, 1020–1021.

Steinbach, J.P., Müller, U., Leist, M., Li, Z.W., Nicotera, P., Aguzzi, A., 1998. Hypersensitivity to seizures in β-amyloid precursor protein deficient mice. Cell Death Differ. 5, 858–866.

Stewart, L.R., White, A.R., Jobling, M.F., Needham, B.E., Maher, F., Thyer, J., Beyreuther, K., Masters, C.L., Collins, S.J., Cappai, R., 2001. Involvement of the 5-lipoxygenase pathway in the neurotoxicity of the prion peptide PrP106–126. J. Neurosci. Res. 65, 565–572.

Sturchler, E., Galichet, A., Weibel, M., Leclerc, E., Heizmann, C.W., 2008. Site-specific blockade of RAGE-Vd prevents amyloid-β oligomer neurotoxicity. J. Neurosci. 28, 5149–5158.

Sultan, A., Nesslany, F., Violet, M., Bégard, S., Loyens, A., Talahari, S., et al., 2011. Nuclear tau, a key player in neuronal DNA protection. J. Biol. Chem. 286, 4566–4575.

Tabaton, M., Zhu, X., Perry, G., Smith, M.A., Giliberto, L., 2010. Signaling effect of amyloid-β_{42} on the processing of AβPP. Exp. Neurol. 221, 18–25.

Tackenberg, C., Brandt, R., 2009. Divergent pathways mediate spine alterations and cell death induced by amyloid-β, wild-type tau, and R406W tau. J. Neurosci. 29, 14439–14450.

Takami, M., Funamoto, S., 2012. γ-Secretase-dependent proteolysis of transmembrane domain of amyloid precursor protein: successive tri- and tetrapeptide release in amyloid β-protein production. Int. J. Alzheimers Dis. 591392.

Tamaki, C., Ohtsuki, S., Iwatsubo, T., Hashimoto, T., Yamada, K., Yabuki, C., et al., 2006. Major involvement of low-density lipoprotein receptor-related protein 1 in the clearance of plasma free amyloid β-peptide by the liver. Pharm. Res. 23, 1407–1416.

Tan, Z., Shi, L., Schreiber, S.S., 2009. Differential expression of redox factor-1 associated with β-amyloid-mediated neurotoxicity. Open Neurosci. 3, 26–34.

Tayeb, H.O., Murray, E.D., Price, B.H., Tarazi, F.I., 2013. Bapineuzumab and solanezumab for Alzheimer's disease: is the 'amyloid cascade hypothesis' still alive? Expert Opin. Biol. Ther. 13, 1075–1084.

Terry, R.D., 1996. The pathogenesis of Alzheimer disease: an alternative to the amyloid hypothesis. J. Neuropathol. Exp. Neurol. 55, 1023–1025.

Thinakaran, G., Koo, E.H., 2008. Amyloid precursor protein trafficking, processing, and function. J. Biol. Chem. 283, 29615–29619.

Tomlinson, B.E., Blessed, G., Roth, M., 1970. Observations on the brains of demented old people. J. Neurol. Sci. 11, 205–242.

Tu, S., Okamoto, S., Lipton, S.A., Xu, H., 2014. Oligomeric Aβ-induced synaptic dysfunction in Alzheimer's disease. Mol. Neurodegener. 9, 48.

Ulrich, J.D., Finn, M.B., Wang, Y., Shen, A., Mahan, T.E., et al., 2014. Altered microglial response to Aβ plaques in APPPS1-21 mice heterozygous for TREM2. Mol. Neurodegener. 9, 20.

van der Flier, W.M., Scheltens, P., 2005. Epidemiology and risk factors of dementia. J. Neurol. Neurosurg. Psychiatry 76, 2–7.

van Duinen, S.G., Castano, E.M., Prelli, F., Bots, G.T., Luyendijk, W., Frangione, B., 1987. Hereditary cerebral hemorrhage with amyloidosis in patients of Dutch origin is related to Alzheimer disease. Proc. Natl. Acad. Sci. U. S. A. 84, 5991–5994.

van Helmond, Z., Miners, J.S., Kehoe, P.G., Love, S., 2010. Higher soluble amyloid β concentration in frontal cortex of young adults than in normal elderly or Alzheimer's disease. Brain Pathol. 20, 787–793.

Vandal, M., White, P.J., Tremblay, C., St-Amour, I., Chevrier, G., Emond, V., et al., 2014. Insulin reverses the high-fat diet-induced increase in brain Aβ and improves memory in an animal model of Alzheimer disease. Diabetes 3, 4291–4301.

Vassar, R., Bennett, B.D., Babu-Khan, S., Kahn, S., Mendiaz, E.A., Denis, P., et al., 1999. β-Secretase cleavage of Alzheimer's amyloid precursor protein by the transmembrane aspartic protease BACE. Science 286, 735–741.

Vazquez, A., 2013. Metabolic states following accumulation of intracellular aggregates: implications for neurodegenerative diseases. PLoS One 8, e63822.

Vekrellis, K., Ye, Z., Qiu, W.Q., Walsh, D., Hartley, D., Chesneau, V., et al., 2000. Neurons regulate extracellular levels of amyloid β-protein via proteolysis by insulin-degrading enzyme. J. Neurosci. 20, 1657–1665.

Velez-Pardo, C., Garcia-Ospina, G., Del Rio, J.M., 2002. Aβ[25–35] peptide and iron promote apoptosis in lymphocytes by a common oxidative mechanism: involvement of hydrogen peroxide (H_2O_2), caspase-3, NF-kappaB, p53 and c-Jun. Neurotoxicology 23, 351–365.

Verdile, G., Fuller, S., Atwood, C.S., Laws, S.M., Gandy, S.E., Martins, R.N., 2004. The role of β amyloid in Alzheimer's disease: still a cause of everything or the only one who got caught? Pharmacol. Res. 2004 (50), 397–409.

Viola, K.L., Klein, W.L., 2015. Amyloid β oligomers in Alzheimer's disease pathogenesis, treatment, and diagnosis. Acta Neuropathol. 129, 183–206.

Viswanathan, A., Greenberg, S.M., 2011. Cerebral amyloid angiopathy in the elderly. Ann. Neurol. 70, 871–880.

von Arnim, C.A., Kinoshita, A., Peltan, I.D., Tangredi, M.M., Herl, L., Lee, B.M., et al., 2005. The low density lipoprotein receptor-related protein (LRP) is a novel β-secretase (BACE1) substrate. J. Biol. Chem. 280, 17777–17785.

Walsh, D.M., Klyubin, I., Fadeeva, J.V., Cullen, W.K., Anwyl, R., Wolfe, M.S., Rowan, M.J., Selkoe, D.J., 2002. Naturally secreted oligomers of amyloid β protein potently inhibit hippocampal long-term potentiation in vivo. Nature 416, 535–539.

Walsh, K.P., Kuhn, T.B., Bamburg, J.R., 2014. Cellular prion protein: a co-receptor mediating neuronal cofilin-actin rod formation induced by β-amyloid and proinflammatory cytokines. Prion 8, 375–380.

Wang, Y., Ha, Y., 2004. The X-ray structure of an antiparallel dimer of the human amyloid precursor protein E2 domain. Mol. Cell 15, 243–253.

Wang, H.Y., Lee, D.H., Davis, C.B., Shank, R.P., 2000. Amyloid peptide $Aβ_{1-42}$ binds selectively and with picomolar affinity to α7 nicotinic acetylcholine receptors. J. Neurochem. 75, 1155–1161.

Wang, S., Wang, R., Chen, L., Bennett, D.A., Dickson, D.W., Wang, D.S., 2010. Expression and functional profiling of neprilysin, insulin-degrading enzyme and endothelin-converting enzyme in prospectively studied elderly and Alzheimer's brain. J. Neurochem. 115, 47–57.

Weller, R.O., Subash, M., Preston, S.D., Mazanti, I., Carare, R.O., 2008. Perivascular drainage of amyloid-β peptides from the brain and its failure in cerebral amyloid angiopathy and Alzheimer's disease. Brain Pathol. 18, 253–266.

Williamson, R., Usardi, A., Hanger, D.P., Anderton, B.H., 2008. Membrane-bound β-amyloid oligomers are recruited into lipid rafts by a fyn-dependent mechanism. FASEB J. 22, 1552–1559.

Wolk, D.A., Price, J.C., Madeira, C., Saxton, J.A., Snitz, B.E., et al., 2012. Amyloid imaging in dementias with atypical presentation. Alzheimers Dement. 8, 389–398.

Wolozin, B., 2004. Cholesterol and the biology of Alzheimer's disease. Neuron 41, 7–10.

Wu, Y., Deng, Y., Zhang, S., Luo, Y., Cai, F., Zhang, Z., Zhou, W., Li, T., Song, W., 2015. Amyloid-β precursor protein facilitates the regulator of calcineurin 1-mediated apoptosis by downregulating proteasome subunit α type-5 and proteasome subunit β type-7. Neurobiol. Aging 36, 169–177.

Yang, L., Wang, Z., Wang, B., Justice, N.J., Zheng, H., 2009. Amyloid precursor protein regulates Cav1.2 L-type calcium channel levels and function to influence GABAergic short-term plasticity. J. Neurosci. 29, 15660–15668.

Yao, Z.X., Papadopoulos, V., 2002. Function of β-amyloid in cholesterol transport: a lead to neurotoxicity. FASEB J. 16, 1677–1679.

Yin, R.H., Tan, L., Jiang, T., Yu, J.T., 2014. Prion-like mechanisms in Alzheimer's disease. Curr. Alzheimer Res. 11, 755–764.

Yu, G., Nishimura, M., Arawaka, S., Levitan, D., Zhang, L., Tandon, A., et al., 2000. Nicastrin modulates presenilin-mediated notch/glp-1 signal transduction and βAPP processing. Nature 407, 48–54.

Zawia, N.H., Lahiri, D.K., Cardozo-Pelaez, F., 2009. Epigenetics, oxidative stress, and Alzheimer disease. Free Radic. Biol. Med. 46, 1241–1249.

Zempel, H., Thies, E., Mandelkow, E., Mandelkow, E.M., 2010. Aβ oligomers cause localized Ca^{2+} elevation, missorting of endogenous Tau into dendrites, Tau phosphorylation, and destruction of microtubules and spines. J. Neurosci. 30, 11938–11950.

Zha, Q., Ruan, Y., Hartmann, T., Beyreuther, K., Zhang, D., 2004. GM1 ganglioside regulates the proteolysis of amyloid precursor protein. Mol. Psychiatry 9, 946–952.

Zhang, C., Saunders, A.J., 2007. Therapeutic targeting of the α-secretase pathway to treat Alzheimer's disease. Discov. Med. 7, 113–117.

Zheng, H., Koo, E.H., 2011. Biology and pathophysiology of the amyloid precursor protein. Mol. Neurodegener. 6, 27.

Zhou, Q., Homma, K.J., Poo, M.M., 2004. Shrinkage of dendritic spines associated with long-term depression of hippocampal synapses. Neuron 44, 749–757.

Zhu, D., Bungart, B.L., Yang, X., Zhumadilov, Z., Lee, J.C., Askarova, S., 2015. Role of membrane biophysics in Alzheimer's-related cell pathways. Front. Neurosci. 9, 186.

Zlokovic, B.V., 2011. Neurovascular pathways to neurodegeneration in Alzheimer's disease and other disorders. Nat. Rev. Neurosci. 12, 723–738.

Zlokovic, B.V., Frangione, B., 2003. Transport-Clearance Hypothesis for Alzheimer's Disease and Potential Therapeutic Implications. Madame Curie Bioscience Database (online) http://www.ncbi.nlm.nih.gov/books/NBK5975/.

Zlokovic, B.V., Deane, R., Sagare, A.P., Bell, R.D., Winkler, E.A., 2010. Low-density lipoprotein receptor related protein-1: a serial clearance homeostatic mechanism controlling Alzheimer's amyloid β-peptide elimination from the brain. J. Neurochem. 115, 1077–1089.

Zou, K., Gong, J.S., Yanagisawa, K., Michikawa, M., 2002. A novel function of monomeric amyloid β-protein serving as an antioxidant molecule against metal-induced oxidative damage. J. Neurosci. 22, 4833–4841.

Risk Factors for Alzheimer's Disease

INTRODUCTION

Alzheimer's disease (AD) is a progressive, irreversible, and multifactorial neurodegenerative disease, which is characterized by neurodegeneration in the frontal cortex as well as the temporal and parietal lobes, including the hippocampus and entorhinal cortex (EC), and the cingulate gyrus, whereas the cerebellum is largely spared (Wenk, 2003). The early stages of AD are accompanied by mild cognitive decline, with progressive memory loss and impairment of other intellectual abilities as the disease progresses, and at the late stages, patients with AD display alteration in their personality and lose their bodily functions (Huang and Mucke, 2012). Currently, AD affects more than 44 million people worldwide and every 4 s, a new patient is diagnosed with AD (Alzheimer's Association, 2012). The incidences of AD are increasing not only due to extended life expectancy but also due to certain lifestyle factors (Farooqui, 2015). Changes from traditional healthy lifestyle to Western lifestyle have occurred between 1951 and 1985. These changes in lifestyle are due to the increased consumption of meat and animal products and lack of physical activity (exercise). Animal products and meat are known to increase the risk of AD because they contain compounds such as excess iron. In addition, the presence of advanced glycation end products (AGEs) and arachidonic acid (ARA) in the Western diet increase the induction of oxidative stress and inflammation in the brain. The number of patients with AD is expected to double every 20 years and is expected to reach 115 million in 2050 (Prince et al., 2013). Majority of AD cases (>93%–95%) are sporadic. These patients are older than 65 years. Only 5%–7% cases appear to have familial form of AD, which arises from the mutations of any of the three genes, e.g., amyloid precursor protein (*APP*), presenilin 1 (*PSEN 1*), and presenilin 2 (*PSEN 2*) and usually appears in the somewhat younger age group than the sporadic form and follows a more aggressive downhill course (van der Flier and scheltens, 2005; Duthey, 2013) (Table 2.1). As stated in Chapter 1, neuropathological hallmarks of AD are senile plaques and tangles in brain regions associated with learning and memory and emotional behaviors such as the entorhinal cortex, hippocampus, basal forebrain, and amygdala. Increasing evidence suggests that the overexpression of APP and subsequent generation of amyloid β (Aβ) are central to neuronal degeneration observed in sporadic and familial forms of AD (Fig. 2.1). Sporadic AD is also accompanied by many metabolic, pathological, and neurochemical alterations, including hypometabolism (Mosconi et al., 2008), disruption of blood–brain barrier (BBB) (Zlokovic, 2011), activation of microglial and astroglial cells (Mrak and Griffin, 2005; Prokop et al., 2013), and onset of oxidative stress, which occurs at early stages of AD before the appearance of amyloid plaques and neurofibrillary tangles (NFTs) and acts to exacerbate the disease progression. The causes of the sporadic form of AD appear be to quite complex and may not only involve age-related alterations in metabolism, repair mechanisms, immune response, and the vascular system but also include exogenous factors such as brain trauma and overall life style (Cohen and Dillin, 2008; Bishop et al., 2010; Farooqui, 2015). Familial AD is driven by the genetic mutations in *PSEN 1* and *PSEN 2* genes along with interactions among numerous environmental risk factors and age. These interactions regulate the generation of Aβ peptide, hyperphosphorylation of Tau protein, induction of neuroinflammation, and other unknown factors, which contribute to the pathogenesis of AD.

TABLE 2.1   Genes Associated With the Pathogenesis of
Alzheimer's Disease

Gene	Chromosomal Location
APP	21
Presenilin 1	14
Presenilin 2	1
CD33	19
CLU	8
Cys C	-
APOE	19
PICALM/CALM	11
SORLI	11
TOMM40	19
BINI	2
LRP6	12
CTNNA3	10
GAB2	11
DNMBP	10
ABCA7	19
CR1	1

*Summarized from Bagyinszky et al. (2014); Lambert et al. (2009); Harold et al.
(2009); Hu et al., 2011; Miyashita et al., 2007.*

Thus it is important to not only unravel the risk factors for AD but also discuss signal transduction processes associated with the effect of these factors on the onset of both forms of AD.

AGING AS A RISK FACTOR FOR ALZHEIMER'S DISEASE

Aging is a complex, inevitable, and undeniable process, which is characterized by a time-dependent progressive decline of physiological function and tissue homeostasis leading to increased vulnerability to degeneration and death. At the molecular and cellular level, aging is not only accompanied by genomic instability, defects in nuclear architecture, telomere attrition, epigenetic alterations, and chromatin remodeling but also by the induction

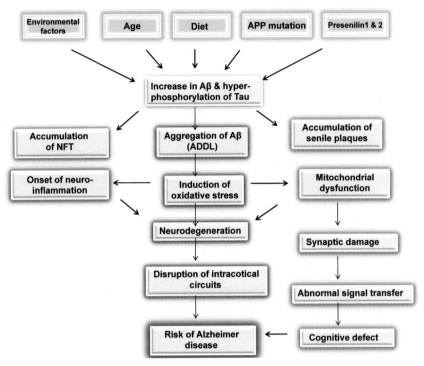

FIGURE 2.1 **Risk factors for the pathogenesis of Alzheimer's disease.** *ADDL*, Aβ-derived diffusible ligand; *Aβ*, amyloid β; *APP*, amyloid precursor protein; *NFT*, neurofibrillary tangles.

of mitochondrial dysfunction, stem cell exhaustion, and altered intercellular communication (Lopez-Otin et al., 2013). In the brain tissue, normal aging is accompanied by not only changes in the brain weight but also by a significant decrease in cognitive abilities (e.g., learning, memory formation, or executive control) leading to gradual constraints in daily activities and independent living in seniors, who are more susceptible to the effects of distracting interference during cognitive tasks (Hasher and Zacks, 1988) and have generally slower processing speed (Healey et al., 2008). Molecular mechanisms associated with cognitive decline are not fully understood. However, it is suggested that cognitive changes in aging brain are mediated and modulated by several conserved mechanisms, which not only control aging process but also are closely associated with the modulation of lifespan through signal transduction processes. These signal transduction processes involve insulin/insulin-like growth (IGF) factor signaling, target of rapamycin signaling, sirtuin signaling, mitochondrial function, and caloric restriction. The cross talk among the signaling pathways may modulate cognitive functions and neural cell longevity in the aging brain (Bishop et al., 2010).

In addition to these mechanisms, telomere shortening, mitochondrial oxidative damage, p53 activation, as well as reduced peroxisome proliferator–activated receptor gamma and coactivator 1 alpha and beta (PGC-1α and PGC-1β) (Sahin et al., 2011; Finck and Kelly, 2006) also modulate the integrity of genome and stability. These are major guarantors of viability and longevity. Furthermore, the aging brain is also modulated by blood flow. Older age is associated with lower blood flow and metabolism at rest (particularly in frontal cortex) probably due to atherosclerosis, whereas in adult brain the blood flow is faster. Reduced regional brain response to challenge tasks among seniors compared to young adults has also been observed in humans (Dennis and Cabeza, 2008).

Aging is the major risk factor for AD. After the age of 65 years the possibility of developing AD doubles every 5 years and reaches 50% in people aged 85 years and over. At the molecular level, aging process not only leads to shortening of telomeres and activation of tumor suppressor genes but also to decrease in levels of interleukin-10 (IL-10) (Ye and Johnson, 2001) and increase in levels of tumor necrosis factor α (TNF-α) and IL-1β

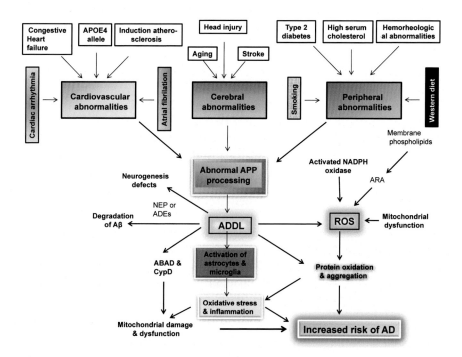

FIGURE 2.2 **Risk factor for Alzheimer's disease (AD) and generation of amyloid β (Aβ) in the pathogenesis of AD.** *APP,* amyloid precursor protein; *ADE,* amyloid degrading enzyme; *NEP,* neprilysin; *ROS,* reactive oxygen species; *ABAD,* Aβ-binding alcohol dehydrogenase; *CypD,* cyclophilin; *BBB,* blood–brain barrier; *ARA,* arachidonic acid.

in the central nervous system (CNS) (Lukiw, 2004) and IL-6 in the plasma (Fig. 2.2) (Godbout and Johnson, 2004). The expression of these cytokines is regulated by nuclear factor-κB, a transcription factor, which is activated by reactive oxygen species (ROS). Under physiological conditions, NF-κB resides in the cytoplasmic compartment as an inactive complex with inhibitor of κB (IκB). Under pathological conditions, high concentrations of ROS promote the breakdown of NF-κB/IκB complex leading to the release of free NF-κB, which then migrates to the nucleus and binds with NF-κB response element. This binding facilitates the expression of TNF-α, IL-1β, and IL-6 (Fig. 2.2). In addition, increase in transforming growth factor β1 messenger RNA (mRNA) in the brain has also been reported (Bye et al., 2001). Furthermore, an age-mediated increase in the number of CD11b$^+$ CD45^{high} cells, compatible with infiltrated monocytes, has been observed in the brain of aged rats (Blau et al., 2012). Similarly, upregulation of chemotactic molecules [interferon (IFN)-inducible protein 10 and monocyte chemotactic protein-1] has been reported in the hippocampus, a region involved in memory formation (Blau et al., 2012). High levels of ROS not only neutralize the reducing capacity of cells and produce changes in signal transduction pathways, but also produce changes in telomeres, the "biological clock" of the cellular aging. One of the most recognized effects of aging is the dysregulation of the immune system as a result of defects in both initiation and resolution of immune responses (*immunosenescence*) and chronic low-grade inflammation (*inflammaging*) (Franceschi et al., 2007). Collective evidence suggests that aging contributes to memory loss and cognitive deficits due to the aforementioned neurochemical changes (Burke and Barnes, 2006; Wimmer et al., 2012). The frequency of age-related cognitive decline is increasing dramatically as human lifespan increased over the past few decades.

At the cellular level, aging also leads to the activation of microglial cells, which are the major resident immune cells in the brain, where they constantly survey the microenvironment and produce factors that influence the surrounding astrocytes and neurons. Under physiological conditions in adult brain, activation of microglial cells protects the brain through much of life through participation and contribution in many processes (Fig. 2.3). The efficacy of microglial protection deteriorates with age. The hallmarks of microglial aging are linked with hyperreactive responses. During aging, the immune system in general undergoes an imbalanced shift toward a proinflammatory status. This process is called as "inflammaging" (Franceschi et al., 2007). If this concept can be applied to microglial cell aging, then microenvironment in microglial cells may shift from a protective role in adult individual to a state in which some protective mechanisms are lost and others are even detrimental to brain health promoting neurodegeneration. The molecular mechanism associated with microbial cell activation is not fully understood. However, it is

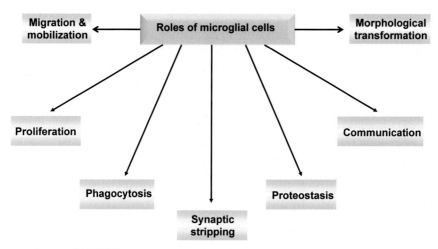

FIGURE 2.3 **Roles of activated microglial cells in the brain.**

suggested that activation of microglial cells (deramification) is companied by the generation of lysophosphatidylcholine (lyso-PtdCho). Activated deramified microglial cells have long processes with secondary branching (Schilling et al., 2004). It is proposed that lyso-PtdCho promotes microglial activation through the modulation of P2X7R signaling (Takenouchi et al., 2007). Under normal conditions in adult brain, activated microglial cells migrate to the area of injured brain tissue and engulf and destroy microbes and cellular debris (Gehrmann et al., 1995). Activated microglia also promote the removal of dysfunctional synapses, a process termed "synaptic stripping" (von Bernhardi, 2010; Kettenmann et al., 2013). In contrast, in aged brain, increased microglial activation due to increased levels of inflammatory mediators (eicosanoids, platelet activating factor, and proinflammatory cytokines), ROS, and NO facilitate neuroinflammation (Fig. 2.2). Increased levels of NO have been shown to suppress neurogenesis under normal and injury conditions (Torroglosa et al., 2007). In AD, chronic microglial activation occurs due to the sustained levels of inflammatory mediators and high oxidative stress, which may contribute to a deleterious microenvironment, such as increase in systemic inflammation, increase in permeability of the BBB, and increase in ROS levels due to degeneration of neurons and other brain cells. The normal BBB integrity is essential in protecting neural cells from systemic toxins and maintaining the necessary level of nutrients and ions for neuronal function. This integrity is mediated by structural BBB components, such as tight junction (TJ) proteins, integrins, annexins, and agrin, of a multicellular system including endothelial cells, astrocytes and pericytes. BBB permeability is significantly increased in aged animals (Blau et al., 2012), facilitating perhaps infiltration by monocytes, leading to mitochondria-mediated generation of ROS. Thus, a compromised BBB can result in neuronal cells to become vulnerable to exposure to circulating potentially proinflammatory macromolecules. This suggestion is supported by observations where capillary leakage of several plasma proteins such as prothrombin, immunoglobulin G, albumin, and lipoproteins have been detected in AD brains (Takechi et al., 2010; Zipser et al., 2007). Moreover, disrupted capillaries can also facilitate blood-to-brain delivery of circulating Aβ, exacerbating inflammatory processes potentially contributing to cerebral amyloid load (Takechi et al., 2010; Zipser et al., 2007; Pallebage-Gamarallage et al., 2015), inflammatory mediators, free radicals, vascular endothelial growth factor, matrix metalloproteinases, microRNAs (miRNAs), etc., compared to normal-aged subjects, supporting the view that alterations in BBB and activated microglial cell–mediated changes may be important risk factors for AD. The onset of neuroinflammation establishes a complex interaction with oxidizing agents through redox sensors present in enzymes, receptors, and sustained activation of NF-κB along with sustained expression and release of these inflammatory cytokines. These processes induce an imbalance in the inflammatory cycle homeostasis promoting further intensification of neuroinflammation (Farooqui, 2010). These factors control and modulate neuron–glia cross talk and neuronal function (Liu et al., 2012), and persistence abnormalities in neuron–glia cross talk may result in loss of cellular homeostasis leading to neurodegeneration in neurodegenerative diseases including AD (Raj et al., 2014). These studies indicate that both aging and AD are multifactorial conditions that share many relevant factors. Therefore, aging is an important risk factor for the development of many neurodegenerative diseases. Furthermore, neuroinflammation and oxidative stress (both reportedly associated with nonpathological aging in humans and animal models) are common features for several disease phenotypes, such as AD and stroke (Farooqui, 2010; Baron et al., 2014; Mosher and Wyss-Coray, 2014; Bachstetter et al., 2015).

DIET AS RISK FACTOR FOR ALZHEIMER'S DISEASE

The diet consumed in Western countries (Western diet) not only contains high amounts of processed macronutrients (saturated fats, cholesterol, proteins, and simple sugars), and too much salt (sodium chloride) but also is low in fiber and deficient in minerals. It is estimated that the present-day Western diet provides about 50% of total daily calories from refined carbohydrates, 35% calories from fat and refined oils, and 15% calories from proteins of animal origin. In the Western diet the ratio between n-6 (ARA) to n-3 fatty acid [docosahexaenoic acid (DHA)] is about 20:1. In contrast, the Paleolithic diet on which our forefathers lived and survived for thousands of years contained unprocessed

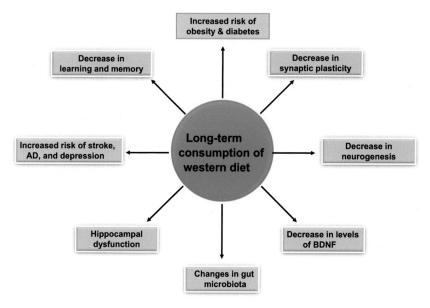

FIGURE 2.4 **Effects of long-term consumption of Western diet on neural and nonneural process.** *BDNF*, brain-derived neurotrophic factor.

carbohydrates (40%), fats (21%), and proteins (39%) (Simopoulos, 2013; Bengmark, 2013). Since 2009, there have been numerous studies reporting inverse associations between diet quality and common neurological disorders, such as depression, anxiety, stroke, AD, and Parkinson's disease (PD) (Jacka et al., 2015; Farooqui, 2013; Farooqui and Farooqui, 2015). The primary mechanisms associated with harmful effects of long-term consumption of Western diet include induction of oxidative stress, onset of inflammation, alterations in gut microbiome, and upregulation of stress responses and immune system (Fig. 2.4) (Jacka et al., 2015; Berk et al., 2013; Moylan et al., 2014; Dash et al., 2015; Jacka, 2015; Farooqui, 2015). As stated, Western diet is low in fiber and lacks fermented foods, which are important for gut health. Soluble fibers, such as psyllium, are probiotics that help nourish beneficial bacteria. These beneficial bacteria assist with digestion and absorption of our food and play a significant role in our immune function (Farooqui, 2015). These days, most grains in Western diet are contaminated with glyphosate, which is now recognized as a probable human carcinogen and has been linked to celiac disease and other gut dysfunction (Samsel and Seneff, 2013), which is the exact converse of what we achieve by adding fiber to our diet. Thus, the consumption of Western diet produces a metabolic state in which energy intake exceeds energy expenditure leading to obesity. This pathological condition is accompanied by induction of cellular oxidative stress and onset of

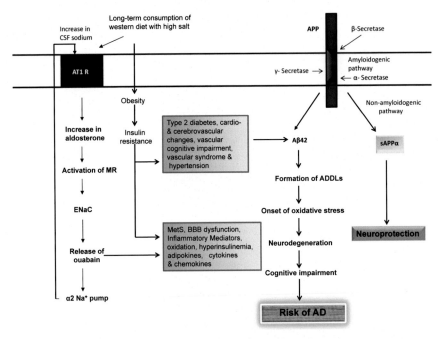

FIGURE 2.5   **Effects of long-term consumption of Western diet on type 2 diabetes, metabolic syndrome (MetS), blood–brain barrier (BBB), and amyloid β (Aβ) generation.** *APP,* amyloid precursor protein; *CSF,* cerebrospinal fluid; *MR,* mineralocorticoid receptor; *ENaC,* epithelial sodium channel; *sAPP,* soluble form of APP.

chronic inflammation. These processes are closely associated with the pathogenesis of cardiovascular, cerebrovascular, neurodevelopmental, neuropsychiatric, and neurodegenerative diseases (O'Keefe et al., 2008; Harris, 2008; Farooqui, 2014, 2015; Jacka, 2015). In addition, presence of high salt in the Western diet may contribute to hypertension (Fig. 2.5). In middle age, hypertension may increase the risk of cognitive impairment, dementia, and AD. Hypertension may increase the risk of AD by decreasing the vascular integrity of the BBB, resulting in protein extravasation into brain tissue (Kalaria, 2010). In turn, protein extravasation may result in cell damage, a reduction in neuronal or synaptic function, apoptosis, and an increase in Aβ accumulation, resulting in cognitive impairment (Deane et al., 2004a,b). Hypertension also produces changes in the structure and function of blood vessels, which modulates perfusion in various areas in the brain leading to hypoperfusion. This may interfere with the delivery of oxygen and nutrients to the brain. Thus hypoperfusion is linked with white matter degradation, gray matter atrophy, and cognitive deficits. Prolonged reduction in blood flow due to aging and atherosclerosis not only results in high hypertension-mediated damage in occipitotemporal, prefrontal, and medial temporal lobe regions, but also impacts blood vessel function (Beason-Held et al., 2007). In fact, a mild chronic cerebrovascular hypoperfusion and hypometabolism caused by decrease in cerebral blood flow may lower metabolic rates of glucose and oxygen. This may be one of the very early events in the pathogenesis of AD (Iadecola, 2004). With increasing age, the effect of hypertension in patients with AD diminishes and may even become inverted, with an increase in blood pressure inducing a protective effect. This observation may be explained by the fact that following the onset of AD, blood pressure begins to decrease, possibly as a result of vessel stiffening, weight loss, and changes in the autonomic regulation of blood flow. Converging evidence suggests that long-term consumption of Western diet not only decreases hippocampal size but also produces negative effects on cognitive performance and learning and memory in animal models of cognition (Wu et al., 2004; Stranahan et al., 2008; Kanoski et al., 2007; Kanoski and Davidson, 2011; Jacka et al., 2015). Hippocampus plays major roles in learning and memory processes. There is a putative causal relationship between Western diet consumption and hippocampal abnormalities related to learning and memory (Stranahan et al., 2008; Kanoski et al., 2007; Kanoski and Davidson, 2011; Kanoski, 2012). The molecular mechanisms by which Western diet impairs cognitive function are not fully understood. However, it is proposed that long-term consumption of Western diet elevates fasting blood glucose, serum cholesterol, and triacylglycerols causing endothelial cell dysfunction and atherosclerotic changes in the cerebrovascular system (Farooqui, 2015). In addition, the consumption of Western diet impairs glutamate metabolism and triggers neurochemical changes leading to desensitization of N-methyl-D-aspartate (NMDA) receptors within the hippocampus, which may contribute to cognitive deficits (Valladolid-Acebes et al., 2012). Western diet also produces changes in hippocampal plasticity by reducing the expression of brain-derived neurotrophic factor (BDNF), a growth factor that is synthesized by neurons in an activity-dependent manner (Stranahan et al., 2008). This growth factor plays an important role in the survival, maintenance, and growth and is essential for learning and memory. It performs its function by interacting and activating high-affinity receptor tyrosine kinase called tropomyosin-related kinase B (trkB). This receptor is located in the plasma membrane, dendrites, and presynaptic terminals. Activation of trkB is coupled with a signaling pathway, which involves phosphatidylinositol (PtdIns) 3-kinase, Akt kinase, and FOXO transcription factors and is crucial for cAMP response element–binding protein (CREB) and CREB-binding protein production that encode proteins involved in β-cell survival. trkB pathway also modulates the expression of genes that enhance synaptic plasticity (glutamate receptor subunits and growth-associated protein 43) and cell survival (due to antioxidant enzyme superoxide dismutase 2, and the antiapoptotic protein Bcl-2, for example) (Koponen et al., 2004; Mattson et al., 2004; Mattson, 2004; Pang and Lu, 2004). Importantly, duration of exposure to Western diet is related to the extent of the decrease in BDNF and the degree of learning and memory impairment in animal models, supporting the view that the longer the exposure to Western diet components, the more profound is the impact on hippocampal functioning and brain plasticity. However, even short-term feeding of Western diet results in cognitive impairment and potentiates the negative impact of mild brain injury on BDNF levels, synaptic plasticity, and cognitive functioning (Farooqui and Farooqui, 2015). In addition, BDNF also stimulates neurogenesis, a process that induces the differentiation of neural stem cells into neurons (Schmidt and Duman, 2007). Converging evidence suggests that decrease in BDNF can reduce synaptic plasticity and neurogenesis in the hippocampus, a brain region that performs two major functions (1) storage and interpretation of spatial information and (2) consolidation of short-term memory into long-term memory. It is reported that the numbers of neuron and neuronal morphologies do not change considerably with normal aging in the hippocampus, supporting the fact that functional weakening of hippocampal neurons with age is the crucial alteration, which may result from defectiveness in synapse functions or decrease in neuronal plasticity. The molecular mechanisms contributing to the regulation of neuronal plasticity and cognitive functions during are not understood. However, it is well known that induction and maintenance of long-term potentiation, a cellular indicator of brain cognitive function, requires gene expression and de novo protein synthesis (Schimanski and Barnes, 2010). Therefore, changes in gene expression/function of neurons may contribute to the functional deterioration of learning, memory, and cognition. Collectively, these studies suggest that synaptic plasticity is modulated by gene expression in the hippocampus and frontal cortex with advancing age and age-related neurodegenerative diseases (AD and PD). The molecular mechanisms underlying altered gene expression in normal aging and neurodegenerative diseases are largely unknown (Xu et al., 2007). Both synaptic plasticity and neurogenesis contribute to neuronal processes that underlie hippocampal-dependent learning and memory (Bergami et al., 2008; Leuner et al., 2006). It is well known that BBB acts to restrict and regulate molecular exchange between the blood and neural tissue or fluid, playing an important role in maintaining a regulated environment for neuronal signaling (Persidsky et al., 2006). Components of Western diet (high fat) have been reported to alter BBB integrity and permeability inducing impairment in hippocampal-dependent tasks such as a feature-negative discrimination problem (Davidson et al., 2012). In contrast, diet enriched in n-3 fatty acids, flavonoids, antioxidant-rich berries and resveratrol, whole grain, as well as fiber stimulate neurogenesis, reduce oxidative stress, and downregulate proinflammatory cytokines and chemokines (Farooqui, 2012).

Epidemiological studies have indicated that long-term consumption of Western diet may contribute to the pathogenesis of AD (Kalmijn, 2000; Bhat, 2010). Western diet significantly increases ROS generation and expression of gp91phox, p22phox, p47phox, and p67phox NADPH oxidase subunits in the cerebral cortex (Zhang et al., 2005). In addition, levels of prostaglandin E_2 (PGE_2) and phospholipase A_2 (PLA_2) and cyclooxygenase-2 (COX-2) activities are also increased in the brain of animals fed with Western diet. Activation of these enzymes produces oxidative stress and neuroinflammation. As stated earlier, ROS stimulates NF-κB, which migrates to the nucleus, where it mediates the expression of proinflammatory cytokines and chemokines [IL-1β, IL-6, and monocyte chemoattractant protein-1 (MCP-1)] (Farooqui, 2010) leading to neuroinflammation in the brain, which is linked with cognitive dysfunction. Furthermore, rodents consuming saturated fatty acids in Western diet have shown elevated markers of brain inflammation and low BDNF relative to standard diet–fed controls (Pistell et al., 2010). Long-term consumption of Western diet also leads to the onset of metabolic syndrome, a condition characterized by a cluster of factors (diabetes, insulin resistance, hypertension, dyslipidemia, visceral obesity). This condition is an important risk factor for AD (Farooqui et al., 2012; Farooqui, 2013). The consumption of Western diet produces insulin resistance and reduces insulin transport across the BBB resulting in lowering insulin levels and brain activity. These effects contribute to not only reduction in insulin levels in the cerebrospinal fluid (CSF) and reduction in brain insulin receptor activity but also to progressive reduction of brain glucose metabolism in AD (Craft et al., 1998). Reduction in brain insulin signaling upregulates the phosphorylation of Tau protein and elevation in Aβ levels in a streptozotocin mouse model of type II diabetes (Jolivalt et al., 2008). Intravenous injections of insulin have been reported to increase plasma levels of Aβ in patients with AD but not in normal subjects, an effect that can be exaggerated in patients with AD with higher body mass index (Kulstad et al., 2006). Furthermore, accumulation of Aβ oligomer has also been reported to induce neuronal insulin resistance in the AD brain not only by inhibiting the insulin network by targeting the insulin/Akt pathway (Townsend et al., 2007) but also by removing insulin receptors after binding at the dendrites of synaptic sites (Zhao et al., 2008). Collective evidence suggests that type II diabetes and metabolic syndrome (MetS) increase the risk of AD. The molecular mechanisms associated with this risk remain unknown. However, it is proposed that interactions among aging brain, insulin resistance, inflammatory cytokines, and oxidative stress may play an important role in the pathogenesis of AD (Bhat, 2010; de la Monte et al., 2009; de la Monte and Tong, 2013). Insulin contributes to the accumulation of Aβ by downregulating the degradation of Aβ through direct competition with the IDE, a 110-kDa thiol zinc-metalloendopeptidase that cleaves small proteins of diverse sequence, many of which share a propensity to form β-pleated-sheet-rich amyloid fibrils. In addition, insulin also contributes to Aβ accumulation by accelerating APP/Aβ trafficking from the trans-Golgi network, a major cellular site for Aβ generation, to the plasma membrane (Gasparini et al., 2002; Farris et al., 2003). Furthermore, consumption of Western diet promotes BACE1/Adaptor protein-2 (AP-2)/clathrin complex formation by increasing AP-2 levels in APP transgenic (APP Tg) mice. In Swedish APP–overexpressing Chinese hamster ovary cells as well as in SH-SY5Y cells, overexpression of AP-2 mediates the formation of BACE1/AP-2/clathrin complex and elevation in the level of the soluble form of APP β (sAPPβ) (Maesako et al., 2015). On the other hand, mutant D495R BACE1, which inhibits formation of this trimeric complex, shows a reduction in the level of sAPPβ. Overexpression of AP-2 promotes the internalization of BACE1 from the cell surface, causing a reduction in the cell surface BACE1 level. Based on these studies, it is suggested that Western diet may induce the formation of the BACE1/AP-2/clathrin complex, leading to the transport of BACE1 from the cell surface to the intracellular compartments supporting the view that these neurochemical processes may be associated with the enhancement of β-site cleavage of APP in APP Tg mice (Maesako et al., 2015). In addition, the consumption of Western diet by 3xTg-AD mice produces defects in insulin production and signaling related to the pathogenesis of type II diabetes and AD (Vandal et al., 2014). In 3xTg-AD mice, Western diet not only enhances glucose intolerance but also promotes soluble Aβ-mediated memory impairment. Importantly, a single insulin injection reverses the deleterious effects of Western diet on memory and soluble Aβ levels, partly through changes in Aβ production and/or clearance. Mechanisms linking type II diabetes and MetS with AD include (1) hyperglycemia-mediated increase in ROS production (Farooqui, 2013; de la Monte and Tong, 2013); (2) induction of insulin resistance and hyperinsulinemia, processes that interfere with Aβ peptides metabolism through the reduction of insulin-degrading enzyme activity (Craft, 2007), leading to its accumulation; and (3) neuroprotective effects of insulin in the brain, leading to the stimulation of dendritic sprouting, regeneration, and stem cell proliferation (Duarte et al., 2013) and impairing insulin signaling promoting the development of AD. Mounting evidence supports the view that cognitive impairment and neurodegeneration in AD are supported by changes in insulin and IGF resistance. Based on these studies, it is suggested that Western diet–mediated neurochemical changes in the brain are associated with a higher risk for developing type II diabetes, metabolic syndrome, and AD as well as other neurodegenerative diseases (Seneff et al., 2011; Farooqui, 2013; Craft, 2007; Talbot et al., 2012). Collectively, many studies in animal models suggest that chronic ingestion of Western diet may alter BBB integrity resulting in inappropriate blood-to-brain extravasation of plasma proteins, including lipid macromolecules that may be enriched in Aβ. Brain parenchymal retention of blood proteins and lipo-protein-bound Aβ may not only contribute to neurovascular inflammation but also induce alterations in redox homeostasis and nitric oxide (NO) metabolism (Farooqui, 2015). Therefore, it is likely that lipid-lowering agents (statins, DHA, and eicosapentaenoic acid) may positively modulate BBB integrity and by extension attenuate risk or progression of AD. In addition to their robust lipid-lowering properties, some lipid-lowering agents also have pleiotropic properties such as modulation of oxidative stress, NO, and Aβ metabolism, which modulate the risk and onset of AD.

SEDENTARY LIFESTYLE AS A RISK FACTOR FOR ALZHEIMER'S DISEASE

Sedentary lifestyle is defined lifestyle with no physical activity. Sedentary lifestyle involves too much sitting for reading, working on a computer, watching TV, playing video games, long deriving time, and lying down during waking hours with little or no vigorous physical exercise (Owen et al., 2000; Basterra-Gortari et al., 2014). These activities require an energy expenditure of 1.0–1.5 basal metabolic rates (Pate et al., 2008).

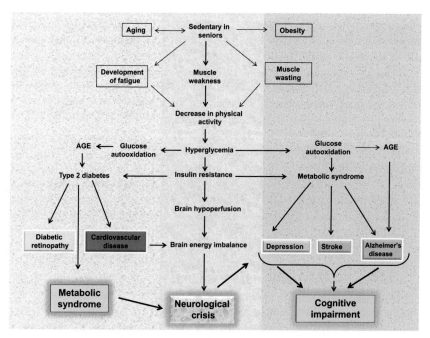

FIGURE 2.6   **Effect of sedentary lifestyle on the pathogenesis of type II diabetes, metabolic syndrome, and neurological disorders.** *AGE,* advanced glycation end product.

Sedentary lifestyle is not only linked with adverse outcome (increased risk of obesity, metabolic syndrome, diabetes, heart disease, and certain types of cancer) but also with early mortality (Fig. 2.6) (Keadle et al., 2015; Grontved and Hu, 2011; de Rezende et al., 2014). Thus sedentary lifestyle increases the risk of developing obesity and type 2 diabetes according to a Nurses' Health Study (Hu et al., 2003). Engaging in moderate-to-vigorous physical activity is an established component of healthy aging, and initiating moderate-to-vigorous physical activity later in life can improve health by reducing risk of obesity metabolic syndrome, diabetes, and heart disease and increasing the longevity (Paffenbarger et al., 1993; Wannamethee et al., 1998; Pahor et al., 2014).

Oxidative stress is closely associated with the pathogenesis of AD (Farooqui, 2010), but attempts to treat AD with vitamins E and C and increase lifespan have failed (Miller et al., 2005; Bjelakovic et al., 2007). Only overexpression of the peroxide and redox active thioredoxin 1 (Mitsui et al., 2002) and mitochondrial-targeted catalase (Schriner et al., 2005) have been reported to increase mouse lifespan. According to the epigenetic oxidative redox shift theory of aging, sedentary lifestyle triggers an oxidized redox shift in the oxidized direction. This process involves an extracellular decrease in the ratio of cysteine/cystine and an intracellular decrease in the ratio between reduced and oxidized glutathione (GSH/GSSG) and NAD(P)H/NAD(P). The oxidized redox shift is initiated by low demand for bursts of energy generated by the mitochondria. The low demand for resting energy results in low levels of physical and mental activity (Brewer, 2010). If there are no needs for the extra energy that is produced by aerobic oxidative phosphorylation, then cells and organs may downregulate the electron transport chain components and survive adequately on glycolysis. Increased consumption of simple sugar in Western diet and soft drinks (Sanchez-Lozada et al., 2008; Johnson et al., 2009) may direct the system to generate energy by glycolysis. Such an oxidative shift has been reported to occur in old rat brain and in plasma from healthy humans (Brewer, 2010). The oxidative redox shift may not only cause changes in the activities of numerous redox-sensitive transcription factors (NF-κB), enzymes, as well as transport and signaling proteins with redox-active cysteines (insulin receptor) but also interfere with the action of NMDA receptor (Lipton, 2008) and the Ca^{+2}-ATPase in the brain (Zaidi et al., 2003) along with increase in glucose level initiating and causing diabetes, metabolic syndrome, and insulin resistance. These conditions increase the risk of AD. Although the nature of link among type II diabetes mellitus, metabolic syndrome, and AD risk remains unclear, several mechanisms including aging, insulin resistance, inflammatory cytokines, and oxidative stress may be associated with this link (Haan, 2006; de la Monte, 2009; Farooqui, 2013). As stated earlier, insulin promotes the accumulation of Aβ by limiting its degradation via direct competition for the

IDE, a 110-kDa thiol zinc-metalloendopeptidase that cleaves small proteins of diverse sequence. Insulin also promotes the Aβ accumulation by accelerating APP/Aβ trafficking from the trans-Golgi network, a major cellular site for Aβ generation, to the plasma membrane (Gasparini et al., 2001). Converging evidence suggests that type II diabetes mellitus and metabolic syndrome are important risk factors for stroke, AD, and depression (Farooqui et al., 2012; Farooqui, 2013).

DISTURBANCE IN SLEEP AS A RISK FACTOR FOR ALZHEIMER'S DISEASE

Sleep is a complex biobehavioral process, which is an essential part of human life. Sleep is required for optimal health and performance. Obstructive sleep apnea is a common sleep disorder, characterized by recurrent partial or complete upper airway obstruction during sleep leading either to stoppage of breathing (apnea) or becomes very shallow breathing (hypopnea). Obstructive sleep apnea not only leads to intermittent hypoxia due to cyclical oxygen desaturation but also to frequent arousals and sleep fragmentation causing fatigue, bad mood, and depression (Levy et al., 2009; Ancoli-Israel, 2007). Due to the presence of polyunsaturated fatty acids and low activities of antioxidant enzymes, the human brain is extremely sensitive to hypoxia, ischemia, and glucose depletion. Impaired delivery of oxygen in obstructive sleep apnea alters neuronal homeostasis, induces pathology, and triggers neuronal degeneration/death leading to neurocognitive impairment, with negative influence on vigilance, attention, executive functioning, and memory (Beebe et al., 2003). Several studies have indicated that sleep apnea–mediated changes occur in the hippocampus, temporal lobe, cerebellum, thalamus, parahippocampal gyrus, anterior cingulate, parietal cortex, frontal cortex, and caudate nucleus (Torelli et al., 2011; Morrell et al., 2010; Joo et al., 2010). Sleep apnea–mediated pathological changes also occur in gray matter loss in the frontal, parietal, temporal, and occipital cortices; the thalamus; the hippocampus; and key brainstem nuclei including the nucleus tractus solitarius (NTS). The neurochemical basis of cognitive impairment in obstructive sleep apnea is not fully understood. However, based on human and animal models studies, it is suggested that intermittent hypoxia in sleep apnea produces increase in aggregation of platelet, metabolic dysregulation, oxidative stress, inflammation, atherosclerosis, endothelial dysfunction, and hypertension (Daulatzai, 2012a,b). Sleep apnea also mediates the activation and translocation of NF-κB in specific brain regions (hippocampus) associated with sleep regulation (Basheer et al., 2001; Ramesh et al., 2007; Gozal et al., 2001), which causes increase in expression of inflammatory cytokines, such as IL-2, IL-4, IL-5, IL-6, IL-8,

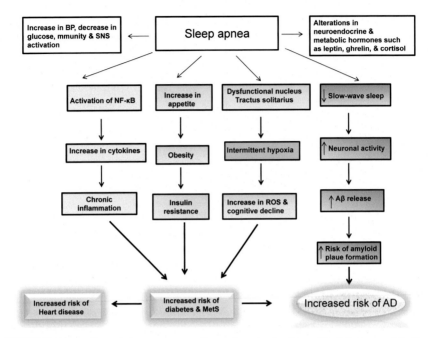

FIGURE 2.7 **Effect of sleep apnea on the pathogenesis of type II diabetes, metabolic syndrome, and neurological disorders.** *Aβ*, amyloid β; *AD*, Alzheimer's disease; *NF-κB*, nuclear factor-κB; *ROS*, reactive oxygen species; *SNS*, sympathetic nervous system.

IFN-γ, and TNF-α (Irwin et al., 2008). Other neurophysiological changes in sleep apnea include reduction in glucose tolerance (Spiegel et al., 1999), increase in blood pressure (Tochikubo et al., 1996; Daulatzai, 2012a,b), activation of the sympathetic nervous system, and alterations in leptin levels (Fig. 2.7) (Spiegel et al., 2004). Changes in leptin (the hormone that signals satiety) may predispose to increased cardiovascular risk including platelet aggregation also (Caro et al., 1996). Independent of body weight, patients with sleep apnea have higher levels of fasting blood glucose, insulin, and glycosylated hemoglobin (Considine et al., 1996). Lack of sleep also down-regulates the satiety hormone leptin (18%) and upregulates the appetite-stimulating hormone ghrelin (28%), a peptide produced predominantly by the stomach and increases hunger and food intake (Morselli et al., 2010). This condition leads to an increase in appetite. Additionally, sleep deprivation tends to lead to food cravings, particularly for sweet and starchy foods. It is becoming increasingly evident that sugar cravings stem from the fact that brain is fueled by glucose (blood sugar); therefore when lack of sleep occurs, the brain is unable to properly respond to insulin, which drives glucose into neural cells leaving brain desperate for carbohydrates to keep going. Too much consumption of simple carbohydrates virtually guarantees weight gain and obesity (Farooqui, 2015). The severity of sleep apnea also correlates with the degree of insulin resistance (Wallace et al., 2001). These processes may contribute to the pathogenesis of metabolic syndrome, a multifactorial condition, which is a risk factor for stroke, AD, and depression (Farooqui, 2013). As stated in Chapter 1, AD is accompanied by the accumulation of Aβ and Tau oligomers several years before the onset of AD. Under normal conditions, the clearance of misfolded proteins is facilitated by proteasome degradation, autophagy, and the newly discovered brain glymphatic system, an astroglial cell–mediated interstitial fluid (ISF) bulk flow (Jessen et al., 2015). It is reported that the activity of the glymphatic system is higher during sleep and disengaged or low during wakefulness (Pistollato et al., 2016). As a consequence, poor sleep quality, which is associated with dementia, may negatively affect glymphatic system activity, thus promoting Aβ accumulation. As stated earlier, diet is another important factor that regulates this complex network. Western diet, which is characterized by high intakes of refined sugars, salt, animal-derived proteins, and fats and by low intakes of fruit and vegetables, contributes to a higher risk of AD. Western diet can perturb the circadian modulation of cortisol secretion, which is associated with poor sleep quality. For this reason, diets and nutritional interventions aimed at restoring cortisol concentrations may ease sleep disorders and may promote brain clearance of Aβ, consequentially reducing the risk of cognitive impairment and dementia (Pistollato et al., 2016).

A novel hypothesis on the contribution of sleep apnea in the pathogenesis of AD has been proposed. According to this hypothesis, persistent neuroinflammation in NTS, a relay nucleus that integrates peripheral chemoreceptor input in response to hypoxia and hence influences the generation of respiratory rhythm as well as producing deleterious effects on nucleus ambiguus, dorsal motor nucleus of vagus, hypoglossal nucleus, parabrachial nucleus, locus coeruleus and many key nuclei in the brainstem, and the hippocampus, entorhinal cortex, prefrontal cortex, amygdala, insula, and basal forebrain in the neocortex leading to the neuronal and synaptic dysfunction associated with interconnected signal transduction pathways, which impact almost the entire brain, promoting cognitive decline and neuropsychiatric symptoms of AD (Daulatzai, 2012a,b). As stated in Chapter 1, AD is a multifactorial disease involving oxidative stress, neuroinflammation, accumulation of Aβ, and hyperphosphorylated Tau protein in the form of senile plaques and NFTs.

AD is associated not only with an increase in fragmentation of the overall sleep–wake pattern, increase in sleep during the daytime, and increase in frequency of nocturnal awakenings but also with a decrease in both slow wave and rapid eye movement sleep (Vitiello and Borson, 2001). Patients with AD often develop irregular sleep–wake rhythm disorder, in which sleep is fragmented into at least three distinct bouts throughout the 24-h period, and individuals also frequently develop "sundowning" whereby confusion and agitation worsen later in the day. These sleep disturbances result in significant functional impairment in individuals with AD (Tractenberg et al., 2005). In addition, patients with AD also have progressive circadian dysfunction, which can in turn result in significant sleep disruption. In the Tg2576 mouse model of AD, which shows age-dependent Aβ accumulation in the brain, Tg2576 mice not only demonstrate a significantly longer circadian period of wheel running rhythms when compared to control mice but also lack the normally observed increase in delta power following sleep deprivation (Wisor et al., 2005). It is also becoming increasingly evident that sleep regulates Aβ toxicity. In a mouse model of AD, knockout of the wake-promoting orexin gene reduces Aβ accumulation, an effect that is reversed by sleep deprivation (Roh et al., 2014). Similarly, in humans the risk of accumulation of Aβ is decreased by consolidated sleep, whereas the risk of developing certain forms of AD is enhanced by poor-quality sleep. Intriguingly, insomnia is common among patients with AD, and the severity of this symptom is correlated with the degree of dementia (Moran et al., 2005) supporting the hypothesis that AD and sleep have a bidirectional relationship (Ju et al., 2014). Converging evidence suggests that sleep apnea in patients with AD may contribute to memory decline, a process that precedes the neuronal death associated with late stages of AD (Lee et al., 2010; Daulatzai, 2012a,b; Farooqui, 2013).

GENES AS RISK FACTOR FOR ALZHEIMER'S DISEASE

As stated earlier, the majority of AD cases are sporadic, with disease onset after 65 years of age. Less than 5%–10% of all AD cases are inherited in an autosomal dominant manner (Bertram and Tanzi, 2005). Aβ peptides are derived from the sequential cleavage of APP by the β-secretase and γ-secretase complexes (Gandy, 2005). Molecular genetic studies indicate that patients with early-onset familial AD contain causative mutations in the *APP*, *PSEN1*, and *PSEN2* genes, and these mutations promote Aβ42 production, aggregation, and stabilization of the protein against clearance (Table 2.1) (Tanzi and Bertram, 2005). Therefore Aβ accumulation in the brain is generally accepted as the cause of AD (Hardy and Selkoe, 2002). However, the mechanism by which the Aβ peptide initiates the pathogenesis of AD still remains elusive. Sporadic AD is also called late-onset AD (LOAD). As stated earlier, onset and symptoms for LOAD typically appear after 65 years of age. LOAD also has a heritable component but has a more genetically complex mechanism than familial AD (Bettens et al., 2010). In the past several years, genome-wide association studies (GWAS) have identified more than 600 genes as susceptibility factors, available in AlzGene database (Wang et al., 2015a,b; Chaudhry et al., 2015). The apolipoprotein E (APOE) gene has

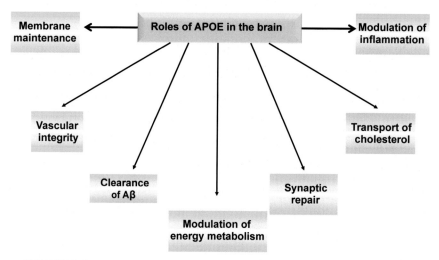

FIGURE 2.8   **Roles of APOE in the brain.** *Aβ*, amyloid β; *APOE*, apolipoprotein E.

been considered as the strongest consistently replicated genetic risk factor for LOAD. This gene is located on chromosome 19 and encodes a glycoprotein that is 299 amino acids long (Frieden, 2015). APOE (~34-kDa protein) is synthesized in various tissues in the body including the liver, brain, and skin and in macrophages (Mahley and Huang, 1999). In the brain, APOE is produced mainly in astrocytes and is a cholesterol-transporting protein and a major determinant of synapse formation and remodeling (Pfrieger, 2003; Bu, 2009). APOE is also a ligand for lipoprotein receptors and thus may have a role in promoting Aβ clearance through the BBB or the blood–CSF barrier. In the blood, APOE protein binds with lipids forming lipoproteins, including very-low-density lipoproteins, which plays an important role in cholesterol and other lipid transport, maintenance of neural membranes, Aβ clearance, and NFT formation (Fig. 2.8). By interacting with its receptors, APOE mediates the clearance of different lipoproteins from the circulation. Absence or structural mutations of APOE cause significant disorders in lipid metabolism and cardiovascular disease. Polymorphisms in this gene define three distinct APOE proteins (the E2, E3, and E4 forms). There are many functional differences between APOE3 and APOE4. APOE4 increases Aβ deposition. APOE4 is the major known genetic risk factor for AD. It increases Aβ deposition and lowers the age of onset of AD (Verghese et al., 2011; Liu et al., 2013). APOE4 carriers account for 65%–80% of all AD cases, highlighting the importance of APOE4 in AD pathogenesis (Fig. 2.9). APOE4 has both Aβ-dependent and Aβ-independent roles in AD pathogenesis (Blacker, 1998). In contrast, APOE3, which differs from APOE4 by a single amino acid change at position 112, poses little or no risk for the development of this

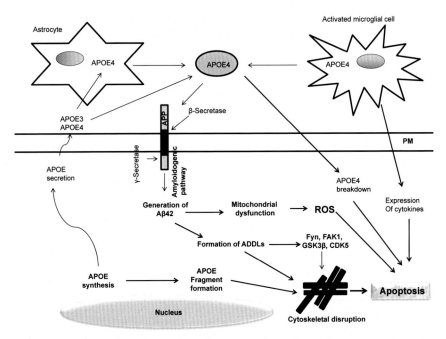

FIGURE 2.9   **Generation of APOE and its role in redistribution of lipids in neural cells.**
Aβ; amyloid β; *APOE*, apolipoprotein E; *APP*, amyloid precursor protein; *Fyn*, tyrosine-specific phosphotransferase; *FAK1*, focal adhesion kinase 1; *GSK3β*, glycogen synthase kinase 3 beta (GSK3β); CDK, cyclin-dependent kinase 5; ROS, reactive oxygen species; PM, plasma membrane.

disease. Amino acid composition studies have indicated that APOE proteins have seven highly conserved regions, representing approximately 47 amino acids, which are distributed throughout the protein and reflect ligand-binding sites as well as regions proposed to be involved in the propagation of the cysteine–arginine change at position 112 to distant regions of the protein in the N- and C-terminal domains. Highly nonconserved regions are at the N- and C-terminal ends of the APOE (Frieden, 2015). APOE2 has cysteines at positions 112 and 158, APOE3 has an arginine at position 158, and APOE4 has arginines at both 112 and 158 (Rall et al., 1982). In brain, APOE is mainly synthesized by astrocytes (either pericytes or microglia) or under certain pathological conditions by neurons (Mahley and Rall, 2000; Zhang et al., 2011). The human brain contains up to 25% of the body's cholesterol, which is not only essential for myelination but also for axonal growth, synaptogenesis, neural plasticity, neurotransmission, dendritogenesis, and remodeling events that are crucial for learning and memory (Dietschy and Turley, 2004; Saher et al., 2005). The depletion of cholesterol in neurons promotes impairment in the synaptic vesicle exocytosis, decrease in the neuronal activity, and abnormalities in neurotransmission. In brain, cholesterol is transported by APOE (Mauch et al., 2001; Pfrieger, 2003). The BBB restricts the exchange of lipoproteins and APOE between the CNS and peripheral system. Brain APOE regulates the clearance of Aβ, which is a common hallmark of AD (Bien-Ly et al., 2012; Kim et al., 2011). In addition, in vitro studies indicate that overload of cholesterol in plasma membrane of primary cultured neurons promote an elevation in Aβ production through increase in α-secretase (BACE1)-mediated APP cleavage (Marquer et al., 2011). Several studies have also indicated that cholesterol-enriched diet increases the levels of hydroxycholesterol (27-hydroxycholesterol) along with elevation in Aβ and phosphorylated Tau levels in rodents. Although the mechanisms by which cholesterol and 27-hydroxycholesterol modulate the production of Aβ and Tau protein phosphorylation is not fully understood, in transgenic mice for AD leptin (an adipocytokine) has been reported to control and modulate Aβ production and Tau protein hyperphosphorylation. Accumulating evidence thus suggests that cholesterol-enriched diets and 27-hydroxycholesterol induce AD-like pathology by altering leptin signaling (Marwarha et al., 2010). 27-Hydroxycholesterol also produces glutathione depletion, ROS generation, inflammation, and apoptosis-mediated cell death (Dasari et al., 2010). Collective evidence suggests that alterations in cholesterol and hydroxycholesterols may be closely associated with pathophysiology of AD. In addition to its function in lipid transport, APOE is an antiinflammatory agent. The deficiency of APOE exacerbates neuroinflammation in several traumatic injury models in rodents (Sheng et al., 1999; Lynch et al., 2001, 2002; Fagan et al., 1998). Furthermore, APOE deficiency is also associated with a reduced clearance of neuronal debris in a model of entorhinal cortex lesion, supporting the view that APOE contributes to the clearance of cholesterol-rich neuronal breakdown products (Sheng et al., 1999; Lynch et al., 2001, 2002; Fagan et al., 1998). In addition, APOE facilitates clearance of Aβ from the brain through two mechanisms: one involving a receptor-mediated process and drainage either through the ISF drainage pathway or through the BBB and the other through endopeptidase-mediated proteolytic degradation. Mounting evidence suggests that APOE isoforms not only differentially regulate Aβ aggregation and clearance in the brain but also mediate distinct functions in regulating brain lipid transport, glucose metabolism, neuronal signaling, neuroinflammation, and mitochondrial and synaptic functions through APOE receptors.

Family history is an important risk factor in the development of AD. There is a ~10%–14% higher risk of AD in patients who have close relatives (parent, brother, sister, or child) suffering from the disease compared to unrelated individuals (Burns and Iliffe, 2009). The expression of APOE, a protein contributing to cholesterol transport in the bloodstream, has also been linked with the incidence of LOAD. There are three isoforms

TABLE 2.2   Genes Associated With Sporadic Alzheimer's Disease, Their Encoded Proteins, and Their Roles

Gene	Protein	Role	References
TREM2	Type 1 transmembrane receptor protein	Stimulation of DAP12	Bakker et al. (1999) and Paloneva et al. (2000)
CD33	Type 1 transmembrane receptor protein	Activation of SHP-1 and SHP-2	Ulyanova et al. (1999) and Walter et al. (2008)
SHIP1	PtdIns 3,4,5-trisphosphatase	Activation of SHIP1	An et al. (2005) and Damen et al. (1996)
CR1	Complementary receptor 1	Regulation of complement system	Khera and Das (2009) and Hazrati et al. (2012)
ABCA7	Member of the ATP-binding cassette transporter	Lipid homeostasis (transport of phospholipid)	Abe-Dohmae et al. (2004) and Ikeda et al. (2003)
APOE	Cholesterol-binding protein	Transport of cholesterol from glial cells to neurons	Xu et al. (2006a,b) and Liu et al. (2013)

DAP12, a 12 kDa transmembrane protein, which is recognized as a key signal transduction receptor element in natural killer cells; *SHIP1*, Src homology 2 domain containing inositol-5-phosphatase 1; *SHP-1*, Src homology region 2 domain-containing phosphatase-1; *SHP-2*, cytoplasmic SH2 domain containing protein tyrosine phosphatase.

of APOE: ε2, ε3 (most common), and ε4. Expression of the ε4 isoform of APOE is associated with increased risk of AD. GWAS have also revealed that other genes such as *CD33* (Bertram et al., 2008; Hollingworth et al., 2011; Naj et al., 2011), *CLU, BIN1, PICALM, CR1, CD2AP, EPHA1, ABCA7, MS4A4A/MS4A6E* (Harold et al., 2009; Seshadri et al., 2010; Bagyinszky et al., 2014), and *TREM2* (Guerreiro et al., 2013; Jonsson et al., 2013) may also contribute to AD (Table 2.2). Two GWAS have identified variants in the clusterin gene *(CLU)*, the phosphatidylinositol-binding clathrin assembly protein gene *(PICALM)* and complement receptor type 1 gene *(CR1)* with AD, but functional data confirming involvement of these genes in AD are still lacking (Harold et al., 2009; Lambert et al., 2009). Clusterin is a lipoprotein that is expressed in mammalian tissues and is incorporated into amyloid plaques. This protein binds to soluble Aβ in CSF, forming complexes that can penetrate the BBB. Clusterin levels are positively correlated with the number of *APOE ε4* alleles, suggesting a compensatory induction of *CLU* in the brains of patients with AD with the *APOE* ε4 allele, who show low brain levels of APOE. It is also reported that single nucleotide polymorphisms (SNPs) may contribute to the risk of AD. Genes controlling SNP-encoded proteins modulate microglial function and neuroinflammatory process. These genes include TREM2, CD33, CR1, ABCA7, and SHIP1 (Harold et al., 2009; Naj et al., 2011; Hollingworth et al., 2011). They contribute to the pathogenesis of AD by stimulating microglial cell activation and increasing the expression of proinflammatory cytokines.

Relatively less data are available on genes encoding for NFTs compared to Aβ. The toxicity of Aβ seems to depend on the presence of microtubule-associated protein Tau (Roberson et al., 2007), the hyperphosphorylated forms of which aggregate and deposit in AD brains as NFTs. The spreading and distribution of NFTs across the brain exhibits high correlation with the cognitive impairment status in AD (Nelson et al., 2012). Studies on whole-genome gene expression analysis using 213 human postmortem brain tissue specimens from three brain regions, the entorhinal, temporal, and frontal cortices, of 71 Japanese brain donor subjects (Murayama and Saito, 2004; Miyashita et al., 2014) indicate that eight genes, *RELN, PTGS2, MYO5C, TRIL, DCHS2, GRB14, NPAS4,* and *PHYHD1,* are associated with the development of NFTs. Genes related to the phosphorylation and dephosphorylation of Tau, such as *GSK3B* and *PPP1CA,* do not show significant association with Braak NFT stages (Bekris et al., 2012; Mondragon-Rodriguez et al., 2013). Three proteins, RELN, PTGS2, and DCHS2, directly interact with COPS5, which directly interacts with APP and thus increases Aβ production (Wang et al., 2013a,b). However, the remaining five genes, *MYO5C, TRIL, GRB14, NPAS4,* and *PHYHD1,* are novel to AD. GWAS have identified few novel risk genes for AD (Hollingworth et al., 2011; Olgiati et al., 2011). These genes include bridging integrator 1 (BIN1), clusterin (CLU, also called apolipoprotein J), ATP-binding cassette transporter A7 (ABCA7), triggering receptor expressed on myeloid cells 2 (TREM2), and phosphatidylinositol-binding clathrin assembly protein (PICALM). Like APOE, CLU contributes to brain cholesterol transport (Yu and Tan, 2012), ABCA7 is involved in lipid homeostasis (Tanaka et al., 2011). Deletion of ABCA7 increases Aβ accumulation in APP Tg mice through the decrease in phagocytic clearance of Aβ (Kim et al., 2013). TREM2 is associated with the microglial cell response through lipid sensing around the senile plaque in a mouse model of AD (Wang et al., 2015a,b). BIN1 not only contributes to endocytosis and membrane trafficking (Itoh and De Camilli, 2006) through PtdIns binding (Kojima et al., 2004) but also modulates APP trafficking in neurons (Chapuis et al., 2013). In microglial cells, BIN1 expression also contributes to Aβ phagocytosis (Chapuis et al., 2013). PICALM plays important roles in clathrin-mediated endocytosis (Dreyling et al., 1996) through the involvement of PtdInsl binding (Ford et al., 2001). It is suggested that PICALM is also associated with modulation of APP trafficking (Xiao et al., 2012). It also promotes endocytosis of γ-secretase (Kanatsu et al., 2014). Collective evidence suggests that many genes contribute to the pathogenesis of AD. On the one hand, genetic variants may prevent development of the neuropathological substrate that underlies susceptibility to cognitive dysfunction. On the other hand, the presence of neuropathology and genetic variants may mitigate the effects of AD-related lesions that otherwise result in cognitive decline.

ENVIRONMENTAL FACTORS AS RISK FACTOR FOR ALZHEIMER'S DISEASE

The pathogenesis of sporadic AD is not only promoted by long-term consumption of high-calorie Western diet, sedentary lifestyle, and interaction of multiple genetic and environmental risk factors but also by the disruption of epigenetic mechanisms controlling gene dynamics and expression (Farooqui, 2015). At the molecular level, long-term consumption of high-calorie Western diet not only contributes to insulin resistance and cerebrovascular dysfunction but also to oxidative damage and mitochondrial dysfunction leading to the loss of synapses as well as axonal pathology promoting dementia (Farooqui, 2010). In addition to the aforementioned changes, long-term consumption of Western diet also alters the synaptic plasticity and neuronal integrity in mature neuronal circuitries, which may modulate the process of neurogenesis (Li et al., 2008; Dong et al., 2004; Jin et al., 2004). These findings support the view that the pathogenesis of AD may not only involve the degeneration of mature neurons but also the disruption of neurogenic niches in the adult brain (Crews et al., 2010). Other risk factors that modulate the pathogenesis of AD include smoking, stroke, heart disease, head injury, impairment of cerebral blood flow, depression, alcohol consumption, and type II diabetes (Ballard et al., 2011; Reitz and Mayeux, 2014). Among these factors, smoking has a special place. Approximately 2 billion people worldwide use tobacco products, mostly in the form of cigarettes, and tobacco smoking–related diseases cause at least 4 million global deaths per year (De Marini, 2004). There are an estimated 44 million active smokers in the United States (Dube et al., 2009); however, the actual prevalence of smoking in the United States may be underestimated, particularly in younger adults and females (Delnevo et al., 2008). Although cardiovascular disease, chronic obstructive pulmonary diseases, and various forms of cancer are the leading causes of smoking-related mortality in the United States (http://www.cdc.gov/tobacco/data_statistics/sgr/2004/), but F^{18} positron emission tomography studies have indicated that smoking is also an important risk factor for LOAD/sporadic AD and AD-related dementia (Durazzo et al., 2014). The molecular mechanism of smoking-mediated neurotoxicity is not fully understood. However, it is known that cigarettes and tobacco smoking enhance the risk of cerebrovascular and neurological disorders (Cataldo et al., 2010; Anstey et al., 2007; National Center for Chronic Disease Prevention and Health Promotion (US) Office on Smoking and Health, 2014) largely due to an increase in ROS generation (Giacco and Brownlee, 2010), proinflammatory activity (Arnson et al., 2010), and BBB impairment (Rosenberg, 2012). Cigarette smoke contains over 4000 chemicals including nicotine, its metabolites, and various forms of ROS [H_2O_2, epoxides, nitrogen dioxide, peroxynitrite (Pryor and Stone, 1993)], which can penetrate through the lung alveolar wall and increase systemic levels of ROS (Yamaguchi et al., 2007).

At the cerebrovascular level, high ROS and NO not only induce oxidative damage, which cause oxidative damage in vascular tissue, but also may exacerbate inflammatory events leading to the breakdown of BBB, modification of TJ, and more activation of proinflammatory pathways (Naik et al., 2014). Cigarette smoke stimulates transcription of inflammatory modulators such as NF-κB, RelB, STAT3, and Nrf-2 (Chen and Han, 2008; Kaisar et al., 2015) and modulates extracellular signals to transcriptionals that control proliferation and immune evasion (Xu et al., 2006a). Pretreatment with α-tocopherol and/or ascorbic acid provides protection against BBB permeability, suggesting that prophylactic administration of antioxidants can reduce cigarette smoking–mediated inflammatory BBB damage (Kaisar et al., 2015). Furthermore, SAA1 (a potent chemoattractant factor) is also responsible for the transcription and upregulation of amyloid A (Xu et al., 2006b) and APOE gene, which is known to promote the pathogenesis of atherosclerosis, ischemic damage, and AD (Abboud et al., 2008). Collective evidence suggests that oxidative stress and inflammation caused by the exposure to cigarette smoke can facilitate the pathogenesis and progression of several neurological disorders including AD and PD (Baldeiras et al., 2010; Seet et al., 2010). It is also known that chronic smokers have a higher incidence of small vessel ischemic disease than nonsmokers (Hossain et al., 2009). Small vessel ischemic disease is characterized by leaky brain microvessels and loss of BBB integrity, which may have secondary effects on cerebral blood flow and vascular tone, further influencing transport across the microvascular endothelium. For example, cigarette smoke produces cerebrovascular vasodilation through sympathetic activation. Nicotine activates nicotine receptors, which leads to the acetylcholine-dependent release of NO from the vascular endothelium (Furchgott, 1993) through activation of endothelial nitric oxide synthase (Bulnes et al., 2010). NO has been reported to play an active role in regulating microvascular tone and the cerebral blood flow under normal and pathological conditions (McCarron et al., 2006). Furthermore, NO also increases vascular permeability at the BBB thus impairing brain homeostasis and facilitating the passage of unwanted substances from the blood into the brain (Bulnes et al., 2010; Kaur and Ling, 2008). Converging evidence suggests that the generation of superoxide and other ROS and NO due to cigarette smoke promotes leaky brain microvessel formation and loss of BBB integrity (Raij et al., 2001; Baldeiras et al., 2010; Seet et al., 2010;

Hossain et al., 2011). NO also reacts with superoxide to generate peroxynitrite, which may promote apoptotic cell death and antioxidant depletion. Thus, smoking-mediated oxidative stress may serve as a fundamental mechanism contributing to the neurobiological abnormalities (e.g., brain atrophy, regional cortical thinning) along with the increased risk for development of Aβ and Tau pathology (Reitz et al., 2007; Sutherland et al., 2013; Ho et al., 2012).

Like the aforementioned pathological conditions, chronic hyperglycemia in type 2 diabetes causes endogenous ROS increase by inhibiting glycolysis and promoting the formation of harmful intermediates such as AGEs and protein kinase C pathway isoforms, which have DNA- and protein-damaging effects (Brownlee, 2005). At the BBB level, chronic hyperglycemia also produces endothelial dysfunction leading to BBB impairment and loss of barrier integrity (Prasad et al., 2014).

EPIGENETIC FACTORS AS RISK FACTOR FOR ALZHEIMER'S DISEASE

Epigenetics is the molecular phenomenon by which phenotypic changes are transmitted from one generation to another with no apparent alterations in structural DNA. Epigenetic mechanisms (DNA methylation, histone modifications, and miRNA regulation) are among the major regulatory elements responsible for the control of metabolic pathways at the molecular level. Epigenetic modifications regulate gene expression transcriptionally, whereas miRNAs suppress gene expression posttranscriptionally (Szulwach and Jin, 2014). Epigenetic mechanisms include the involvement of DNA methylases, histone acetyltransferases and histone deacetylases as well as by the processes such as phosphorylation and ubiquitination (Szulwach and Jin, 2014). Histone modifications and DNA methylation are the two major epigenetic dysregulations that appear to result in excess Aβ and NFTs of Tau protein, which contribute to neurodegeneration. Epigenetic modifications control and regulate expression of gene transcriptionally, whereas the miRNAs suppresses gene expression posttranscriptionally (Szulwach and Jin, 2014). Epigenetic mechanisms modulate brain development, maturation and aging, puberty-related changes, mental disorders, addictive behaviors, and neurodegeneration (Wang et al., 2013a,b; Cacabelos and Torrellas, 2014; Cacabelos et al., 2014). As stated earlier, AD is a multifactorial disease, in which aging as well as genetic, metabolic, nutritional, vascular, and environmental factors, along with diabetes, inflammation, and traumatic brain injury (TBI) are associated with onset and progression of the pathology and pathogenesis of AD (Marques et al., 2011). As stated in Chapter 1, AD is strongly linked with the accumulation of Aβ and amyloid plaque formation. The fact that TBI may cause rapid elevations of brain Aβ along with early deposition of amyloid plaques, even in younger individuals, has provided a suggested pathobiological linkage between TBI and AD (Johnson et al., 2010). Moreover, certain pathological features also differ, and some controlled epidemiologic studies have indicated that TBI, stroke, and vascular factors are better associated with delayed-onset dementias other than AD (Wang et al., 2012). How do the aforementioned factors promote and induce epigenetic changes that mediate the network genes involved in this disease is a question that remains to be answered. Among these factors, aging is the most important risk factor in which epigenetic changes may contribute to the pathogenesis of AD (Wu et al., 2008). Epigenetic mechanisms, which may contribute to AD, range from DNA methylation to altered posttranslational marking of histones to regulatory actions of noncoding RNAs, with particularly rich and challenging studies published on miRNAs (Millan, 2013), along with the modification of the proteins that package the DNA, the histones. DNA hypomethylation may lead to upregulation of the proinflammatory gene, NF-κB, as well as encodes COX-2, which catalyzes the generation of prostaglandins, leukotrienes, and thromboxanes (Rao et al., 2012; Gu et al., 2013). These observations support the view that aberrant DNA methylation may contribute to the neuroinflammation in AD. Conversely, the hypermethylation of promoters for *BDNF* and *CREB* has been reported to interfere with synaptic plasticity (Rao et al., 2012). Many genes (*APP, PSEN1, APOE, BACE*) contributing to the pathogenesis of AD contain methylated CpG sites. The promoter region of the *APP* gene is hypomethylated. This hypomethylation results in enhancement of Aβ production. Some investigators report no relevant changes in *APP* methylation but an epigenetic drift in brain samples from patients with LOAD (Wang et al., 2008). It is interesting to note that the intracellular domain of APP acts as a key epigenetic regulator of gene expression, controlling a diverse range of genes, including the *APP* itself, the amyloid-degrading enzyme (ADE) neprilysin (NEP), and aquaporin-1 (Octave et al., 2013). Abnormal processing of neuronal cell membrane APP is accompanied by elevated levels of 24-hydroxycholesterol in human serum and CSF. This lipid mediator is an endogenous ligand for liver X receptor (LXR-α). There is an epigenomic pathway that connects LXR-α activation with genes associated with the regulation of aberrant Aβ production, leading to apoptotic cell death. LXR-α activation contributes to the overexpression of the *PAR-4* gene and the suppression of the *AATF* gene. Overexpression of the *PAR-4* gene is accompanied by aberrant Aβ production followed by ROS generation and subsequently ROS-mediated neuronal death. Aβ-induced heme oxygenase-1 has been reported not only to contribute to cholesterol oxidation but also to production of endogenous ligands for the sustained activation of neuronal LXR-α-dependent epigenomic pathways, causing neuronal death in AD (Raina and Kaul, 2010). In addition, folate deprivation–mediated hypomethylation increases the expression of *BACE* and *PSEN1* and restoration of folate deficiency by supplementation of S-adenosylmethionine leads to normal expression of these genes. Aβ-mediated hypomethylation also contributes to the upregulation of genes involved in neuroinflammation (TNF-α) and apoptosis (caspase-3), which intensifies the production of more Aβ (Wang et al., 2013a,b). Aberrant epigenetic changes in fully methylated 3′-CpG island region of APOE have been reported to contribute to LOAD pathology (Wang et al., 2008, 2013a,b). The APOE4 sequence may alter the epigenetic function of the methylated 3′-CpG island, since the APOE4 allele induces a C to T transition contributing to the loss of a methylatable CpG unit (Wang et al., 2013a,b). APOE4 carriers show a dose-dependent risk, and the relative mRNA level of APOE4 is increased in AD compared to controls, indicating that variability in the neuronal expression of APOE contributes to disease risk (Caesar and Gandy, 2012).

Converging evidence suggests that risk factors for AD are classified into three groups, namely, (1) heart-related abnormalities, (2) cerebral abnormalities, and (3) peripheral abnormalities (Fig. 2.2). Heart abnormalities include cardiac arrhythmia, atrial fibrillation, congestive heart failure, APOE4, and induction of atherosclerosis. Cerebral abnormalities include normal aging, TBI, and stroke-mediated signal transduction alteration in brain metabolism, and peripheral abnormalities are associated with metabolic changes induced by type 2 diabetes, high cholesterol levels, smoking, and long-term consumption of Western diet along with hemorheological abnormalities. These metabolic changes not only induce abnormal APP processing and increased production of Aβ but also promote the production of ROS. Increased production of ROS; oligomerization of Aβ; reduction in levels of NEP, a zinc metalloendopeptidase; and ADEs in specific brain areas during aging along with activation of microglial cells and inability of the BBB to regulate the transendothelial transport and clearance of the Aβ may be associated with the development of AD (Nalivaeva et al., 2014; Marr and Hafez, 2014; Farooqui, 2014).

CONCLUSION

Aging, APOE4, diabetes, metabolic syndrome, and long-term consumption of Western diet are the major risk factors for AD (Fig. 2.10). Other risk factors for AD include physical inactivity, hypertension, smoking and obesity, heart disease, head injury, impairment of cerebral blood flow, and alcohol consumption. Except aging and genetic factors, many of factors listed previously are potentially modifiable. These factors usually coexist in metabolic syndrome, a condition that is not only characterized by insulin resistance but also by diabetes, hypertension, dyslipidemia, vascular syndrome, and increased production of Aβ. These processes increase the likelihood of developing AD and dementia. Among genetic factors (APP, AOPE4, PSEN 1, and PSEN 2), a single APOE-4 allele is associated with

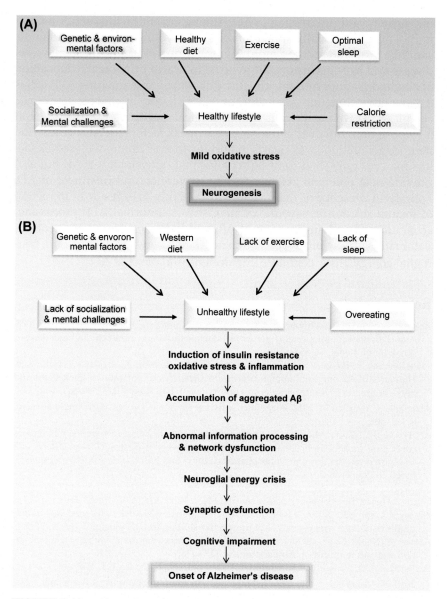

FIGURE 2.10 **Effects of healthy (A) and unhealthy lifestyles (B) on the brain and onset of Alzheimer's disease.** *Aβ*, amyloid β.

a two- to threefold increase in risk of developing AD, and having two copies is associated with a fivefold or more increase. In addition, each inherited APOE4 allele lowers the age at onset by 6–7 years. APOE4 also lowers cognitive performance, in particular the memory domain, supporting the view that this gene makes a major contribution to cognitive impairment and progression in AD. Among genetic factors, abnormalities in *PS 1* and *PS 2* usually appear in somewhat younger age group than the sporadic form and follow a more aggressive downhill course. Small vessel disease and hypoperfusion and hypoxia may also promote the deposition of amyloid plaques by damaging the BBB and by upregulating APP. TBI is a major environmental risk factor for AD, TBI, viral infections, neurotoxic chemicals, as well as various immunological and hormonal disorders. It is important to discover and identify early risk factors for AD because the neurodegeneration in AD starts in midlife. Identification of these risk factors may not only shed some light on the pathophysiology of AD but also provide new potential avenues for the prevention and treatment of AD. Long-term presence of these factors in the body may result in induction of insulin resistance, severe oxidative stress, abnormal APP processing, accumulation and aggregation of Aβ, and induction of chronic neuroinflammation. These processes lead to neuroglial network dysfunction, neuroglial energy crisis, cognitive impairment, and onset of AD.

References

Abboud, S., Viiri, L.E., Lutjohann, D., Goebeler, S., Luoto, T., Friedrichs, S., et al., 2008. Associations of apolipoprotein E gene with ischemic stroke and intracranial atherosclerosis. Eur. J. Hum. Genet. 16, 955–960.

Abe-Dohmae, S., Ikeda, Y., Matsuo, M., Hayashi, M., Okuhira, K., Ueda, K., et al., 2004. Human ABCA7 supports apolipoprotein-mediated release of cellular cholesterol and phospholipid to generate high density lipoprotein. J. Biol. Chem. 279, 604–611.

Alzheimer's Association, 2012. Alzheimer's disease facts and figures. Alzheimers Dement. 8, 131–168.

An, H., Xu, H., Zhang, M., Zhou, J., Feng, T., Qian, C., et al., 2005. Src homology 2 domain-containing inositol-5-phosphatase 1 (SHIP1) negatively regulates TLR4-mediated LPS response primarily through a phosphatase activity- and PI-3 K-independent mechanism. Blood 105, 4685–4692.

Ancoli-Israel, S., 2007. Sleep apnea in older adults—is it real and should age be the determining factor in the treatment decision matrix? Sleep. Med. Rev. 2007 (11), 83–85.

Anstey, K.J., von, S.C., Salim, A., O'Kearney, R., 2007. Smoking as a risk factor for dementia and cognitive decline: a meta-analysis of prospective studies. Am. J. Epidemiol. 166, 367–378.

Arnson, Y., Shoenfeld, Y., Amital, H., 2010. Effects of tobacco smoke on immunity, inflammation and autoimmunity. J. Autoimmun. 34, J258–J265.

Bachstetter, A.D., Van Eldik, L.J., Schmitt, F.A., Neltner, J.H., Ighodaro, E.T., Webster, S.J., et al., 2015. Disease-related microglia heterogeneity in the hippocampus of Alzheimer's disease, dementia with Lewy bodies and hippocampal sclerosis of aging. Acta Neuropathol. Commun. 3, 32.

Bagyinszky, E., Youn, Y.C., An, S.S., Kim, S., 2014. The genetics of Alzheimer's disease. Clin. Interv. Aging 9, 535–551.

Bakker, A.B., Baker, E., Sutherland, G.R., Phillips, J.H., Lanier, L.L., 1999. Myeloid DAP12-associating lectin (MDL)-1 is a cell surface receptor involved in the activation of myeloid cells. Proc. Natl. Acad. Sci. U.S.A. 96, 9792–9796.

Baldeiras, I., Santana, I., Proenca, M.T., Garrucho, M.H., Pascoal, R., Rodrigues, A., Duro, D., Oliveira, C.R., 2010. Oxidative damage and progression to Alzheimer's disease in patients with mild cognitive impairment. J. Alzheimers Dis. 165, 165–177.

Ballard, C., Gauthier, S., Corbett, A., Brayne, C., Aarsland, D., Jones, E., 2011. Alzheimer's disease. Lancet 377, 1019–1031.

Baron, R., Babcock, A.A., Nemirovsky, A., Finsen, B., Monsonego, A., 2014. Accelerated microglial pathology is associated with Abeta plaques in mouse models of Alzheimer's disease. Aging Cell 13, 584–595.

Basheer, R., Rainnie, D.G., Porkka-Heiskanen, T., Ramesh, V., McCarley, R.W., 2001. Adenosine, prolonged wakefulness, and A1-activated NF-kappaB DNA binding in the basal forebrain of the rat. Neuroscience 104731–104739.

Basterra-Gortari, F.J., Bes-Rastrollo, M., Gea, A., Nunez-Cordoba, M., Toledo, E., 2014. Television viewing, computer use, time driving and all-cause mortality: the SUN cohort. J. Am. Assoc. 3, e000864.

Beebe, D.W., Groesz, L., Wells, C., Nichols, A., McGee, K., 2003. The neuropsychological effects of obstructive sleep apnea: a meta-analysis of norm-referenced and case-controlled data. Sleep 26, 298–307.

Beason-Held, L.L., Moghekar, A., Zonderman, A.B., Kraut, M.A., Resnick, S.M., 2007. Longitudinal changes in cerebral blood flow in the older hypertensive brain. Stroke 38, 1766–1773.

Bekris, L.M., Millard, S., Lutz, F., Li, G., Galasko, D.R., Farlow, M.R., et al., 2012. Tau phosphorylation pathway genes and cerebrospinal fluid tau levels in Alzheimer's disease. Am. J. Med. Genet. B Neuropsychiatr. Genet. 159B, 874–883.

Bengmark, S., 2013. Processed foods, Dysbiosis, systemic inflammation, and poor health. Curr. Nutri Food Sci. 9, 113–143.

Bergami, M., Rimondini, R., Santi, S., Blum, R., Gotz, M., Canossa, M., 2008. Deletion of TrkB in adult progenitors alters newborn neuron integration into hippocampal circuits and increases anxiety-like behavior. Proc. Natl. Acad. Sci. U.S.A. 105, 15570–15575.

Berk, M., Williams, L.J., Jacka, F., O'Neil, A., Pasco, J.A., Moylan, S., et al., 2013. So depression is an inflammatory disease, but where does the inflammation come from? BMC Med. 11, 200.

Bertram, L., Tanzi, R.E., 2005. The genetic epidemiology of neurodegenerative disease. J. Clin. Invest. 115, 1449–1457.

Bertram, L., Lange, C., Mullin, K., Parkinson, M., Hsiao, M., Hogan, M.F., Schjeide, B.M., Hooli, B., Divito, J., Ionita, I., et al., 2008. Genome-wide association analysis reveals putative Alzheimer's disease susceptibility loci in addition to APOE. Am. J. Hum. Genet. 83, 623–632.

Bettens, K., Sleegers, K., Van Broeckhoven, C., 2010. Current status on Alzheimer disease molecular genetics: from past, to present, to future. Hum. Mol. Genet. 19 (1), R4–R11.

Bhat, N.R., 2010. Linking cardiometabolic disorders to sporadic Alzheimers disease: a perspective on potential mechanisms and mediators. J. Neurochem. 115, 551–562.

Bien-Ly, N., Gillespie, A.K., Walker, D., Yoon, S.Y., Huang, Y., 2012. Reducing human apolipoprotein E levels attenuates age-dependent Aβ accumulation in mutant human amyloid precursor protein transgenic mice. J. Neurosci. 32, 4803–4811.

Bishop, N.A., Lu, T., Yankner, B.A., 2010. Neural mechanisms of ageing and cognitive decline. Nature 8, 529–535.

Bjelakovic, G., Nikolova, D., Gluud, L.L., Simonetti, R.G., Gluud, C., 2007. Mortality in randomized trials of antioxidant supplements for primary and secondary prevention: systematic review and meta-analysis. JAMA 297, 842–857.

Blacker, D., Tanzi, R.E., 1998. The genetics of Alzheimer disease: current status and future prospects. Arch. Neurol. 55, 294–296.

Blau, C.W., Cowley, T.R., O'sullivan, J., Grehan, B., Browne, T.C., et al., 2012. The age-related deficit in LTP is associated with changes in perfusion and blood–brain barrier permeability. Neurobiol. Aging 33, 1005.e1023–1005.e1035.

Brewer, G.J., 2010. Epigenetic oxidative redox shift (EORS) theory of aging unifies the free radical and insulin signaling theories. Exp. Gerontol. 45, 173–179.

Brownlee, M., 2005. The pathobiology of diabetic complications: a unifying mechanism. Diabetes 54, 1615–1625.

Bu, G., 2009. Apolipoprotein E and its receptors in Alzheimer's disease: pathways, pathogenesis and therapy. Nat. Rev. Neurosci. 10, 333–344.

Bulnes, S., Argandona, E.G., Bengoetxea, H., Leis, O., Ortuzar, N., Lafuente, J.V., 2010. The role of eNOS in vascular permeability in ENU-induced gliomas. Acta Neurochir. Suppl. 106, 277–282.

Burke, S.N., Barnes, C.A., 2006. Neural plasticity in the ageing brain. Nat. Rev. Neurosci. 7, 30–40.

Burns, A., Iliffe, S., 2009. Alzheimer's disease. BMJ 338, b158.

Bye, N., Zieba, M., Wreford, N.G., Nichols, N.R., 2001. Resistance of the dentate gyrus to induced apoptosis during ageing is associated with increases in transforming growth factor-beta1 messenger RNA. Neuroscience 105, 853–862.

Cacabelos, R., Torrellas, C., 2014. Epigenetic drug discovery for Alzheimer's disease. Expert Opin. Drug Discov. 9, 1059–1086.

Cacabelos, R., Cacabelos, P., Torrellas, C., Tellado, I., Carril, J.C., 2014. Pharmacogenomics of Alzheimer's disease: novel therapeutic strategies for drug development. Methods Mol. Biol. 1175, 323–356.

Caesar, I., Gandy, S., 2012. Evidence that an APOE 4 "double whammy" increases risk for Alzheimer's disease. BMC Med. 10.

Caro, J.F., Sinha, M.K., Kolaczynski, J.W., Zhang, P.L., Considine, R.V., 1996. Leptin: the tale of an obesity gene. Diabetes 45, 1455–1462.

Cataldo, J.K., Prochaska, J.J., Glantz, S.A., 2010. Cigarette smoking is a risk factor for Alzheimer's disease: an analysis controlling for tobacco industry affiliation. J. Alzheimers Dis. 19, 465–480.

Chapuis, J., Hansmannel, F., Gistelinck, M., Mounier, A., Van Cauwenberghe, C., Kolen, K.V., et al., 2013. Increased expression of BIN1 mediates Alzheimer genetic risk by modulating tau pathology. Mol. Psychiatry 18, 1225–1234.

Chaudhry, M., Wang, X., Bamne, M.N., Hasnain, S., Demirci, F.Y., Lopez, O.L., Kamboh, M.I., 2015. Genetic variation in imprinted genes is associated with risk of late-onset Alzheimer's disease. J. Alzheimers Dis. 44, 989–994.

Chen, Z., Han, Z.C., 2008. STAT3: a critical transcription activator in angiogenesis. Med. Res. Rev. 28, 185–200.

Cohen, E., Dillin, A., 2008. The insulin paradox: aging, proteotoxicity and neurodegeneration. Nat. Rev. Neurosci. 8, 759–767.

Considine, R.V., Sinha, M.K., Heiman, M.L., et al., 1996. Serum immunoreactive-leptin concentrations in normal-weight and obese humans. N. Engl. J. Med. 334, 292–295.

Craft, S., Peskind, E., Schwartz, M.W., Schellenberg, G.D., Raskind, M., Porte Jr., D., 1998. Cerebrospinal fluid and plasma insulin levels in Alzheimer's disease: relationship to severity of dementia and apolipoprotein E genotype. Neurology 50, 164–168.

Craft, S., 2007. Insulin resistance and Alzheimer's disease pathogenesis: potential mechanisms and implications for treatment. Curr. Alzheimer Res. 4, 147–152.

Crews, L., Rockenstein, E., Masliah, E., 2010. APP transgenic modeling of Alzheimer's disease: mechanisms of neurodegeneration and aberrant neurogenesis. Brain Struct. Funct. 214, 111–126.

Damen, J.E., Liu, L., Rosten, P., Humphries, R.K., Jefferson, A.B., Majerus, P.W., et al., 1996. The 145-kDa protein induced to associate with Shc by multiple cytokines is an inositol tetraphosphate and phosphatidylinositol 3,4,5-triphosphate 5-phosphatase. Proc. Natl. Acad. Sci. U.S.A. 93, 1689–1693.

Dasari, B., Prasanthi, J.R., Marwarha, G., Singh, B.B., Ghribi, O., 2010. The oxysterol 27-hydroxycholesterol increases β-amyloid and oxidative stress in retinal pigment epithelial cells. BMC 10, 22.

Dash, S., Clarke, G., Berk, M., Jacka, F.N., 2015. The gut microbiome and diet in psychiatry: focus on depression. Curr. Opin. Psychiatry 28, 1–6.

Daulatzai, M.A., 2012a. Quintessential risk factors: their role in promoting cognitive dysfunction and Alzheimer's disease. Neurochem. Res. 37, 2627–2658.

Davidson, T.L., Monnot, A., Neal, A.U., Martin, A.A., Horton, J.J., Zheng, W., 2012. The effects of a high-energy diet on hippocampal-dependent discrimination performance and blood–brain barrier integrity differ for diet-induced obese and diet-resistant rats. Physiol. Behav. 107, 26–33.

de la Monte, S.M., 2009. Insulin resistance and Alzheimer's disease. BMB Rep. 42, 475–481.

de la Monte, S.M., Tong, M., 2013. Insulin resistance and metabolic failure underlie Alzheimer disease. In: Farooqui, T., Farooqui, A.A. (Eds.), Metabolic Syndrome and Neurological Disorders, Oxford UK. John Wiley & Sons, Inc, Oxford, UK, pp. 1–30.

Daulatzai, M.A., 2012b. Pathogenesis of cognitive dysfunction in patients with obstructive sleep apnea: a hypothesis with emphasis on the nucleus tractus solitaries. Sleep Disord. 2012, 251096.

Delnevo, C.D., Gundersen, D.A., Hagman, B.T., 2008. Declining estimated prevalence of alcohol drinking and smoking among young adults nationally: artifacts of sample undercoverage? Am. J. Epidemiol. 167, 15–19.

Deane, R., Wu, Z., Sagare, A., Davis, J., Du Yan, S., et al., 2004a. LRP/amyloid beta-peptide interaction mediates differential brain efflux of Abeta isoforms. Neuron 43, 333–344.

DeMarini, D.M., 2004. Genotoxicity of tobacco smoke and tobacco smoke condensate: a review. Mutat. Res. 567, 447–474.

Deane, R., Wu, Z., Zlokovic, B.V., 2004b. RAGE (yin) versus LRP (yang) balance regulates Alzheimer amyloid β-peptide clearance through transport across the blood–brain barrier. Stroke 35, 2628–2631.

Dennis, N.A., Cabeza, R., 2008. Neuroimaging of healthy cognitive aging. In: Craik, F.I.M., Salthouse, T.A. (Eds.), The Handbook of Aging and Cognition, third ed. Psychology Press, New York, pp. 1–54.

de Rezende, L.F., Rodrigues Lopes, M., Rey-Lopez, J.P., Matsudo, V.K., Luiz, O.C., 2014. Sedentary behavior and health outcomes: an overview of systematic reviews. PLoS One 9, e105620.

Dietschy, J.M., Turley, S.D., 2004. Thematic review series: brain Lipids. Cholesterol metabolism in the central nervous system during early development and in the mature animal. J. Lipid Res. 45, 1375–1397.

Dong, H., Goico, B., Martin, M., Csernansky, C.A., Bertchume, A., Csernansky, J.G., 2004. Modulation of hippocampal cell proliferation, memory, and amyloid plaque deposition in APPsw (Tg2576) mutant mice by isolation stress. Neuroscience 127, 601–609.

Dreyling, M.H., Martinez-Climent, J.A., Zheng, M., Mao, J., Rowley, J.D., Bohlander, S.K., 1996. The t(10;11)(p13;q14) in the U937 cell line results in the fusion of the AF10 gene and CALM, encoding a new member of the AP-3 clathrin assembly protein family. Proc. Natl. Acad. Sci. U.S.A. 93, 4804–4809.

Duarte, A.I., Candeias, E., Correia, S.C., Santos, R.X., Carvalho, C., Cardoso, S., Plácido, A., Santos, M.S., Oliveira, C.R., Moreira, P.I., 2013. Crosstalk between diabetes and brain: glucagon-like peptide-1 mimetics as a promising therapy against neurodegeneration. Biochim. Biophys. Acta 1832, 527–541.

Durazzo, T.C., Mattsson, N., Weiner, M.W., 2014. Alzheimer's disease neuroimaging initiative. Smoking and increased Alzheimer's disease risk: a review of potential mechanisms. Alzheimers Dement. 10 (Suppl. 3), S122–S145.

Dube, S.R., McClave, A., James, C., Santos, K., Cardoso, G., et al., 2009. Vital signs: current cigarette smoking among adults aged ≥18 years – United States. MMWR Morb. Mortal. Wkly. Rep. 2010 (59), 1135–1140.

Duthey, B., 2013. Background Paper 6.11. Alzheimer Disease and Other Dementias, pp. 4–74.

Fagan, A.M., Murphy, B.A., Patel, S.N., Kilbridge, J.F., Mobley, W.C., Bu, G., Holtzman, D.M., 1998. Evidence for normal aging of the septo-hippocampal cholinergic system in apoE (−/−) mice but impaired clearance of axonal degeneration products following injury. Exp. Neurol. 151, 314–325.

Farooqui, A.A., 2010. Neurochemical Aspects of Neurotraumatic and Neurodegenerative Diseases. Springer. New.

Farooqui, A.A., 2012. Phytochemicals, Signal Transduction, and Neurological Disorders. Springer, New York.

Farooqui, A.A., Farooqui, T., Panza, F., Frisardi, V., 2012. Metabolic syndrome as a risk factor for neurological disorders. Cell Mol. Life Sci. 69, 741–746.

Farooqui, A.A., 2013. Metabolic Syndrome: An Important Risk Factor for Stroke, Alzheimer, and Depression. Springer, New York.

Farooqui, A.A., 2014. Inflammation and Oxidative Stress in Neurological Disorders: Effect of Lifestyle, Genes and Age. Springer, New York.

Farooqui, A.A., Farooqui, T., 2015. Neurochemical effects of western diet consumption on human brain. In: Farooqui, T., Farooqui, A.A. (Eds.), Diet and Exercise in Cognitive Function and Neurological Diseases. John Wiley and Sons, Inc, New York, pp. 15–28.

Farooqui, A.A., 2015. High Calorie Diet and the Human Brain. Springer, International Publishing, Switzerland.

Farris, W., Mansourian, S., Chang, Y., Lindsley, L., Eckman, E.A., Frosch, M.P., Eckman, C.B., Tanzi, R.E., Selkoe, D.J., Guenette, S., 2003. Insulin-degrading enzyme regulates the levels of insulin, amyloid beta-protein, and the beta-amyloid precursor protein intracellular domain in vivo. Proc. Natl. Acad. Sci. U.S.A. 100, 4162–4167.

Finck, B.N., Kelly, D.P., 2006. PGC-1 coactivators: inducible regulators of energy metabolism in health and disease. J. Clin. Invest. 116, 615–622.

Ford, M.G., Pearse, B.M., Higgins, M.K., Vallis, Y., Owen, D.J., Gibson, A., et al., 2001. Simultaneous binding of PtdIns(4,5)P2 and clathrin by AP180 in the nucleation of clathrin lattices on membranes. Science 29, 1051–1055.

Franceschi, C., Capri, M., Monti, D., Giunta, S., Olivieri, F., Sevini, F., et al., 2007. Inflammaging and anti-inflammaging: a systemic perspective on aging and longevity emerged from studies in humans. Mech. Ageing Dev. 128, 92–105.

Frieden, C., 2015. ApoE: the role of conserved residues in defining function. Protein Sci. 24, 138–144.

Furchgott, R.F., 1993. Introduction to EDRF research. J. Cardiovasc. Pharmacol. 22 (Suppl. 7), S1–S2.

Gandy, S., 2005. The role of cerebral amyloid beta accumulation in common forms of Alzheimer disease. J. Clin. Invest. 115, 1121–1129.

Gasparini, L., Gouras, G.K., Wang, R., Gross, R.S., Beal, M.F., Greengard, P., Xu, H., 2001. Stimulation of beta-amyloid precursor protein trafficking by insulin reduces intraneuronal beta-amyloid and requires mitogen-activated protein kinase signaling. J. Neurosci. 21, 2561–2570.

Gasparini, L., Netzer, W.J., Greengard, P., Xu, H., 2002. Does insulin dysfunction play a role in Alzheimer's disease? Trends Pharmacol. Sci. 23, 288–293.

Gehrmann, J., Matsumoto, Y., Kreutzberg, G.W., 1995. Microglia: intrinsic immuneffector cell of the brain. Brain Res. Brain Res. Rev. 20, 269–287.

Giacco, F., Brownlee, M., 2010. Oxidative stress and diabetic complications. Circ. Res. 107, 1058–1070.

Godbout, J.P., Johnson, R.W., 2004. Interleukin-6 in the aging brain. J. Neuroimmunol. 147, 141–144.

Gozal, D., Daniel, J.M., Dohanich, G.P., 2001. Behavioural and anatomical correlates of chronic episodic hypoxia during sleep in rat. J. Neurol. Sci. 21, 2442–2450.

Grontved, A., Hu, F.B., 2011. Television viewing and risk of type 2 diabetes, cardiovascular disease, and all-cause mortality: a meta-analysis. JAMA 305, 2448–2455.

Gu, X., Sun, J., Li, S., Wu, X., Li, L., 2013. Oxidative stress induces DNA demethylation and histone acetylation in SH-SY5Y cells: potential epigenetic mechanisms in gene transcription in Aβ production. Neurobiol. Aging 34, 1069–1079.

Guerreiro, R., Wojtas, A., Bras, J., Carrasquillo, M., Rogaeva, E., Majounie, E., Cruchaga, C., Sassi, C., Kauwe, J.S., Younkin, S., et al., 2013. TREM2 variants in Alzheimer's disease. N. Engl. J. Med. 368, 117–127.

Haan, M.N., 2006. Therapy Insight: type 2 diabetes mellitus and the risk of late-onset Alzheimer's disease. Nat. Clin. Pract. Neurol. 2, 159–166.

Hasher, L., Zacks, R.T., 1988. In: Bower, G.H. (Ed.), The Psychology of Learning and Motivation, Vol. 22. Academic Press, pp. 193–225.

Hardy, J., Selkoe, D.J., 2002. The amyloid hypothesis of Alzheimer's disease: progress and problems on the road to therapeutics. Science 297, 353–356.

Harris, W.S., 2008. The omega-3 index as a risk factor for coronary heart disease. Am. J. Clin. Nutr. 87, 1997S–2002S.

Harold, D., Abraham, R., Hollingworth, P., Sims, R., Gerrish, A., Hamshere, M.L., 2009. Genome-wide association study identifies variants at CLU and PICALM associated with Alzheimer's disease. Nat. Genet. 41, 1088–1093.

Hazrati, L.N., Van Cauwenberghe, C., Brooks, P.L., Brouwers, N., Ghani, M., Sato, C., et al., 2012. Genetic association of CR1 with Alzheimer's disease: a tentative disease mechanism. Neurobiol. Aging 33, 2949e5–2949e12.

Healey, M.K., Campbell, K.L., Hasher, L., 2008. Progress in brain research. In: Sossin, W., Lacaille, J.C., Castellucci, V.F., Belleville, S. (Eds.), The Essence of Memory, Vol. 169. Elsevier.

Ho, Y.S., Yang, X., Yeung, S.C., Chiu, K., Lau, C.F., Tsang, A.W., Mak, J.C., Chang, R.C., 2012. Cigarette smoking accelerated brain aging and induced pre-Alzheimer-like neuropathology in rats. PLoS One 7, e36752.

Hollingworth, P., Harold, D., Sims, R., Gerrish, A., Lambert, J.C., Carrasquillo, M.M., et al., 2011. Common variants at ABCA7, MS4A6A/MS4A4E, EPHA1, CD33 and CD2AP are associated with Alzheimer's disease. Nat. Genet. 43, 429–435.

Hossain, M., Sathe, T., Fazio, V., Mazzone, P., Weksler, B., Janigro, D., Rapp, E., Cucullo, L., 2009. Tobacco smoke: a critical etiological factor for vascular impairment at the blood–brain barrier. Brain Res. 1287, 192–205.

Hossain, M., Mazzone, P., Tierney, W., Cucullo, L., 2011. In vitro assessment of tobacco smoke toxicity at the BBB: do antioxidant supplements have a protective role? BMC Neurosci. 12, 92.

Hu, F.B., Li, T.Y., Colditz, G.A., Willett, W.C., Manson, J.E., 2003. Television watching and other sedentary behaviors in relation to risk of obesity and type 2 diabetes mellitus in women. JAMA 289, 1785–1791.

Hu, X., Pickering, E., Liu, Y.C., Hall, S., Fournier, H., et al., 2011. Alzheimer's disease neuroimaging initiative meta-analysis for genome-wide association study identifies multiple variants at the BIN1 locus associated with late-onset Alzheimer's disease. PLoS One 6, e16616.

Huang, Y., Mucke, L., 2012. Alzheimer mechanisms and therapeutic strategies. Cell 148, 1204–1222.

Iadecola, C., 2004. Neurovascular regulation in the normal brain and in Alzheimer's disease. Nat. Rev. Neurosci. 5, 347–360.

Ikeda, Y., Abe-Dohmae, S., Munehira, Y., Aoki, R., Kawamoto, S., Furuya, A., et al., 2003. Posttranscriptional regulation of human ABCA7 and its function for the apoA-I-dependent lipid release. Biochem. Biophys. Res. Commun. 311, 313–318.

Irwin, M.R., Wang, M., Ribeiro, D., Cho, H.J., Olmstead, R., Breen, E.C., et al., 2008. Sleep loss activates cellular inflammatory signaling. Biol. Psychiatry 64, 538–540.

Itoh, T., De Camilli, P., 2006. BAR, F-BAR (EFC) and ENTH/ANTH domains in the regulation of membrane-cytosol interfaces and membrane curvature. Biochim. Biophys. Acta 1761, 897–912.

Jacka, F.N., Cherbuin, N., Anstey, K.J., Sachdev, P., Butterworth, P., 2015. Western diet is associated with a smaller hippocampus: a longitudinal investigation. BMC Med. 13, 215.

Jacka, F.N., 2015. Lifestyle factors in preventing mental health disorders: an interview with Felice Jacka. BMC Med. 13, 264.

Jessen, N.A., Munk, A.S., Lundgaard, I., Nedergaad, M., 2015. Associations between sleep, cortisol regulation, and diet: possible implications for the risk of Alzheimer disease. Neurochem. Res. 40, 2583–2599.

Jin, K., Peel, A.L., Mao, X.O., Xie, L., Cottrell, B.A., Henshall, D.C., Greenberg, D.A., 2004. Increased hippocampal neurogenesis in Alzheimer's disease. Proc. Natl. Acad. Sci. U.S.A. 101, 343–347.

Johnson, R.J., Perez-Pozo, S.E., Sautin, Y.Y., Manitius, J., Sanchez-Lozada, L.G., Feig, D.I., Shafiu, M., Segal, M., Glassock, R.J., Shimada, M., Roncal, C., Nakagawa, T., 2009. Hypothesis: could excessive fructose intake and uric acid cause type 2 diabetes? Endocr. Rev. 30, 96–116.

Johnson, V.E., Stewart, W., Smith, D.H., 2010. Traumatic brain injury and amyloid-beta pathology: a link to Alzheimer's disease? Nat. Rev. 11, 361–370.

Jolivalt, C.G., Lee, C.A., Beiswenger, K.K., Smith, J.L., Orlov, M., Torrance, M.A., Masliah, E., 2008. Defective insulin signaling pathway and increased glycogen synthase kinase-3 activity in the brain of diabetic mice: parallels with Alzheimer's disease and correction by insulin. J. Neurosci. Res. 86, 3265–3274.

Jonsson, T., Stefansson, H., Steinberg, S., Jonsdottir, I., Jonsson, P.V., Snaedal, J., Bjornsson, S., Huttenlocher, J., Levey, A.I., Lah, J.J., et al., 2013. Variant of TREM2 associated with the risk of Alzheimer's disease. N. Engl. J. Med. 368, 107–116.

Joo, E.Y., Tae, W.S., Lee, M.J., Suh, M., Hong, S.B., et al., 2010. Reduced brain gray matter concentration in patients with obstructive sleep apnea syndrome. Sleep 33, 235–241.

Ju, Y.E., Lucey, B.P., Holtzman, D.M., 2014. Sleep and Alzheimer disease pathology—a bidirectional relationship. Nat. Rev. Neurol. 10, 115–119.

Kaisar, M.A., Prasad, S., Cucullo, L., 2015. Protecting the BBB endothelium against cigarette smoke-induced oxidative stress using popular antioxidants: are they really beneficial? Brain Res. 19 (1627), 90–100.

Kalaria, R.N., 2010. Vascular basis for brain degeneration: faltering controls and risk factors for dementia. Nutr. Rev. 68, S74–S87.

Kalmijn, S., 2000. Fatty acid intake and the risk of dementia and cognitive decline: a review of clinical and epidemiological studies. J. Nutr. Health Aging 4, 202–207.

Kanatsu, K., Morohashi, Y., Suzuki, M., Kuroda, H., Watanabe, T., Tomita, T., et al., 2014. Decreased CALM expression reduces Aβ42 to total Aβ ratio through clathrin-mediated endocytosis of γ-secretase. Nat. Commun. 5, 3386.

Kanoski, S.E., Meisel, R.L., Mullins, A.J., Davidson, T.L., 2007. The effects of energy-rich diets on discrimination reversal learning and on BDNF in the hippocampus and prefrontal cortex of the rat. Behav. Brain Res. 182, 57–66.

Kanoski, S.E., Davidson, T.L., 2011. Western diet consumption and cognitive impairment: links to hippocampal dysfunction and obesity. Physiol. Behav. 103, 59–68.

Kanoski, S.E., 2012. Cognitive and neuronal systems underlying obesity. Physiol. Behav. 106, 337–344.

Kaur, C., Ling, E.A., 2008. Blood–brain barrier in hypoxic-ischemic conditions. Curr. Neurovasc. Res. 5, 71–81.

Keadle, S.K., Arem, H., Moore, S.C., Sampson, J.N., Matthews, C.E., 2015. Impact of changes in television viewing time and physical activity on longevity: a prospective cohort study. Int. J. Behav. Nutr. Phys. Act. 12, 156.

Kettenmann, H., Kirchhoff, F., Verkhratsky, A., 2013. Microglia: new roles for the synaptic stripper. Neuron 77, 10–18.

Khera, R., Das, N., 2009. Complement receptor 1: disease associations and therapeutic implications. Mol. Immunol. 46, 761–772.

Kim, J., Jiang, H., Park, S., et al., 2011. Haploinsufficiency of human APOE reduces amyloid deposition in a mouse model of amyloid-beta amyloidosis. J. Neurosci. 31, 18007–18012.

Kim, W.S., Li, H., Ruberu, K., Chan, S., Elliott, D.A., Low, J.K., et al., 2013. Deletion of Abca7 increases cerebral amyloid-β accumulation in the J20 mouse model of Alzheimer's disease. J. Neurosci. 33, 4387–4394.

Kojima, C., Hashimoto, A., Yabuta, I., Hirose, M., Hashimoto, S., Kanaho, Y., et al., 2004. Regulation of Bin1 SH3 domain binding by phosphoinositides. EMBO J. 23, 4413–4422.

Koponen, E., Lakso, M., Castrén, E., 2004. Overexpression of the full-length neurotrophin receptor trkB regulates the expression of plasticity-related genes in mouse brain. Brain Res. Mol. Brain Res. 130, 81–94.

Kulstad, J.J., Green, P.S., Cook, D.G., Watson, G.S., Reger, M.A., Baker, L.D., et al., 2006. Differential modulation of plasma beta-amyloid by insulin in patients with Alzheimer disease. Neurology 66, 1506–1510.

Lambert, J.C., Heath, S., Even, G., Campion, D., Sleegers, K., Hiltunen, M., Combarros, O., Zelenika, D., 2009. Genome-wide association study identifies variants at *CLU* and *CR1* associated with Alzheimer's disease. Nat. Genet. 41, 1094–1099.

Lee, Y.J., Han, S.B., Nam, S.Y., Oh, K.W., Hong, J.T., 2010. Inflammation and Alzheimer's disease. Arch. Pharm. Res. 33, 1539–1556.

Levy, P., Bonsignore, M.R., Eckel, J., 2009. Sleep, sleep-disordered breathing and metabolic consequences. Eur. Respir. J. 34, 243–260.

Leuner, B., Gould, E., Shors, T.J., 2006. Is there a link between adult neurogenesis and learning? Hippocampus 16, 216–224.

Li, B., Yamamori, H., Tatebayashi, Y., Shafit-Zagardo, B., Tanimukai, H., Chen, S., Iqbal, K., Grundke-Iqbal, I., 2008. Failure of neuronal maturation in Alzheimer disease dentate gyrus. J. Neuropathol. Exp. Neurol. 67, 78–84.

Lipton, S.A., 2008. NMDA receptor activity regulates transcription of antioxidant pathways. Nat. Neurosci. 11, 381–382.

Liu, X., Wu, Z., Hayashi, Y., Nakanishi, H., 2012. Age-dependent neuroinflammatory responses and deficits in long-term potentiation in the hippocampus during systemic inflammation. Neuroscience 216, 133–142.

Liu, C.C., Kanekiyo, T., Xu, H., Bu, G., 2013. Apolipoprotein E and Alzheimer disease: risk, mechanisms and therapy. Nat. Rev. Neurol. 9, 106–118.

Lopez-Otin, C., Blasco, M.A., Partridge, L., Serrano, M., Kroemer, G., 2013. The hallmarks of aging. Cell 153, 1194–1217.

Lukiw, W.J., 2004. Gene expression profiling in fetal, aged and Alzheimer hippocampus: a continuum of stress-related signaling. Neurochem. Res. 29, 1287–1297.

Lynch, J.R., Morgan, D., Mance, J., Matthew, W.D., Laskowitz, D.T., 2001. Apolipoprotein E modulates glial activation and the endogenous central nervous system inflammatory response. J. Neuroimmunol. 114, 107–113.

Lynch, J.R., Pineda, J.A., Morgan, D., Zhang, L., Warner, D.S., Benveniste, H., Laskowitz, D.T., 2002. Apolipoprotein E affects the central nervous system response to injury and the development of cerebral edema. Ann. Neurol. 51, 113–117.

Maesako, M., Uemura, M., Tashiro, Y., Sasaki, K., Watanabe, K., et al., 2015. High fat diet enhances β-site cleavage of amyloid precursor protein (APP) via promoting β-site APP cleaving enzyme 1/Adaptor protein 2/Clathrin complex formation. PLoS One 10 (9), e0131199.

Mahley, R.W., Huang, Y., 1999. Apolipoprotein E: from atherosclerosis to Alzheimer's disease and beyond. Curr. Opin. Lipidol. 10, 207–217.

Mahley, R.W., Rall, S.C., 2000. Apolipoprotein E: far more than a lipid transport protein. Annu. Rev. Genomics Hum. Genet. 1, 507–537.

Marquer, C., Devauges, V., Cossec, J.C., Liot, G., Lecart, S., Saudou, F., Duyckaerts, C., Leveque-Fort, S., Potier, M.C., 2011. Local cholesterol increase triggers amyloid precursor protein-BACE1 clustering in lipid rafts and rapid endocytosis. FASEB J. 25, 1295–1305.

Marques, J.F., Cappa, S.F., Sartori, G., 2011. Naming from definition, semantic relevance and feature type: the effects of aging and Alzheimer's disease. Neuropsychology 25, 105–113.

Marwarha, G., Dasari, B., Prasanthi, J.R., Schommer, J., Ghribi, O., 2010. Leptin reduces the accumulation of Abeta and phosphorylated tau induced by 27-hydroxycholesterol in rabbit organotypic slices. J. Alzheimers Dis. 19, 1007–1019.

Marr, R.A., Hafez, D.M., 2014. Amyloid-beta and Alzheimer's disease: the role of neprilysin-2 in amyloid-beta clearance. Front. Aging Neurosci. 6, 187.

Mattson, M.P., Maudsley, S., Martin, B., 2004. BDNF and 5-HT: a dynamic duo in age-related neuronal plasticity and neurodegenerative disorders. Trends Neurosci. 27, 589–594.

Mattson, M.P., 2004. Pathways towards and away from Alzheimer's disease. Nature 430, 631–639.

Mauch, D.H., Nägler, K., Schumacher, S., Göritz, C., Müller, E.C., Otto, A., Pfrieger, F.W., 2001. CNS synaptogenesis promoted by glia-derived cholesterol. Science 294, 1354–1357.

McCarron, R.M., Chen, Y., Tomori, T., Strasser, A., Mechoulam, R., Shohami, E., Spatz, M., 2006. Endothelial-mediated regulation of cerebral microcirculation. J. Physiol. Pharmacol. 57 (Suppl. 11), 133–144.

Millan, M.J., 2013. An epigenetic framework for neurodevelopmental disorders: from pathogenesis to potential therapy. Neuropharmacology 68, 2–82.

Miller III, E.R., Pastor-Barriuso, R., Dalal, D., Riemersma, R.A., Appel, L.J., Guallar, E., 2005. Meta-analysis: high-dosage vitamin E supplementation may increase all-cause mortality. Ann. Intern. Med. 142, 37–46.

Mitsui, A., Hamuro, J., Nakamura, H., Kondo, N., Hirabayashi, Y., Ishizaki-Koizumi, S., Hirakawa, T., Inoue, T., Yodoi, J., 2002. Overexpression of human thioredoxin in transgenic mice controls oxidative stress and life span. Antioxid. Redox Signal. 4, 693–696.

Miyashita, A., Arai, H., Asada, T.A., et al., 2007. Japanese genetic study consortium for Alzheimer's disease. Genetic association of CTNNA3 with late-onset Alzheimer's disease in females. Hum. Mol. Genet. 16, 2854–2869.

Miyashita, A., Hatsuta, H., Kikuchi, M., Nakaya, A., Saito, Y., Tsukie, T., et al., 2014. Japanese Alzheimer's disease neuroimaging initiative. Genes associated with the progression of neurofibrillary tangles in Alzheimer's disease. Transl. Psychiatry 4, e396.

Mondragón-Rodríguez, S., Perry, G., Zhu, X., Moreira, P.I., Acevedo-Aquino, M.C., Williams, S., 2013. Phosphorylation of tau protein as the link between oxidative stress, mitochondrial dysfunction, and connectivity failure: implications for Alzheimer's disease. Oxid. Med. Cell Longev. 2013, 940603.

Moran, M., Lynch, C.A., Walsh, C., Coen, R., Coakley, D., et al., 2005. Sleep disturbance in mild to moderate Alzheimer's disease. Sleep Med. 6, 347–352.

Morrell, M.J., Jackson, M.L., Twigg, G.L., Ghiassi, R., McRobbie, D.W., et al., 2010. Changes in brain morphology in patients with obstructive sleep apnoea. Thorax 65, 908–914.

Morselli, L., Leproult, R., Balbo, M., Spiegel, K., 2010. Role of sleep duration in the regulation of glucose metabolism and appetite. Best. Pract. Res. Clin. Endocrinol. Metab. 24, 687–702.

Mosconi, L., Pupi, A., De Leon, M.J., 2008. Brain glucose hypometabolism and oxidative stress in preclinical Alzheimer's disease. Ann. N.Y. Acad. Sci. 1147, 180–195.

Mosher, K.I., Wyss-Coray, T., 2014. Microglial dysfunction in brain aging and Alzheimer's disease. Biochem. Pharmacol. 88, 594–604.

Moylan, S., Berk, M., Dean, O.M., Samuni, Y., Williams, L.J., O'Neil, A., et al., 2014. Oxidative & nitrosative stress in depression: why so much stress? Neurosci. Biobehav. Rev. 45C, 46–62.

Mrak, R.E., Griffin, W.S., 2005. Potential inflammatory biomarkers in Alzheimer's disease. J. Alzheimers Dis. 8, 369–375.

Murayama, S., Saito, Y., 2004. Neuropathological diagnostic criteria for Alzheimer's disease. Neuropathology 24, 254–260.

Naik, P., Fofaria, N., Prasad, S., Sajja, R.K., Weksler, B., Couraud, P.O., et al., 2014. Oxidative and pro-inflammatory impact of regular and denicotinized cigarettes on blood–brain barrier endothelial cells: is smoking reduced or nicotine-free products really safe? BMC Neurosci. 15, 51.

Naj, A.C., Jun, G., Beecham, G.W., Wang, L.S., Vardarajan, B.N., Buros, J., et al., 2011. Common variants at MS4A4/MS4A6E, CD2AP, CD33 and EPHA1 are associated with late-onset Alzheimer's disease. Nat. Genet. 43, 436–441.

Nalivaeva, N.N., Belyaev, N.D., Kerridge, C., Turner, A.J., 2014. Amyloid-clearing proteins and their epigenetic regulation as a therapeutic target in Alzheimer's disease. Front. Aging Neurosci. 6, 235.

National Center for Chronic Disease Prevention and Health Promotion (US) Office on Smoking and Health, 2014. The Health Consequences of Smoking—50 Years of Progress: A Report of the Surgeon General. Centers for Disease Control and Prevention (US), Atlanta (GA).

Nelson, P.T., Alafuzoff, I., Bigio, E.H., Bouras, C., Braak, H., Cairns, N.J., et al., 2012. Correlation of Alzheimer disease neuropathologic changes with cognitive status: a review of the literature. J. Neuropathol. Exp. Neurol. 71, 362–381.

Octave, J.N., Pierrot, N., Ferao Santos, S., Nalivaeva, N.N., Turner, A.J., 2013. From synaptic spines to nuclear signaling: nuclear and synaptic actions of the amyloid precursor protein. J. Neurochem. 126, 183–190.

O'Keefe, J.H., Gheewala, N.M., O'Keefe, J.O., 2008. Dietary strategies for improving postprandial glucose, lipids, inflammation, and cardiovascular health. J. Am. Coll. Cardiol. 51, 249–255.

Olgiati, P., Politis, A.M., Papadimitriou, G.N., De Ronchi, D., Serretti, A., 2011. Genetics of late-onset Alzheimer's disease: update from the alzgene database and analysis of shared pathways. Int. J. Alzheimers Dis. 2011, 832379.

Owen, N., Leslie, E., Salmon, J., Fotheringham, M.J., 2000. Environmental determinants of physical activity and sedentary behavior. Exerc. Sport Sci. Rev. 28, 153–158.

Paffenbarger, R., Hyde, R., Wing, A., Lee, I., Jung, D., Kampert, J., 1993. The association of changes in physical activity level and other lifestyle characteristics with mortality among men. New Eng. J. Med. 328, 538–545.

Pahor, M., Guralnik, J.M., Ambrosius, W.T., Blair, S., Bonds, D.E., Church, T.S., et al., 2014. Effect of structured physical activity on prevention of major mobility disability in older adults: the LIFE study randomized clinical trial. JAMA 311, 2387–2396.

Paloneva, J., Kestila, M., Wu, J., Salminen, A., Bohling, T., Ruotsalainen, V., et al., 2000. Loss-of-function mutations in TYROBP (DAP12) result in a presenile dementia with bone cysts. Nat. Genet. 25, 357–361.

Pallebage-Gamarallage, M., Takechi, R., Lam, V., Elahy, M., Mamo, J., 2015. Pharmacological modulation of dietary lipid-induced cerebral capillary dysfunction: considerations for reducing risk for Alzheimer's disease. Crit. Rev. Clin. Lab. Sci. 18, 1–18.

Pang, P.T., Lu, B., 2004. Mechanisms of late-phase LTP and long-term memory in normal and aging hippocampus. Aging Res. Rev. 4, 407–430.

Pate, R.R., O'Neill, J.R., Lobelo, F., 2008. The evolving definition of "sedentary". Exerc. Sport Sci. Rev. 36, 173–178.

Persidsky, Y., Ramirez, S.H., Haorah, J., Kanmogne, G.D., 2006. Blood–brain barrier: structural components and function under physiologic and pathologic conditions. J. Neuroimmun. Pharmacol. 1, 223–236.

Pfrieger, F.W., 2003. Cholesterol homeostasis and function in neurons of the central nervous system. Cell Mol. Life Sci. 60, 1158–1171.

Pistell, P.J., Morrison, C.D., Gupta, S., Knight, A.G., Keller, J.N., Ingram, D.K., et al., 2010. Cognitive impairment following high fat diet consumption is associated with brain inflammation. J. Neuroimmunol. 219, 25–32.

Pistollato, F., Sumalla Cano, S., Elio, I., Masias Vergara, M., et al., 2016. Associations between sleep, cortisol regulation, and diet: possible implications for the risk of Alzheimer disease. Adv. Nutr. 7, 679–689.

Prasad, S., Sajja, R.K., Naik, P., Cucullo, L., 2014. Diabetes mellitus and blood–brain barrier dysfunction: an overview. J. Pharmacovigil. 2, 125.

Prince, M., Bryce, R., Albanese, E., Wimo, A., Ribeiro, W., et al., 2013. The global prevalence of dementia: a systematic review and metaanalysis. Alzheimers Dement. J. Alzheimer Assoc. 9, 63–75.

Prokop, S., Miller, K.R., Heppner, F.L., 2013. Microglia actions in Alzheimer's disease. Acta Neuropathol. 126, 461–477.

Pryor, W.A., Stone, K., 1993. Oxidants in cigarette smoke. Radicals, hydrogen peroxide, peroxynitrate, and peroxynitrite. Ann. N.Y. Acad. Sci. 686, 12–27.

Raj, D.D., Jaarsma, D., Holtman, I.R., Olah, M., Ferreira, F.M., Schaafsma, W., 2014. Priming of microglia in a DNA-repair deficient model of accelerated aging. Neurobiol. Aging 35, 2147–2160.

Raij, L., Demaster, E.G., Jaimes, E.A., 2001. Cigarette smoke-induced endothelium dysfunction: role of superoxide anion. J. Hypertens. 19, 891–897.

Raina, A., Kaul, D., 2010. LXR-α genomics programmes neuronal death observed in Alzheimer's disease. Apoptosis 15, 1461–1469.

Rall Jr., S.C., Weisgraber, K.H., Mahley, R.W., 1982. Human apolipoprotein E. The complete amino acid sequence. J. Biol. Chem. 257, 4171–4178.

Ramesh, V., Thatte, H.S., McCarley, R.W., Basheer, R., 2007. Adenosine and sleep deprivation promote NF-kappaB nuclear translocation in cholinergic basal forebrain. J. Neurochem. 100, 1351–1363.

Rao, J.S., Keleshian, V.L., Klein, S., Rapoport, S.I., 2012. Epigenetic modifications in frontal cortex from Alzheimer's disease and bipolar disorder patients. Transl. Psychiatry 2, e132.

Reitz, C., Mayeux, R., 2014. Alzheimer disease: epidemiology, diagnostic criteria, risk factors and biomarkers. Biochem. Pharmacol. 88, 640–651.

Roberson, E.D., Scearce-Levie, K., Palop, J.J., Yan, F., Cheng, I.H., Wu, T., Gerstein, H., Yu, G.Q., Mucke, L., 2007. Reducing endogenous tau ameliorates amyloid β-induced deficits in an Alzheimer's disease mouse model. Science 316, 750–754.

Reitz, C., den Heijer, T., van Duijn, C., Hofman, A., Breteler, M.M., 2007. Relation between smoking and risk of dementia and Alzheimer disease: the Rotterdam Study. Neurology 69, 998–1005.

Roh, J.H., Jiang, H., Finn, M.B., Stewart, F.R., Mahan, T.E., et al., 2014. Potential role of orexin and sleep modulation in the pathogenesis of Alzheimer's disease. J. Exp. Med. 211, 2487–2496.

Rosenberg, G.A., 2012. Neurological diseases in relation to the blood–brain barrier. J. Cereb. Blood Flow. Metab. 32, 1139–1151.

Saher, G., Brügger, B., Lappe-Siefke, C., Möbius, W., Tozawa, R., Wehr, M.C., Wieland, F., Ishibashi, S., Nave, K.A., 2005. High cholesterol level is essential for myelin membrane growth. Nat. Neurosci. 8, 468–475.

Sahin, E., Colla, S., Liesa, M., Moslehi, J., Muller, F.L., Guo, M., 2011. Telomere dysfunction induces metabolic and mitochondrial compromise. Nature 470, 359–365.

Samsel, A., Seneff, S., 2013. Glyphosate, pathways to modern diseases II: celiac sprue and gluten intolerance. Interdiscip. Toxicol. 6, 159–184.

Sanchez-Lozada, L.G., Le, M., Segal, M., Johnson, R.J., 2008. How safe is fructose for persons with or without diabetes? Am. J. Clin. Nutr. 88, 1189–1190.

Schilling, T., Lehmann, F., Ruckert, B., Eder, C., 2004. Physiological mechanisms of lysophosphatidylcholine-induced de-ramification of murine microglia. J. Physiol. (London) 557, 105–120.

Schimanski, L.A., Barnes, C.A., August 6, 2010. Neural protein synthesis during aging: effects on plasticity and memory. Front. Aging Neurosci. 2, 26.

Schmidt, H.D., Duman, R.S., 2007. The role of neurotrophic factors in adult hippocampal neurogenesis, antidepressant treatments and animal models of depressive-like behavior. Behav. Pharmacol. 18, 391–418.

Schriner, S.E., Linford, N.J., Martin, G.M., Treuting, P., Ogburn, C.E., Emond, M., et al., 2005. Extension of murine lifespan by overexpression of catalase targeted to mitochondria. Science 308, 1909–1911.

Seet, R.C., Lee, C.Y., Lim, E.C., Tan, J.J., Quek, A.M., Chong, W.L., Looi, W.F., Huang, S.H., Wang, H., Chan, Y.H., et al., 2010. Oxidative damage in Parkinson disease: measurement using accurate biomarkers. Free Radic. Biol. Med. 48, 560–566.

Seneff, S., Wainwright, G., Mascitelli, L., 2011. Nutrition and Alzheimer's disease: the detrimental role of a high carbohydrate diet. Eur. J. Intern. Med. 22, 134–140.

Seshadri, S., Fitzpatrick, A.L., Ikram, M.A., DeStefano, A.L., Gudnason, V., Boada, M., Bis, J.C., Smith, A.V., Carassquillo, M.M., Lambert, J.C., et al., 2010. Genome-wide analysis of genetic loci associated with Alzheimer disease. JAMA 303, 1832–1840.

Sheng, H., Laskowitz, D.T., Mackensen, G.B., Kudo, M., Pearlstein, R.D., Warner, D.S., 1999. Apolipoprotein E deficiency worsens outcome from global cerebral ischemia in the mouse. Stroke 30, 1118–1124.

Simopoulos, A.P., 2013. Dietary omega-3 fatty acid deficiency and high fructose intake in the development of metabolic syndrome, brain metabolic abnormalities, and non-alcoholic fatty liver disease. Nutrients 5, 2901–2923.

Spiegel, K., Leproult, R., Van Cauter, E., 1999. Impact of sleep debt on metabolic and endocrine function. Lancet 354, 1435–1439.

Spiegel, K., Tasali, E., Penev, P., Van Cauter, E., 2004. Brief communication: sleep curtailment in healthy young men is associated with decreased leptin levels, elevated ghrelin levels, and increased hunger and appetite. Ann. Intern. Med. 141, 846–850.

Stranahan, A.M., Norman, E.D., Lee, K., Cutler, R.G., Telljohann, R.S., Egan, J.M., Mattson, M.P., 2008. Diet-induced insulin resistance impairs hippocampal synaptic plasticity and cognition in middle-aged rats. Hippocampus 18, 1085–1088.

Sutherland, G.T., Chami, B., Youssef, P., Witting, P.K., 2013. Oxidative stress in Alzheimer's disease: primary villain or physiological by-product? Redox Rep. 18, 134–141.

Szulwach, K.E., Jin, P., 2014. Integrating DNA methylation dynamics into a framework for understanding epigenetic codes. Bioessays 36, 107–117.

Takechi, R., Galloway, S., Pallebage-Gamarallage, M.M.S., et al., 2010. Dietary fats, cerebrovasculature integrity and Alzheimer's disease risk. Prog. Lipid Res. 49, 159–170.

Takenouchi, T., Sato, M., Kitani, H., 2007. Lysophosphatidylcholine potentiates $Ca2+$ influx, pore formation and p44/42 MAP kinase phosphorylation mediated by P2X7 receptor activation in mouse microglial cells. J. Neurochem. 102, 1518–1532.

Talbot, K., Wang, H.Y., Kazi, H., Han, L.Y., Bakshi, K.P., Stucky, A., et al., 2012. Demonstrated brain insulin resistance in Alzheimer's disease patients is associated with IGF-1 resistance, IRS-1 dysregulation, and cognitive decline. J. Clin. Invest. 122, 1316–1338.

Tanaka, N., Abe-Dohmae, S., Iwamoto, N., Yokoyama, S., 2011. Roles of ATP-binding cassette transporter A7 in cholesterol homeostasis and host defense system. J. Atheroscler. Thromb. 18, 274–281.

Tanzi, R.E., Bertram, L., 2005. Twenty years of the Alzheimer's disease amyloid hypothesis: a genetic perspective. Cell 120, 545–555.

Tochikubo, O., Ikeda, A., Miyajima, E., Ishii, M., 1996. Effects of insufficient sleep on blood pressure monitored by a new multibiomedical recorder. Hypertension 27, 1318–1324.

Torelli, F., Moscufo, N., Garreffa, G., Placidi, F., Romigi, A., Zannino, S., et al., 2011. Cognitive profile and brain morphological changes in obstructive sleep apnea. NeuroImage. 54, 787–793.

Torroglosa, A., Murillo-Carretero, M., Romero-Grimaldi, C., Matarredona, E.R., Campos-Caro, A., Estrada, C., 2007. Nitric oxide decreases subventricular zone stem cell proliferation by inhibition of epidermal growth factor receptor and phosphoinositide-3-kinase/Akt pathway. Stem Cells 25, 88–97.

Townsend, M., Mehta, T., Selkoe, D.J., 2007. Soluble Abeta inhibits specific signal transduction cascades common to the insulin receptor pathway. J. Biol. Chem. 282, 33305–33312.

Tractenberg, R.E., Singer, C.M., Kaye, J.A., 2005. Symptoms of sleep disturbance in persons with Alzheimer's disease and normal elderly. J. Sleep. Res. 14, 177–185.

Ulyanova, T., Blasioli, J., Woodford-Thomas, T.A., Thomas, M.L., 1999. The sialoadhesin CD33 is a myeloid-specific inhibitory receptor. Eur. J. Immunol. 29, 3440–3449.

Valladolid-Acebes, I., Merino, B., Principato, A., Fole, A., Barbas, C., Lorenzo, M.P., García, A., Del Olmo, N., Ruiz-Gayo, M., Cano, V., 2012. High-fat diets induce changes in hippocampal glutamate metabolism and neurotransmission. Am. J. Physiol. Endocrinol. Metab. 302, E396–E402.

von Bernhardi, R., 2010. Immunotherapy in Alzheimer's disease: where do we stand? Where should we go? J. Alzheimers Dis. 19, 405–421.

van der Flier, W.M., Scheltens, P., 2005. Epidemiology and risk factors of dementia. J. Neurol. Neurosurg. Psychiatry 76, 2–7.

Vandal, M., White, P.J., Tremblay, C., St-Amour, I., Chevrier, G., et al., 2014. Insulin reverses the high-fat diet-induced increase in brain $A\beta$ and improves memory in an animal model of Alzheimer disease. Diabetes 63, 4291–4301.

Verghese, P.B., Castellano, J.M., Holtzman, D.M., 2011. Apolipoprotein E in Alzheimer's disease and other neurological disorders. Lancet Neurol. 10, 241–252.

Vitiello, M.V., Borson, S., 2001. Sleep disturbances in patients with Alzheimer's disease: epidemiology, pathophysiology and treatment. CNS Drugs 15, 777–796.

Wang, H., Dey, D., Carrera, I., Minond, D., Bianchi, E., Xu, S., et al., 2013a. COPS5 (Jab1) Protein increases β site processing of amyloid precursor protein and amyloid β peptide generation by stabilizing RanBP9 protein levels. J. Biol. Chem. 288, 26668–26677.

Wallace, A.M., McMahon, A.D., Packard, C.J., et al., 2001. Plasma leptin and the risk of cardiovascular disease in the west of Scotland coronary prevention study (WOSCOPS). Circulation 104, 3052–3056.

Walter, R.B., Raden, B.W., Zeng, R., Hausermann, P., Bernstein, I.D., Cooper, J.A., 2008. ITIM-dependent endocytosis of CD33-related Siglecs: role of intracellular domain, tyrosine phosphorylation, and the tyrosine phosphatases, Shp1 and Shp2. J. Leukoc. Biol. 83, 200–211.

Wang, S.C., Oelze, B., Schumacher, A., 2008. Age-specific epigenetic drift in late-onset Alzheimer's disease. PLos One 3, 32698.

Wang, H.K., Lin, S.H., Sung, P.S., Wu, M.H., Hung, K.W., Wang, L.C., et al., 2012. Population based study on patients with traumatic brain injury suggests increased risk of dementia. J. Neurol. Neurosurg. Psychiatry 83, 1080–1085.

Wang, J., Yu, J.T., Tan, M.S., Jiang, T., Tan, L., 2013b. Epigenetic mechanisms in Alzheimer's disease: implications for pathogenesis and therapy. Ageing Res. Rev. 12, 1024–1041.

Wang, X., Lopez, O.L., Sweet, R.A., Becker, J.T., DeKosky, S.T., Barmada, M.M., Demirci, F.Y., Kamboh, M.I., 2015a. Genetic determinants of disease progression in Alzheimer's disease. J. Alzheimers Dis. 43, 649–655.

Wang, Y., Cella, M., Mallinson, K., Ulrich, J.D., Young, K.L., Robinette, M.L., et al., 2015b. TREM2 lipid sensing sustains the microglial response in an Alzheimer's disease model. Cell 2014 (160), 1061–1071.

Wannamethee, S.G., Shaper, A.G., Walker, M., 1998. Changes in physical activity, mortality, and incidence of coronary heart disease in older men. Lancet 351, 1603–1608.

Wenk, G.L., 2003. Neuropathologic changes in Alzheimer's disease. J. Clin. Psychiatry 64 (Suppl. 9), 7–10.

Wimmer, M.E., Hernandez, P.J., Blackwell, J., Abel, T., 2012. Aging impairs hippocampus-dependent long-term memory for object location in mice. Neurobiol. Aging 33, 2220–2224.

Wisor, J.P., Edgar, D.M., Yesavage, J., et al., 2005. Sleep and circadian abnormalities in a transgenic mouse model of Alzheimer's disease: a role for cholinergic transmission. Neuroscience 131, 375–385.

Wu, A., Ying, Z., Gomez-Pinilla, F., 2004. The interplay between oxidative stress and brain-derived neurotrophic factor modulates the outcome of a saturated fat diet on synaptic plasticity and cognition. Eur. J. Neurosci. 19, 1699–1707.

Wu, J., Basha, M.R., Brock, B., Cox, D.P., Cardozo-Pelaez, F., McPherson, C.A., et al., 2008. Alzheimer's disease (AD)-like pathology in aged monkeys after infantile exposure to environmental metal lead (Pb): evidence for a developmental origin and environmental link for AD. J. Neurosci. 28, 3–9.

Xiao, Q., Gil, S.C., Yan, P., Wang, Y., Han, S., Gonzales, E., et al., 2012. Role of phosphatidylinositol clathrin assembly lymphoid-myeloid leukemia (PICALM) in intracellular amyloid precursor protein (APP) processing and amyloid plaque pathogenesis. J. Biol. Chem. 287, 21279–21289.

Xu, Q., Bernardo, A., Walker, D., Kanegawa, T., Mahley, R.W., Huang, Y., 2006a. Profile and regulation of apolipoprotein E (ApoE) expression in the CNS in mice with targeting of green fluorescent protein gene to the ApoE locus. J. Neurosci. 26, 4985–4994.

Xu, Y., Yamada, T., Satoh, T., Okuda, Y., 2006b. Measurement of serum amyloid A1 (SAA1), a major isotype of acute phase SAA. Clin. Chem. Lab. Med. 44, 59–63.

Xu, X., Zhan, M., Duan, W., Prabhu, V., Brenneman, R., Wood, W., Firman, J., 2007. Gene expression atlas of the mouse central nervous system: impact and interactions of age, energy intake and gender. Genome Biol. 8, R234.

Yamaguchi, Y., Nasu, F., Harada, A., Kunitomo, M., 2007. Oxidants in the gas phase of cigarette smoke pass through the lung alveolar wall and raise systemic oxidative stress. J. Pharmacol. Sci. 103, 275–282.

Ye, S.M., Johnson, R.W., 2001. An age-related decline in interleukin-10 may contribute to the increased expression of interleukin-6 in brain of aged mice. Neuroimmunomodulation 9, 183–192.

Yu, J.T., Tan, L., 2012. The role of clusterin in Alzheimer's disease: pathways, pathogenesis, and therapy. Mol. Neurobiol. 45, 314–326.

Zaidi, A., Barron, L., Sharov, V.S., Schoneich, C., Michaelis, E.K., Michaelis, M.L., 2003. Oxidative inactivation of purified plasma membrane Ca(2+)-ATPase by hydrogen peroxide and protection by calmodulin. Biochemistry 42, 12001–12010.

Zhang, X., Dong, F., Ren, J., Driscoll, M.J., Culver, B., 2005. High dietary fat induces NADPH oxidase-associated oxidative stress and inflammation in rat cerebral cortex. Exp. Neurol. 191, 318–325.

Zhang, H., Wu, L.M., Wu, J., 2011. Cross-talk between apolipoprotein E and cytokines. Mediat. Inflamm. 949072.

Zhao, W.Q., De Felice, F.G., Fernandez, S., Chen, H., Lambert, M.P., Quon, M.J., Krafft, G.A., Klein, W.L., 2008. Amyloid beta oligomers induce impairment of neuronal insulin receptors. FASEB J. 22, 246–260.

Zipser, B.D., Johanson, C.E., Gonzalez, L., et al., 2007. Microvascular injury and blood–brain barrier leakage in Alzheimer's disease. Neurobiol. Aging 28, 977–986.

Zlokovic, B.V., 2011. Neurovascular pathways to neurodegeneration in Alzheimer's disease and other disorders. Nat. Rev. Neurosci. 12, 723–738.

Contribution of Neural Membrane Phospholipids, Sphingolipids, and Cholesterol in the Pathogenesis of Alzheimer's Disease

INTRODUCTION

Neural membranes are dynamic structures that separate the extracellular and intracellular environments. Neural membranes contain glycerophospholipids, sphingolipids, cholesterol, and proteins. In neural membranes, lipids are organized in bilayers with the amine-containing

Neurochemical Aspects of Alzheimer's Disease
http://dx.doi.org/10.1016/B978-0-12-809937-7.00003-3

93

phospholipids enriched on the cytoplasmic side of the plasma membrane, whereas the choline-containing phospholipids and sphingolipids enriched on the outer surface. The two lipid bilayers of neural membranes are held together by hydrophobic, coulombic, and van der Waal forces and hydrogen bonding. This organization of lipid bilayer is spontaneous, meaning it is a natural process and does not require energy. The distribution of lipids in the two leaflets of the lipid bilayer is asymmetric with the hydrophobic tails on the interior of the membrane and the hydrophilic heads pointing outward (Ikeda et al., 2006; Yamaji-Hasegawa and Tsujimoto, 2006). A high degree of asymmetry in neural membranes controls various cellular functions, such as signal transduction, membrane fusion, and neural cell apoptosis (Boon and Smith, 2002; Fadok et al., 2001). The protein composition of each bilayer is different. Glycerophospholipids and sphingolipids contribute to lipid asymmetry, whereas cholesterol and sphingolipids form lipid microdomains or lipid rafts. Lipid rafts are characterized by a differentiated lipid and protein composition that segregate within the plasma membrane (Brown and Jessup, 2009; Hicks et al., 2012). These chemical features are responsible for many physicochemical properties of rafts including closer lipid packing, rather restricted lateral movement, higher viscosity, and differential thermodynamic properties, compared to the surrounding nonraft regions (Brown and Jessup, 2009; Klymchenko and Kreder, 2014; Zajchowski and Robbins, 2002; Schroeder et al., 1998). Lipid rafts are enriched in cholesterol, a key element, which maintains the stability of these domains compared with their surroundings area. In addition, sphingolipids and saturated fatty acids are notably augmented compared to nonraft membrane domains, they contribute to the higher density and degree of packing of lipid rafts (Simons and Ehehalt, 2002; Lucero and Robbins, 2004; Róg and Vattulainen, 2014). Lipid rafts also contain a high amount of sphingomyelin (SM), which is enriched in the outer leaflet of the plasma membrane, indicating that some transbilayer translocation must occur to form and stabilize these domains (van Meer, 2011). In addition to lipids, lipid rafts also contain anchored and signal transduction proteins. Lipid rafts are involved in intracellular trafficking of lipids and proteins and lipids. They modulate signal transduction processes associated with neural cell functions (Edidin, 2003). In neurons, membrane rafts have been detected at synapses, where they are thought to contribute to pre- and postsynaptic function (Suzuki, 2002).

An important property of the lipid bilayer is its fluidity. This property allows the structural components mobility within the lipid bilayer. Membrane fluidity not only depends on specific structure of the fatty acid chains but also on temperature (fluidity increases at lower temperatures). Membrane fluidity plays an important role in membrane transport. In neural cell membranes, lipid asymmetry is introduced by selective synthesis of lipids on one side of the neural membrane and maintained by three classes of lipid translocases, P-type ATPases, ABC transporters, and scramblases (Daleke, 2003; Pomorski and Menon, 2006). The disruption of asymmetry not only results in neural cell activation but also plays an important role in neurodegeneration.

Glycerophospholipids, sphingolipids, and cholesterol are key regulators of brain function and have been increasingly implicated in pathogenesis of AD. A major risk factor for late-onset AD is the ε4 allelic variant of apolipoprotein E (APOE), which encodes a protein involved in cholesterol metabolism and lipid transport (Lambert et al., 2009). Importantly, a variety of genes have been linked to late-onset AD through genome-wide association studies and are directly or indirectly connected to lipid metabolism or cellular membrane dynamics (Hollingworth et al., 2011; Naj et al., 2011). In support of a role for glycerophospholipids, sphingolipids, and cholesterol in AD, analysis of the prefrontal cortex, entorhinal cortex (ENT), and cerebellum of patients with late-onset AD has indicated an elevation of diacylglycerol and sphingolipids in the prefrontal cortex of patients with AD (Chan et al., 2012). Enrichment of lysobisphosphatidic acid, SM, the ganglioside GM3, and cholesterol esters has also been observed in the affected ENT, but no change in lipids occur within the cerebellum. Mapstone and his group have identified a fingerprint of 10 plasma lipids that defines AD in older adults. However, it remains unknown whether these lipids play a mechanistic role in the development and progression of AD (Mapstone et al., 2014).

GLYCEROPHOSPHOLIPIDS, SPHINGOLIPIDS, AND CHOLESTEROL AS PRECURSORS FOR LIPID MEDIATORS

Lipid mediators are a diverse family of endogenous bioactive molecules, which are derived from fatty acids either enzymatically or nonenzymatically. Neural membrane phospholipids contain arachidonic acid (ARA, 20:4n-6) and docosahexaenoic acid (DHA; 22:6n-3) belonging to n-6 and n-3 family of fatty acids, respectively. In phospholipids, these fatty acids are located at the sn-2 position of glycerol moiety (Farooqui et al., 2000). They provide neural membranes with many physicochemical properties such as fluidity, stability, ion permeability, fusion, rapid flip-flopping, packing, and elastic compressibility. The majority of ARA is associated with phosphatidylcholine (PtdCho), whereas phosphatidylethanolamine and ethanolamine plasmalogen (PlsEtn) contain both ARA and DHA. Phosphatidylserine is enriched in DHA (Farooqui et al., 2000). ARA is liberated from glycerophospholipids by the action of cytosolic phospholipase A_2 (cPLA$_2$), whereas DHA is released by the action of Ca^{2+}-independent PLA$_2$ (Farooqui and Horrocks, 2007). Free ARA and DHA are precursors for

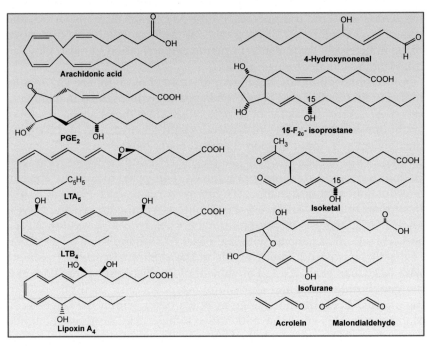

FIGURE 3.1   Chemical structures of arachidonic acid–derived enzymatic and nonenzymatic lipid mediators.

phospholipid-derived lipid mediators. ARA is oxidized by enzymatic and nonenzymatic mechanisms into oxygenated metabolites. Enzymatically derived lipid mediators of ARA metabolism are prostaglandins (PGs), leukotrienes (LTs), thromboxanes (TXs), and lipoxins (LXs) (Fig. 3.1). These metabolites are collectively known as eicosanoids. They modulate neural function by acting through eicosanoid receptors (EP1, EP3, and EP4). These receptors are localized not only on neural cell surface but also on nuclear membranes of neural cells. Among ARA-derived metabolites PGs, LTs, and TXs induce proinflammatory effects, whereas LXs act as "signal braking" metabolite, which inhibits inflammatory processes and promotes resolution of inflammation. In addition to endogenous generation of LXs, aspirin can also trigger the biosynthesis of a group of LXs known as aspirin-triggered lipoxin A_4 (LXA$_4$), which mimic the antiinflammatory actions of LXs but are more resistant to degradation (McMahon et al., 2001). The nonenzymatic lipid mediators of ARA metabolism include 4-hydroxynonenal, isoprostanes (IsoP), isoketal, isofuran, acrolein, and malondialdehyde (Fig. 3.1) (Esterbauer et al., 1991). These mediators are strong oxidant and biomarkers for lipid peroxidation and oxidative stress (Farooqui, 2011).

Free DHA is metabolized by 15-lipoxygenase (15-LOX). Enzymically lipid mediators of DHA metabolism include resolvins (Rvs), protectin

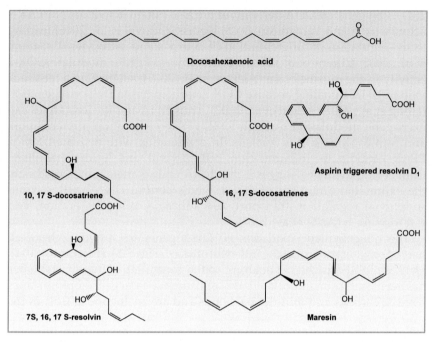

FIGURE 3.2 Chemical structures of docosahexaenoic acid–derived enzymatic lipid mediators.

D1 (PD1) or neuroprotectins (NPDs), and maresins (MaRs) (Fig. 3.2). These mediators induce antiinflammatory and proresolutionary effects (Serhan and Petasis, 2011; Serhan et al., 2008; Farooqui, 2012). In the brain, NPD1 not only stimulates the expression of antiapoptotic Bcl-2 proteins (Bcl-2 and BclxL) and inhibits the expression of proapoptotic proteins (pro-Baxds and pro-Bad) (Bazan, 2005) but also retards oxidative stress. In the presence of aspirin, the action of cyclooxygenase (COX)-2 on DHA leads to the synthesis of aspirin-triggered forms of Rvs. A growing body of evidence indicates that Rvs not only possess potent antiinflammatory properties but also mediate immunoregulatory actions by inhibiting the production of proinflammatory mediators and regulating trafficking of leukocytes (Serhan et al., 2004). Specifically, Rvs stop polymorphonuclear leukocytes infiltration in vivo and transmigration (Serhan et al., 2002). They reduce cytokine expression by isolated microglia cells (Hong et al., 2003). DHA is an endogenous ligand for retinoid X receptors (RXR), which form heterodimers with peroxisome proliferator-activated receptors (PPARs) to generate nuclear transcription factors RXR/PPAR, which modulate gene expression in different cell types (Bordoni et al., 2006; Dyall et al., 2010). In addition, NPD1, but not DHA, activates the PPAR-γ isoform in human neuronal and glial cells (Zhao et al.,

2011). Oxygenated DHA derivatives are also potent activators of PPAR-γ activity (Itoh and Yamamoto, 2008; Heras-Sandoval et al., 2016) leading to the regulation of differentiation of neurons and astrocytes (Cristiano et al., 2005; Cimini et al., 2005). Furthermore, PPARs modulate COX-2 activity in the murine brain (Aleshin et al., 2009), suggesting a feedback loop in DHA signaling because COX-2 participates in DHA oxygenation (Hong et al., 2007; Serhan et al., 2002). Thus DHA and its derivatives can exert gene modulation through RXR/PPAR-γ activation, dimerization, and translocation to the nucleus. In conjunction with activated RXRs, PPAR-γ modulates inflammation, cell survival, and lipid metabolism. Converging evidence suggests that DHA and its metabolites modulate the expression a number of genes, which control DNA binding, transcriptional regulation, transport, cell adhesion, cell proliferation, and raft formation (Arita et al., 2007).

The nonenzymatic oxidation of DHA generates 4-hydroxyhexanal, neuroprostane, neuroketal, and neurofuran (Fig. 3.3) (Farooqui, 2011). DHA is highly enriched in neurons, so the generation of these metabolites has been used as an important index of neuronal damage. Neuroprostanes have 22 carbons and four double bonds and are analogous to IsoP. In the

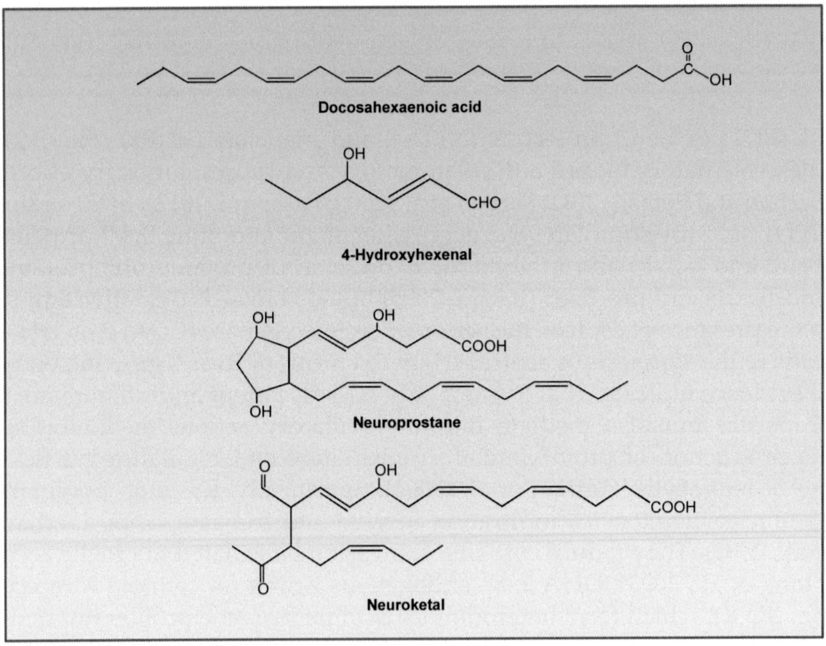

FIGURE 3.3   Chemical structures of docosahexaenoic acid–derived nonenzymatic lipid mediators.

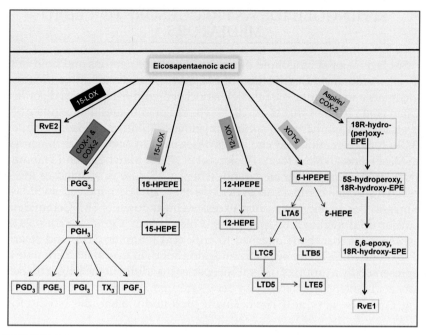

FIGURE 3.4 **Generation of eicosapentaenoic acid (EPA)–derived lipid mediators in neural and nonneural tissues.** *5-LOX*, lipoxygenase; *12-HETE*, 12-hydroxy-5,8,10,14-eicosatetraenoic acid; *12-HPEPE*, 12-hydroperoxy-5,8,11,13,12-eicosapentaenoic acid; *15-HETE*, 15-hydroxy-5,8,10,14-eicosatetraenoic acid; *15-HPEPE*, 15-hydroperoxy-5,8,11,13,15-eicosapentaenoic acid; *LTB5, LTC5, and LTD5*, 5-series leukotrienes; *PGE3 and PGI3*, 3-series of prostaglandins and thromboxanes; *RvE1*, resolvin E_1; *RvE2*, resolvin E_2.

synthesis of neuroprostanes, nonenzymatic oxidation of DHA also generates peroxyl radicals, which may contribute to alterations in neural membrane fluidity and permeability of neural membranes leading to neuronal dysfunction (Farooqui, 2011).

Mammalian brain neural membrane contains small amounts of eicosapentaenoic acid (EPA), which is released from EPA containing phospholipids by the action of calcium independent phospholipase A_2 (iPLA$_2$). Free EPA is oxidized by COXs and LOXs into 3-series PGs (PGE3) and TXs (TX3) and the 5-series LTs (LTB5) (Farooqui, 2011). These metabolites of EPA metabolism are less active than ARA-derived PGs, LTs, and TXs (Calder, 2011). Enzymatic oxidation of EPA by 15-LOX produces E-series Rvs (RvE$_1$ and RvE$_2$) (Fig. 3.4). These metabolites interact with specific G-protein-coupled ChemR23 receptors and mediate antiinflammatory effect in vivo (Ohira et al., 2010).

SPHINGOLIPIDS AS PRECURSORS FOR LIPID MEDIATORS

SM is a major sphingolipid of myelin sheaths of neurons and lipid rafts of cell membranes. Degradation of SM may cause disturbances in action potentials and synaptic dysfunction associated with neurological disorders (Chan et al., 2012; Muhle et al., 2013). The major enzymes that regulate SM levels are SM synthase and sphingomyelinases (SMases). SMases degrade SM to form ceramide and choline. SMases exist in acid (aSMase), neutral (nSMase), and alkaline forms (Jenkins et al., 2009; Marchesin and Hannun, 2004). nSMase is membrane-bound, whereas aSMase is secreted or localizes to lysosomal compartments (Jenkins et al., 2011). An increase in SMase expression or activity may cause excessive breakdown of SM. Ceramide is a major lipid mediator of sphingolipid metabolism. Ceramide consists of a sphingosine backbone attached to fatty acid [palmitic (C16) and stearic (C18) nonhydroxy fatty acids] by an amide bond. In mammals, ceramide is represented by a group of several subspecies that differ in the length of their fatty acid moiety, which can vary between 2 and 28 carbon atoms in length (Fig. 3.5). The fatty acid chain length profoundly alters the biophysical

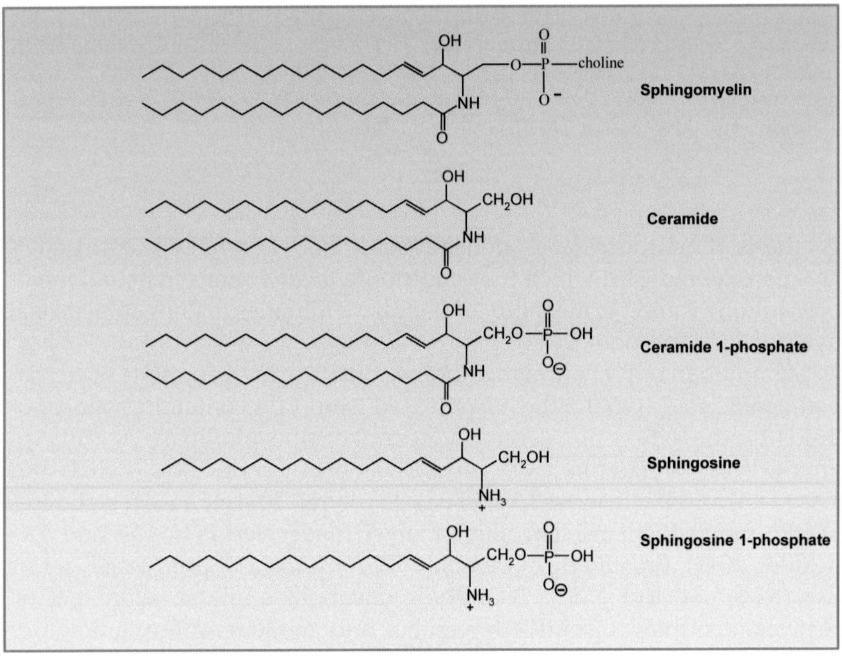

FIGURE 3.5   Chemical structures of sphingomyelin, ceramide, ceramide 1-phosphate, sphingosine, and sphingosine 1-phosphate.

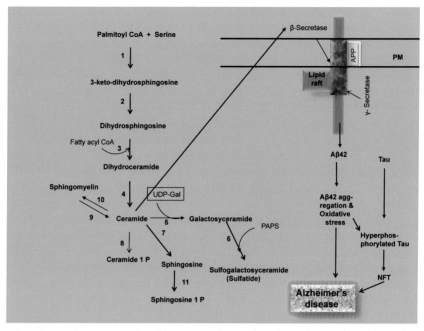

FIGURE 3.6 **De novo synthesis of ceramide.** Serine palmitoyltransferase (1); ketodihydrosphingosine reductase (2); ceramide synthase (3); dihydroceramide desaturase (4); UDP-Gal transferase (5); cerebroside sulfotransferase (6); ceramidase (7); ceramide kinase (8); sphingomyelinase (9); sphingomyelin synthase (10); ceramide synthase (11); and 3'-phosphoadenosine-5'-phosphosulfate (PAPS).

properties of ceramides. Short-chain ceramides with fatty acyl chains of fewer than 12 carbons can be easily dispersed in water and serve as detergents (Sot et al., 2005). In contrast, most ceramides found in mammalian cellular membranes contain long fatty acyl chains of 16–28 carbon atoms rendering them hydrophobic lipids lacking detergent properties. In the brain, synthesis of ceramide occurs through the de novo route (Fig. 3.6). In the brain, ceramide metabolism is highly compartmentalized. Hydrophobic ceramides are synthesized by membrane-associated enzymes and exert their effects either in close proximity to the generation site or require specific transport mechanisms to reach their targets in another intracellular compartment (Futerman and Riezman, 2005). Beside their essential role in signal transduction, ceramides also act as regulators of synaptic function, contributing to the maintenance of synapses, dendritic spines, and neuronal transmission in association with other sphingolipids and cholesterol in lipid rafts (Haughey et al., 2010; Mencarelli and Martinez-Martinez, 2013). Long-chain ceramides have the ability to flip-flop across the membrane (López-Montero et al., 2005). However, spontaneous interbilayer transfer is extremely slow (Contreras et al., 2010). Therefore, the transfer of ceramide between intracellular compartments is performed by vesicular transport pathways (Perry and Ridgway, 2005). Thus ceramide is synthesized by the condensation of serine and palmitoyl coenzyme A (CoA) in the presence of the enzyme serine palmitoyltransferase in the endoplasmic reticulum (Fig. 3.6). From the endoplasmic reticulum, ceramide is transported by ceramide transport protein to the Golgi apparatus, where it is required for the synthesis of sphingomyelin (CerPCho or SM) (Hanada et al., 2003). In addition to de novo biosynthesis, ceramide is generated by the action of SMases on SM (Ong et al., 2015). Ceramide functions as a second messenger in a variety of cellular events, including proliferation, differentiation, growth arrest, inflammation, stress responses, cell signaling, synaptic activity, and apoptosis (Hannun and Obeid, 2002; Yu et al., 2000; El Alwani et al., 2006). Levels of plasma ceramides are increased in obese children (Lopez et al., 2013) and in human adults with diabetic and metabolic syndrome (Haus et al., 2009; Brozinick et al., 2013). In these pathological conditions, levels of ceramide correlate with the severity of insulin resistance (Haus et al., 2009). Studies on the effects of insulin-sensitizing drug (pioglitazone) on plasma ceramides indicate a correlation between the decrease in plasma ceramides and improved insulin sensitivity (Warshauer et al., 2015). Collectively, the majority of human studies have demonstrated the role of ceramides in insulin resistance associated with obesity, type 2 diabetes, and metabolic syndrome. These pathological conditions are risk factor for AD, stroke, and depression (Farooqui et al., 2012; Farooqui, 2013).

Ceramide is phosphorylated into ceramide 1-phosphate (C1P) by ceramide kinase (Fig. 3.7). C1P is synthesized in the trans-Golgi network (Pettus et al., 2004). C1P is known to activate $cPLA_2\alpha$. The enzyme hydrolyzes neural membrane PtdCho and generates lyso-PtdCho and ARA, which is converted into eicosanoid by the actions of COXs and 5-LOXs. As stated previously, eicosanoids induce neuroinflammation and increased levels of eicosanoids are associated with the pathogenesis of many inflammatory disorders; understanding the role of C1P in neuroinflammation is of utmost importance (Fig. 3.7). Simanshu et al. have characterized a unique cytosolic C1P-specific transfer protein called lipid transfer protein (CPTP) (Simanshu et al., 2013). Depletion of CPTP increases C1P in the Golgi and nucleus, but decreases it in the plasma membrane. This suggests that CPTP controls the levels of C1P, which is synthesized by CERK in the Golgi by transferring it to the plasma membrane, thereby suppressing $cPLA_2\alpha$ activity and reducing ARA release and eicosanoid production (Simanshu et al., 2013). The C1P is transferred to the plasma membrane and is rapidly degraded by lipid phosphatases to ceramide. Another important piece of information is that changes in C1P levels are directly correlated with changes in sphingosine and sphingosine 1-phosphate (S1P), raising the possibility that C1P may be a precursor of S1P that is mediated by deacylase activity, which is yet to be characterized (Simanshu et al., 2013) (see later).

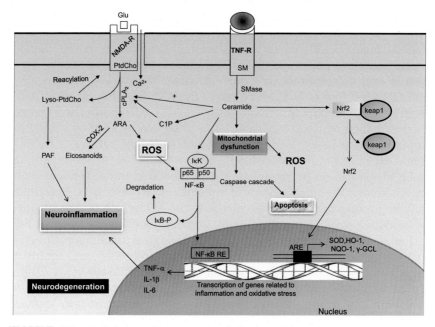

FIGURE 3.7   **Modulation of NF-κB and Nrf2 by ceramide.** *ARE*, antioxidant response element; *C1P*, ceramide 1-phosphate; *COX-2*, cyclooxygenase-2; *cPLA₂*, cytosolic phospholipase A_2; *Glu*, glutamate; *HO-1*, heme oxygenase; *IL-1β*, interleukin-1β; *IL-6*, interleukin-6; *Keap1*, kelch-like erythroid Cap'n'Collar homologue-associated protein 1; *lyso-PtdCho*, lyso-phosphatidylcholine; *NF-κB*, nuclear factor-κB; *NF-κB-RE*, nuclear factor-κB response element; *NMDA-R*, N-Methyl-ᴅ-aspartate receptor; *NQO-1*, NADPH quinine oxidoreductase; *Nrf2*, nuclear factor erythroid-2-related factor 2; *PAF*, platelet-activating factor; *PtdCho*, phosphatidylcholine; *ROS*, reactive oxygen species; *SM*, sphingomyelin; *SMase*, sphingomyelinase; *TNF-R*, tumor necrosis factor-α-receptor; *TNF-α*, tumor necrosis factor-α; *γ-GCL*, γ-glutamate cysteine ligase.

Levels of endogenous ceramide are tightly controlled by various mechanisms, particularly activation of SMase and de novo ceramide synthesis (Marchesini and Hannun, 2004). Ceramide activates a number of serine/threonine protein phosphatases, including PP1, PP2A, and PP2C (Canals et al., 2012; Saddoughi et al., 2013). The activation of PP2A results in dephosphorylation of Akt, a potent promoter of cell survival. This may partly explain the proapoptotic actions of ceramide. In addition, PP1 is an important target for ceramide, which is generated at the plasma membrane. Stimulation of the atypical protein kinase Cζ by ceramide, probably concomitant with its membrane recruitment, also leads to the inhibition of Akt (Fox et al., 2007).

In rat astrocytes, C2 ceramide inhibits hydrogen peroxide–mediated reactive oxygen species (ROS) generation and subsequent cell death. C2 ceramide upregulates the expression of phase II antioxidant enzymes, such as heme oxygenase-1 (HO-1), NAD(P)H:quinine oxidoreductase 1, and superoxide dismutase. These enzymes are under the control of nuclear factor-E2-related factor 2 (Nrf2)/antioxidant response element (ARE) signaling pathways (Fig. 3.7). Detailed mechanistic studies have shown that C2 ceramide not only increases the nuclear translocation and DNA binding of Nrf2 and c-Jun to the ARE but also enhances ARE-mediated transcriptional activity. Furthermore, C2 ceramide also increases the interaction between Nrf2 and c-Jun as shown by antibody coimmunoprecipitation assay. Further analysis of signaling pathways indicate that AMPK and mitogen-activated protein kinases (MAPKs) contribute to HO-1 expression by modulating ARE-mediated transcriptional activity. Therefore, the upregulation of antioxidant enzymes by C2 ceramide may be a potential therapeutic modality for AD, which is accompanied by oxidative stress (Jung et al., 2016). Converging evidence suggests that C1P plays important roles in a number of biological functions such as cell growth, survival and mediation of macrophage migration, and control of inflammatory responses (Maceyka and Spiegel, 2014; Arana et al., 2010).

The degradation product of ceramide is sphingosine. Phosphorylation of sphingosine in the presence of ATP and sphingosine kinases (SphKs) results in the synthesis of S1P (Gómez-Muñoz, 2006) (Figs. 3.6 and 3.7). This metabolite is exported out of cells via transporters such as spinster homolog 2 (Spns2). S1P regulates diverse physiological processes by binding to specific G-protein-binding receptors, S1P receptors (S1PRs) 1–5, through a process termed "inside-out signaling." The S1P concentration gradient between various tissues promotes S1PR1-dependent migration of T cells from secondary lymphoid organs into the lymphatic and blood circulation (Spiegel and Milstien, 2011). S1P suppresses T-cell egress from and promotes retention in inflamed peripheral tissues (Cyster and Schwab, 2012), whereas S1PR2 inhibits motility to promote retention of B cells in germinal centers (Green et al., 2011). S1PR1 in T and B cells as well as Spns2 in endothelial cells contributes to lymphocyte trafficking. FTY720 (fingolimod) is a functional antagonist of S1PRs that induces systemic lymphopenia by suppression of lymphocyte egress from lymphoid organs. Very little is known about the intracellular targets of S1P. It has been shown that S1P is synthesized by SphK1 in response to tumor necrosis factor (TNF)-α or interleukin-1 (IL-1). TNF-α interacts not only with TNF-α receptor-associated factor 2 but also with cellular inhibitor of apoptosis 2 (cIAP2). These interactions enhance their lysine-63-linked polyubiquitination activities (Alvarez et al., 2010; Harikumar et al., 2014). Similarly, in response to IL-1, which plays an important role in autoinflammatory diseases, cIAP2 and SphK1 form a complex with interferon regulatory factor-1 (IRF1). This results in polyubiquitination and activation. Consequently, IRF1 enhances expression of the chemokines CXCL10 and CCL5 that recruit mononuclear cells to sites of sterile inflammation

(Harikumar et al., 2014). Thus in addition to canonical S1P receptor signaling, these new intracellular functions of S1P may contribute to the regulation of inflammatory responses of these potent cytokines. In the nucleus, S1P is synthesized by SphK2 or by inhibition of sphingolipid. It binds and inhibits histone deacetylase (HDAC) 1 and HDAC2, linking sphingolipid metabolism to inflammatory and metabolic gene expression (Hait et al., 2009; Ihlefeld et al., 2012). Thus, unlike ceramide, which induces apoptosis, C1P induces opposite effects. C1P produces mitogenic effects and has prosurvival properties (Arana et al., 2010). Furthermore, S1P promotes proliferation, cell survival, and angiogenesis. It also contributes to neuritogenesis and immune function (Spiegel and Milstien, 2003; Alvarez et al., 2007). Converging evidence suggests that ceramide, C1P and S1P, the second messengers of sphingolipid metabolism perform different functions in the brain and are closely associated with neuroinflammation, oxidative stress, neural cell survival, and apoptosis.

At the molecular level, ceramides may contribute to the progression of AD and its pathology not only via β-amyloid (Aβ) generation but also through the induction of oxidative stress (Fig. 3.6) (Jazvinšćak Jembrek et al., 2015). Thus increase in levels of ceramides may directly increase the stabilization of β-secretase. As stated in Chapter 1, this enzyme is associated amyloidogenic processing of Aβ precursor protein (APP). As a positive feedback loop, oligomeric and fibrillar Aβ promote further increase in ceramide levels by activating SMases. These enzymes catalyze the catabolic breakdown of SM to ceramide (Jazvinšćak Jembrek et al., 2015). Furthermore, ceramides also facilitate neuronal apoptosis. Ceramides may initiate a cascade of biochemical alterations, which ultimately leads to neuronal death by diverse biochemical events, such as depolarization and permeabilization of mitochondria, elevation in production of ROS, cytochrome c release, Bcl-2 depletion, and caspase-3 activation, mainly by modulating intracellular signaling, particularly through the pathways related to Akt/PKB kinase and MAPKs.

CHOLESTEROL AS A PRECURSOR FOR LIPID MEDIATORS

The human brain contains approximately 25% of the body's cholesterol. This lipid is essential for normal brain function. It is a major component of neuronal cell membranes, where it not only plays an essential role in determining the fluidity and electrical permeability but also regulates activities of membrane-bound enzymes, receptors, and ion channels (Simons and Ikonen, 2000). In the adult brain, cholesterol is mostly present in its nonesterified form. In the brain, cholesterol is present in two pools: one major pool associated with the myelin membrane mediating the propagation of the electrical signals along the axons and the minor pool associated with the plasma membrane of astrocytes and neurons (Dietschy and Turley, 2004). Blood circulation in the brain is separated from peripheral circulation by the blood–brain barrier (BBB), which prevents the entry of dietary cholesterol from the circulation to the brain, because lipoproteins do not cross the BBB. This suggests that nearly all the brain cholesterol is synthesized in the brain through de novo synthesis, which starts with the conversion of acetyl-CoA to 3-hydroxy-3-methylglutaryl (HMG)-CoA by HMG-CoA synthase. The latter constitutes the only known rate-limiting and irreversible step in cholesterol synthesis (Dietschy, 2009). This is followed by a long series of enzymatic reactions that convert mevalonate into 3-isopenenyl pyrophosphate, farnesyl pyrophosphate, squalene, lanosterol and, through a 19-step process, to cholesterol (Dietschy, 2009). Cholesterol is synthesized in the endoplasmic reticulum from where it is distributed to neural membrane compartments by vesicular and nonvesicular transport mechanisms. Because the synthesis of cholesterol in neurons does not occur efficiently, neurons depend on astrocytes for the supply of cholesterol as an external source. Astrocytes meet neuronal cholesterol demands by secreting APOE–cholesterol complexes, which are transported to the neurons for their development and function (Nieweg et al., 2009; Pfrieger, 2003). There is constant cholesterol turnover in the brain, which has mechanisms to control and excrete cholesterol across the BBB. Four major processes control the cholesterol content in the metabolically active pool of the neural and nonneural cells. These cells express abundant messenger RNA for acetyl-CoA acetyltransferase 2 (ACAT2), sterol regulatory element-binding protein (SREBP), 3-hydroxy-3-methylglutaryl-CoA reductase (HMGR), and low-density lipoprotein receptor (LDLR). In addition, in the mouse brain, there is also significant expression of the nuclear receptors LXRα, LXRβ, RXRα, and RXRγ, but little or no pregnane X receptor (PXR) or farnesoid X receptor (FXR) (Repa et al., 2007; Wang et al., 2002). Deleting the function of LXRα and LXRβ leads to diminished expression of a number of liver X receptor (LXR) target genes (Wang et al., 2002). However, the administration of an LXR agonist enhances the expression of genes controlling the synthesis of ABCA1, ABCG1, and apo D (Repa et al., 2007). Dysregulation of cholesterol homeostasis may result in high cholesterol levels in the brain. Potential sources of excess cholesterol in the brain are likely to include excessive synthesis in astrocytes, plasma membrane breakdown, myelin breakdown, and neuronal loss. Hypercholesterolemia in the blood can come from de novo synthesis and dietary sources, whereas hypercholesterolemia in the brain is likely to be only affected by de novo synthesis and metabolism (Björkhem et al., 2009). The role of cholesterol in the brain remained unknown until the discovery of APOE, which is known to be the primary mechanism of cholesterol transport within the brain. Later on, clinical retrospective studies in humans demonstrated that hypercholesterolemia predisposes to cognitive deficits, including dementia of the Alzheimer type (Zambón et al., 2010; Fantini and Yahi, 2010), and that chronic treatment with cholesterol-lowering drugs (statins) results in retardation of AD (Wolozin et al., 2000). A multisite, medical center–based analysis of patients with early-onset AD has confirmed a correlation between high low-density lipoprotein cholesterol, low high-density lipoprotein cholesterol, and high Pittsburgh compound B (PIB) index, which measures the levels of cerebral Aβ with carbon C11-labeled PIB (Reed et al., 2014). In addition, large consortium studies have also identified several genes involved in cholesterol transport and metabolism that increase AD susceptibility (Di Paolo and Kim, 2011). Cholesterol homoeostasis in the brain is maintained primarily through the de novo synthesis, transport, and clearance. These processes are tightly regulated to prevent the accumulation of cholesterol in the brain.

In neural and nonneural tissues, oxysterols are derived from either enzymatic or nonenzymatic oxidation of cholesterol (Popp et al., 2012; Hughes et al., 2013). The chemical structures of oxysterols vary depending upon the number and position of oxygenated functional groups, and include keto-, hydroxyperoxy-, and epoxy forms. Enzymic pathways of oxysterol production mainly involve cytochrome P450 family enzymes. This family includes 22-, 24-, 25-, and 27-cholesterol hydroxylases (Popp et al., 2012; Hughes et al., 2013). Certain oxysterols are synthesized by nonenzymatic oxidation (auto-oxidation) of cholesterol. This process involves ROS and

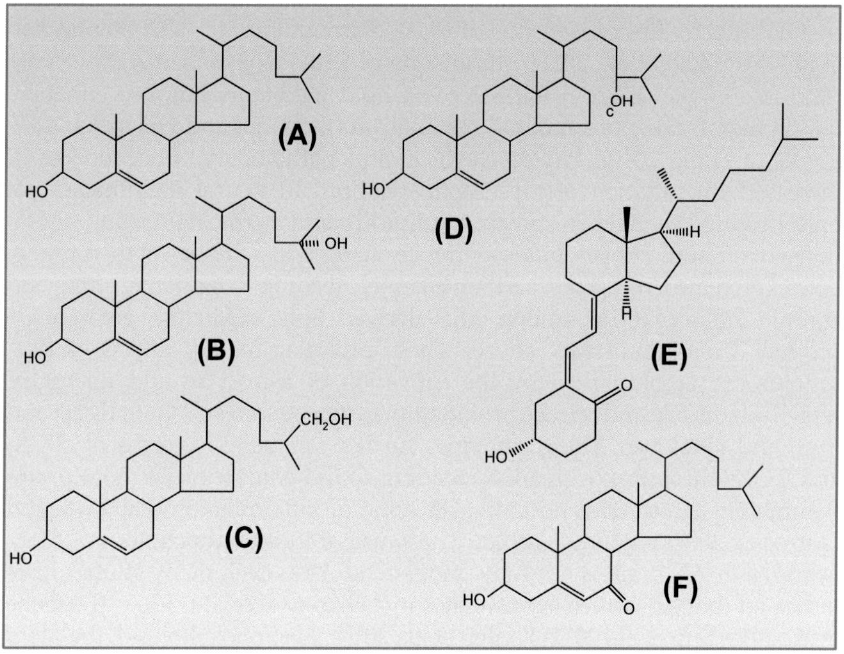

FIGURE 3.8 Conversion of cholesterol into hydroxycholesterols. Cholesterol (A); 25-hydroxycholesterol (B); 27-hydroxycholesterol (C); and 24-hydroxycholesterol (D); vitamin D3 (E); and 7-ketocholesterol (F).

reactive nitrogen species. Examples of ROS that participate in nonenzymatic oxidation of cholesterol include hydroxyl radical ($\cdot$OH), superoxide anion radical (O_2^-), hydrogen peroxide (H_2O_2), and singlet oxygen ($^1O_2$).

Increasing evidence suggests that cholesterol serves as a precursor for the synthesis of hydroxycholesterols (24-hydroxycholesterol, 27-hydroxycholesterol, 22-hydroxycholesterol, and 25-hydroxycholesterol), steroid hormones, and vitamin D in the brain (Fig. 3.8). Changes in levels of cholesterol-derived products may modulate brain cholesterol homeostasis and function in normal brain and brain from patients with neurological disorders (Farooqui, 2011).

ALTERATIONS IN LEVELS OF PHOSPHOLIPIDS, SPHINGOLIPIDS, AND CHOLESTEROL, AND INVOLVEMENT OF THEIR LIPID MEDIATORS IN ALZHEIMER'S DISEASE

Significant decrease in glycerophospholipids levels is observed in different regions in patients with AD compared with age-matched controls (Söderberg et al., 1991; Wells et al., 1995; Guan et al., 1999; Han et al., 2002). This decrease in neural membrane phospholipid levels is caused by the stimulation of activities of isoforms of PLA_2 (Farooqui et al., 1997; Stephenson et al., 1999; Rao et al., 2011). Stimulation of PLA_2 isoforms in AD not only produces an elevation in ARA-derived lipid mediators but also enhances neural membrane phospholipid metabolism (Farooqui and Horrocks, 2007; Farooqui et al., 2010). Physicochemical and pathological consequences of increased degradation of glycerophospholipid in neural membranes not only produce changes in membrane fluidity and permeability, and alterations in ion homeostasis, but also induce changes in activities of membrane-bound enzymes, receptors, and ion channels leading to oxidative stress and chronic inflammation. Among ARA-derived lipid mediators, eicosanoids produce proinflammatory effects. These proinflammatory effects of eicosanoids are accompanied by the activation of astrocytes and microglia, which synthesize and release proinflammatory cytokines (TNF-α, IL-1β, and IL-6) and cytokines. These cytokines further stimulate isoforms of $cPLA_2$ and COXs through positive loop leading to the generation of more proinflammatory eicosanoids and intensification of inflammation (Farooqui and Horrocks, 2007; Farooqui, 2010a). The cause of increased activities of PLA_2 isoforms in AD brain is not fully understood. However, many studies have indicated that Aβ, which accumulates in AD, may activate $cPLA_2$ (Desbène et al., 2012; Sun et al., 2012; Fonteh et al., 2013). Lysophospholipid, the other product of $cPLA_2$-catalyzed reaction, is transformed into platelet-activating factor, a lipid mediator, which also promotes and supports neuroinflammation (Farooqui et al., 2000). Finally, ceramide, a metabolite of sphingolipid metabolism, which accumulates in AD brain (Han et al., 2002), has been reported to stimulate isoforms of PLA$_2$ (Farooqui, 2010b). Under pathological conditions (such as AD), PlsEtn-PLA$_2$ may be the first PLA$_2$ isoform that initiates neural injury by altering neural membrane permeability due to the loss of plasmalogens allowing slow Ca^{2+} influx. Low levels of Ca^{2+} in the presence of ceramide may facilitate translocation of cPLA$_2$ from the cytosol to neural membranes and its activation resulting in the hydrolysis of neural membrane PtdCho. As concentration of Ca^{2+} reaches millimolar levels, secretory phospholipase A$_2$ (sPLA$_2$) may be activated promoting neural cell injury and death. Thus during neural injury process sequence, PlsEtn-PLA$_2$ is situated at the proximal end, cPLA$_2$ in the middle, and sPLA$_2$ at the distal end (Farooqui, 2010b). Studies on analysis of the ENT from patients with AD indicate that the levels of MaR1, PD1, and resolvin D5 (RvD$_5$) are decreased in ENT of patients with AD as compared to age-matched controls, whereas levels of the proinflammatory prostaglandin D$_2$ (PGD$_2$) are markedly increased in AD patients. In vitro studies have also shown that lipoxin A$_4$ (LXA$_4$),

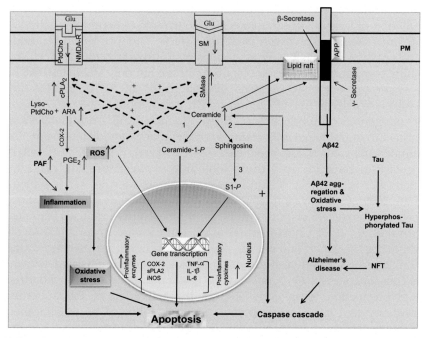

FIGURE 3.9 Interactions among phospholipid, sphingolipid, and β-amyloid metabolism. APP, amyloid precursor protein; ARA, arachidonic acid; Aβ, beta-amyloid; C1P, ceramide 1-phosphate; COX-2, cyclooxygenase-2; cPLA$_2$, cytosolic phospholipase A$_2$; Glu, glutamate; lyso-PtdCho, lyso-phosphatidylcholine; NMDA-R, N-methyl-D-aspartate receptor; PAF, platelet-activating factor; PGE$_2$, prostaglandin E$_2$; PtdCho, phosphatidylcholine; ROS, reactive oxygen species; S1P, sphingosine 1-phosphate; SM, sphingomyelin; SMase, sphingomyelinase.

MaR1, RvD_1, and protectin DX produce neuroprotective effects by blocking oxidative stress and neuroinflammation. In addition, in human microglial cells, MaR1 and RvD_1 also downregulate Aβ-mediated neuroinflammation. Furthermore, MaR1 exerts a stimulatory effect on microglial cell–mediated uptake of Aβ. These studies support the view that AD is accompanied with a disturbance in the resolution pathway (Zhu et al., 2016; Wang et al., 2015).

Unlike glycerophospholipids among sphingolipids, levels of ceramide are increased and cerebroside 3-sulfate (sulfatide) are decreased in brains from patients with AD (Han et al., 2002; Cutler et al., 2004). Increase in ceramide levels are probably caused by increased in activities of ASMase and NSMase in AD (Fig. 3.9) (He et al., 2010). Furthermore, gene expression abnormalities of enzymes associated with sphingolipid metabolism are observed in different stages of AD progression. In particular, expression of genes related to ceramide homeostasis is altered in a way that enzymes participating in the de novo synthesis of ceramide and in SM hydrolysis are increased as a function of disease progression (Katsel et al., 2007; Filippov et al., 2012). Alterations in sphingolipid metabolism during normal brain aging and in the brain of patients AD that result in accumulation of long-chain ceramides may contribute to neurotoxic action of Aβ and exacerbate progression of the disease. The increase in ceramide levels may be responsible for alterations in multiple enzymes and cell signaling components. For example, C6 ceramide not only stabilizes BACE1 (β-secretase), but also increases the biogenesis of Aβ by modulating β-secretase activity in AD. Fumonisin B1, which inhibits the biosynthesis of endogenous ceramide, reduces the production of Aβ. Addition of exogenous C6 ceramide not only restores intracellular ceramide levels but also induces Aβ production in fumonisin B1-treated cells (Puglielli et al., 2003). Increase in ceramide may cause inhibition of the Phosphatidylinositol 3-kinase/Akt pathway leading to neurodegeneration in AD (Arboleda et al., 2009). Collective evidence suggests that ceramide and Aβ produce their neurotoxic effects through the production of free radicals (Rosales-Corral et al., 2012; Chakrabarti et al., 2013). Increase in ceramide levels stimulates the generation of ROS in a dose-dependent manner, indicating a link between oxidative stress and ceramide metabolism, with detrimental consequences for neuronal survival (Cutler et al., 2004). Mitochondrial ROS production is an early step in the activation of the apoptotic SM-dependent transduction pathway. This is followed by translocation of the transcription factor nuclear factor-κB (NF-κB) from cytosol to the nucleus in differentiated PC12 cells. Complex 1 of the mitochondrial electron transport chain is the primary site of the generation of free radicals. Antioxidant treatment blocks ROS production almost completely and prevents both the NF-κB translocation and neuronal death (Darios et al., 2003). It is also becoming increasingly evident that there occurs a communication between neurons and glia related to sphingolipid metabolism. Thus deregulation of ceramide metabolism in astrocytes promotes the secretion of Aβ, as well as Tau phosphorylation, in neurons (Patil et al., 2006). These observations support the view that interactions between ceramide and Aβ may promote neuronal death in AD. Based on microarray analysis, it is proposed that there is upregulation of gene expression of the enzymes associated with the de novo synthesis of ceramide and downregulation of the enzymes involved in glycosphingolipid synthesis in early AD progression (Katsel et al., 2007).

APOE is involved in sulfatide transport and contributes to sulfatide trafficking and homeostasis in the brain through lipoprotein metabolism pathways (Han, 2007). The depletion of sulfatides is not only tightly associated with Aβ pathology in AD (Han, 2010) but also linked with the white matter abnormality in AD (Xie et al., 2006; Zhang et al., 2009). The molecular mechanism of sulfatide-mediated changes in Aβ pathology in AD is not fully understood. However, lipidation status of APOE modulates the metabolism of Aβ peptide accumulation and deposition in the neural parenchyma and cerebrovasculature of patients with AD. It should be noted that APOE not only inhibits the transport of Aβ from blood to the brain through the BBB but also facilitates the proteolytic degradation of Aβ by neprilysin and insulin-degrading enzyme, which are enhanced when APOE is lipidated (Martins et al., 2009). It is also reported that reduction in S1P levels in the AD brain, together with elevated ceramide, and decrease in sulfatide along with changes in Apo E metabolism may contribute to the pathogenesis of AD (Bu, 2009).

Alterations in ganglioside levels have also been reported to occur in AD. For example, ganglioside levels are reduced by 58%–70% of that of control brains in gray matter, and to 81% in frontal white matter, in early-onset or familial AD cases. Other studies have also indicated that levels of gangliosides are also significantly reduced in the temporal cortex, hippocampus, and frontal white matter in late-onset cases (Svennerholm and Gottfries, 1994; Kalanj et al., 1991). Several studies have indicated that Aβ specifically binds to GM1 that occurs in clusters, but not when it is uniformly distributed. The mechanism of the abnormal aggregation of Aβ in vivo is not fully understood. However, it is proposed that ganglioside clusters in lipid rafts mediate the formation of Aβ fibrils. The toxicity and physicochemical properties of Aβ fibrils are different in the presence of cholesterol than in amyloids formed in solution. Detailed investigations have indicated that at lower protein densities in the membrane (Aβ:GM1 ratio of less than ~0.013), only the helical species are formed. At intermediate protein densities (Aβ:GM1 ratio between ~0.013 and ~0.044), the helical species and aggregated β-sheets (~15-mer) are generated. However, the β-structure is stable and does not form larger aggregates. At Aβ:GM1 ratios above ~0.044, the β-structure is converted to a second, seed-prone β-structure. These seeds recruit monomers from the aqueous phase to form amyloid fibrils. These results may shed some light on a molecular mechanism for the pathogenesis of AD (Matsuzaki et al., 2010, 2014). In contrast, amyloid fibrils formed in aqueous solution are less toxic and have parallel β-sheets. The less polar environments of GM1 clusters play an important role in the formation of these toxic fibrils. Membranes that contain GM1 clusters not only accelerate the aggregation of Aβ by locally concentrating Aβ molecules but also generate amyloid fibrils with unique structures and significant cytotoxicity (Matsuzaki et al., 2010). Because GM1 possesses neurotrophic properties, administration of GM1 may exert beneficial effects in AD. Additionally, GM1 infusions may be useful for the sequestration of excess Aβ in patients with AD to alleviate their toxic effects (Ariga et al., 2008).

It is well known that brain is the richest source of cholesterol in the body. Two pools of cholesterol are present in the brain. One pool, which accounts for ~70% of the total cholesterol, is metabolically stable. This pool is present in myelin membranes of white matter (Davison, 1965). The second pool, which represents ~30% of total cholesterol, is associated with the plasma and subcellular membranes of neurons and glial cells of gray matter. This is metabolically active and contribute to the formation of "lipid rafts," which are not only found in the plasma membrane and Golgi apparatus but also in endosomal membranes. As stated earlier, lipid rafts are involved in protein sorting, signaling complex formation, and the initiation of signal transduction pathways. The roles of cholesterol and oxysterols in AD were largely unknown until the discovery of APOE as the primary genetic risk factor for AD (Hughes et al., 2013). APOE transports cholesterol within the brain; however, its direct involvement in AD pathogenesis is still unknown. Excessive levels of cholesterol may enhance amyloidogenesis and Aβ production in the brain in patients with AD. In vitro studies have shown that plasma membranes containing high levels of cholesterol are more likely to lead to a primary cleavage of the C99 terminal of the Aβ protein precursor (AβPP) by γ-secretase to release Aβ polypeptides (Fassbender et al., 2002). Barrett and his colleagues have reported identification of a cholesterol-binding site on C99 terminal of AβPP (Barrett et al., 2012). While the functional consequences of cholesterol binding to AβPP remains unknown, this mechanism may provide further insight into the role of cholesterol in amyloidogenesis (Barrett et al., 2012). The deposition and aggregation of Aβ as neuritic plaques is believed to lead to widespread loss of neurons and synapses (Brown and Jessup, 2009). It is also possible that degeneration of neurons and synapses in AD may destroy myelin, releasing excess cholesterol and other myelin constituents into the extracellular space (Brown and Jessup, 2009). This process may promote more aggregation of Aβ.

Impaired cholesterol metabolism may also contribute to Tau hyperphosphorylation (Michikawa, 2006; Maccioni et al., 2010). Emerging evidence suggests that bridging integrator 1 (BIN1) modulates Tau pathology in addition to Aβ (Zhou et al., 2014). BIN colocalizes and interacts with Tau (Chapuis et al., 2013; Zhou et al., 2014). It is proposed that BIN1 levels may correlate with number of neurofibrillary tangles (NFTs) in AD (Glennon et al., 2013; Holler et al., 2014). In addition to their effects on Aβ metabolism, statins suppress Tau hyperphosphorylation induced by excess cholesterol in the brain and also reduce NFTs in a Tau pathology model (Boimel et al., 2009), indicating that the regulation and dysregulation of cholesterol metabolism affect Tau pathology in the brain. As stated in Chapter 2, cholesterol-enriched diet increases the levels of cholesterol-derived lipid mediators such as oxysterols (24S-hydroxycholesterol; 25-hydroxycholesterol; 27-hydroxycholesterol; keto, hydroperoxy, epoxy, and cholesterol oxides; and cholesterol esters) (Mast et al., 2003; Olkkonen et al., 2012). Oxysterols have been reported to exert several in vitro and in vivo biochemical activities of both physiologic and pathologic relevance and they have been implicated in the pathogenesis of various age-related chronic diseases, including atherosclerosis and AD, where hypercholesterolemia represents a primary risk factor along with impairment of redox state and induction of neuroinflammation.

Free cholesterol cannot cross the BBB. However, some cholesterol oxidation products (27-hydroxycholesterol and 24S-hydroxycholesterol) (Björkhem et al., 2009) can diffuse into the brain. These cholesterol metabolites not only exhibit strong proapoptotic and proinflammatory effects (Gamba et al., 2014) but also affect the renin–angiotensin system (Mateos et al., 2009; Mateos et al., 2011). This system in addition to its roles in salt and water homeostasis and the regulation of blood pressure regulates multiple brain functions such as learning and memory, processing of sensory information, and regulation of emotional responses (von Bohlen und Halbach and Albrecht, 2006). Significant increase in 27-hydroxycholesterol and 24S-hydroxycholesterol levels in the frontal cortex and cerebrospinal fluid (CSF) samples from patients with AD (Gamba et al., 2014, 2015). Studies in rodent model of AD indicate that cholesterol and 27-hydroxycholesterol upregulate the production of Aβ and Tau protein phosphorylation (Fig. 3.10). Similarly, treatment of hippocampal organotypic slices with leptin retards the 27-hydroxycholesterol-mediated increase in Aβ and phosphorylated Tau protein by decreasing the levels of β-secretase (BACE1; also called Asp2 and memapsin2) and glycogen synthase kinase-3β (GSK-3β), respectively (Marwarha et al., 2010; Dasari et al., 2010; Mateos et al., 2009). In human neuroblastoma SH-SY5Y cells, 27-hydroxycholesterol increases BACE1 and Aβ levels (Marwarha et al., 2013). This increase in BACE1 involves a crosstalk between the two transcription factors NF-κB and the endoplasmic reticulum stress marker, the growth arrest and DNA damage–induced gene-153 (gadd153, also called CHOP). Treatment with 27-hydroxycholesterol stimulates the binding of NF-κB

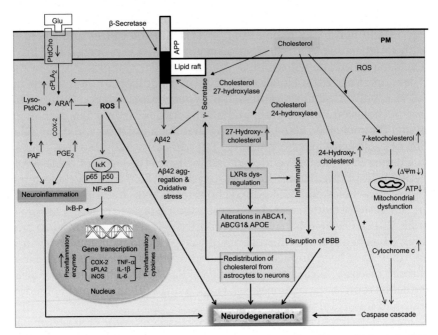

FIGURE 3.10 **Interactions among phospholipid, cholesterol, and β-amyloid metabolism.** *APP*, amyloid precursor protein; *ARA*, arachidonic acid; *Aβ*, β-amyloid; *COX-2*, cyclooxygenase-2; *cPLA₂*, cytosolic phospholipase A₂; *Glu*, glutamate; *IL-1β*, interleukin-1β; *IL-6*, interleukin-6; *iNOS*, inducible nitric oxide synthase; *lyso-PtdCho*, lyso-phosphatidylcholine; *NMDA-R*, N-methyl-ᴅ-aspartate receptor; *PAF*, platelet-activating factor; *PGE₂*, prostaglandin E₂; *PtdCho*, phosphatidylcholine; *ROS*, reactive oxygen species; *sPLA₂*, secretory phospholipase A₂; *TNF-α*, tumor necrosis factor-α.

with the BACE1 promoter and subsequent increases in BACE1 transcription and Aβ production. The NF-κB inhibitor, sc514, significantly blocks the 27-hydroxycholesterol-coupled increase in NF-κB-mediated BACE1 expression and Aβ genesis (Marwarha et al., 2013). Similarly, 27-hydroxycholesterol also increases the expression of gadd153. Silencing gadd153 expression with small interfering RNA retards the 27-hydroxycholesterol-mediated increase in NF-κB activation. Based on detailed investigation, it is proposed that gadd153 and NF-κB work in concert to regulate BACE1 expression and production of Aβ (Marwarha et al., 2013). It is interesting to note that in AD transgenic mice, leptin, a hormone mainly synthesized by adipocytes, produces beneficial effects. Leptin circulates in the plasma in proportion to fat mass. It acts as a regulator of body energy expenditure and food intake (Elmquist et al., 1998). Circulating leptin is taken into the brain, across the BBB, where it mainly acts on the hypothalamus. However, leptin receptors are widely expressed in other brains regions including hippocampus. Accumulating evidence from cell and animal models indicates its positive effects on memory processes and neuroprotection

(Guo et al., 2008). In addition, leptin also participates in the immune functions, including inflammation (Paz-Filho et al., 2013), and plays a major role in the chronic proinflammatory state that is seen in obesity and atherosclerosis (Conde et al., 2011). The importance of leptin in AD is supported by several studies. In vitro, leptin treatment reduces Aβ levels in neurons by inhibiting β-secretase activity and reducing Tau phosphorylation through the modulation of Tau kinases (Greco et al., 2009). Leptin treatment in murine models of AD also results in significant reductions of Aβ and phosphorylated Tau (p-tau) and cognitive deficits (Greco et al., 2009, 2010; Maioli et al., 2015).

27-Hydroxycholesterol also causes glutathione depletion, ROS generation, inflammation, and apoptosis (Dasari et al., 2010). Similarly, cell culture and animal studies have indicated that 24S-hydroxycholesterol may also contribute to amyloidogenesis by increasing sAPP production (Lukiw, 2006). Lowering cholesterol levels has been shown to reduce BACE1 activity and Aβ production (Ehehalt et al., 2003; Xiong et al., 2008). Furthermore, other experimental studies revealed an influence of the Aβ levels on de novo synthesis of cholesterol and its concentrations (Grimm et al., 2005; Canepa et al., 2011).

Cholesterol 24-hydroxylase catalyzes the synthesis of 24S-hydroxycholesterol (Lund et al., 1999). This enzyme has been identified in specific regions of neurons (endoplasmic reticulum, cell body, dendrites) and brain areas (e.g., hippocampal CA1 pyramidal cells, hippocampal and cerebellar interneurons) (Ramirez et al., 2008). There are reports that 24S-hydrocholesterol levels are also elevated in CSF and plasma of patients with AD as compared with control subjects (Papassotiropoulos et al., 2002; Popp et al., 2012, 2013). These observations support the view that alterations in cholesterol and hydroxycholesterols may be closely associated with pathophysiology of kainic acid neurotoxicity (Ong et al., 2010). Like 27-hydroxycholesterol, 24S-hydroxycholesterol may also promote alterations in inflammatory signaling, apoptotic genetic responses, and AD-type change. In clinical studies, cholesterol-lowering statins, nonsteroidal antiinflammatory drugs, cholesterol absorption/transport inhibitors, and related modulators of cholesterol trafficking have demonstrated some pharmacological benefit for the treatment of AD. In neural cells, the metabolism of glycerophospholipid, sphingolipid, and cholesterol is interrelated and interconnected. Neurotransmitters, cytokines, and growth factors modulate more than one enzyme of phospholipid, sphingolipid, and cholesterol metabolism at the same time. This process produces complexity in regulatory processes associated with glycerophospholipid, sphingolipid, and cholesterol metabolism. Under normal physiological conditions, homeostasis in neurotransmitters, cytokines, and growth factor–mediated glycerophospholipid, sphingolipid, and cholesterol metabolism is based not only on optimal levels of lipid mediators and organization of signaling network but also on the complexity and interconnectedness of their metabolism. In contrast, under pathological condition such as AD marked enhancement in levels of glycerophospholipid-, sphingolipid-, and cholesterol-derived lipid mediators disturbs the signaling networks, resulting in loss of communication among neurons, astrocytes, oligodendrocytes, and microglial cells. This process threatens the integrity of neural cell lipid homeostasis among the aforementioned neural cells leading to neurodegeneration through intense interactions among glycerophospholipid-, sphingolipid-, and cholesterol-derived lipid mediators (Farooqui, 2011; Farooqui and Horrocks, 2006, 2007; Farooqui, 2011).

As stated in Chapter 2 aging is an important risk factor for AD. Other risk factors are associated with vascular disease, including hypercholesterolemia, hypertension, atherosclerosis, coronary heart disease, smoking, and diabetes. Converging evidence suggests that both aging and AD are accompanied by the generation of high levels of several stress factors such as increase in glycerophospholipid-, sphingolipid-, and cholesterol-derived lipid mediators; activation of isoforms of PLA_2s, SMases, and cholesterol hydroxylases; and induction of oxidative stress and neuroinflammation (Farooqui, 2011). The difference between neurochemical changes in normal aging and AD is the intensity and levels of glycerophospholipid-, sphingolipid-, and cholesterol-derived lipid mediators. Levels of these mediators are higher in the AD brain than normal-aged human brain (Farooqui, 2011). AD is the most common form of age-related dementia, which is characterized not only by the onset of oxidative stress and neuroinflammation, mitochondrial dysfunction, and accumulation of Aβ peptides in the brain but also by the deposition of NFTs. These processes may be responsible not only for neuronal death in neural cell culture in vitro but also for cognitive dysfunction in patients with AD. In addition, due to the involvement of receptors and ion channels, metabolism of neural membrane components and generation of lipid mediators may not only modulate the formation of Aβ but also contribute to neurodegeneration in AD (Farooqui et al., 2007a, 2007b; Farooqui, 2011; Lukiw, 2013). In addition, enhanced metabolism of neural membrane glycerophospholipid, sphingolipid, and cholesterol may alter the physicochemical properties of neural membranes; fluidity and permeability causes dysregulation of ion homeostasis in AD brain. All these changes may also contribute to the pathogenesis of AD (Yahi et al., 2010).

CONCLUSION

Presence of various combinations of polar head groups, fatty acyl chains, and backbone structures makes neural membrane glycerophospholipids, sphingolipids, and cholesterol and their lipid mediators to perform receptor-mediated diverse functions. Phospholipid-derived polyunsaturated fatty acids (DHA and ARA) are precursors for lipid mediators, such as docosanoids (RVs, PDs, NPD, and MaRs) and eicosanoids (PGs, LTs, TXs, and LXs). Generation of these fatty acids may markedly influence neural membrane properties such as fluidity, flexibility, elasticity, permeability, and gene expression, but modulate onset and resolution of neuroinflammation through the production of lipid mediators in AD. Enhanced levels of ceramides directly increase Aβ levels through stabilization of β-secretase, the key enzyme in the amyloidogenic processing of Aβ precursor protein (APP). It is also shown that increase in production of Aβ not only may increase the levels of ceramide by activating SMases through a positive feedback loop but also may induce depolarization and permeabilization of mitochondria, increase in ROS production, release of cytochrome c, depletion of Bcl-2, and activation of caspase-3 activation, along the pathways related to Akt/PKB kinase and MAPKs.

Ceramide and ganglioside are complex classes of signaling molecules, which are essential part of neural cell membranes. Their cellular levels regulate proliferation, apoptosis, and inflammation depending on the specific sphingolipid species, cell and receptor type, and different intracellular targets. Ceramides are the backbone of more complex sphingolipids and the precursor of the versatile signaling molecules such as C1P and S1P. Gangliosides are sialylated glycosphingolipids composed of a hydrophobic ceramide base anchored within the cellular membrane and a carbohydrate moiety that extends into the extracellular space. Levels of ceramide and ganglioside are markedly increased in AD. Production of high levels of ceramide from SM in AD may be responsible for ceramide-mediated apoptotic cell death. Finally, increase in 27-hydroxycholesterol and 24-hydroxycholesterol from cholesterol may contribute to oxidative stress and neuroinflammation in AD. Oxidative stress and neuroinflammation are interlinked processes, which play an important role in the pathophysiology of AD and other neurological disorders. However, it is difficult to establish the temporal sequence in the relationship between oxidative stress and neuroinflammation. It is interesting to note that many proinflammatory transcription factors [NF-κB and activator protein 1 (AP-1)] are redox sensitive; therefore, ROS may stimulate NF-κB and trigger the release of inflammatory cytokines, which in turn enhance ROS production through the stimulation of cPLA$_2$ and generation of ROS, thus establishing a vicious circle. In addition, the chronic and sustained nature of neuroinflammation produces alterations in BBB permeability leading to increase in infiltration of peripheral macrophages into the brain parenchyma to further perpetuate the neuroinflammation. Collective evidence suggests that neural membrane glycerophospholipids, sphingolipids, and cholesterol lipids may modulate the biogenesis of Aβ from APP. In particular, increase in plasma membrane cholesterol augments Aβ generation and shows a strong positive correlation with AD progression. Furthermore,

APOE, which transports cholesterol in the CSF and interacts with Aβ or competes with it for the lipoprotein receptor binding, significantly influences Aβ clearance in an isoform-specific manner and is the major genetic risk factor for the onset of AD. Aβ is an amphiphilic peptide that interacts with various lipids, proteins, and their assemblies, which can lead to variation in Aβ aggregation in vitro and in vivo. Upon interaction with the lipid raft components, such as cholesterol, gangliosides, and glycerophospholipids, Aβ can aggregate on the neural membrane and thereby disrupt it. This leads to perturbed cellular calcium homeostasis, supporting the view that Aβ-lipid interactions at the neural cell membrane probably trigger the neurotoxic cascade in AD.

References

Aleshin, S., Grabeklis, S., Hanck, T., Sergeeva, M., Reiser, G., 2009. Peroxisome proliferator-activated receptor (PPAR)-γ positively controls and PPARα negatively controls cyclooxygenase-2 expression in rat brain astrocytes through a convergence on PPARβ/δ via mutual control of PPAR expression levels. Mol. Pharmacol. 76, 414–424.

Alvarez, S.E., Milstien, S., Spiegel, S., 2007. Autocrine and paracrine roles of sphingosine-1-phosphate. Trends Endocrinol. Metab. 18, 300–307.

Alvarez, S.E., Harikumar, K.B., Hait, N.C., Allegood, J., Strub, G.M., Kim, E.Y., et al., 2010. Sphingosine-1-phosphate is a missing cofactor for the E3 ubiquitin ligase TRAF2. Nature 465, 1084–1088.

Arana, L., Gangoiti, P., Ouro, A., Trueba, M., Gómez-Muñoz, A., 2010. Lipids Health Dis. 9, 1–12.

Arboleda, G., Morales, L.C., Benítez, B., Arboleda, H., 2009. Regulation of ceramide-induced neuronal death: cell metabolism meets neurodegeneration. Brain Res. Rev. 59, 333–346.

Ariga, T., McDonald, M.P., Yu, R.K., 2008. Thematic Review Series: Sphingolipids. Role of ganglioside metabolism in the pathogenesis of Alzheimer's disease—a review. J. Lipid Res. 49, 1157–1175.

Arita, M., Ohira, T., Sun, Y.P., Elangovan, S., Chiang, N., Serhan, C.N., 2007. Resolvin E1 selectively interacts with leukotriene B4 receptor BLT1 and ChemR23 to regulate inflammation. J. Immunol. 178, 3912–3917.

Barrett, P., Song, Y., Van Horn, W., Hustedt, E., Schafer, J., Hadziselimovic, A., et al., 2012. The amyloid precursor protein has a flexible transmembrane domain and binds cholesterol. Science 336, 1168–1171.

Bazan, N.G., 2005. Lipid signaling in neural plasticity, brain repair, and neuroprotection. Mol. Neurobiol. 32, 89–103.

Björkhem, I., Cedazo-Minguez, A., Leoni, V., Meaney, S., 2009. Oxysterols and neurodegenerative diseases. Mol. Asp. Med. 30, 171–179.

Boimel, M., Grigoriadis, N., Lourbopoulos, A., Touloumi, O., Rosenmann, D., Abramsky, O., et al., 2009. Statins reduce the neurofibrillary tangle burden in a mouse model of tauopathy. J. Neuropathol. Exp. Neurol. 68, 314–325.

Boon, J.M., Smith, B.D., 2002. Chemical control of phospholipid distribution across bilayer membranes. Med. Res. Rev. 22, 251–281.

Bordoni, A., Di Nunzio, M., Danesi, F., Biagi, P.L., 2006. Polyunsaturated fatty acids: from diet to binding to PPARs and other nuclear receptors. Genes Nutr. 1, 95–106.

Brown, A., Jessup, W., 2009. Oxysterols: sources, cellular storage and metabolism, and new insights into their roles in cholesterol hoeostasis. Mol. Asp. Med. 30, 111–122.

Brozinick, J.T., Hawkins, E., Hoang Bui, H., Kuo, M.S., Tan, B., et al., 2013. Plasma sphingo-lipids are biomarkers of metabolic syndrome in non-human primates maintained on a Western-style diet. Int. J. Obes. (Lond.) 37, 1064–1070.

Bu, G., 2009. Apolipoprotein E and its receptors in Alzheimer's disease: pathways, pathogen-esis and therapy. Nat. Rev. Neurosci. 10, 333–344.

Calder, P.C., 2011. Fatty acids and inflammation: the cutting edge between food and pharma. Eur. J. Pharmacol. 668 (Suppl. 1), S50–S58.

Canals, D., Roddy, P., Hannun, Y.A., 2012. Protein phosphatase 1α mediates ceramide-induced ERM protein dephosphorylation: a novel mechanism independent of phospha-tidylinositol 4,5-biphosphate (PIP2) and myosin/ERM phosphatase. J. Biol. Chem. 287, 10145–10155.

Canepa, E., Borghi, R., Vina, J., Traverso, N., Gambini, J., Domenicotti, C., et al., 2011. Cholesterol and amyloid-β: evidence for a cross-talk between astrocytes and neuronal cells. J. Alzheimers Dis. 2011 (25), 645–653.

Chakrabarti, S., Sinha, M., Thakurta, I.G., Banerjee, P., Chattopadhyay, M., 2013. Oxidative stress and amyloid β toxicity in Alzheimer's disease: intervention in a complex relation-ship by antioxidants. Curr. Med. Chem. 20, 4648–4664.

Chan, R.B., Oliveira, T.G., Cortes, E.P., Honig, L.S., Duff, K.E., Small, S.A., et al., 2012. Comparative lipidomic analysis of mouse and human brain with Alzheimer disease. J. Biol. Chem. 287, 2678–2688.

Chapuis, J., Hansmannel, F., Gistelinck, M., Mounier, A., Van Cauwenberghe, C., Kolen, K.V., et al., 2013. Increased expression of BIN1 mediates Alzheimer genetic risk by modulating tau pathology. Mol. Psychiatry 18, 1225–1234.

Cimini, A., Benedetti, E., Cristiano, L., Sebastiani, P., D'Amico, M.A., D'Angelo, B., et al., 2005. Expression of peroxisome proliferator-activated receptors (PPARs) and retinoic acid receptors (RXRs) in rat cortical neurons. Neuroscience 130, 325–337.

Conde, J., Scotece, M., Gomez, R., Lopez, V., Gomez-Reino, J.J., Lago, F., et al., 2011. Adipokines: biofactors from white adipose tissue. A complex hub among inflammation, metabolism, and immunity. BioFactors 37, 413–420.

Contreras, F.X., Sánchez-Magraner, L., Alonso, A., Goñi, F.M., 2010. Transbilayer (flip-flop) lipid motion and lipid scrambling in membranes. FEBS Lett. 584, 1779–1786.

Cristiano, L., Cimini, A., Moreno, S., Ragnelli, A.M., Paola, C.M., 2005. Peroxisome prolifera-tor-activated receptors (PPARs) and related transcription factors in differentiating astro-cyte cultures. Neuroscience 131, 577–587.

Cutler, R.G., Kelly, J., Storie, K., Pedersen, W.A., Tammara, A., Hatanpaa, K., et al., 2004. Involvement of oxidative stress-induced abnormalities in ceramide and cholesterol metabolism in brain aging and Alzheimer's disease. Proc. Natl. Acad. Sci. U. S. A. 17, 2070–2075.

Cyster, J.G., Schwab, S.R., 2012. Sphingosine-1-phosphate and lymphocyte egress from lym-phoid organs. Annu. Rev. Immunol. 30, 69–94.

Daleke, D.L., 2003. Regulation of transbilayer plasma membrane phospholipid asymmetry. J. Lipid Res. 44, 233–242.

Darios, F., Lambeng, N., Troadec, J.-D., Michel, P.P., Ruberg, M., 2003. Ceramide increases mitochondrial free calcium levels via caspase 8 and Bid: role in initiation of cell death. J. Neurochem. 84, 643–654.

Dasari, B., Prasanthi, J.R., Marwarha, G., Singh, B.B., Ghribi, O., 2010. The oxysterol 27-hydroxycholesterol increases β-amyloid and oxidative stress in retinal pigment epi-thelial cells. BMC 10, 22.

Davison, A.N., 1965. Brain sterol metabolism. Adv. Lipid Res. 3, 171–196.

Desbène, C., Malaplate-Armand, C., Youssef, I., Garcia, P., Stenger, C., Sauvée, M., et al., 2012. Critical role of cPLA2 in Aβ oligomer-induced neurodegeneration and memory deficit. Neurobiol. Aging 33, 1123.e17–1123.e29.

Di Paolo, G., Kim, T.-W., 2011. Linking lipids to Alzheimer's disease: cholesterol and beyond. Nat. Rev. Neurosci. 12, 284–296.

Dietschy, J.M., 2009. Central nervous system: cholesterol turnover, brain development and neurodegeneration. Biol. Chem. 390, 287–293.

Dietschy, J.M., Turley, S.D., 2004. Thematic review series: brain lipids. Cholesterol metabolism in the central nervous system during early development and in the mature animal. J. Lipid Res. 45, 1375–1397.

Dyall, S.C., Michael, G.J., Michael-Titus, A.T., 2010. Omega-3 fatty acids reverse age-related decreases in nuclear receptors and increase neurogenesis in old rats. J. Neurosci. Res. 88, 2091–2102.

Edidin, 2003. Lipids on the frontier: a century of cell-membrane bilayers. Nat. Rev. Mol. Cell Biol. 4, 414–418.

Ehehalt, R., Keller, P., Haass, C., Thiele, C., Simons, K., 2003. Amyloidogenic processing of the Alzheimer β-amyloid precursor protein depends on lipid rafts. J. Cell Biol. 160, 113–123.

El Alwani, M., Wu, B.X.J., Obeid, L.M., Hannun, Y.A., 2006. Bioactive sphingolipids in the modulation of the inflammatory response. Pharmacol. Ther. 112, 171–183.

Elmquist, J.K., Maratos-Flier, E., Saper, C.B., Flier, J.S., 1998. Unraveling the central nervous system pathways underlying responses to leptin. Nat. Neurosci. 1, 445–450.

Esterbauer, H., Schaur, R.J., Zollner, H., 1991. Chemistry and biochemistry of 4-hydroxynonenal, malonaldehyde and related aldehydes. Free Radic. Biol. Med. 11, 81–128.

Fadok, V.A., de Cathelineau, A., Daleke, D.L., Henson, P.M., Bratton, D.L., 2001. Loss of phospholipid asymmetry and surface exposure of phosphatidylserine is required for phagocytosis of apoptotic cells by macrophages and fibroblasts. J. Biol. Chem. 276, 1071–1077.

Fantini, J., Yahi, N., 2010. Molecular insights into amyloid regulation by membrane cholesterol and sphingolipids: common mechanisms in neurodegenerative diseases. Exp. Rev. Mol. Med. 12, e27.

Farooqui, A.A., 2013. Metabolic Syndrome: An Important Risk Factor for Stroke, Alzheimer Disease, and Depression. Springer, New York.

Farooqui, A.A., 2010a. Neurochemical Aspects of Neurotraumatic and Neurodegenerative Diseases. Springer, New York.

Farooqui, A.A., 2010b. Studies on plasmalogen-selective phospholipase A_2 in brain. Mol. Neurobiol. 41, 267–273.

Farooqui, A.A., 2011. Lipid Mediators and Their Metabolism in the Brain. Springer, New York.

Farooqui, A.A., 2012. n-3 Fatty acid-derived lipid mediators in the brain; new weapons against oxidative stress and inflammation. Curr. Med. Chem. 19, 532–543.

Farooqui, A.A., Horrocks, L.A., 2006. Phospholipase A_2-generated lipid mediators in the brain: the good, the bad, and the ugly. Neuroscientist 12, 245–260.

Farooqui, A.A., Horrocks, L.A., 2007. Glycerophospholipids in Brain. Springer, New York.

Farooqui, A.A., Rapoport, S.I., Horrocks, L.A., 1997. Membrane phospholipid alterations in Alzheimer's disease: deficiency of ethanolamine plasmalogens. Neurochem. Res. 22, 523–527.

Farooqui, A.A., Horrocks, L.A., Farooqui, T., 2000. Glycerophospholipids in brain: their metabolism, incorporation into membranes, functions, and involvement in neurological disorders. Chem. Phys. Lipids 106, 1–29.

Farooqui, A.A., Horrocks, L.A., Farooqui, T., 2007a. Modulation of inflammation in brain: a matter of fat. J. Neurochem. 101, 577–599.

Farooqui, A.A., Horrocks, L.A., Farooqui, T., 2007b. Interactions between neural membrane glycerophospholipid and sphingolipid mediators: a recipe for neural cell survival or suicide. J. Neurosci. Res. 85, 1834–1850.

Farooqui, A.A., Ong, W.Y., Farooqui, T., 2010. Lipid mediators in the nucleus: their potential contribution to Alzheimer's disease. Biochim. Biophys. Acta 1801, 906–916.

Farooqui, A.A., Farooqui, T., Panza, F., Frisardi, V., 2012. Metabolic syndrome as a risk factor for neurological disorders. Cell. Mol. Life Sci. 69, 741–762.

Fassbender, K., Stoick, M., Bertsch, T., Ragoschke, A., Kuehl, S., Walter, S., et al., 2002. Effects of statins on cerebral cholesterol metabolism and secretion of Alzheimer amyloid peptide. Neurology 59, 1257–1258.

Filippov, V., Song, M.A., Zhang, K., Vinters, H.V., Tung, S., et al., 2012. Increased ceramide in brains with Alzheimer's and other neurodegenerative diseases. J. Alzheimers Dis. 29, 537–547.

Fonteh, A.N., Chiang, J., Cipolla, M., Hale, J., Diallo, F., Chirino, A., Arakaki, X., Harrington, M.G., 2013. Alterations in cerebrospinal fluid glycerophospholipids and phospholipase A_2 activity in Alzheimer's disease. J. Lipid Res. 54, 2884–2897.

Fox, T.E., Houck, K.L., O'Neill, S.M., Nagarajan, M., Stover, T.C., Pomianowski, P.T., et al., 2007. Ceramide recruits and activates protein kinase C ζ (PKC ζ) within structured membrane microdomains. J. Biol. Chem. 282, 12450–12457.

Futerman, A.H., Riezman, H., 2005. The ins and outs of sphingolipid synthesis. Trends Cell Biol. 15, 312–318.

Gamba, P., Guglielmotto, M., Testa, G., Monteleone, D., Zerbinati, C., Gargiulo, S., 2014. Up-regulation of β-amyloidogenesis in neuron-like human cells by both 24- and 27-hydroxycholesterol: protective effect of N-acetyl-cysteine. Aging Cell 13, 561–572.

Gamba, P., Testa, G., Gargiulo, S., Staurenghi, E., Poli, G., Leonarduzzi, G., 2015. Oxidized cholesterol as the driving force behind the development of Alzheimer's disease. Front. Aging Neurosci. 7, 119.

Glennon, E.B., Whitehouse, I.J., Miners, J.S., Kehoe, P.G., Love, S., Kellett, K.A., et al., 2013. BIN1 is decreased in sporadic but not familial Alzheimer's disease or in aging. PLoS One 8, e78806.

Gómez-Muñoz, A., 2006. Ceramide 1-phosphate/ceramide, a switch between life and death. Biochim. Biophys. Acta 1758, 2049–2056.

Greco, S.J., Sarkar, S., Casadesus, G., Zhu, X., Smith, M.A., Ashford, J.W., et al., 2009. Leptin inhibits glycogen synthase kinase-3β to prevent tau phosphorylation in neuronal cells. Neurosci. Lett. 455, 191–194.

Greco, S.J., Bryan, K.J., Sarkar, S., Zhu, X., Smith, M.A., Ashford, J.W., Johnston, J.M., Tezapsidis, N., Casadesus, G., 2010. Leptin reduces pathology and improves memory in a transgenic mouse model of Alzheimer's disease. J. Alzheimers Dis. 19, 1155–1167.

Green, J.A., Suzuki, K., Cho, B., Willison, L.D., Palmer, D., Allen, C.D., et al., 2011. The sphingosine 1-phosphate receptor $S1P_2$ maintains the homeostasis of germinal center B cells and promotes niche confinement. Nat. Immunol. 12, 672–680.

Grimm, M.O., Grimm, H.S., Patzold, A.J., et al., 2005. Regulation of cholesterol and sphingomyelin metabolism by amyloid-β and presenilin. Nat. Cell Biol. 7, 1118–1123.

Guan, Z., Wang, Y., Cairns, N.J., Lantos, P.L., Dallner, G., Sindelar, P.J., 1999. Decrease and structural modifications of phosphatidylethanolamine plasmalogen in the brain with Alzheimer disease. J. Neuropathol. Exp. Neurol. 58, 740–747.

Guo, Z., Jang, M.H., Otani, K., Bai, Z., Umemoto, E., Matsumoto, M., et al., 2008. $CD4^+CD25^+$ regulatory T cells in the small intestinal lamina propria show an effector/memory phenotype. Int. Immunol. 20, 307–315.

Hait, N.C., Allegood, J., Maceyka, M., Strub, G.M., Harikumar, K.B., Singh, S.K., et al., 2009. Regulation of histone acetylation in the nucleus by sphingosine-1-phosphate. Science 325, 1254–1257.

Han, X., 2007. Potential mechanisms contributing to sulfatide depletion at the earliest clinically recognizable stage of Alzheimer's disease: a tale of shotgun lipidomics. J. Neurochem. 103 (Suppl. 1), 171–179.

Han, X., 2010. The pathogenic implication of abnormal interaction between apolipoprotein E isoforms, amyloid-β peptides, and sulfatides in Alzheimer's disease. Mol. Neurobiol. 41, 97–106.

Han, X., Holtzman, D.M., McKeel Jr., D.W., Kelley, J., Morris, J.C., 2002. Substantial sulfatide deficiency and ceramide elevation in very early Alzheimer's disease: potential role in disease pathogenesis. J. Neurochem. 82, 809–818.

Hanada, K., Kumagai, K., Yasuda, S., Miura, Y., Kawano, M., Fukasawa, M., Nishijima, M., 2003. Molecular machinery for non-vesicular trafficking of ceramide. Nature 426, 803–809.

Hannun, Y.A., Obeid, L.M., 2002. The ceramide-centric universe of lipid-mediated cell regulation: stress encounters of the lipid kind. J. Biol. Chem. 277, 25847–25850.

Harikumar, K.B., Yester, J.W., Surace, M.J., Oyeniran, C., Price, M.M., Huang, W.C., 2014. K63-linked polyubiquitination of transcription factor IRF1 is essential for IL-1-induced production of chemokines CXCL10 and CCL5. Nat. Immunol. 15, 231–238.

Haughey, N.J., Bandaru, V.V.R., Bae, M., Mattson, M.P., 2010. Roles for dysfunctional sphingolipid metabolism in Alzheimer's disease neuropathogenesis. Biochim. Biophys. Acta 1801, 878–886.

Haus, J.M., Kashyap, S.R., Kasumov, T., Zhang, R., Kelly, K.R., 2009. Plasma ceramides are elevated in obese subjects with type 2 diabetes and correlate with the severity of insulin resistance. Diabetes 58, 337–343.

He, X., Huang, Y., Li, B., Gong, C.X., Schuchman, E.H., 2010. Deregulation of sphingolipid metabolism in Alzheimer's disease. Neurobiol. Aging 31, 398–408.

Heras-Sandoval, D., Pedraza-Chaverri, J., Pérez-Rojas, J.M., 2016. Role of docosahexaenoic acid in the modulation of glial cells in Alzheimer's disease. J. Neuroinflammation 13, 61.

Hicks, D.A., Nalivaeva, N.N., Turner, A.J., 2012. Lipid rafts and Alzheimer's disease: protein-lipid interactions and perturbation of signaling. Front. Physiol. 3, 189.

Holler, C.J., Davis, P.R., Beckett, T.L., Platt, L., Webb, R.L., Head, E., et al., 2014. Bridging integrator 1 (BIN1) protein expression increases in the Alzheimer's disease brain and correlates with neurofibrillary tangle pathology. J. Alzheimers Dis. 42, 1221–1227.

Hollingworth, P., Harold, D., Sims, R., Gerrish, A., Lambert, J.C., et al., 2011. Common variants at ABCA7, MS4A6A/MS4A4E, EPHA1, CD33 and CD2AP are associated with Alzheimer's disease. Nat. Genet. 43, 429–435.

Hong, S., Gronert, K., Devchand, P., Moussignac, R.L., Serhan, C.N., 2003. Novel docosatrienes and 17S-resolvins generated from docosahexaenoic acid in murine brain, human blood and glial cells: autacoids in anti-inflammation. J. Biol. Chem. 278, 14677–14687.

Hong, S., Lu, Y., Yang, R., Gotlinger, K.H., Petasis, N.A., Serhan, C.N., 2007. Resolvin D1, protectin D1, and related docosahexaenoic acid-derived products: analysis via electrospray/low energy tandem mass spectrometry based on spectra and fragmentation mechanisms. J. Am. Soc. Mass Spectrom. 18, 128–144.

Hughes, T.M., Rosano, C., Evans, R.W., Kuller, L.H., 2013. Brain cholesterol metabolism, oxysterols, and dementia. J. Alzheimers Dis. 33, 891–911.

Ihlefeld, K., Claas, R.F., Koch, A., Pfeilschifter, J.M., Meyer, Z., Heringdorf, D., 2012. Evidence for a link between histone deacetylation and Ca^{2+} homoeostasis in sphingosine-1-phosphate lyase-deficient fibroblasts. Biochem. J. 447, 457–464.

Ikeda, M., Kihara, A., Igarashi, Y., 2006. Lipid asymmetry of the eukaryotic plasma membrane: functions and related enzymes. Biol. Pharm. Bull. 29, 1542–1546.

Itoh, T., Yamamoto, K., 2008. Peroxisome proliferator activated receptor γ and oxidized docosahexaenoic acids as new class of ligand. Naunyn Schmiedebergs Arch. Pharmacol. 377, 541–547.

Jazvinšćak Jembrek, M., Hof, P.R., Šimić, G., 2015. Ceramides in Alzheimer's disease: key mediators of neuronal apoptosis induced by oxidative stress and Aβ accumulation. Oxid. Med. Cell Longev. 346783.

Jenkins, R.W., Canals, D., Hannun, Y.A., 2009. Roles and regulation of secretory and lysosomal acid sphingomyelinase. Cell. Signal. 21, 836–846.

Jenkins, R.W., Idkowiak-Baldys, J., Simbari, F., Canals, D., Roddy, P., Riner, C.D., et al., 2011. A novel mechanism of lysosomal acid sphingomyelinase maturation: requirement for carboxyl-terminal proteolytic processing. J. Biol. Chem. 286, 3777–3788.

Jung, J.S., Choi, M.J., Ko, H.M., Kim, H.S., 2016. Short-chain C2 ceramide induces heme oxygenase-1 expression by upregulating AMPK and MAPK signaling pathways in rat primary astrocytes. Neurochem. Int. 94, 39–47.

Kalanj, S., Kracun, I., Rosner, H., Cosovic, C., 1991. Regional distribution of brain gangliosides in Alzheimer's disease. Neurol. Croat. 40, 269–281.

Katsel, P., Li, C., Haroutunian, V., 2007. Gene expression alterations in the sphingolipid metabolism pathways during progression of dementia and Alzheimer's disease: a shift toward ceramide accumulation at the earliest recognizable stages of Alzheimer's disease? Neurochem. Res. 32, 845–856.

Klymchenko, A.S., Kreder, R., 2014. Fluorescent probes for lipid rafts: from model membranes to living cells. Chem. Biol. 21, 97–113.

Lambert, J., Heath, S., Even, G., Campion, D., Sleegers, K., et al., 2009. Genome-wide association study identifies variants at CLU and CR1 associated with Alzheimer's disease. Nat. Genet. 41, 1094–1099.

Lopez, X., Goldfine, A.B., Holland, W.L., Gordillo, R., Scherer, P.E., 2013. Plasma ceramides are elevated in female children and adolescents with type 2 diabetes. J. Pediatr. Endocrinol. Metab. 26, 995–998.

López-Montero, I., Rodriguez, N., Cribier, S., Pohl, A., Vélez, M., Devaux, P.F., 2005. Rapid transbilayer movement of ceramides in phospholipid vesicles and in human erythrocytes. J. Biol. Chem. 280, 25811–25819.

Lucero, H.A., Robbins, P.W., 2004. Lipid rafts-protein association and the regulation of protein activity. Arch. Biochem. Biophys. 426, 208–224.

Lukiw, W.J., 2006. Cholesterol and 24S-hydroxycholesterol trafficking in Alzheimer's disease. Expert Rev. Neurother. 6, 683–693.

Lukiw, W.J., 2013. Alzheimer's disease as a disorder of the plasma membrane. Front. Physiol. 4, 24.

Lund, E.G., Guileyardo, J.M., Russell, D.W., 1999. cDNA cloning of cholesterol 24-hydroxylase, a mediator of cholesterol homeostasis in the brain. Proc. Natl. Acad. Sci. U. S. A. 96, 7238–7243.

Maccioni, R.B., Farías, G., Morales, I., Navarrete, L., 2010. The revitalized tau hypothesis on Alzheimer's disease. Arch. Med. Res. 41, 226–231.

Maceyka, M., Spiegel, S., 2014. Sphingolipid metabolites in inflammatory disease. Nature 510, 58–67.

Maioli, S., Lodeiro, M., Merino-Serrais, P., Falahati, F., Khan, W., Puerta, E., Alzheimer's Disease Neuroimaging Initiative, et al., 2015. Alterations in brain leptin signalling in spite of unchanged CSF leptin levels in Alzheimer's disease. Aging Cell 14, 122–129.

Mapstone, M., Cheema, A.K., Fiandaca, M.S., 2014. Plasma phospholipids identify antecedent memory impairment in older adults. Nat. Med. 20, 415–418.

Marchesini, N., Hannun, Y.A., 2004. Acid and neutral sphingomyelinases: roles and mechanisms of regulation. Biochem. Cell Biol. 82, 27–44.

Martins, I.J., Berger, T., Sharman, M.J., Verdile, G., Fuller, S.J., Martins, R.N., 2009. Cholesterol metabolism and transport in the pathogenesis of Alzheimer's disease. J. Neurochem. 111, 1275–1308.

Marwarha, G., Dasari, B., Prasanthi, J.R., Schommer, J., Ghribi, O., 2010. Leptin reduces the accumulation of Aβ and phosphorylated tau induced by 27-hydroxycholesterol in rabbit organotypic slices. J. Alzheimers Dis. 19, 1007–1019.

Marwarha, G., Raza, S., Prasanthi, J.R., Ghribi, O., 2013. Gadd153 and NF-κB crosstalk regulates 27-hydroxycholesterol-induced increase in BACE1 and β-amyloid production in human neuroblastoma SH-SY5Y cells. PLoS One 8, e70773.

Mast, N., Norcross, R., Andersson, U., Shou, M., Nakayama, L., Bjorkhem, I., et al., 2003. Broad substrate specificity of human cytochrome P450 46A1 which initiates cholesterol degradation in the brain. Biochemistry 42, 14284–14292.

Mateos, L., Akterin, S., Gil-Bea, F.J., Spulber, S., Rahman, A., Björkhem, I., et al., 2009. Activity-regulated cytoskeleton-associated protein in rodent brain is down-regulated by high fat diet in vivo and by 27-hydroxycholesterol in vitro. Brain Pathol. 19, 69–80.

Mateos, L., Ismail, M.A., Gil-Bea, F.J., Schüle, R., Schöls, L., Heverin, M., et al., 2011. Side chain-oxidized oxysterols regulate the brain renin-angiotensin system through a liver X receptor-dependent mechanism. J. Biol. Chem. 286, 25574–25585.

Matsuzaki, K., 2014. How do membranes initiate Alzheimer's disease? Formation of toxic amyloid fibrils by the amyloid β-protein on ganglioside clusters. Acc. Chem. Res. 47, 2397–2404.

Matsuzaki, K., Kato, K., Yanagisawa, K., 2010. Aβ polymerization through interaction with membrane gangliosides. Biochim. Biophys. Acta 1801, 868–877.

McMahon, B., Mitchell, S., Brady, H.R., Godson, C., 2001. Lipoxins: revelations on resolution. Trends Pharmacol. Sci. 22, 391–395.

Mencarelli, C., Martinez-Martinez, P., 2013. Ceramide function in the brain: when a slight tilt is enough. Cell. Mol. Life Sci. 70, 181–203.

Michikawa, M., 2006. Role of cholesterol in amyloid cascade: cholesterol-dependent modulation of tau phosphorylation and mitochondrial function. Acta Neurol. Scand. Suppl. 185, 21–26.

Muhle, C., Reichel, M., Gulbins, E., Kornhuber, J., 2013. Sphingolipids in psychiatric disorders and pain syndromes. Handb. Exp. Pharmacol. 431–456.

Naj, A.C., Jun, G., Beecham, G.W., Wang, L.S., Vardarajan, B.N., et al., 2011. Common variants at MS4A4/MS4A6E, CD2AP, CD33 and EPHA1 are associated with late-onset Alzheimer's disease. Nat. Genet. 43, 436–441.

Nieweg, K., Schaller, H., Pfrieger, F.W., 2009. Marked differences in cholesterol synthesis between neurons and glial cells from postnatal rats. J. Neurochem. 109, 125–134.

Ohira, T., Arita, M., Omori, K., Recchiuti, A., Van Dyke, T.E., Serhan, C.N., 2010. J. Biol. Chem. 285, 3451–3461.

Olkkonen, V.M., Béaslas, O., Nissilä, E., 2012. Oxysterols and their cellular effectors. Biomolecules 2, 76–103.

Ong, W.Y., Kim, J.-H., He, X., Chen, P., Farooqui, A.A., Jenner, A.M., 2010. Changes in brain cholesterol metabolome after kainate excitotoxicity. Mol. Neurobiol. 41, 299–313.

Ong, W.Y., Herr, D.R., Farooqui, T., Ling, E.A., Farooqui, A.A., 2015. Role of sphingomyelinases in neurological disorders. Expert Opin. Ther. Targets 19, 1725–1742.

Papassotiropoulos, A., Lutjohann, D., Bagli, M., Locatelli, S., Jessen, F., et al., 2002. 24S-hydroxycholesterol in cerebrospinal fluid is elevated in early stages of dementia. J. Psychiatr. Res. 36, 27–32.

Patil, S., Sheng, L., Masserang, A., Chan, C., 2006. Palmitic acid-treated astrocytes induce BACE1 upregulation and accumulation of C-terminal fragment of APP in primary cortical neurons. Neurosci. Lett. 406, 55–59.

Paz-Filho, G., Mastronardi, C., Franco, C.B., Wang, K.B., Wong, M.L., Licinio, J., 2013. Leptin: molecular mechanisms, systemic pro-inflammatory effects, and clinical implications. Arq. Bras. Endocrinol. Metabol. 56, 597–607.

Perry, R.J., Ridgway, N.D., 2005. Molecular mechanisms and regulation of ceramide transport. Biochim. Biophys. Acta 1734, 220–2234.

Pettus, B.J., Bielawska, A., Subramanian, P., Wijesinghe, D.S., Maceyka, M., Leslie, C.C., et al., 2004. Ceramide 1-phosphate is a direct activator of cytosolic phospholipase A$_2$. J. Biol. Chem. 279, 11320–11326.

Pfrieger, F.W., 2003. Outsourcing in the brain: do neurons depend on cholesterol delivery by astrocytes? Bioessays 25, 72–78.

Pomorski, T., Menon, A.K., 2006. Lipid flippases and their biological functions. Cell. Mol. Life Sci. 63, 2908–2921.

Popp, J., Lewczuk, P., Kölsch, H., Meichsner, S., Maier, W., Kornhuber, J., et al., 2012. Cholesterol metabolism is associated with soluble amyloid precursor protein production in Alzheimer's disease. J. Neurochem. 123, 310–316.

Popp, J., Meichsner, S., Kölsch, H., Lewczuk, P., Maier, W., Kornhuber, J., et al., 2013. Cerebral and extracerebral cholesterol metabolism and CSF markers of Alzheimer's disease. Biochem. Pharmacol. 86, 37–42.

Puglielli, L., Ellis, B.C., Saunders, A.J., Kovacs, D.M., 2003. Ceramide stabilizes β-site amyloid precursor protein-cleaving enzyme 1 and promotes amyloid β-peptide biogenesis. J. Biol. Chem. 278, 19777–19783.

Ramirez, D.M.O., Andersson, S., Russel, D.W., 2008. Neuronal expression and subcellular localization of cholesterol 24-hydroxylase in the mouse brain. J. Comp. Neurol. 507, 1676–1693.

Rao, J.S., Rapoport, S.I., Kim, H.W., 2011. Altered neuroinflammatory, arachidonic acid cascade and synaptic markers in postmortem Alzheimer's disease brain. Transl. Psychiatry 1, e31.

Reed, B., Villeneuve, S., Mack, W., DeCarli, C., Chui, H.C., Jagust, W., 2014. Associations between serum cholesterol levels and cerebral amyloidosis. JAMA Neurol. 71, 195–200.

Repa, J.J., Li, H., Frank-Cannon, T.C., Valasek, M.A., Turley, S.D., Tansey, M.G., Dietschy, J.M., 2007. Liver X receptor activation enhances cholesterol loss from the brain, decreases neuroinflammation, and increases survival of the NPC1 mouse. J. Neurosci. 27, 14470–14480.

Róg, T., Vattulainen, I., 2014. Cholesterol, sphingolipids, and glycolipids: what do we know about their role in raft-like membranes? Chem. Phys. Lipids 184, 82–104.

Rosales-Corral, S., Acuna-Castroviejo, D., Tan, D.X., et al., 2012. Accumulation of exogenous amyloid-β peptide in hippocampal mitochondria causes their dysfunction: a protective role for melatonin. Oxidative Med. Cell. Longev. 2012, 2015.

Saddoughi, S.A., Gencer, S., Peterson, Y.K., Ward, K.E., Mukhopadhyay, A., Oaks, J., Bielawski, J., et al., 2013. Sphingosine analogue drug FTY720 targets I2PP2A/SET and mediates lung tumour suppression via activation of PP2A-RIPK1-dependent necroptosis. EMBO Mol. Med. 5, 105–121.

Schroeder, R.J., Ahmed, S.N., Zhu, Y., London, E., Brown, D.A., 1998. Cholesterol and sphingolipid enhance the Triton X-100 insolubility of glycosylphosphatidylinositol-anchored proteins by promoting the formation of detergent-insoluble ordered membrane domains. J. Biol. Chem. 273, 1150–1157.

Serhan, C.N., Petasis, N.A., 2011. Resolvins and protectins in inflammation resolution. Chem. Rev. 111, 5922–5943.

Serhan, C.N., Hong, S., Gronert, K., Colgan, S.P., Devchand, P.R., Mirick, G., Moussignac, R.L., 2002. Resolvins: a family of bioactive products of omega-3 fatty acid transformation circuits initiated by aspirin treatment that counter pro-inflammation signals. J. Exp. Med. 196, 1025–1037.

Serhan, C.N., Arita, M., Hong, S., Gotlinger, K., 2004. Resolvins, docosatrienes, and neuroprotectins, novel omega-3-derived mediators, and their endogenous aspirin-triggered epimers. Lipids 39, 1125–1132.

Serhan, C.N., Chiang, N., Van Dyke, T.E., 2008. Resolving inflammation: dual anti-inflammatory and pro-resolution lipid mediators. Nat. Rev. Immunol. 8, 249–261.

Simanshu, D.K., Kamlekar, R.K., Wijesinghe, D.S., Zou, X., Zhai, X., Mishra, S.K., et al., 2013. Non-vesicular trafficking by a ceramide-1-phosphate transfer protein regulates eicosanoids. Nature 500, 463–467.

Simons, K., Ehehalt, R., 2002. Cholesterol, lipid rafts, and disease. J. Clin. Invest. 110, 597–603.

Simons, K., Ikonen, E., 2000. How cells handle cholesterol. Science 290, 1721–1726 51.

Söderberg, M., Edlund, C., Kristensson, K., Dallner, G., 1991. Fatty acid composition of brain phospholipids in aging and in Alzheimer's disease. Lipids 26, 421–425.

Sot, J., Goni, F.M., Alonso, A., 2005. Molecular associations and surface-active properties of short-and long-N-acyl chain ceramides. Biochim. Biophys. Acta 1711, 12–19.

Spiegel, S., Milstien, S., 2003. Exogenous and intracellularly generated sphingosine 1-phosphate can regulate cellular processes by divergent pathways. Biochem. Soc. Trans. 31, 1216–1219.

Spiegel, S., Milstien, S., 2011. The outs and the ins of sphingosine-1-phosphate in immunity. Nat. Rev. Immunol. 11, 403–415.

Stephenson, D., Rash, K., Smalstig, B., Roberts, E., Johnstone, E., Sharp, J., et al., 1999. Cytosolic phospholipase A2 is induced in reactive glia following different forms of neurodegeneration. Glia 27, 110–128.

Sun, G.Y., He, Y., Chuang, D.Y., Lee, J.C., Gu, Z., Simonyi, A., et al., 2012. Integrating cytosolic phospholipase A_2 with oxidative/nitrosative signaling pathways in neurons: a novel therapeutic strategy for AD. Mol. Neurobiol. 46, 85–95.

Suzuki, T., 2002. Lipid rafts at postsynaptic sites: distribution, function and linkage to postsynaptic density. Neurosci. Res. 44, 1–9.

Svennerholm, L., Gottfries, C.G., 1994. Membrane lipids, selectively diminished in Alzheimer brains, suggest synapse loss as a primary event in early-onset form (type I) and demyelination in late-onset form (type II). J. Neurochem. 62, 1039–1047.

van Meer, G., 2011. Dynamic transbilayer lipid asymmetry. Cold Spring Harb. Perspect. Biol. 3, a004671.

von Bohlen und Halbach, O., Albrecht, D., 2006. The CNS renin-angiotensin system. Cell Tissue Res. 326, 599–616.

Wang, L., Schuster, G.U., Hultenby, K., Zhang, Q., Andersson, S., Gustafsson, J.-Å., 2002. Liver X receptors in the central nervous system: from lipid homeostasis to neuronal degeneration. Proc. Natl. Acad. Sci. U.S.A. 99, 13878–13883.

Wang, X., Zhu, M., Hjorth, E., Cortés-Toro, V., Eyjolfsdottir, H., Graff, C., et al., 2015. Resolution of inflammation is altered in Alzheimer's disease. Alzheimers Dement. 11, 40–50.

Warshauer, J.T., Lopez, X., Gordillo, R., Hicks, J., Holland, W.L., et al., 2015. Effect of pioglitazone on plasma ceramides in adults with metabolic syndrome. Diabetes Metab. Res. Rev. 31, 734–744.

Wells, K., Farooqui, A.A., Liss, L., Horrocks, L.A., 1995. Neural membrane phospholipids in Alzheimer disease. Neurochem. Res. 20, 1329–1333.

Wolozin, B., Kellman, W., Ruosseau, P., Celesia, G.G., Siegel, G., 2000. Decreased prevalence of Alzheimer disease associated with 3-hydroxy-3-methyglutaryl coenzyme A reductase inhibitors. Arch. Neurol. 57, 1439–1443.

Xie, S., Xiao, J.X., Gong, G.L., Zang, Y.F., Wang, Y.H., Wu, H.K., Jiang, X.X., 2006. Voxel-based detection of white matter abnormalities in mild Alzheimer disease. Neurology 66, 1845–1849.

Xiong, H., Callaghan, D., Jones, A., et al., 2008. Cholesterol retention in Alzheimer's brain is responsible for high β- and γ-secretase activities and Aβ production. Neurobiol. Dis. 29, 422–437.

Yahi, N., Aulas, A., Fantini, J., 2010. How cholesterol constrains glycolipid conformation for optimal recognition of Alzheimer's β amyloid peptide (Aβ1-40). PLoS One 5, e9079.

Yamaji-Hasegawa, A., Tsujimoto, M., 2006. Asymmetric distribution of phospholipids in biomembranes. Biol. Pharm. Bull. 29, 1547–1553.

Yu, Z.F., Nikolova-Karakashian, M., Zhou, D.H., Cheng, G.J., Schuchman, E.H., Mattson, M.P., 2000. Pivotal role for acidic sphingomyelinase in cerebral ischemia-induced ceramide and cytokine production, and neuronal apoptosis. J. Mol. Neurosci. 15, 85–97.

Zajchowski, L.D., Robbins, S.M., 2002. Lipid rafts and little caves–compartmentalized signalling in membrane microdomains. Eur. J. Biochem. 269, 737–752.

Zambón, D., Quintana, M., Mata, P., Alonso, R., Benavent, J., Cruz-Sánchez, F., et al., 2010. Higher incidence of mild cognitive impairment in familial hypercholesterolemia. Am. J. Med. 123, 267–274.

Zhang, Y., Schuff, N., Du, A.T., Rosen, H.J., Kramer, J.H., Gorno-Tempini, M.L., et al., 2009. White matter damage in frontotemporal dementia and Alzheimer's disease measured by diffusion MRI. Brain 132, 2579–2592.

Zhao, Y., Calon, F., Julien, C., Winkler, J.W., Petasis, N.A., Lukiw, W.J., et al., 2011. Docosahexaenoic acid-derived neuroprotectin D1 induces neuronal survival via secretase- and PPARγ-mediated mechanisms in Alzheimer's disease models. PLoS One 6, e15816.

Zhou, Y., Hayashi, I., Wong, J., Tugusheva, K., Renger, J.J., Zerbinatti, C., 2014. Intracellular clusterin interacts with brain isoforms of the bridging integrator 1 and with the microtu- bule-associated protein Tau in Alzheimer's disease. PLoS One 9, e103187.

Zhu, M., Wang, X., Hjorth, E., Colas, R.A., Schroeder, L., Granholm, A.C., Serhan, C.N., Schultzberg, M., 2016. Pro-resolving lipid mediators improve neuronal survival and increase Aβ42 phagocytosis. Mol. Neurobiol. 53, 2733–2749. http://www.nature.com/articles/srep32721.

CHAPTER

4

Contribution of Nucleic Acids in the Pathogenesis of Alzheimer's Disease

OUTLINE

INTRODUCTION

Alzheimer's disease (AD) is a common age-dependent neurodegenerative disorder that impairs memory and causes cognitive deficits. It is characterized by the accumulation of senile plaques [enriched in β-amyloid peptide (Aβ)] and neurofibrillary tangles (enriched in hyperphosphorylated Tau protein). The accumulation of these hallmarks leads to massive neuronal loss in the brain (Farooqui, 2010; Kumar et al., 2015). AD is also accompanied by reduction in mitochondrial membrane potential, increase in permeability, and impaired cerebral glucose metabolism ("hypometabolism") along with insulin resistance and excessive production of reactive oxygen species (ROS), which damages proteins, lipids, and nucleic acids. All these processes have been reported to cause membrane defects (Farooqui, 2010; Butterfield and Lashuel, 2010), disruption of neuronal networks (Shankar et al., 2007; Kuperstein et al., 2010), neuronal dysfunction (Lacor et al., 2007; Shankar et al., 2008), impairment of long-term potentiation (Walsh et al., 2002), and changes in animal behavior (Ford et al., 2015). Induction of these processes may contribute to the pathogenesis of neurodegeneration in AD (Farooqui, 2014). Increasing age is a major risk factor for sporadic forms of AD. While some progress has been made in the pathogenesis of familial form of AD, we still know very little about the molecular mechanisms associated with the progression of this neurodegenerative disease. As the number of seniors in the world continues to increase, the prevalence of AD cases will increase remarkably throughout the world, and among seniors, AD will become one of the leading cause of disability and death (Kalaria et al., 2008). Currently, AD affects 5 million Americans. It is estimated that ~3% of Americans aged between 65 and 74 years, 19% aged between 75 and 84 years, and ~47% older than 85 years are victims of AD (Hebert et al., 2003) and that ~60% of nursing home patients older than 65 years have AD. As the baby boom generation ages and without preventive strategies, it is estimated that there may be 14 million Americans with AD by 2040 (Alzheimer's Association, 2010; Alzheimer's Association, 2012; Alzheimer's Association, 2013). As stated in Chapter 1, soluble Aβ oligomers (ADDLs) are widely regarded as instigating neuronal damage leading to AD. The molecular mechanism associated with ADDL-mediated neurotoxicity is not fully understood. However, it is proposed that ADDL binding with its receptors triggers a redistribution of critical synaptic proteins and induces hyperactivity in metabotropic and ionotropic glutamate receptors. This results in Ca^{2+} influx and overload and instigates major facets of AD neuropathology, including Tau hyperphosphorylation, insulin resistance, oxidative stress, neuroinflammation, and synapse loss (Ong et al., 2013; Viola and Klein, 2015). It is also proposed that ADDL and Tau protein work together, independently of their accumulation into plaques and tangles, changing healthy neurons into the diseased neurons. Thus acute as well as delayed neurodegenerations are triggered by ADDL species and Tau protein.

OXIDATIVE STRESS IN THE BRAIN

The brain tissue is highly susceptible to oxidative imbalance due to its high energy demand, high oxygen consumption, rich abundance of easily peroxidizable polyunsaturated fatty acids, high level of iron, and relative paucity of antioxidants and related enzymes [superoxide dismutase (SOD), catalase, glutathione peroxidase (GPx), and heme oxygenase]. Oxidative

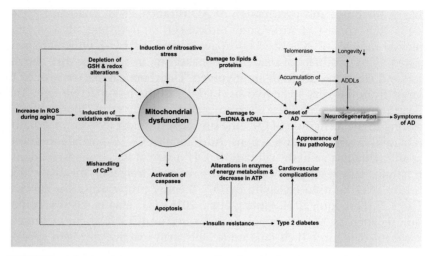

FIGURE 4.1 **Involvement of age-mediated mitochondrial dysfunction in the pathogenesis of Alzheimer's disease.** *AD*, Alzheimer's disease; *ADDL*, soluble Aβ oligomers; *Aβ*, β-amyloid; *GSH*, reduced glutathione; *mtDNA*, mitochondrial DNA; *nDNA*, nuclear DNA; *ROS*, reactive oxygen species.

imbalance and subsequent oxidative stress–mediated damage to neural biomolecules are extensively reported in AD, and increasing evidence suggest that oxidative imbalance plays a critical role in AD (Nunomura et al., 2001). As the main source of ROS generation and also a major target of oxidative damage, progressive impairment of mitochondrial function has also been implicated in aging and AD (Fig. 4.1) (Swerdlow, 2011).

Aging contributes to mitochondrial dysfunction, which causes increase in oxidative stress. Current evidence also indicates there is an elevated inflammatory profile in the aging brain consisting of an increased population of reactive microglia cells (Alonso-Fernandez and De la Fuente, 2011). These factors set the stage for the pathogenesis of AD and other neurodegenerative diseases. During aging, oxidative stress negatively impacts mitochondrial integrity and function. Mitochondrial dysfunction results in decrease in ATP production, impairment in calcium buffering, and increase in ROS generation (Beal, 2005). Alterations in mitochondrial dynamics, either due to the response to genetic deficits or metabolic or environmental alterations exacerbate mitochondrial dysfunction. Onset of this process results in a vicious downward spiral that produces alterations in mitochondrial energy metabolism; decrease in ATP; increase in levels of phospholipid-, sphingolipid-, and cholesterol-derived lipid mediators; and induction of insulin resistance. These deficits play an important role in the pathogenesis of AD. In addition, numerous studies have indicated that oxidative stress, an imbalance between free radicals and the antioxidant system, increases the production of Aβ and plays an important role in the onset and progression of AD (Praticò, 2008; Zhao and Zhao, 2013). ROS consist of superoxide radical ($O_2{}^{\cdot-}$); hydroxyl, alkoxyl, and peroxyl radicals ($ROO\cdot$); and hydrogen peroxide (H_2O_2), which are generated as by-products of oxidative metabolism, in which energy activation and electron reduction are involved. The chemical reactivity of ROS varies from the very toxic hydroxyl ($\cdot OH$) to the less reactive $O_2{}^{\cdot-}$. The reduction of oxygen by one electron generates fairly stable intermediates leading to the formation of $O_2{}^{\cdot-}$, the precursor of most ROS and mediator in oxidative stress chain reactions. Additionally, $O_2{}^{\cdot-}$ is partially reduced by antioxidants to a hydroxyl radical ($OH\cdot$), one of the strongest oxidants in nature. This reaction is catalyzed by reduced transition metals, which in turn may be reduced again by $O_2{}^{\cdot-}$ thereby propagating the process (Fig. 4.2) (Farooqui, 2014; Huang et al., 2016). $O_2{}^{\cdot-}$ also reacts with other radicals such as $NO\cdot$ to generate peroxynitrite ($OONO^-$), an extremely powerful oxidant, which falls under the term reactive nitrogen species (RNS). ROS and/or RNS play a dual role in intracellular signaling, cell proliferation, and survival (Farooqui, 2014). However, slight changes in the steady-state levels of ROS and RNS result in free radical–mediated chain reactions that indiscriminately target lipids, proteins, polysaccharides, and DNA (Farooqui, 2014).

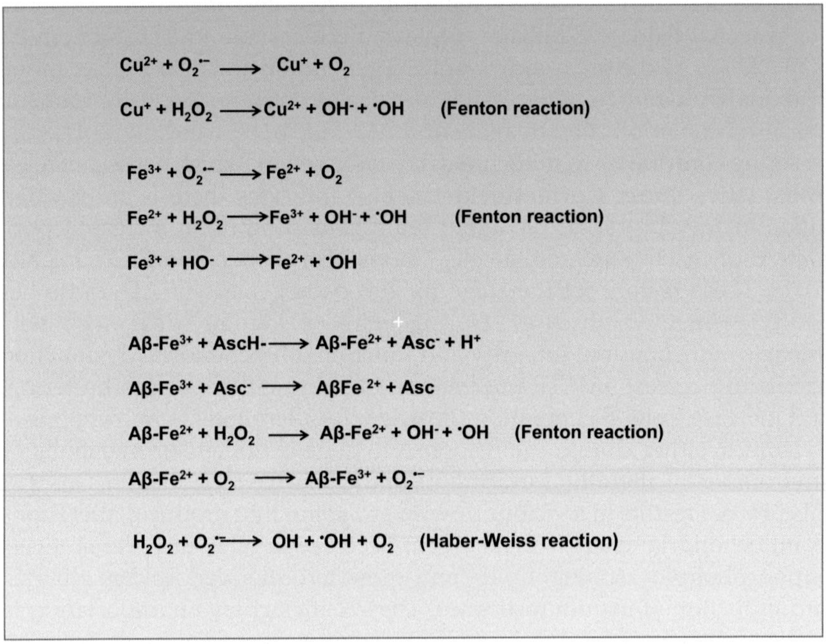

$$Cu^{2+} + O_2{}^{\cdot-} \longrightarrow Cu^+ + O_2$$

$$Cu^+ + H_2O_2 \longrightarrow Cu^{2+} + OH^- + {}^\cdot OH \quad \text{(Fenton reaction)}$$

$$Fe^{3+} + O_2{}^{\cdot-} \longrightarrow Fe^{2+} + O_2$$

$$Fe^{2+} + H_2O_2 \longrightarrow Fe^{3+} + OH^- + {}^\cdot OH \quad \text{(Fenton reaction)}$$

$$Fe^{3+} + HO^- \longrightarrow Fe^{2+} + {}^\cdot OH$$

$$A\beta\text{-}Fe^{3+} + AscH\text{-} \longrightarrow A\beta\text{-}Fe^{2+} + Asc^{\cdot-} + H^+$$

$$A\beta\text{-}Fe^{3+} + Asc^- \longrightarrow A\beta Fe^{2+} + Asc$$

$$A\beta\text{-}Fe^{2+} + H_2O_2 \longrightarrow A\beta\text{-}Fe^{2+} + OH^- + {}^\cdot OH \quad \text{(Fenton reaction)}$$

$$A\beta\text{-}Fe^{2+} + O_2 \longrightarrow A\beta\text{-}Fe^{3+} + O_2{}^{\cdot-}$$

$$H_2O_2 + O_2{}^{\cdot-} \longrightarrow OH + {}^\cdot OH + O_2 \quad \text{(Haber-Weiss reaction)}$$

FIGURE 4.2 Contribution of copper and iron in the generation of superoxide and hydroxyl radicals through Fenton and Haber–Weiss reactions.

The detoxification of ROS is a major prerequisite for the survival of aerobic life. The detoxification is performed by several enzymatic and nonenzymatic antioxidant mechanisms, which are available to the cells in different cellular compartments (Droge, 2002; Farooqui, 2010). Accumulation of Aβ in AD damages neurons through the production of ROS indicating Aβ peptides are probably oxidant peptides. ROS are mainly (90%) generated in mitochondria, in particular from mitochondrial dysfunction. Complexes I and II of mitochondrial electron transport chain (ETC) produce ROS only into the matrix, whereas complex III produces ROS on both sides of the mitochondrial inner membrane (Murphy, 2009; Muller et al., 2004). Between 1% and 5% of all O_2 used in complexes I and III of the ETC escapes as $O_2^{\cdot-}$. In response, SODs in the mitochondria (Mn-SOD or SOD2) and cytosol (Cu-Zn-SOD or SOD1) catalyze a reaction changing $O_2^{\cdot-}$ to diatomic oxygen and hydrogen peroxide. GPxs and catalase, in turn, act as additional antioxidant defenses by converting H_2O_2 to water. H_2O_2 is not a radical; thus it can cross membranes and rapidly propagate throughout the cell.

ROS are also generated through uncontrolled arachidonic acid (ARA) cascade. From neural membrane glycerophospholipids, ARA is released by cytosolic phospholipase A_2 and oxidized by cyclooxygenases (COXs), lipoxygenases, and epoxygenases leading to the formation of prostaglandins, leukotriene, lipoxins, and thromboxanes, as well as hydroxyeicosatetraenoic acid and epoxyeicosatetraenoic acids, and dihydroxyeicosatrienoic acids. These metabolites are collectively known as eicosanoids. Eicosanoids produce a wide range of biological actions including potent effects on neuroinflammation, vasodilation, vasoconstriction, apoptosis, and immune responses (Phillis et al., 2006).

Nonenzymatic oxidation of ARA produces ROS. In neurons and neuroblastoma cells, NADPH oxidase also contribute to ROS production (Sun et al., 2007). The ability of NADPH oxidase inhibitors to inhibit ROS-mediated cytotoxicity provides strong support for the role of this enzyme in regulation of ROS generation. As stated earlier, in the presence of metal ions, such as Fe^{2+} and Cu^{2+}, generation of ROS is one of the leading cause of oxidative stress in AD (da Silva et al., 2009). In AD brain, Fe^{3+}, Zn^{2+}, and Al^{3+} are colocalized with Aβ42 peptides in the senile plaque cores (Pogue et al., 2009). Excessive Fe^{3+} and Cu^{2+} ions binding with Aβ produces deleterious effects not only on the aggregation of Aβ but also on the generation of elevated levels of ROS. In the presence of metal ions (Fe^{2+} and Cu^{2+}) H_2O_2 is also transformed into $\cdot OH$ through the Fenton reaction (Fig. 4.2). High concentrations of zinc are found in the neocortex and amygdala, and hippocampus. These regions are not only affected in AD pathology but also play important roles in memory formation and maintenance of cognitive function (Huang et al., 2004). Consequently, the immunological/inflammatory response to nonsoluble Aβ plaques involves the disruption of zinc homeostasis followed by uncontrolled cerebral zinc release, which typically occurs on the onset of oxidative stress. Thus the uncontrolled accumulation of zinc or Aβ leads to zinc-induced and Aβ-mediated oxidative stress and cytotoxicity (Pal et al., 2013).

Although AD is a multifactorial neurodegenerative disease, oxidative stress plays a major role in the onset and progression of AD. Alterations caused by ROS on DNA and RNA integrity are particularly harmful in differentiated neurons particularly in the old age. Nonrepaired ROS-mediated nucleic acid damage may trigger transcriptional and translational deregulation, which may lead to reduction in protein synthesis, protein mutation, production of truncated proteins, and genomic instability. Fe^{3+} and Cu^{2+} can also cause DNA damage by direct binding. The metal-induced genotoxic damage has been shown by the wide spectrum of damaged bases, oxidized apurinic/apyrimidinic (AP) sites, as well as single-strand breaks (SSBs) in cultured cells and animals exposed to pro-oxidant metals (Mitra et al., 2014). Furthermore, the activity of DNA repair proteins such as DNA-dependent protein kinase (Shackelford, 2006), DNA polymerase β (Pol β) (Weisman et al., 2007), and 8-oxoguanine DNA glycosylase (OGG1) (Lovell et al., 2000; Iida et al., 2002) are impaired in AD-affected brains. The decrease in activities of the aforementioned enzymes and accumulation of DNA damage in affected brain regions in AD correlates with metal overload, but the mechanism of selective metal accumulation in affected regions vs. other regions is not understood. ROS generate a wide range of nucleic acid lesions including base modifications, deletions, and strand breaks in DNA and RNA. Neurons in the brain continuously face the harmful effects of ROS due to high oxygen consumption. Therefore, the preservation of nucleic acid integrity from oxidative damage is essential to maintain neuronal functionality and ensure their longevity (Englander, 2008; Mantha et al., 2013).

Neural cells have multiple defense mechanisms to protect neuronal DNA integrity in normal brain. For example, neural cells control and maintain Fe^{2+} and Cu^{2+} homeostasis to prevent the formation of toxic hydroxyl radicals. It is important to note that the changes in H_2O_2 that are required for signaling do not cause significant changes in intracellular ratio of oxidized glutathione/reduced glutathione (GSH) or NADPH/NADP+ (Morgan et al., 2011). Hydroxyl radicals can not only attack nucleic acids but also damage polyunsaturated fatty acids forming the ROO· and then propagate the chain reaction of lipid peroxidation. During normal aerobic metabolism, ROS production is kept under tight control through the activities of antioxidant defense systems. The antioxidant defense systems of the brain include low-molecular-weight antioxidants like glutathione, uric acid, lipoic acid, and vitamins E and C (Fig. 4.3) and high-molecular-weight antioxidant enzymes such as SOD, catalase, transferrin, and GPx. ROS act by damaging lipids, proteins, and DNA, resulting in damage or

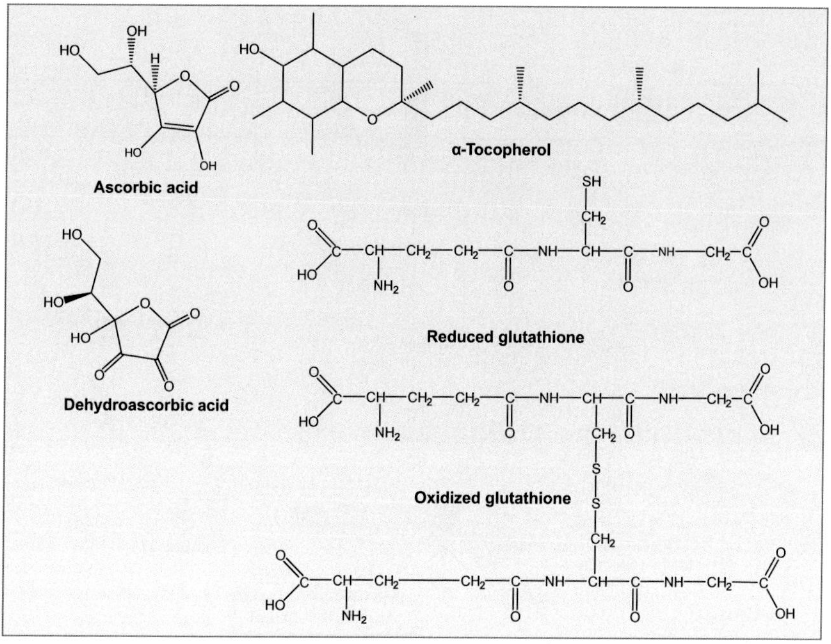

FIGURE 4.3 Chemical structures of ascorbic acid, dehydroascorbic acid, α-tocopherol, and glutathione.

genomic instability (Dizdaroglu and Jaruga, 2012). However, from the past two decades, it is becoming increasingly apparent that ROS may also serve as signaling molecules to regulate biological and physiological processes (Farooqui, 2014).

Levels of ROS and Aβ are markedly increased in the brains of patients with AD (Fig. 4.4) (Williams et al., 2006; Farooqui, 2010). Furthermore, the activities of antioxidants and antioxidant enzymes are decreased in the early phase of AD (Cervellati et al., 2014). These processes elevate levels of ROS mainly due to the production of superoxide anion by mitochondria along with decrease in ATP. ROS are potentially toxic for neurons and oligodendrocytes and may induce toxicity by damaging lipids, glycerophospholipids, proteins, and nucleic acids (Farooqui, 2010). ROS-mediated damage to glycerophospholipids, sphingolipids, and cholesterol has been described in Chapter 3. ROS-mediated damage to proteins in aging and in age-related neurodegenerative diseases is accompanied by oxidative modification of proteins. This is a nonreversible phenomenon, which requires clearance systems for removal of oxidized/dysfunctional proteins (Perluigi et al., 2014). In general, oxidation of proteins not only affects protein expression and protein turnover but also affects neural cell signaling, eventually leading to neuronal cell death (Butterfield et al., 2012).

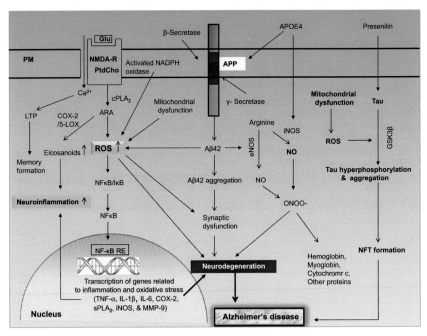

FIGURE 4.4 **Hypothetical diagram showing the contribution of β-amyloid in neurodegeneration in Alzheimer's disease.** *5-LOX*, 5-lipoxygenase; *APP*, amyloid precursor protein; *ARA*, arachidonic acid; *Aβ*, β-amyloid; *COX-2*, cyclooxygenase; *cPLA₂*, cytosolic phospholipase A₂; *eNOS*, endothelial nitric oxide synthase; *Glu*, glutamate; *IL-1β*, interleukin-1β; *IL-6*, interleukin-6; *iNOS*, inducible nitric oxide synthase; *LTP*, long-term potentiation; *NF-κB*, nuclear factor-κB; *NF-κB-RE*, nuclear factor-κB response element; *NFT*, neurofibrillary tangle; *NMDA-R*, N-methyl-D-aspartate receptor; *NO*, nitric oxide; *ONOO⁻*, peroxynitrite; *PM*, plasma membrane; *PtdCho*, phosphatidylcholine; *ROS*, reactive oxygen species; *sPLA₂*, secretory phospholipase A₂; *TNF-α*, tumor necrosis factor- α.

Another major consequence of protein oxidation is the formation of large protein aggregates, which are often toxic to cells if allowed to accumulate. Insoluble aggregates can be formed as a result of covalent cross-linking among peptide chains, as in the case of Aβ and hyperphosphorylated Tau in AD. Deposits of aggregated, misfolded, and oxidized proteins accumulate normally over time in cells and tissues and are often present in increased amounts in a range of age-related diseases. Two major pathways have been reported to play important roles in the removal of intracellular proteins. These pathways are the ubiquitin–proteasome system (UPS) (Grune et al., 2005) and the autophagy–lysosome pathway (Metcalf et al., 2012). The UPS is located in the cytosol and the nucleus, and it is responsible for the degradation of more than 70%–80% of intracellular proteins. Most of the proteins are targeted for proteasomal degradation after being tagged with a polyubiquitin chain, which, in turn, is recognized by the proteasome. Experimental evidence suggests that failure of the UPS may contribute to neurodegeneration. However, additional factors, including other protein degradation pathways and mitochondrial dysfunction associated with a decline in ATP levels, may contribute to cellular viability.

In addition, in AD and other neurological diseases, activation of endothelial nitric oxide synthase (eNOS) and inducible nitric oxide synthase (iNOS) results in generation of nitric oxide (NO), which reacts with O_2^- to form peroxynitrite (ONOO⁻). ONOO⁻ interacts with proteins, such as hemoglobin, myoglobin, and cytochrome c, oxidizing ferrous heme into the corresponding ferric forms (Fig. 4.4) (Pacher et al., 2007). ONOO⁻ also mediates neural cell damage via lipid peroxidation as well as inactivation of proteins by oxidation and nitration. ONOO⁻ also acts on mitochondria to decrease the membrane potential (Ψ) and triggers the release of proapoptotic factors such as cytochrome c (Cyt c) and apoptosis-inducing factor (AIF). These factors mediate caspase-dependent and caspase-independent apoptotic death pathways. Peroxynitrite, in concert with other oxidants (e.g., H_2O_2), promotes breaking of DNA strands through the activation of nuclear enzyme poly(ADP-ribose) polymerase-1 (PARP-1). Mild damage to DNA activates the DNA repair machinery. By contrast, once excessive oxidative stress– and nitrosative stress–induced DNA damage occurs, overactivation of PARP-1 initiates an energy-consuming cycle by transferring ADP-ribose units (small red spheres) from NAD⁺ to nuclear proteins, resulting in rapid depletion of the intracellular NAD⁺ and ATP pools, slowing the rate of glycolysis and mitochondrial respiration, and eventually leading to cellular dysfunction and death (Islam et al., 2015). Poly(ADP-ribose) glycohydrolase degrades poly(ADP-ribose) (PAR) polymers, generating free PAR polymer and ADP-ribose, which signals to the mitochondria to induce AIF release. PARP-1 activation also leads to the inhibition of cellular glyceraldehyde-3-phosphate dehydrogenase activity which, in turn, promotes the activation of protein kinase C (Islam et al., 2015). Furthermore, ONOO⁻ inactivates iNOS by oxidative modification of its heme group (Huhmer et al., 1997), a reaction that may be responsible for negative feedback negative control of ONOO⁻ generation under inflammatory conditions (Pacher et al., 2007). Peroxynitrite also not only reacts with iron-sulfur-cluster-containing proteins causing inactivation of mitochondrial aconitase and phosphogluconate dehydratase but also inactivates Zn^{2+} sulfur motifs containing eNOS and alcohol dehydrogenase (Pacher et al., 2007).

Generation of low levels (physiological concentrations) of ROS is necessary for many fundamental neurochemical processes, such as cell growth, adhesion of cells toward other cells, adaptation responses, senescence, and optimal working of immune system (Thannickal and Fanburg, 2000). ROS-mediated signal transduction processes involve the participation of several transcription factors, including nuclear factor-κB (NF-κB), activator protein-1 (AP-1), NF-E2-related factor 2 (Nrf2), and hypoxia-inducible

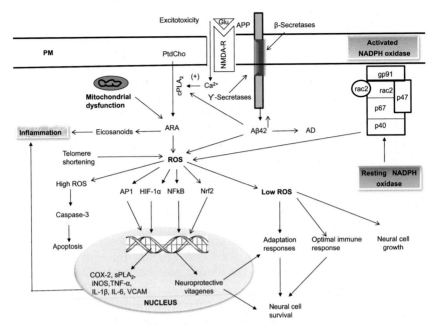

FIGURE 4.5 **Hypothetical model showing generation of ROS from various sources and modulation of transcription factors by ROS.** *AP-1*, activator protein 1; *APP*, amyloid precursor protein; *ARA*, arachidonic acid; *Aβ*, β-amyloid; *COX-2*, cyclooxygenase; *cPLA₂*, cytosolic phospholipase A₂; *Glu*, glutamate; *gp91, rac1, rac2, p47, p67, and p40*, various subunits of NADPH oxidase; *HIF-1α*, hypoxia-inducible transcription factor-1α; *IL-1β*, interleukin-1β; *IL-6*, interleukin-6; *lyso-PtdCho*, lyso-phosphatidylcholine; *NF-κB*, nuclear factor-κB; *NF-κB-RE*, nuclear factor-κB response element; *NMDA-R*, N-methyl-ᴅ-aspartate receptor; *Nrf2*, NF-E2-related factor 2; *PM*, plasma membrane; *PtdCho*, phosphatidylcholine; *ROS*, reactive oxygen species; *sPLA₂*, secretory phospholipase A₂; *TNF-α*, tumor necrosis factor-alpha.

factor (HIF)-1α (Valko et al., 2007) (Fig. 4.5). These transcription factors reside in cytoplasm. Production of high ROS in neural cells promotes the translocation of the aforementioned transcription factors to the nucleus, where they modulate gene transcriptions. Thus in the presence of high ROS levels, NF-κB translocates from cytoplasm to the nucleus, where it interacts with NF-κB response element to facilitate the expression of pro-inflammatory enzymes (secretory phospholipase A₂, COX-2, iNOS), cytokines [tumor necrosis factor-α, interleukin (IL)-1β, IL-6, IL-12], chemokines [macrophage inflammatory protein (MIP-1α), meta-chlorophenylpiperazine (MCPP1)], growth factors, cell cycle regulatory molecules, adhesion molecule leading to inflammation (intercellular adhesion molecule, vascular cell adhesion molecule, and E-selectin), and antiinflammatory molecules and adhesion molecules (Fig. 4.5). The DNA-binding ability of NF-κB is modulated by redox status in the cell (Nishi et al., 2002). It is also shown that redox factor protein, Ref-1 reduces cysteine 62 in NF-κB in the nucleus and this reaction is required for NF-κB to bind to DNA (Nishi et al., 2002). Conversely, oxidation of this residue inhibits binding to DNA (Toledano and Leonard, 1991). In addition, glutathionylation of NF-κB in the presence of ROS results in a decrease in its DNA-binding ability and downstream transcriptional activity (Pineda-Molina et al., 2001).

AP-1 is a redox-sensitive inducible intracellular transcription factor, which is composed of proteins belonging to the c-FBJ murine osteosarcoma viral oncogene homolog, c-Jun proto-oncogene, and activating transcription factor families (Pronk et al., 2014). The activation of AP-1 is mediated by sulfhydryl modification of critical cysteine residues found on this protein and/or other upstream redox-sensitive molecular targets in response to proinflammatory cytokines or oxidative stimuli (Angel and Karin, 1991). AP-1 plays an important role in the control and expression of genes associated with eukaryotic cellular behavior such as cell cycle proliferation, development, stress responses, and apoptosis (Shaulian and Karin, 2001, 2002; Ozanne et al., 2007). Activated AP-1 translocates from cytosol to the nucleus, recognizes 12-O-tetradecanoylphorbol-13-acetate response elements (5′-TGAG/CTCA-3′), and induces the expression of a variety of genes (Choi, 2014). AP-1 contributes to the chronic inflammation in response to oxidative and electrophilic stress. It is demonstrated that the phophatidylinositol 3-kinase (PtdIns 3K)/serine/threonine kinase (Akt1) pathway plays an important role in the transcriptional regulation of AP-1 expression. Although the histone post-translational modifications are assumed to affect AP-1 transcriptional regulation by the PtdIns 3K/Akt pathway, the detailed mechanisms are completely unknown.

Like NF-κB, Nrf2 is retained in the cytoplasm along with Kelch-like ECH-associated protein 1 (Keap1), which promotes its proteosomal degradation under physiological conditions (Itoh et al., 2003). In response to ROS, Nrf2 dissociates from Keap1 and translocates into the nucleus, where it dimerizes mainly with the MAF (Maf-G, Maf-F, and Maf-K), JUN (c-Jun, Jun-B, and Jun-D), and ATF (ATF-4) families of bZIP proteins and transactivates a network of genes encoding cytoprotective and antioxidative enzymes containing the antioxidant response element, 5′-TGAG/CnnnGC-3′ in their promoters such as Gpx, NAD(P)H:quinone oxidoreductase, and heme oxygenase-1. HIF-1 is a basic helix-loop-helix transcription factor, which mediates the adaptation of cells to low oxygen tensions. It is composed of an HIF-1β subunit and one of three α subunits (HIF-1α, HIF-2α, or HIF-3α). Under normoxic conditions, HIF-1α has an extremely short half-life of less than 5 min. It is continuously synthesized and degraded (Huang et al., 1996). The degradation of HIF-1α is accompanied by the hydroxylation of two prolyl residues (402 and 564) in the oxygen-dependent degradation domain by the specific prolyl hydroxylases (Jaakkola et al., 2001), which require oxygen and 2-oxoglutarate, as cosubstrates, and iron (Fe^{2+}) and ascorbate, as cofactors (Weidemann and Johnson, 2008). Prolyl hydroxylation of HIF-1α is required for binding of the von Hippel–Lindau protein (VHL), which forms part of the E3 ligase complex (a ubiquitin ligase complex) that targets HIF-1α for polyubiquitination and subsequent proteasomal degradation (Maxwell et al., 1999). Besides this canonical pathway, there is evidence to demonstrate that the degradation of HIF-1α may also occur in oxygen- or VHL-independent ways (Liu et al., 2007). While under hypoxic conditions, the hydroxylation of prolyl residues is inhibited, thus HIF-1α evades VHL-mediated proteasomal destruction leading to accumulation.

EFFECTS OF OXIDATIVE STRESS ON DEOXYRIBONUCLEIC ACID

As stated in Chapter 1, aging is a major uncontrollable risk factor for AD. Age-mediated changes in the brain not only increase oxidative damage to neural cell components such as lipids, proteins, and nucleic acids (Fig. 4.6)

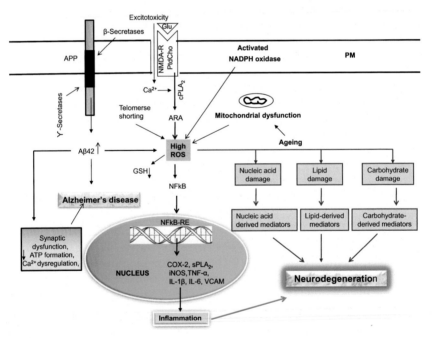

FIGURE 4.6 **Hypothetical diagram showing the effect of high ROS on neural cell components in neurodegeneration.** *ARA*, arachidonic acid; *COX-2*, cyclooxygenase; *cPLA$_2$*, cytosolic phospholipase A$_2$; *Glu*, glutamate; *IL-1β*, interleukin-1β; *IL-6*, interleukin-6; *lyso-PtdCho*, lyso-phosphatidylcholine; *NF-κB*, nuclear factor-κB; *NF-κB-RE*, nuclear factor-κB response element; *NMDA-R*, N-methyl-D-aspartate receptor; *Nrf2*, NF-E2-related factor 2; *PM*, plasma membrane (PM); *PtdCho*, phosphatidylcholine; *ROS*, reactive oxygen species; *sPLA$_2$*, secretory phospholipase A$_2$; *TNF-α*, tumor necrosis factor-α.

(Mattson, 2004) but also reduce neurotrophic factor signaling (Zuccato and Cattaneo, 2009) and induce dysregulation of neuronal calcium homeostasis (Bezprozvanny and Mattson, 2008). Mammalian neurons are highly vulnerable to DNA damage due to their high metabolic rate for energy production, generating cytotoxic ROS that can produce oxidative DNA damage (Sedelnikova et al., 2004; von Figura et al., 2009; Mata-Garrido et al., 2016). DNA damage not only influences neuron glial network activity, neuronal excitability through the inhibition of ion-motive ATPases, modification of ligand, and alterations in voltage-gated ion channel activity but also changes neurotransmitter signaling pathways (Gleichmann and Mattson, 2011). Very little is known about oxidative stress–mediated alterations in neuronal and glial nucleic acids (DNA and RNA). However, many different types of alterations in DNA bases have been reported (Cooke et al., 2003; Cadet et al., 2014). Guanine has the lowest standard reduction potential. Thus, it represents the major site of oxidative damage in DNA (Fleming et al., 2015). The most frequently detected lesions are 8-hydroxyguanine, 8-hydroxyadenine, 2,6-diamino-4-hydroxy-5-formamidoguanine, 4,6-diamino-5-formamidoadenine, and cytosine glycol (Gajewski et al., 1990). In the brain, H_2O_2-mediated oxidative damage to DNA occurs continuously. Under pathological conditions (AD and stroke), excessive amounts of H_2O_2 escape from antioxidant systems and eventually reach the nucleus due to its highly diffusible characteristic. Chromatin-bound iron interacts with H_2O_2, producing hydroxyl radicals in situ (Meneghini, 1997). Hydroxyl radicals attack a nearby DNA residue producing DNA modifications. These DNA modifications are investigated by examining the following events: (1) base modifications, (2) covalent binding of bases within DNA or DNA-protein cross-links, (3) abasic sites, and (4) strand breaks (Yu and Anderson, 1997). This DNA damage results in breakage of DNA strands along with production of damaged nucleotides. It is estimated that 10,000 oxidative interactions occur between DNA and endogenously generated free radicals per human cell per day. 8-Hydroxy-2′-deoxyguanosine (8-OHdG) is one of the most widely studied base lesions. The levels of 8-OHdG are elevated (fourfold) in aged brains, when compared with the young brains (Hamilton et al., 2001). In mitochondria, higher levels of oxidative damage and DNA mutations have been ascribed to the location of the DNA near the inner mitochondrial membrane sites where superoxide anions are mainly formed supporting the view that the degree of DNA damage is higher in mitochondrial DNA than in nuclear DNA (Shen et al., 2000). In brains from patients with AD, levels of multiple oxidized bases from nuclear and mitochondrial DNA are significantly higher in the frontal, parietal, and temporal lobes as compared to age-matched control human brain regions (Wang et al., 2005). Moreover, as stated earlier, damage to mitochondrial DNA has approximately 10-fold higher levels of oxidized bases than nuclear DNA. These data are consistent with higher levels of oxidative stress in mitochondria suggesting that oxidative damage to mitochondrial DNA may contribute to the neurodegeneration observed in AD. High levels of DNA breaks have been reported in neurons from patients with AD (Coppedè and Migliore, 2015). Moreover, neurons from patients with AD accumulate oxidized DNA bases both in nuclear and mitochondrial DNA (Coppedè and Migliore, 2015). Similarly, high levels of the oxidized DNA bases 8-oxoG in both nuclear and mitochondrial DNA from brains of patients with mild cognitive impairment (MCI), the phase between normal aging and early dementia (Wang et al., 2006). The symptoms of MCI include: (1) memory impairment observed over time corrected for age and education, (2) normal general cognitive function, (3) many daily living activities, and (d) the subject not meeting criteria for dementia (DeCarli, 2003; Maynard et al., 2010). Thus patients with MCI exhibit characteristics of both clinically normal individuals as well as patients with AD. The oxidation of DNA bases and levels of 8-oxoG are an early event in MCI and AD pathology and may play a role in neurodegeneration. Analysis of peripheral leukocytes derived from patients with AD and MCI also indicate increased levels of ROS-mediated DNA damage (Migliore et al., 2005). Furthermore, treatment of human primary fibroblasts with oxidizing agents induces a gene expression pattern typical of fibroblasts from patients with AD (Ramamoorthy et al., 2012) suggesting the possible role of oxidative stress in the gene expression profile seen in AD. It is essential that damage to DNA does not persist. Brain tissue has several mechanisms that repair DNA damage. This DNA repair is accomplished mostly through nucleotide or base excision pathways. Base excision repair (BER) occurs through the involvement of enzymes called glycosylases. These enzymes cleave the N-glycosylic bond between the target base and deoxyribose, thus releasing a free base and leaving an AP site. In addition, several DNA glycosylases (DGs) are bifunctional, since they also display a lyase activity that cleaves the phosphodiester backbone 3′ to the AP site generated by the glycosylase activity (Chung et al., 1991) indicating that these enzymes selectively cleave phosphodiester bonds 5′ and 3′ to damaged bases, producing free hydroxyl and phosphate groups at the respective termini (Chung et al., 1991). The capacity to repair DNA damage not only varies with age but also with different regions of the brain, and that pathological features of AD are higher in brain regions where more DNA damage is detected (Weissman et al., 2007). DNA repair mechanisms play a critical role in protecting the genome. Several studies have also indicated that there also occurs a decline in the DNA repair mechanisms in AD (Weissman et al., 2007; Markesbery and Lovell, 2006). Accumulation of oxidized DNA and its products may render the cell dysfunctional and facilitate disease progression. In the case of DNA and proteins, repair and degradation of oxidized macromolecules provide further defenses for the cells against any deleterious effects.

In nonneural tissues (liver), more than 100 oxidative DNA lesions have been reported (Dizdaroglu and Jaruga, 2012). The 8-oxo-2-deoxyguanosine

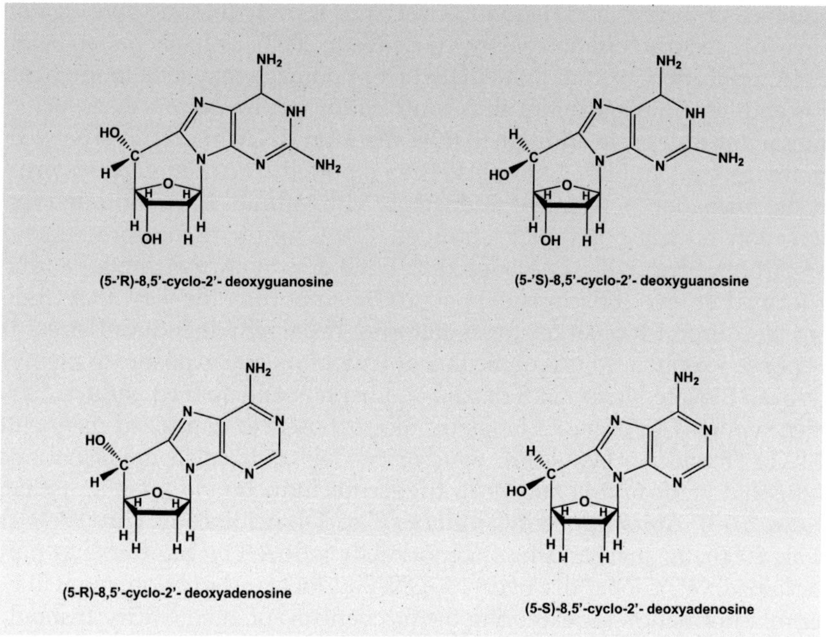

FIGURE 4.7   Chemical structures of oxidation products of DNA.

(oxo8dG) lesion has been studied most extensively not only because it is the major oxidative lesion (oxo8dG accounts for ~5% of the total oxidized bases known to occur in DNA) (Fig. 4.7) but also because it can be detected using a variety of procedures (Yen et al., 1991). The attack of ROS (hydroxyl radicals) on liver DNA not only produces strand breaks but also results in DNA–DNA interactions and DNA–protein cross-linking, with the formation of at least 20 modified base adducts (Barciszewski et al., 1999; Lovell and Markesbery, 2007). Among DNA bases, the guanine residues are most readily oxidized by the hydroxyl radical ($\cdot$OH) and singlet oxygen ($^1O_2$) (Steenken and Jovanovic, 1997). Formation of 5-R0-8,5-cyclo-2′-deoxyadenosine (5′-ScdA) has also been reported (Dizdaroglu and Jaruga, 2012). Oxo8dG is the most abundant ROS-related product of DNA oxidation, which has been implicated in mutagenesis (Lu and Liu, 2010; Damsma and Cramer, 2009). OGG1 is a bifunctional DG that removes oxidized bases such as 8-oxo-G, 2,6-diamino-4-hydroxy-5-formamidopyrimidine (FaPyG) and 7,8-dihydro-8-oxoadenine (8-oxo-A) from the DNA (Morales-Ruiz et al., 2003). This enzyme is specific for incising 8-oxoG to avoid the transversion of GC→TA, or DNA damage–induced apoptosis. Liu et al. have reported that OGG1 can rescue neurons subjected to ischemic conditions (Liu et al., 2011). It is known that acetylation of OGG1 can increase the activity of the enzyme nearly 10-fold in a cell culture (Bhakat et al., 2006), but the presence of in vivo acetylation has yet to be reported. Many investigators have focused on oxidation of DNA (Friedberg, 2003). Unless repaired before DNA replication, oxidation-mediated DNA damage may lead to mutagenesis and genomic instability, thus contributing to disease processes including carcinogenesis. In addition to ROS attack on DNA, interactions between methylglyoxal and DNA have also been reported. These interactions result in the formation of covalent adducts. In vitro studies have indicated that glycation adducts of double-stranded DNA result in the formation of N^2-carboxyethyl-2'-deoxyguanosine (CEdG). It is likely that similar adduct is formed in vivo (Frischmann et al., 2005) supporting the view that CEdG can be a useful biomarker for monitoring oxoaldehyde-induced stress in response to enhanced glycolytic flux or environmental exposure to methylglyoxal. Elevated levels of 8-oxodG and heptanoetheno (Hε) adducts have been reported to occur in lungs of mice exposed to nanosized-magnetite (MGT) (Totsuka et al., 2014), which is widely utilized in medicinal and industrial fields and is known to trigger for induction of gene mutations (Kew, 2013). Although Hε-dC induces C to T transitions in vitro (Pollack et al., 2006), the mechanisms of genotoxicity induced by MGT are not fully explained yet. Global discovery of DNA adducts in target organs will be useful information for exploring the mechanisms of genotoxicity. It should be noted that in liver, 5',8-cyclo-2'-deoxypurine (cdPu) DNA lesions contain an extra covalent bond formed between the C5 of the 2'-deoxyribose and the C8 of the purine. cdPus occur in two diastereomeric forms with a 5'R or 5'S configuration. It is reported that 180–320 cdPus per cell can be detected in fetal and postnatal rat liver per day (Chatgilialoglu et al., 2011). Because the extra covalent bond of a cdPu retards the breakdown of the glycosidic bond by DG, the lesion can only be repaired by nucleotide excision repair (NER) (Das et al., 2012). However, the rate of hydrolysis of a cdPu by NER is two- to fourfold less than that of removal of other bulky lesions such as a cis-B[α]P-N^2-dG adduct. This process results in the accumulation of a high level of cdPu lesions in genomic DNA, causing inhibition of DNA synthesis by human pol δ/ε (Kropachev et al., 2014), replication fork stalling, and inefficient lesion bypass DNA synthesis by translesion DNA polymerases (Kuraoka et al., 2000, 2001). This can subsequently lead to the accumulation of DNA strand breaks that can ultimately cause cell death. Because the bypass of a cdPu by translesion DNA polymerases, such as pol ζ and pol κ, can also produce A to T and G to T transversions as well as a G to A transition (You et al., 2013), accumulation of cdPus in the genome may result in mutagenesis and cell death, which is associated with aging, cancer, and neurodegeneration (Wang et al., 2012).

As stated earlier, neural cell response to DNA damage involves execution of DNA repair and activation of a repertoire of DNA damage signaling molecules [DNA damage response (DDR)]. As stated previously, the main DNA repair pathways involve the participation of NER, BER, mismatch repair (MMR), homologous recombination (HR), and nonhomologous end joining (NHEJ). These pathways are devoted to the repair of specific DNA alterations and complementary in some respects. NER is a multistep process that repairs the damage caused by UV-induced damage and bulky (Marteijn et al., 2014). BER corrects DNA damage from oxidation, deamination, and alkylation including SSBs, which are all lesions that cause little distortion to the DNA helix structure (Bauer et al., 2015). MMR corrects mismatches generated during DNA replication and escape proofreading (Jiricny, 2006). Recombinational repair deals with the most lethal form of DNA damage, double-strand breaks (DSB), by using a homologous DNA sequence as in the case of HR or requiring little or no sequence homology for efficient repair as in the case of NHEJ (Chapman et al., 2012). In the brain, BER is the main DNA repair mechanism in the removal of oxidized DNA bases and oxidized DNA break termini that are formed at high frequency in neuronal cells. BER involves five steps: (1) base removal by a specific DG, (2) incision at the resulting abasic site by an AP endonuclease, (3) formation of the gap, (4) gap filling by a DNA polymerase, and (5) resealing of the damaged DNA strand by a DNA ligase (Hegde et al., 2008, 2012).

Collective evidence suggests that oxidative damage to DNA, in particular DNA DSBs, is a potent inducer of a DDR. DDR is characterized by the activation of sensor kinases, including DNA protein kinase catalytic subunit and regions of DNA damage incorporating phosphorylated histone γH2AX (Shrivastav et al., 2008). Chronic oxidative stress, a persistent DDR, and telomere dysfunction may contribute to irreversible arrest of the cell cycle leading to cellular senescence (Rodier and Campisi, 2011). Studies on brain aging have also indicated that marked age-related changes occur in human astrocytes (Bhat et al., 2012) and mouse neurons (Jurk et al., 2012). At present, very little is known about the relationship among oxidative stress, DDR, and age-related signaling pathways in postmitotic neurons in the aged human brain and in patient with AD. Future studies on characterization of oxidative DNA damage on the neuronal molecular profile and any interaction with similar processes in the glial cell populations may potentially reveal novel dementia biomarkers, cellular pathways, and specific therapeutic targets.

EFFECTS OF OXIDATIVE STRESS ON MITOCHONDRIAL DEOXYRIBONUCLEIC ACID

Mitochondria are highly dynamic subcellular organelles involved in ATP production and cell survival in the face of environmental stresses. Subsequently, it is shown that electron transfer along mitochondrial respiratory chain also leads to the formation of free radicals. Change in overall mitochondrial mass is associated with alterations in balance between mitochondrial biogenesis and rates of mitophagy (Kelly and Scarpulla, 2004; Youle and Narendra, 2011). Mitochondria are composed of a smooth outer membrane surrounding an inner membrane of significantly larger surface area, which in turn, surrounds a protein-rich core, the matrix. Mitochondria contain 2–10 molecules of DNA called the mitochondrial DNA (mtDNA) (Di Mauro and Schon, 2003). Human mtDNA is a circular, double-stranded molecule of 16,569 base pairs. It contains 37 genes, including 13 protein-encoding genes, 22 transfer RNA (tRNA) genes, and two ribosomal RNA (rRNA) genes. All 13 protein-encoding genes are components of the mitochondrial respiratory chain, which is located in the inner membrane (DiMauro and Schon, 2003; DiMauro and Davidzon, 2005; Beal, 2005). They are encoded using a nonstandard genetic code, which requires its own translational machinery separate from that of the nucleus. Two rRNAs and 22 tRNAs involved in this mitochondrial protein synthesis are also encoded by mtDNA. mtDNA is densely packed into nucleoids, each containing as few as 1–2 mtDNA molecules (Kukat et al., 2011). mtDNA not only has a short half-life (Wallace, 1992) but also encodes few to no noncoding sequences between one another. Aging-related mtDNA damage results in decreased expression of mitochondrial electron transfer complexes, especially in complexes I and IV, because they contain a relatively large number of mtDNA-encoded subunits. The reduced activity of complex I further promotes the generation of ROS (Lin et al., 2002), establishing a vicious cycle (Kang et al., 2007). Most nonneural cells contain protective mechanisms to either breakdown or scavenge ROS. However, antioxidant systems are less active in the brain compared to other tissues. This leads in a persistent increase in levels of ROS reacting with the various target molecules (lipids, proteins, and nucleic acids) (Halliwell, 2006). Due to the lack of protective histones, low activity of antioxidant system, diminished DNA repair capability, and the proximity to the site of generation of ROS, mtDNA is more vulnerable to oxidative damage compared to nuclear DNA (Wang et al., 2014). The main products of mtDNA base damage are thymine glycol among pyrimidines (Wang et al., 1998). This metabolite has low mutagenicity. In contrast, among purines, the main mtDNA product is 7,8-dihydro-8-oxo-2′-deoxyguanosine (De Bont and van Larebeke, 2004; Bohr, 2002; Hirai et al., 2001), which can cause characteristic G→T transversions upon replication (Wang et al., 1998; Bohr, 2002). Large-scale sporadic mtDNA rearrangements are predominantly deletions up to 9 kb in size with one particular 5-kb deletion being the most common deletion. The molecular mechanisms associated with ROS-mediated mtDNA damage are not known. However, it is suggested that oxidative damage–associated SSBs or DSBs may be closely involved in the ROS-mediated mtDNA deletions. Indeed, numerous studies have indicated that somatic mutations in mtDNA progressively accumulate with age in a variety of tissues in humans, and importantly, terminally differentiated tissues with active oxidative metabolism such as the brain accumulate relatively higher levels of mutant mtDNA during the aging process, starting from the mid-30s in humans (Lee et al., 1994). Collective evidence suggests that oxidative stress markedly influences mtDNA in the aging brain producing mtDNA deletions. Unlike damage to other macromolecules, mtDNA damage can lead to mutations and deletions, which can be transmitted to and accumulated in daughter molecules through the process of replication, enabling deterioration of the integrity of hereditary information over time. This suggestion is based on mitochondrial theory of aging, which consist of the following basic tenets: (1) mitochondria are a significant source of ROS in the cell, (2) mitochondrial ROS inflict damage on mtDNA, (3) oxidative mtDNA damage results in mutations, (4) mtDNA mutations lead to the synthesis of defective polypeptide components of the ETC, and (5) incorporation of these defective subunits into the ETC results in a further increase in ROS production, initiating a "vicious" cycle of ROS production, mtDNA mutations, and mitochondrial dysfunction (Shokolenko et al., 2014). Although, the molecular mechanisms are not fully understood, oxidative damage–associated SSBs or DSBs may contribute to the deletions of mtDNA. As a result, these gradually accumulated mtDNA mutations may potentially produce a reduction in the efficiency of the ETC complex, decrease in ATP formation, and increase in ROS production. The increase in ROS production may produce more mtDNA mutations and create a positive feedback loop of increasing mutations and ROS production, followed by eventual cell death (Hsieh et al., 1994). Mutations in mtDNA may also contribute to the pathogenesis of depression, which is the leading risk factor for AD (Coskun et al., 2004). In addition, increased levels of 4-HNE in AD inhibit DNA synthesis (Farooqui, 2009). In AD, Tau dysfunction is accompanied by hyperphosphorylation and aggregation of Tau protein leading to neuron loss in the brain (Bloom, 2014). Tau depletion has been reported to protect against Aβ-associated neuron death (Leroy et al., 2012; Nussbaum et al., 2012). Furthermore, Tau has also been shown to induce chromatin relaxation, which subsequently leads to mtDNA damage and global changes in transcription (Frost et al., 2014). It should be noted that mtDNA deletions that are characteristic of diseases due to defects in mtDNA maintenance and accumulate in postmitotic tissues of aging humans are found at increased levels in many neurodegenerative diseases including AD (Krishnan et al., 2008). Collectively, these studies indicate that the accumulation of Aβ in AD promotes the formation of ROS, which may contribute to ROS-mediated mitochondrial dysfunction (Abramov et al., 2004) in large vulnerable neurons of the hippocampus and neocortex in AD brains. This observation has been confirmed by in situ hybridization (Hirai et al., 2001). A high incidence of ROS-mediated mtDNA base changes is also found in patients with Down syndrome who progress to have an AD-like dementia (Arbuzova et al., 2002). Presence of many more sporadic mutations is also found in the mtDNA control region in patients with AD compared to age-matched human control subjects (Coskun et al., 2004). These mutations occur at sites of known mtDNA transcription and replication regulatory elements; so, they may be involved in reduction in transcript levels of and deleterious functional consequences for mitochondrial homeostasis once they reach a critical mass in postmitotic cells in the brain. Many studies have also indicated that mtDNA is implicated in longevity (Brand et al., 1992). However, other investigators indicate that associations between mtDNA and longevity are weak (Castri et al., 2009). Investigations on mtDNA and aging support the view that the relationship between mtDNA variants and longevity may be very complex. Overall, these studies indirectly support the view that mtDNA variations may contribute to longevity.

In AD, elevated nitroxidative stress affects mtDNA due to nitrosation and/or nitration. Increased nitrosative stress may also result in deletion and/or mutation through oxidative and nitrosative modifications of mtDNA, which are particularly important, since they encode 13 polypeptides, all of which are subunits of the four mitochondrial ETC proteins (i.e., complexes I, III, IV, and V) (Chandra and Singh, 2011). mtDNA is more susceptible to oxidative/nitrative damage due to the absence of protective antioxidant protein catalase, histones, or polyamines and a relatively low activity of DNA repair enzyme in mitochondria compared to nuclei (Wan et al., 2004). It is also reported that the rate of mutation in mtDNA is 10-fold higher than that in the nuclear DNA (Monsalve et al., 2007). In addition, the level of GSH in mitochondria is relatively low compared to that in cytosol because it is imported into mitochondria through the specific GSH transporter protein due to the absence of its synthesis in mitochondria. Thus it is likely that oxidative and nitrosative damage and/or deletion of mtDNA may lead to reduction in expression and function of mitochondrial ETC proteins, contributing to greater ROS production.

Mitochondria have been reported to interact with the nucleus (Ryan and Hoogenraad, 2007). Although the precise details of these interactions remain elusive and controversial, there is ample evidence for crosstalk between the mitochondria and nucleus in humans. For example, many defects in mtDNA maintenance are caused by mutations in nuclear genes of the replisome (Kaukonen et al., 2000; Spelbrink et al., 2001) and the mitochondrial dysfunction, which results from mtDNA depletion or damage activates responses in a large number of nuclear genes (Hansson et al., 2004). Moreover, the interactions between mitochondria and nuclear DNA are important for longevity and aging in humans because nonrandom associations between mtDNA and nuclear variability have been reported to occur in centenarians (De Benedictis et al., 2000). While mitochondrial biogenesis occurs on a regular basis in healthy neural cells where mitochondria constantly divide and fuse with each other (Liu and Hajnóczky, 2011), it also occurs in response not only to ROS generation (oxidative stress) and increased energy demand but also to exercise training and certain neurodegenerative and neurotraumatic diseases (Farooqui, 2014). The status of mitochondrial biogenesis in AD neurons is unclear (Onyango et al., 2010). However, mitochondrial biogenesis is induced by the peroxisome proliferator-activated receptor γ coactivator-1α (PGC-1α), a master transcription factor that regulates cellular oxidant–antioxidant homeostasis by stimulating the gene expression of SOD2, catalase, GPx1, and uncoupling protein. PGC-1α activates different transcription factors, including nuclear respiratory factor (NRF)-1 and NRF-2 proteins and the mitochondrial transcription factor A (TFAM), which leads to increased mitochondrial DNA replication and gene transcription (Yin et al., 2008; Medeiros, 2008; Onyango et al., 2010). NRF-1 and NRF-2 regulate transcription of nuclear and mitochondrial genes involved in oxidative phosphorylation, electron transport (complexes I–V), mtDNA transcription/replication, heme biosynthesis, protein import/assembly, ion channels, shuttles, and translation (Kelly and Scarpulla, 2004). Collectively, these studies indicate that mitochondrial biogenesis is a complex process supported by the regulatory network of transcription factors that provides a dynamic coordinate control to nuclear and mitochondrial genes during development, adulthood, and neurodegenerative processes. New experimental approaches such as the delineation of tissue-specific mitochondrial proteomes (Mootha et al., 2003) can provide an excellent framework for future studies aimed at understanding the molecular events involved in defining the mitochondrial dysfunction in AD.

EFFECTS OF OXIDATIVE STRESS ON RIBONUCLEIC ACID

Oxidative damage to RNA occurs more frequently than DNA, not only because RNA molecules are mostly single stranded but also because RNA is less associated with proteins. Moreover, its bases are not protected by hydrogen bonding and are thus more easily accessible to ROS attack. RNA has an extensive cytoplasmic distribution in close proximity to mitochondria (Nunomura et al., 1999, 2009; Nunomura, 2013), where the majority of ROS is generated. In neural cells, RNA is more abundant than DNA; RNA probably carries most of the oxidized nucleotides under any condition. This presents a great challenge to RNA function; however, information about the potential effects of RNA oxidation is scarce. Earlier studies have indicated that normal turnover can take care of oxidized RNA. However, oxidation of RNA occurs in minutes and the half-life of human messenger

RNA (mRNA) is in the order of hours; there are many chances for the oxidized RNA to induce deleterious effects. RNA oxidation is also an important event in the pathogenesis of AD as up to 50% of mRNAs purified from AD frontal cortices were oxidized (Shan and Lin, 2006) and RNA oxidation occurred to a large extent in those neurons that are especially vulnerable to degeneration in AD (Nunomura et al., 1999).

As stated previously, damage to RNA is caused by HO· rising from $O_2^{·-}$ and H_2O_2 by the Fenton and Haber–Weiss reactions. In cytoplasm, cellular RNA is located close to mitochondria, which are the primary generator of ROS. ROS oxidizes purine and pyrimidine bases in RNA leading to the formation of 8-hydroxygauanosine (8-OHG) (Fig. 4.8). This metabolite serves as a sensitive biomarker for RNA oxidation (Wamer et al., 1997). The highly reactive hydroxyl radical first reacts with guanine to form a C8-OH adduct radical. Then the loss of an electron (e^-) and proton (H^+) generates 8-OHG (Wurtmann and Wolin, 2009). 8-OHG generation can alter genetic information by incorrectly pairing with adenine at similar or higher efficiency than with cytosine in RNA and producing mutations at the level of transcription (Neeley and Essigmann, 2006). Alterations of ribose, base excision, and strand break can represent other modifications induced by ROS on RNA (Rhee et al., 1995). In addition, presence of 8-OHG in RNA may interfere with the interaction between RNA and other cellular molecules.

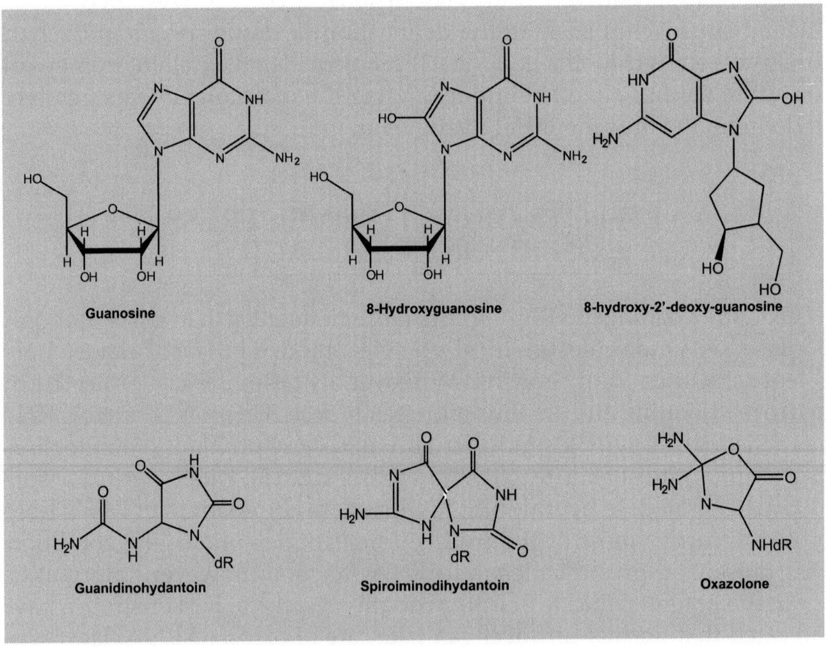

FIGURE 4.8 Chemical structures of oxidation products of RNA.

As an example, oxidative damage to RNA template may not only produce the block of reverse transcription (Rhee et al., 1995) but also may produce reduction in translation efficiency and abnormal protein production in mRNA (Shan and Lin, 2006) leading to ribosomal dysfunction (Ding et al., 2005). Quantitative analysis in AD brain indicates that some mRNA species are more susceptible to oxidative damage (Shan et al., 2003). Oxidative modification not only occurs in protein-coding RNAs but also in noncoding RNAs (rRNAs) and small nuclear RNAs. Damage to coding and noncoding RNAs may induce errors in proteins and alter the regulation of gene expression in AD (Nunomura et al., 2009). Significant mRNA oxidation (30%–70%) occurs in the frontal cortex of AD brain (Shan and Lin, 2006). In earlier stages of AD, levels of 8-OHG are elevated in the cytoplasm of AD hippocampus, frontal, and occipital neocortex, which is correlated with the $A\beta$ load supporting the view that perhaps increase in oxidative RNA damage occurs in the early stage of AD (Lovell and Markesbery, 2008; Nunomura et al., 1999, 2009). It is also reported that levels of 8-OHG in cerebrospinal fluid (CSF) of patients with AD is approximately fivefold higher than that in age-matched controls subjects. The levels of 8-OHG in CSF are decreased significantly with the duration of illness and the progression of cognitive dysfunctions (Abe et al., 2002). rRNA is also affected by oxidative damage. A significant decline in protein synthesis occurs in areas of the brain experiencing oxidative damage due to ribosomal dysfunction, featured by increased oxidation of rRNA (Ding et al., 2005). Another study suggests that the high oxidation potential of ribosomes from vulnerable hippocampal neurons in patients with AD may be related to the rRNA's high affinity for redox iron (Ding et al., 2005). This supports the view that RNA interacts with iron and RNA-bound iron plays a pivotal role in RNA oxidation. Thus cytoplasm of hippocampal neurons displays a significantly higher redox activity and iron staining in the brains of patients with AD than that in the brains of age-matched controls. Importantly, both iron staining and redox activity are susceptible to RNase treatment, suggesting a possible physical association between iron and RNA. When oxidized ribosomes were used for translation, protein synthesis was significantly reduced (Honda et al., 2005). These results are also supported by studies on old rats with memory loss (Shen et al., 2000) suggesting that introduction of 8-OHG in RNA may be associated with neurodegeneration and aging.

NITROSATIVE DAMAGE TO NUCLEIC ACIDS

$ONOO^-$ reacts with DNA leading to both sugar (Szabó and Ohshima, 1997; Kennedy et al., 1997) and nucleobase damage. Among four DNA bases, guanine, with the lowest oxidation potential, is the most reactive (Burney et al., 1999; Douki and Cadet, 1996). Major oxidation products arising from the reaction of dG with ONOO⁻ include 8-oxodG (Niles et al., 1999) oxazolone, and its precursor, imidazolone (Raoul et al., 1996) (Fig. 4.8). In addition, 8-oxodG is at least 1000 times more reactive toward ONOO⁻ than is deoxyguanine (Uppu et al., 1996), resulting in the formation of secondary oxidation products such as oxazolone (Tretyakova et al., 1999), spiroiminodihydantoin (Niles et al., 2006; Sawa and Ohshima, 2006; Yu et al., 2005), and guanidinohydantoin (Niles et al., 2004; Niles et al., 2006). These four DNA oxidation products are all promutagenic. 8-OxodG is a well-known DNA oxidation product and has been shown to cause G→T transversions in both prokaryotic and eukaryotic cells (Moriya et al., 1991). Onset of these reactions trigger a wide array of cellular responses ranging from subtle modulations of cell signaling to overwhelming oxidative injury, committing cells to necrosis or apoptosis. Thus, ONOO⁻-mediated DNA damage with the formation of 8-nitroguanine and 8-oxoguanine and breakage of DNA strand may cause cell death. Low concentrations of ONOO⁻ produce apoptotic death, whereas higher concentrations cause necrosis with alterations in cellular energetics (Abraham and Rabi, 2009; Chen et al., 2011). It is proposed that DNA damage mediated by ONOO⁻ triggers the activation of DNA repair systems. A DNA nick–sensing enzyme PARP-1 becomes activated upon detecting DNA breakage, and it cleaves NAD⁺ into nicotinamide and ADP-ribose and polymerizes the latter on nuclear acceptor proteins. Overactivation of PARP induced by ONOO⁻ consumes NAD⁺ and consequently ATP decreases, culminating in cell dysfunction, apoptosis, or necrosis. This mechanism may contribute to the pathogenesis of not only diabetes and cardiovascular diseases but also neurodegenerative diseases including AD (Szabo et al., 2001; Islam et al., 2015). In addition to oxidative damage, peroxynitrite can also nitrate guanine to form 8-nitroguanine and 5-guanidino-4-nitroimidazole (Niles et al., 2006; Sawa and Ohshima, 2006).

CONCLUSION

AD is characterized by a progressive decline in cognitive function, which typically begins with deterioration in memory. Neuropathologically, AD is characterized by the accumulation of senile plaques derived from the proteolytic cleavage of amyloid precursor protein and the presence of neurofibrillary tangles composed of an abnormally phosphorylated and aggregated microtubule-associated Tau protein. As a consequence of neurofibrillary tangles, AD is accompanied by the progressive loss of neurons and induction of oxidative stress and onset of inflammation around the senile plaques with reactive the accumulation of microglial cells. Nucleic acids are vulnerable to oxidative damage by ROS and ONOO⁻. ROS-mediated DNA damage results in formation of 8-hydroxyguanine, 8-hydroxyadenine,

2,6-diamino-4-hydroxy-5-formamidoguanine, and 4,6-diamino-5-formami-doadenine, whereas $ONOO^-$-mediated DNA damage produces 8-oxodG. Formation of the aforementioned nucleic acid-derived products in AD may contribute to neurodegeneration in AD. Similarly, ROS-mediated RNA damage has also been reported. A major product of RNA oxidation is 8-OHG. Oxidative damage to RNA may also aid neurodegenerative process in AD brain. Information on oxidation of DNA and RNA is still in a developing state. It remains an open question that how do cells handle oxidatively damaged DNA and RNA under normal physiological conditions? Are oxidized DNA and RNA degraded, repaired, or both? What are the repair/degradation processes and what proteins are involved in the processes? These are very important and challenging questions that require further research to move forward.

References

Abe, T., Tohgi, H., Isobe, C., Murata, T., Sato, C., 2002. Remarkable increase in the concentration of 8-hydroxyguanosine in cerebrospinal fluid from patients with Alzheimer's disease. J. Neurosci. Res. 70, 447–450.

Abraham, P., Rabi, S., 2009. Protein nitration, PARP activation and NAD+ depletion may play a critical role in the pathogenesis of cyclophosphamide-induced hemorrhagic cystitis in the rat. Cancer Chemother. Pharmacol. 64, 279–285.

Abramov, A.Y., Canevari, L., Duchen, M.R., 2004. β-Amyloid peptides induce mitochondrial dysfunction and oxidative stress in astrocytes and death of neurons through activation of NADPH oxidase. J. Neurosci. 24, 565–575.

Alonso-Fernandez, P., De la Fuente, M., 2011. Role of the immune system in aging and longevity. Curr. Aging Sci. 4, 78–100.

Alzheimer's Association, 2012. 2011 Alzheimer's diseases facts and figures: prevalence. Alzheimers Dement. 7, 12–13.

Alzheimer's Association, 2010. Alzheimer's disease facts and figures. Alzheimers Dement. 6, 158–194.

Alzheimer's Association, 2013. Alzheimer's disease facts and figures. Alzheimers Dement. 9, 208–245.

Angel, P., Karin, M., 1991. The role of Jun, Fos and the AP-1 complex in cell-proliferation and transformation. Biochim. Biophys. Acta 1072, 129–157.

Arbuzova, S., Hutchin, T., Cuckle, H., 2002. Mitochondrial dysfunction and Down's syndrome. Bioessays 24, 681–684.

Barciszewski, J., Barciszewska, M.Z., Siboska, G., Rattan, S.I.S., Clark, B.F.C., 1999. Some unusual nucleic acid bases are products of hydroxyl radical oxidation of DNA and RNA. Mol. Biol. Rep. 26, 231–238.

Bauer, N.C., Corbett, A.H., Doetsch, P.W., 2015. The current state of eukaryotic DNA base damage and repair. Nucleic Acids Res. 43, 10083–10101.

Beal, M.F., 2005. Mitochondria take center stage in aging and neurodegeneration. Ann. Neurol. 58, 495–505.

Bezprozvanny, I., Mattson, M.P., 2008. Neuronal calcium mishandling and the pathogenesis of Alzheimer's disease. Trends Neurosci. 31, 454–463.

Bhakat, K.K., Mokkapati, S.K., Boldogh, I., Hazra, T.K., Mitra, S., 2006. Acetylation of human 8-oxoguanine-DNA glycosylase by p300 and its role in 8-oxoguanine repair in vivo. Mol. Cell Biol. 26, 1654–1665.

Bhat, R., Crowe, E.P., Bitto, A., Moh, M., Katsetos, C.D., et al., 2012. Astrocyte senescence as a component of Alzheimer's disease. PLoS One 7, e45069.

Bloom, G.S., 2014. Amyloid-β and tau. JAMA Neurol. 71, 505.

Bohr, V.A., 2002. Repair of oxidative DNA damage in nuclear and mitochondrial DNA, and some changes with aging in mammalian cells. Free Radic. Biol. Med. 32, 804–812.

Brand, F.N., Kiely, D.K., Kannel, W.B., Myers, R.H., 1992. Family patterns of coronary heart disease mortality: the Framingham Longevity Study. J. Clin. Epidemiol. 45, 169–174.

Burney, S., Niles, J.C., Dedon, P.C., Tannenbaum, S.R., 1999. DNA damage in deoxynucleosides and oligonucleotides treated with peroxynitrite. Chem. Res. Toxicol. 12, 513–520.

Butterfield, S.M., Lashuel, H.A., 2010. Amyloidogenic protein-membrane interactions: mechanistic insight from model systems. Angew. Chem. 49, 5628–5654.

Butterfield, D.A., Perluigi, M., Reed, T., Muharib, T., Hughes, C.P., et al., 2012. Redox proteomics in selected neurodegenerative disorders: from its infancy to future applications. Antioxid. Redox Signal. 17, 1610–1655.

Cadet, J., Wagner, J.R., Shafirovich, V., Geacintov, N.E., 2014. One-electron oxidation reactions of purine and pyrimidine bases in cellular DNA. Int. J. Radiat. Biol. 90, 423–432.

Castri, L., Melendez-Obando, M., Villegas-Palma, R., Barrantes, R., Raventos, H., et al., 2009. Mitochondrial polymorphisms are associated both with increased and decreased longevity. Hum. Hered. 67, 147–153.

Cervellati, C., Romani, A., Seripa, D., Cremonini, E., Bosi, C., Magon, S., et al., 2014. Systemic oxidative stress and conversion to dementia of elderly patients with mild cognitive impairment. Biomed. Res. Int. 309507.

Chandra, D., Singh, K.K., 2011. Genetic insights into OXPHOS defect and its role in cancer. Biochim. Biophys. Acta 1807, 620–625.

Chapman, J.R., Taylor, M.R.G., Boulton, S.J., 2012. Playing the end game: DNA double-strand break repair pathway choice. Mol. Cell 47, 497–510.

Chatgilialoglu, C., Ferreri, C., Terzidis, M.A., 2011. Purine 5′,8-cyclonucleoside lesions: chemistry and biology. Chem. Soc. Rev. 40, 1368–1382.

Chen, W., Li, Y., Li, J., Han, Q., Ye, L., Li, A., 2011. Myricetin affords protection against peroxynitrite-mediated DNA damage and hydroxyl radical formation. Food Chem. Toxicol. 49, 2439–2444.

Choi, W.J., 2014. The Heterochromatin-1 phosphorylation contributes to TPA-Induced AP-1 expression. Biomol. Ther. (Seoul) 22, 308–313.

Chung, M.H., Kim, H.S., Ohtsuka, E., Kasai, H., Yamamoto, F., et al., 1991. An endonuclease activity in human polymorphonuclear neutrophils that removes 8-hydroxyguanine residues from DNA. Biochem. Biophys. Res. Commun. 178, 1472–1478.

Cooke, M.S., Evans, M.D., Dizdaroglu, M., Lunec, J., 2003. Oxidative DNA damage: mechanisms, mutation, and disease. FASEB J. 17, 1195–1214.

Coppedè, F., Migliore, L., 2015. DNA damage in neurodegenerative diseases. Mutat. Res. 776, 84–97.

Coskun, P.E., Beal, M.F., Wallace, D.C., 2004. Alzheimer's brains harbor somatic mtDNA control-region mutations that suppress mitochondrial transcription and replication. Proc. Natl. Acad. Sci. U. S. A. 101, 10726–10731.

da Silva, G.F., Lykourinou, V., Angerhofer, A., Ming, L.J., 2009. Methionine does not reduce Cu(II)-β-amyloid!—rectification of the roles of methionine-35 and reducing agents in metal-centered oxidation chemistry of Cu(II)-β-amyloid. Biochim. Biophys. Acta 1792, 49–55 67.

Damsma, G.E., Cramer, P., 2009. Molecular basis of transcriptional mutagenesis at 8-oxoguanine. J. Biol. Chem. 284, 31658–31663.

Das, R.S., Samaraweera, M., Morton, M., Gascon, J.A., Basu, A.K., 2012. Stability of N-glycosidic bond of (5′S)-8,5′-cyclo-2′-deoxyguanosine. Chem. Res. Toxicol. 25, 2451–2461.

De Benedictis, G., Carrieri, G., Garasto, S., Rose, G., Varcasia, O., et al., 2000. Does a retrograde response in human aging and longevity exist? Exp. Gerontol. 35, 795–801.

De Bont, R., van Larebeke, N., 2004. Endogenous DNA damage in humans: a review of quantitative data. Mutagenesis 19, 169–185.

DeCarli, C., 2003. Mild cognitive impairment: prevalence, prognosis, aetiology, and treatment. Lancet Neurol. 2, 15–21.

DiMauro, S., Davidzon, G., 2005. Mitochondrial DNA and disease. Ann. Med. 37, 222–232.

DiMauro, S., Schon, E.A., 2003. Mitochondrial respiratory-chain diseases. N. Engl. J. Med. 348, 2656–2668.

Ding, Q., Markesbery, W.R., Chen, Q., Li, F., Keller, J.N., 2005. Ribosome dysfunction is an early event in Alzheimer's disease. J. Neurosci. 25, 9171–9175.

Dizdaroglu, M., Jaruga, P., 2012. Mechanisms of free radical-induced damage to DNA. Free Radic. Res. 46, 382–419.

Douki, T., Cadet, J., 1996. Peroxynitrite mediated oxidation of purine bases of nucleosides and isolated DNA. Free Radic. Res. 24, 369–380.

Droge, W., 2002. Free radicals in the physiological control of cell function. Physiol. Rev. 82, 47–95.

Englander, E.W., 2008. Brain capacity for repair of oxidatively damaged DNA and preservation of neuronal function. Mech. Ageing Dev. 129, 475–482.

Farooqui, A.A., 2009. Hot Topics in Neural Membrane Lipidology. Springer, New York, NY.

Farooqui, A.A., 2010. Neurochemical Aspects of Neurotraumatic and Neurodegenerative Diseases. Springer, New York.

Farooqui, A.A., 2014. Inflammation and Oxidative Stress in Neurological Disorders. Springer, New York, NY.

Fleming, A.M., Alshykhly, O.R., Zhu, J., Muller, J.G., Burrows, C.J., 2015. Rates of chemical cleavage of DNA and RNA oligomers containing guanine oxidation products. Chem. Res. Toxicol. 28, 1292–1300.

Ford, L., Crossley, M., Williams, T., Thorpe, J.R., Serpell, L.C., et al., 2015. Effects of Aβ exposure on long-term associative memory and its neuronal mechanisms in a defined neuronal network. Sci. Rep. 5, 10614.

Friedberg, E.C., 2003. DNA damage and repair. Nature 421, 436–440.

Frischmann, M., Bidmon, C., Angerer, J., Pischetsrieder, M., 2005. Identification of DNA adducts of methylglyoxal. Chem. Res. Toxicol. 18, 1586–1592.

Frost, B., Hemberg, M., Lewis, J., Feany, M.B., 2014. Tau promotes neurodegeneration through global chromatin relaxation. Nat. Neurosci. 17, 357–366.

Gajewski, E., Rao, G., Nackerdien, Z., Dizdaroglu, M., 1990. Modification of DNA bases in mammalian chromatin by radiation-generated free radicals. Biochemistry 29, 7876–7882.

Gleichmann, M., Mattson, M.P., 2011. Neuronal calcium homeostasis and dysregulation. Antioxid. Redox Signal. 14, 1261–1273.

Grune, T., Merker, K., Jung, T., Sitte, N., Davies, K.J., 2005. Protein oxidation and degradation during postmitotic senescence. Free Radic. Biol. Med. 39, 1208–1215.

Halliwell, B., 2006. Oxidative stress and neurodegeneration: where are we now? J. Neurochem. 97, 1634–1658.

Hamilton, M.L., Van Remmen, H., Drake, J.A., Yang, H., Guo, Z.M., et al., 2001. Does oxidative damage to DNA increase with age? Proc. Natl. Acad. Sci. U. S. A. 98, 10469–10474.

Hansson, A., Hance, N., Dufour, E., Rantanen, A., Hultenby, K., Clayton, D.A., et al., 2004. A switch in metabolism precedes increased mitochondrial biogenesis in respiratory chain-deficient mouse hearts. Proc. Natl. Acad. Sci. U. S. A. 101, 3136–3141.

Hebert, L.E., Scherr, P.A., Bienias, J.L., Bennett, D.A., Evans, D.A., 2003. Alzheimer disease in the US population: prevalence estimates using the 2000 census. Arch. Neurol. 60, 1119–1122.

Hegde, M.L., Hazra, T.K., Mitra, S., 2008. Early steps in the DNA base excision/single-strand interruption repair pathway in mammalian cells. Cell Res. 18, 27–47.

Hegde, M.L., Mantha, A.K., Hazra, T.K., Bhakat, K.K., Mitra, S., et al., 2012. Oxidative genome damage and its repair: implications in aging and neurodegenerative diseases. Mech. Ageing Dev. 133, 157–168.

Hirai, K., Aliev, G., Nunomura, A., Fujioka, H., Russell, R.L., Atwood, C.S., et al., 2001. Mitochondrial abnormalities in Alzheimer's disease. J. Neurosci. 21, 3017–3023.

Honda, K., Smith, M.A., Zhu, X., Baus, D., Merrick, W.C., Tartakoff, A.M., et al., 2005. Ribosomal RNA in Alzheimer disease is oxidized by bound redox-active iron. J. Biol. Chem. 280, 20978–20986.

Hsieh, R.H., Hou, J.H., Hsu, H.S., Wei, Y.H., 1994. Age-dependent respiratory function decline and DNA deletions in human muscle mitochondria. Biochem. Mol. Biol. Int. 32, 1009–1022.

Huang, L.E., Arany, Z., Livingston, D.M., Bunn, H.F., 1996. Activation of hypoxia-inducible transcription factor depends primarily upon redox-sensitive stabilization of its alpha subunit. J. Biol. Chem. 271, 32253–32259.

Huang, X., Moir, R.D., Tanzi, R.E., Bush, A.I., Rogers, J.T., 2004. Redox-active metals, oxidative stress, and Alzheimer's disease pathology. Ann. N. Y. Acad. Sci. 1012, 153–163.

Huang, W.J., Zhang, X., Chen, W.W., 2016. Role of oxidative stress in Alzheimer's disease. Biomed. Rep. 4, 519–522.

Huhmer, A.F., Nishida, C.R., Ortiz de Montellano, P.R., Schoneich, C., 1997. Inactivation of the inducible nitric oxide synthase by peroxynitrite. Chem. Res. Toxicol. 10, 618–626.

Iida, T., Furuta, A., Nishioka, K., Nakabeppu, Y., Iwaki, T., 2002. Expression of 8-oxoguanine DNA glycosylase is reduced and associated with neurofibrillary tangles in Alzheimer's disease brain. Acta Neuropathol. 103, 20–25.

Islam, B.U., Habib, S., Ahmad, P., Allarakha, S., Moinuddin, A.A., 2015. Pathophysiological role of peroxynitrite induced DNA damage in human diseases: a special focus on poly(ADP-ribose) polymerase (PARP). Indian J. Clin. Biochem. 30, 368–385.

Itoh, K., Wakabayashi, N., Katoh, Y., Ishii, T., O'Connor, T., Yamamoto, M., 2003. Keap1 regulates both cytoplasmic-nuclear shuttling and degradation of Nrf2 in response to electrophiles. Genes Cells 8, 379–391.

Jaakkola, P., Mole, D.R., Tian, Y.M., Wilson, M.I., Gielbert, J., Gaskell, S.J., et al., 2001. Targeting of HIF-1α to the von Hippel-Lindau ubiquitylation complex by O_2-regulated prolyl hydroxylation. Science 292, 468–472.

Jiricny, J., 2006. The multifaceted mismatch-repair system. Nat. Rev. Mol. Cell Biol. 7, 335–346.

Jurk, D., Wang, C., Miwa, S., Maddick, M., Korolchuk, V., et al., 2012. Postmitotic neurons develop a p21-dependent senescence-like phenotype driven by a DNA damage response. Aging Cell 11, 996–1004.

Kalaria, R.N., Maestre, G.E., Arizaga, R., et al., 2008. Alzheimer's disease and vascular dementia in developing countries: prevalence, management, and risk factors. Lancet Neurol. 7, 812–826.

Kang, D., Kim, S.H., Hamasaki, N., 2007. Mitochondrial transcription factor A (TFAM): roles in maintenance of mtDNA and cellular functions. Mitochondrion 7, 39–44.

Kaukonen, J., Juselius, J.K., Tiranti, V., Kyttälä, A., Zeviani, M., et al., 2000. Role of adenine nucleotide translocator 1 in mtDNA maintenance. Science 289, 782–785.

Kelly, D.P., Scarpulla, R.C., 2004. Transcriptional regulatory circuits controlling mitochondrial biogenesis and function. Genes Dev. 18, 357–368.

Kennedy, L.J., Moore Jr., K., Caulfield, J.L., Tannenbaum, S.R., Dedon, P.C., 1997. Quantitation of 8-oxoguanine and strand breaks produced by four oxidizing agents. Chem. Res. Toxicol. 10, 386–392.

Kew, M.C., 2013. Aflatoxins as a cause of hepatocellular carcinoma. J. Gastrointestin. Liver Dis. 22, 305–310.

Krishnan, K.J., Reeve, A.K., Samuels, D.C., Chinnery, P.F., Blackwood, J.K., Taylor, R.W., et al., 2008. What causes mitochondrial DNA deletions in human cells? Nat. Genet. 40, 275–279.

Kropachev, K., Ding, S., Terzidis, M.A., Masi, A., Liu, Z., et al., 2014. Structural basis for the recognition of diastereomeric 5',8-cyclo-2'-deoxypurine lesions by the human nucleotide excision repair system. Nucleic Acids Res. 42, 5020–5032.

Kukat, C., Wurm, C.A., Spåhr, H., Falkenberg, M., Larsson, N.G., et al., 2011. Super-resolution microscopy reveals that mammalian mitochondrial nucleoids have a uniform size and frequently contain a single copy of mtDNA. Proc. Natl. Acad. Sci. U. S. A. 108, 13534–13539.

Kumar, A., Singh, A., Ekavali, M., 2015. A review on Alzheimer's disease pathophysiology and its management: an update. Pharmacol. Rep. 67, 195–203.

Kuperstein, I., Broersen, K., Benilova, I., Rozenski, J., Jonckheere, W., et al., 2010. Neurotoxicity of Alzheimer's disease $A\beta$ peptides is induced by small changes in the $A\beta_{42}$ to $A\beta_{40}$ ratio. EMBO J. 29, 3408–3420.

Kuraoka, I., Bender, C., Romieu, A., Cadet, J., Wood, R.D., Lindahl, T., 2000. Removal of oxygen free-radical-induced 5',8-purine cyclodeoxynucleosides from DNA by the nucleotide excision-repair pathway in human cells. Proc. Natl. Acad. Sci. U. S. A. 97, 3832–3837.

Kuraoka, I., Robins, P., Masutani, C., Hanaoka, F., Gasparutto, D., Cadet, J., et al., 2001. Oxygen free radical damage to DNA. Translesion synthesis by human DNA polymerase eta and resistance to exonuclease action at cyclopurine deoxynucleoside residues. J. Biol. Chem. 276, 49283–49288.

Lacor, P.N., Buniel, M.C., Furlow, P.W., Clemente, A.S., Velasco, P.T., et al., 2007. $A\beta$ oligomer-induced aberrations in synapse composition, shape, and density provide a molecular basis for loss of connectivity in Alzheimer's disease. J. Neurosci. 27, 796–807.

Lee, H.C., Pang, C.Y., Hsu, H.S., Wei, Y.H., 1994. Differential accumulations of 4,977bp deletion in mitochondrial DNA of various tissues in human ageing. Biochim. Biophys. Acta 1226, 37–43.

Leroy, K., Ando, K., Laporte, V., Dedecker, R., Suain, V., Octave, J., et al., 2012. Lack of Tau proteins rescues neuronal cell death and decreases amyloidogenic processing of APP in APP/PS1 mice. AJPA 181, 1928–1940.

Lin, M.T., Simon, D.K., Ahn, C.H., Kim, L.M., Beal, M.F., 2002. High aggregate burden of somatic mtDNA point mutations in aging and Alzheimer's disease brain. Hum. Mol. Genet. 11, 133–145.

Liu, X., Hajnóczky, G., 2011. Altered fusion dynamics underlie unique morphological changes in mitochondria during hypoxia-reoxygenation stress. Cell Death Differ. 18, 1561–1572.

Liu, Y.V., Baek, J.H., Zhang, H., Diez, R., Cole, R.N., Semenza, G.L., 2007. RACK1 competes with HSP90 for binding to HIF-1α and is required for O_2-independent and HSP90 inhibitor-induced degradation of HIF-1 α. Mol. Cell 25, 207–217.

Liu, D., Croteau, D.L., Souza-Pinto, N., Pitta, M., Tian, J., et al., 2011. Evidence that OGG1 glycosylase protects neurons against oxidative DNA damage and cell death under ischemic conditions. J. Cereb. Blood Flow Metab. 31, 680–692.

Lovell, M.A., Markesbery, W.R., 2007. Oxidative DNA damage in mild cognitive impairment and late-stage Alzheimer's disease. Nucleic Acids Res. 35, 7497–7504.

Lovell, M.A., Markesbery, W.R., 2008. Oxidative DNA damage in mild cognitive impairment and late-stage Alzheimer's disease. Nucleic Acids Res. 35, 7497–7504.

Lovell, M.A., Xie, C., Markesbery, W.R., 2000. Decreased base excision repair and increased helicase activity in Alzheimer's disease brain. Brain Res. 855, 116–123.

Lu, J., Liu, Y., 2010. Deletion of Ogg1 DNA glycosylase results in telomere base damage and length alteration in yeast. EMBO J. 29, 398–409.

Mantha, A.K., Sarkar, B., Tell, G., 2013. A short review on the implications of base excision repair pathway for neurons: relevance to neurodegenerative diseases. Mitochondrion 2013 (16), 38–49.

Markesbery, W.R., Lovell, M.A., 2006. DNA oxidation in Alzheimer's disease. Antioxid. Redox Signal. 8, 2039–2045.

Marteijn, J.A., Lans, H., Vermeulen, W., Hoeijmakers, J.H.J., 2014. Understanding nucleotide excision repair and its roles in cancer and ageing. Nat. Rev. Mol. Cell Biol. 15, 465–481.

Mata-Garrido, J., Casafont, I., Tapia, O., Berciano, M.T., Lafarga, M., 2016. Neuronal accumulation of unrepaired DNA in a novel specific chromatin domain: structural, molecular and transcriptional characterization. Acta Neuropathol. Commun. 4, 41.

Mattson, M.P., 2004. Metal-catalyzed disruption of membrane protein and lipid signaling in the pathogenesis of neurodegenerative disorders. Ann. N. Y. Acad. Sci. 1012, 37–50.

Maxwell, P.H., Wiesener, M.S., Chang, G.W., Clifford, S.C., Vaux, E.C., Cockman, M.E., et al., 1999. The tumor suppressor protein VHL targets hypoxia-inducible factors for oxygen-dependent proteolysis. Nature 399, 271–275.

Maynard, S., de Souza-Pinto, N.C., Scheibye-Knudsen, M., Bohr, V.A., 2010. Mitochondrial base excision repair assays. Methods 51, 416–425.

Medeiros, D.M., 2008. Assessing mitochondria biogenesis. Methods 46, 288–294.

Meneghini, R., 1997. Iron homeostasis, oxidative stress, and DNA damage. Free Radic. Biol. Med. 23, 783–792.

Metcalf, D.J., Garcia-Arencibia, M., Hochfeld, W.E., Rubinsztein, D.C., 2012. Autophagy and misfolded proteins in neurodegeneration. Exp. Neurol. 238, 22–28.

Migliore, L., Fontana, I., Trippi, F., Colognato, R., Coppedè, F., et al., 2005. Oxidative DNA damage in peripheral leukocytes of mild cognitive impairment and AD patients. Neurobiol. Aging 26, 567–573.

Mitra, J., Vasquez, V., Hegde, P.M., Boldogh, I., Mitra, S., Mitra, J., Vasquez, V., Hegde, P.M., Boldogh, I., Mitra, S., Kent, T.A., Rao, K.S., Hegde, M.L., et al., 2014. Neurol. Res. Ther. 1 (2) pii: 107.

Monsalve, M., Borniquel, S., Valle, I., Lamas, S., 2007. Mitochondrial dysfunction in human pathologies. Front. Biosci. 12, 1131–1153.

Mootha, V.K., Bunkenborg, J., Olsen, J.V., Hjerrid, M., Wisniewski, J.R., et al., 2003. Integrated analysis of protein composition, tissue diversity, and gene regulation in mouse mitochondria. Cell 2003 (115), 629–640.

Morales-Ruiz, T., Birincioglu, M., Jaruga, P., Rodriguez, H., Roldan-Arjona, T., Dizdaroglu, M., 2003. Arabidopsis thaliana Ogg1 protein excises 8-hydroxyguanine and 2,6-diamino-4-hydroxy-5-formamidopyrimidine from oxidatively damaged DNA containing multiple lesions. Biochemistry 42, 3089–3095.

Morgan, B., Sobotta, M.C., Dick, T.P., 2011. Measuring E(GSH) and H_2O_2 with roGFP2-based redox probes. Free Radic. Biol. Med. 51, 1943–1951.

Moriya, M., Ou, C., Bodepudi, V., Johnson, F., Takeshita, M., Grollman, A.P., 1991. Site-specific mutagenesis using a gapped duplex vector: a study of translesion synthesis past 8-oxodeoxyguanosine in E. coli. Mutat. Res. 254, 281–288.

Muller, F.L., Liu, Y., Van Remmen, H., 2004. Complex III releases superoxide to both sides of the inner mitochondrial membrane. J. Biol. Chem. 279, 49064–49073.

Murphy, M.P., 2009. How mitochondria produce reactive oxygen species. Biochem. J. 417, 1–13.

Neeley, W.L., Essigmann, J.M., 2006. Mechanisms of formation, genotoxicity, and mutation of guanine oxidation products. Chem. Res. Toxicol. 19, 491–505.

Niles, J.C., Burney, S., Singh, S., Wishnok, J.S., Tannenbaum, S.R., 1999. Peroxynitrite reaction products of 3',5'-di-O-acetyl-8-oxo-7,8-dihydro-2'-deoxyguanosine. Proc. Natl. Acad. Sci. U. S. A. 96, 11729–11734.

Niles, J.C., Wishnok, J.S., Tannenbaum, S.R., 2004. Spiroiminodihydantoin and guanidinohydantoin are the dominant reaction products of 8-oxoguanosine oxidation at low fluxes of peroxynitrite: mechanistic studies with ^{18}O. Chem. Res. Toxicol. 17, 1510–1519.

Niles, J.C., Wishnok, J.S., Tannenbaum, S.R., 2006. Peroxynitrite-induced oxidation and nitration products of guanine and 8-oxoguanine: structures and mechanisms of product formation. Nitric Oxide 14, 109–121.

Nishi, T., Shimizu, N., Hiramoto, M., Sato, I., Yamaguchi, Y., Hasegawa, M., Aizawa, S., Tanaka, H., Kataoka, K., Watanabe, H., Handa, H., 2002. Spatial redox regulation of a critical cysteine residue of NF-kappa B in vivo. J. Biol. Chem. 277, 44548–44556.

Nunomura, A., Perry, G., Pappolla, M.A., et al., 1999. RNA oxidation is a prominent feature of vulnerable neurons in Alzheimer's disease. J. Neurosci. 19, 1959–1964.

Nunomura, A., Perry, G., Aliev, G., Hirai, K., Takeda, A., et al., 2001. Oxidative damage is the earliest event in Alzheimer disease. J. Neuropathol. Exp. Neurol. 60, 759–767.

Nunomura, A., Hofer, T., Moreira, P.T., Castellani, R.J., Smith, M.A., Perry, G., 2009. RNA oxidation in Alzheimer disease and related neurodegenerative disorders. Acta Neuropathol. 118, 151–166.

Nunomura, A., 2013. Role of oxidative RNA damage in aging and neurodegenerative disorders. Brain Nerve 65, 179–194.

Nussbaum, J.M., Schilling, S., Cynis, H., Silva, A., Swanson, E., Wangsanut, T., et al., 2012. Prion-like behaviour and tau-dependent cytotoxicity of pyroglutamylated amyloid β. Nature 485, 651–655.

Ong, W.Y., Tanaka, K., Dawe, G.S., Ittner, L.M., Farooqui, A.A., 2013. Slow excitotoxicity in Alzheimer's disease. J. Alzheimers Dis. 35, 643–668.

Onyango, I.G., Lu, J., Rodova, M., Lezi, E., Crafter, A.B., et al., 2010. Regulation of neuron mitochondrial biogenesis and relevance to brain health. Biochim. Biophys. Acta 1802, 228–234.

Ozanne, B.W., Spence, H.J., McGarry, L.C., Hennigan, R.F., 2007. Transcription factors control invasion: AP-1 the first among equals. Oncogene 26, 1–10.

Pacher, P., Beckman, J.S., Liaudet, L., 2007. Nitric oxide and peroxynitrite in health and disease. Physiol. Rev. 87, 315–424.

Pal, A., Badyal, R.K., Vasishta, R.K., Attri, S.V., Thapa, B.R., et al., 2013. Biochemical, histological, and memory impairment effects of chronic copper toxicity: a model for non-Wilsonian brain copper toxicosis in Wistar rat. Biol. Trace Elem. Res. 153, 257–268.

Perluigi, M., Swomley, A.M., Butterfield, D.A., 2014. Redox proteomics and the dynamic molecular landscape of the aging brain. Ageing Res. Rev. 13, 75–89.

Phillis, J.W., Horrocks, L.A., Farooqui, A.A., 2006. Cyclooxygenases, lipoxygenases, and epoxygenases in CNS: their role and involvement in neurological disorders. Brain Res. Rev. 52, 201–243.

Pineda-Molina, E., Klatt, P., Vazquez, J., Marina, A., Garcia de Lacoba, M., Perez-Sala, D., Lamas, S., 2001. Glutathionylation of the p50 subunit of NF-kappaB: a mechanism for redox-induced inhibition of DNA binding. Biochemistry 40, 14134–14142.

Pogue, A.I., Li, Y.Y., Cui, J.G., Zhao, Y., Kruck, T.P., Percy, M.E., et al., 2009. Characterization of an NF-kappaB-regulated, miRNA-146a-mediated down-regulation of complement factor H (CFH) in metal-sulfate-stressed human brain cells. J. Inorg. Biochem. 103, 1591–1595.

Pollack, M., Yang, I.Y., Kim, H.Y., Blair, I.A., et al., 2006. Translesion DNA synthesis across the heptanone-etheno-2'-deoxycytidine adduct in cells. Chem. Res. Toxicol. 19, 1074–1079.

Praticò, D., 2008. Oxidative stress hypothesis in Alzheimer's disease: a reappraisal. Trends Pharmacol. Sci. 29, 609–615.

Pronk, T.E., van der Veen, J.W., Vandebriel, R.J., van Loveren, H., de Vink, E.P., Pennings, J.L., 2014. Comparison of the molecular topologies of stress-activated transcription factors HSF1, AP-1, NRF2 and NF-κB in their induction kinetics of HMOX1. Biosystems 124, 75–85.

Ramamoorthy, M., Sykora, P., Scheibye-Knudsen, M., Dunn, C., Kasmer, C., et al., 2012. Sporadic Alzheimer disease fibroblasts display an oxidative stress phenotype. Free Radic. Biol. Med. 53, 1371–1380.

Raoul, S., Berger, M., Buchko, G.W., Joshi, P.C., Morin, B., Weinfeld, M., Cadet, J., 1996. Novel oxidation products of deoxyguanosine:oxazolone and imidazolone nucleosides. J. Chem. Soc. Perkin Trans. 2 2, 371–381.

Rhee, Y., Valentine, M.R., Termini, J., 1995. Oxidative base damage in RNA detected by reverse transcriptase. Nucleic Acids Res. 23, 3275–3282.

Rodier, F., Campisi, J., 2011. Four faces of cellular senescence. J. Cell Biol. 192, 547–556.

Ryan, M.T., Hoogenraad, N.J., 2007. Mitochondrial-nuclear communications. Annu. Rev. Biochem. 76, 701–722.

Sawa, T., Ohshima, H., 2006. Nitrative DNA damage in inflammation and its possible role in carcinogenesis. Nitric Oxide 14, 91–100.

Sedelnikova, O.A., Horikawa, I., Zimonjic, D.B., Popescu, N.C., Bonner, W.M., Barrett, J.C., 2004. Senescing human cells and ageing mice accumulate DNA lesions with unrepairable double-strand breaks. Nat. Cell Biol. 6, 168–170.

Shackelford, D.A., 2006. DNA end joining activity is reduced in Alzheimer's disease. Neurobiol. Aging 7, 596–605.

Shan, X., Lin, C.L.G., 2006. Quantification of oxidized RNAs in Alzheimer's disease. Neurobiol. Aging 27, 657–662.

Shan, X., Tashiro, H., Lin, C.L., 2003. The identification and characterization of oxidized RNAs in Alzheimer's disease. J. Neurosci. 23, 4913–4921.

Shankar, G.M., Bloodgood, B.L., Townsend, M., Walsh, D.M., Selkoe, D.J., et al., 2007. Natural oligomers of the Alzheimer amyloid-β protein induce reversible synapse loss by modulating an NMDA-type glutamate receptor-dependent signaling pathway. J. Neurosci. 27, 2866–2875.

Shankar, G.M., Li, S., Mehta, T.H., Garcia-Munoz, A., Shepardson, N.E., et al., 2008. Amyloid-β protein dimers isolated directly from Alzheimer's brains impair synaptic plasticity and memory. Nat. Med. 14, 837–842.

Shaulian, E., Karin, M., 2001. AP-1 in cell proliferation and survival. Oncogene 20, 2390–2400.

Shaulian, E., Karin, M., 2002. AP-1 as a regulator of cell life and death. Nat. Cell Biol. 4, E131–E136.

Shen, Z., Wu, W., Hazen, S.L., 2000. Activated leukocytes oxidatively damage DNA, RNA, and the nucleotide pool through halide-dependent formation of hydroxyl radical. Biochemistry 39, 5474–5482.

Shokolenko, I.N., Wilson, G.L., Alexeyev, M.F., 2014. Aging: a mitochondrial DNA perspective, critical analysis and an update. World J. Exp. Med. 4, 46–57.

Shrivastav, M., De Haro, L.P., Nickoloff, J.A., 2008. Regulation of DNA double-strand break repair pathway choice. Cell Res. 18, 134–147.

Spelbrink, J., Li, F.Y., Tiranti, V., Nikali, K., Yuan, Q.P., et al., 2001. Human mitochondrial DNA deletions associated with mutations in the gene for twinkle, a phage t7 gene 4-like protein localized to mitochondrial nucleoids. Nat. Genet. 28, 223–231.

Steenken, S., Jovanovic, S.V., 1997. How easily oxidizable is DNA? One-electron reduction potentials of adenosine and guanosine radicals in aqueous solution. J. Am. Chem. Soc. 119, 617–618.

Sun, G.Y., Horrocks, L.A., Farooqui, A.A., 2007. The role of NADPH oxidase and phospholipases A_2 in mediating oxidative and inflammatory responses in neurodegenerative diseases. J. Neurochem. 103, 1–16.

Swerdlow, R.H., 2011. Brain aging, Alzheimer's disease, and mitochondria. Biochim. Biophys. Acta 1812, 1630–1639.

Szabó, C., Ohshima, H., 1997. DNA damage induced by peroxynitrite: subsequent biological effects. Nitric Oxide 1, 373–385.

Szabo, E., Virag, L., Bakondi, E., Gyure, L., Hasko, G., Bai, P., Hunyadi, J., Gergely, P., Szabo, C., 2001. Peroxynitrite production, DNA breakage, and poly(ADP-ribose) polymerase activation in a mouse model of oxazolone-induced contact hypersensitivity. J. Invest. Dermatol. 117, 74–80.

Thannickal, V.J., Fanburg, B.L., 2000. Reactive oxygen species in cell signaling. Am. J. Physiol. Lung Cell. Mol. Physiol. 279, L1005–L1028.

Toledano, M.B., Leonard, W.J., 1991. Modulation of transcription factor NF-kappa B binding activity by oxidation-reduction in vitro. Proc. Natl. Acad. Sci. U. S. A. 88, 4328–4332.

Totsuka, Y., Ishino, K., Kato, T., Goto, S., Tada, Y., Nakae, D., et al., 2014. Magnetite nanoparticles induce genotoxicity in the lungs of mice via inflammatory response. Nanomaterials 4, 175–188.

Tretyakova, N.Y., Niles, J.C., Burney, S., Wishnok, J.S., Tannenbaum, S.R., 1999. Peroxynitrite-induced reactions of synthetic oligonucleotides containing 8-oxoguanine. Chem. Res. Toxicol. 12, 459–466.

Uppu, R.M., Cueto, R., Squadrito, G.L., Salgo, M.G., Pryor, W.A., 1996. Competitive reactions of peroxynitrite with 2′-deoxyguanosine and 7,8-dihydro-8-oxo-2′-deoxyguanosine (8-oxodG): relevance to the formation of 8-oxodG in DNA exposed to peroxynitrite. Free Radic. Biol. Med. 21, 407–411.

Valko, M., Leibfritz, D., Moncol, J., Cronin, M.T., Mazur, M., Telser, J., 2007. Free radicals and antioxidants in normal physiological functions and human disease. Int. J. Biochem. Cell Biol. 39, 44–84.

Viola, K.L., Klein, W.L., 2015. Amyloid β oligomers in Alzheimer's disease pathogenesis, treatment, and diagnosis. Acta Neuropathol. 129, 183–206.

von Figura, G., Hartmann, D., Song, Z., Rudolph, K.L., 2009. Role of telomere dysfunction in aging and its detection by biomarkers. J. Mol. Med. 87, 1165–1171.

Wallace, D.C., 1992. Mitochondrial genetics: a paradigm for aging and degenerative diseases? Science 256, 628–632.

Walsh, D.M., Klyubin, I., Fadeeva, J.V., Cullen, W.K., Anwyl, R., et al., 2002. Naturally secreted oligomers of amyloid beta protein potently inhibit hippocampal long-term potentiation in vivo. Nature 416, 535–539.

Wan, J., Bae, M.A., Song, B.J., 2004. Acetoaminophen-induced accumulation of 8-oxodeoxy-guanosine through reduction of Ogg1 DNA repair enzyme in C6 glioma cells. Exp. Mol. Med. 36, 71–77.

Warner, W.G., Yin, J.J., Wei, R.R., 1997. Oxidative damage to nucleic acids photosensitized by titanium dioxide. Free Radic. Biol. Med. 23, 851–858.

Wang, D., Kreutzer, D.A., Essigmann, J.M., 1998. Mutagenicity and repair of oxidative DNA damage: insights from studies using defined lesions. Mutat. Res. 400, 99–115.

Wang, J., Xiong, S., Xie, C., Markesbery, W.R., Lovell, M.A., 2005. Increased oxidative damage in nuclear and mitochondrial DNA in Alzheimer's disease. J. Neurochem. 93, 953–962.

Wang, J., Markesbery, W.R., Lovell, M.A., 2006. Increased oxidative damage in nuclear and mitochondrial DNA in mild cognitive impairment. J. Neurochem. 96, 825–832.

Wang, J., Clauson, C.L., Robbins, P.D., Niedernhofer, L.J., Wang, Y., 2012. The oxidative DNA lesions 8,5′-cyclopurines accumulate with aging in a tissue-specific manner. Aging Cell 11, 714–716.

Wang, X., Wang, W., Li, L., Perry, G., Lee, H.-G., Zhu, X., 2014. Oxidative stress and mito-chondrial dysfunction in Alzheimer's disease. Biochim. Biophys. Acta 1842, 1240–1247.

Weidemann, A., Johnson, R.S., 2008. Biology of HIF-1α. Cell Death Differ. 15, 621–627.

Weissman, L., Jo, D.G., Sorensen, M.M., de Souza-Pinto, N.C., Markesbery, W.R., et al., 2007. Defective DNA base excision repair in brain from individuals with Alzheimer's disease and amnestic mild cognitive impairment. Nucleic Acids Res. 35, 5545–5555.

Williams, T.I., Lynn, B.C., Markesbery, W.R., Lovell, M.A., 2006. Increased levels of 4-hydroxynonenal and acrolein, neurotoxic markers of lipid peroxidation, in the brain in Mild Cognitive Impairment and early Alzheimer's disease. Neurobiol. Aging 27, 1094–1099.

Wurtmann, E.J., Wolin, S.L., 2009. RNA under attack: cellular handling of RNA damage. Crit. Rev. Biochem. Mol. Biol. 44, 34–49.

Yen, T.C., Su, J.H., King, K.L., Wei, Y.H., 1991. Ageing-associated 5kb deletion in human liver mitochondrial DNA. Biochem. Biophys. Res. Commun. 178, 124–131.

Yin, W., Signore, A.P., Iwai, M., Cao, G., Gao, Y., Chen, J., 2008. Rapidly increased neuronal mitochondrial biogenesis after hypoxic-ischemic brain injury. Stroke 39, 3057–3063.

You, C., Swanson, A.L., Dai, X., Yuan, B., Wang, J., Wang, Y., 2013. Translesion synthesis of 8,5′-cyclopurine-2′-deoxynucleosides by DNA polymerases eta, iota, and zeta. J. Biol. Chem. 288, 28548–28556.

Youle, R.J., Narendra, D.P., 2011. Mechanisms of mitophagy. Nat. Rev. Mol. Biol. 12, 9–14.

Yu, T.W., Anderson, D., 1997. Reactive oxygen species-induced DNA damage and its modification: a chemical investigation. Mutat. Res. 379, 201–210.

Yu, H., Venkatarangan, L., Wishnok, J.S., Tannenbaum, S.R., 2005. Quantitation of four guanine oxidation products from reaction of DNA with varying doses of peroxynitrite. Chem. Res. Toxicol. 18, 1849–1857.

Zhao, Y., Zhao, B., 2013. Oxidative stress and the pathogenesis of Alzheimer's disease. Oxidative Med. Cell. Longev. 316523.

Zuccato, C., Cattaneo, E., 2009. Brain-derived neurotrophic factor in neurodegenerative diseases. Nat. Rev. Neurol. 5, 311–322.

Type II Diabetes and Metabolic Syndrome as Risk Factors for Alzheimer's Disease

INTRODUCTION

Type II diabetes (diabetes mellitus) is a group of complex endocrine and metabolic disorders characterized by chronic hyperglycemia with disturbances in carbohydrate, fat, and protein metabolism resulting from defects in insulin secretion and action (American Diabetes Association,

2010; Centers for Medicare and Medicaid Services, 2013). Thus, type II diabetes results in elevated blood glucose levels due to reduction in insulin production by pancreas and induction of "insulin resistance," a condition in which insulin responsiveness is decreased due to abnormal expression of the insulin receptors (IRs) (American Diabetes Association, 2010; Centers for Medicare and Medicaid Services, 2013). Hyperglycemia-mediated alterations in lipid metabolism bring about a series of adverse effects, including enhancement in high-density lipoprotein (HDL) clearance, decrease in apoA-1 transcription, and accelerated HDL glycation (Drew et al., 2012). These dramatic changes promote the onset of atherosclerosis (Mooradian, 2009). Pathogenesis of type II diabetes involves impaired secretion of insulin, a pancreatic hormone, which regulates glucose uptake and utilization by cells, and free fatty acid levels in peripheral blood. These factors along with vascular changes play the central role in regulating energy metabolism in the body. Insulin resistance is defined by reduction of insulin capacity to stimulate glucose utilization, either by insulin deficiency or by impairment in its secretion and/or utilization (Stumvoll et al., 2005). There are two types of diabetes, namely, type I diabetes (diminished production of insulin) and type II diabetes (impaired response to insulin and b-cell dysfunction). Both conditions are associated with hyperglycemia, excessive urine production, compensatory thirst, blurred vision, unexplained weight loss, lethargy, and changes in energy metabolism along with increased risk of diabetic neuropathy, blood–brain barrier (BBB) alterations, stroke, Alzheimer's disease (AD), and depression (Fig. 5.1). In type II diabetes hyperglycemic or hypoglycemic changes in blood glucose levels produce marked alterations not only in transport of glucose, insulin, choline, and amino acids transport

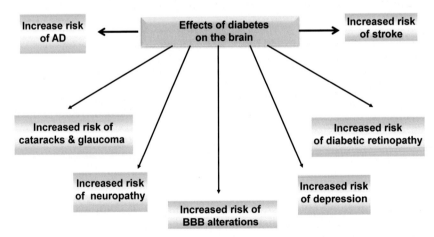

FIGURE 5.1   **Effects of diabetes on neurological diseases.** *AD*, Alzheimer's disease; *BBB*, blood–brain barrier.

but also cause disruption in tight junction leading to oxidative stress in the central nervous system microcapillaries. The transport of glucose into most cells is controlled by insulin. This hormone also stimulates lipogenesis and the synthesis of protein in hepatic and adipose tissues, while reducing lipolysis and proteolysis (Rorsman and Braun, 2013). The liver is a vital organ, which not only controls metabolic homeostasis, but also contributes to glucose uptake-storage-generation and processing. About one-third of consumed glucose is utilized by the liver, and it is a key target for insulin action. Insulin controls lipogenesis (fatty acid and triglycerides biosynthesis) and restrains hepatic gluconeogenesis (glucose production) in the liver suggesting that insulin sensitivity is closely associated with rates of hepatic gluconeogenesis and lipid accumulation (Bechmann et al., 2012). As stated previously, insulin resistance is defined as the decrease in insulin capacity to stimulate glucose utilization, either due to insulin deficiency or due to impairment in its secretion and/or utilization. The excess of glucose not only promotes hyperinsulinemia but also supports insulin resistance (Parillo and Riccardi, 2004; Hwu et al., 2009). Insulin resistance can also be induced by different environmental factors, including dietary habits. Thus, vitamin D deficiency (i.e., hypovitaminosis D) is associated with increased insulin resistance, impaired insulin secretion, and poorly controlled glucose homeostasis (Leung, 2016). The consumption of energy-dense/high-fat and high-carbohydrate diets is strongly and positively associated with obesity, which negatively effects insulin sensitivity, particularly when the excess of body fat is located in the abdominal region. However, the link between consumption of high-fat diet and obesity is not limited to the high energy content of fatty foods. In some individuals, the ability to oxidize and metabolize dietary fat under hyperglycemic conditions is impaired genetically predisposing the subjects to obesity and insulin resistance (Farooqui, 2013). The consumption of energy-dense/high-fat and high-carbohydrate diets results in stimulation of storage mechanisms. Adipose tissue is an important energy sink, which stores the energy that cannot be used otherwise. However, adipose tissue growth also has limits, and the excess of energy induces inflammation, promoted by the ineffective intervention of the immune system. However, even under this acute situation, the presence of excess glucose remains, favoring its final conversion to fat. In addition to energy storage, adipose tissue plays an active role in many homeostatic processes including energy expenditure, appetite regulation, and glucose regulation. Fat tissue is critical for thyroid function, immune response, bone health maintenance, reproduction, and blood clotting. The adipose tissue is an active endocrine organ secreting free fatty acids, leptin, adiponectin, adipsin, complement factor 3, interleukin-6 (IL-6), tumor necrosis factor (TNF)-α, angiotensinogen, and plasminogen activation inhibitor-1, among others. Abnormal signaling and the deficiency of the aforementioned hormones result in deleterious effects (Kissebah et al., 1982; Xing et al., 1998). Collective evidence suggests that insulin resistance not only contributes to the defects in IR function, abnormalities in insulin signaling, alterations in glucose metabolism, and induction of hyperinsulinemia, hyperglycemia and inflammation, but also increases blood pressure (Wang and Jin, 2009).

Type II diabetes markedly effects the brain at the cellular and subcellular levels. Brain is a highly metabolic organ, which utilizes glucose-derived energy for its optimal function. Among various subcellular structures, mitochondria are the major bioenergetic subcellular structures, which are highly susceptible to type II diabetes–mediated metabolic damage. Type II diabetes not only lowers the expression of mitochondrial genes but also induces abnormal mitochondrial morphology with the onset of oxidative stress. In addition, type II diabetes disrupts insulin and brain-derived neurotrophic factor (BDNF) receptor signaling, thereby affecting the cotranscriptional regulator peroxisome proliferator-activated receptor gamma coactivator 1-alpha (PGC-1α) signaling (Fig. 5.2). Impairment in PGC-1α function not only decreases mitochondrial biogenesis by suppressing mitochondrial DNA transcription but also induces mitochondrial proliferation

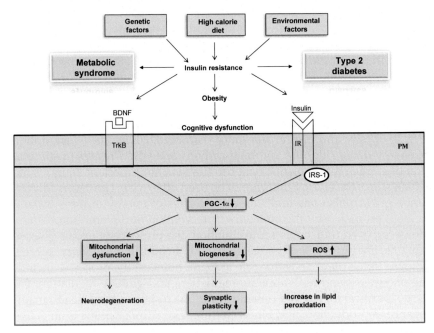

FIGURE 5.2 **Effect of insulin resistance on mitochondrial biogenesis and synaptic plasticity.** *BDNF,* brain-derived neurotrophic factor; *IR,* insulin receptor; *PGC-1α,* coactivator-1alpha; *PM,* plasma membrane; *ROS,* reactive oxygen species; *TrkB,* tyrosine kinase B receptor. *Upward arrows* indicate increase and *downward arrow* indicates decrease.

through mitochondrial transcription factor A (TFAM) (Ventura-Clapier et al., 2008; Campbell et al., 2012). Reactive oxygen species (ROS), a by-product of mitochondrial electron transport chain, leads to the generation of 4-hydroxynonenal (4-HNE), isoprostanes, and oxidative stress, producing metabolites via lipid peroxidation. In turn, 4-HNE disrupts mitochondrial function and overall neuronal functions not only through the inhibition of DNA and RNA synthesis and disturbance in calcium homeostasis (Mark et al., 1997; Keller and Mattson, 1998) but also through impairment of glucose transport and decrease in cellular ATP levels (Keller et al., 1997). 4-HNE contributes to the pathogenesis of insulin resistance in metabolic syndrome (MetS). Levels of 4-HNE are increased in the blood and muscle tissue of obese subjects compared to those of normal weight subjects (Vincent et al., 2001; Russell et al., 2003). Converging evidence suggests that in the brain 4-HNE produces dysfunction and degeneration of neurons by modifying membrane-associated glucose and glutamate transporters, ion-motive ATPases, and cytoskeletal proteins (Farooqui, 2011).

Several targets of the insulin dysfunction have been reported to influence brain function in neurological disorders at the molecular level. Thus it is well known that insulin and insulin-like growth factor (IGF)-1 receptor (IGF-1R) are capable of regulating key processes such as energy homeostasis, neuronal survival, longevity, learning, and memory (Plum et al., 2005; Freude et al., 2009). Insulin and IGF-1 interacts with the tyrosine kinase receptors, IR, and IGF-1R, which share a high degree of identity in their structure and function (Plum et al., 2005; Freude et al., 2009). IR and IGF-1R are selectively distributed in the brain with a higher density in the olfactory bulb, hypothalamus, as well as in two of the main brain regions, which are markedly affected in AD pathology, i.e., hippocampus and cerebral cortex (Plum et al., 2005; Freude et al., 2009). Binding of insulin or IGF-1 induces a conformational change in the receptor leading to their autophosphorylation on specific tyrosine residues on the β-subunit with the consequent recruitment of the insulin receptor substrate-1 (IRS-1) (Plum et al., 2005; Freude et al., 2009). The latter, in turn, activates two main signaling pathways. They are (1) the phosphoinositide 3-kinase (PtdIns 3K) pathway, which activates Akt and glycogen synthase kinase-3β (GSK-3β), and PtdIns 3K signaling also contributes to the maintenance of synaptic plasticity and memory consolidation (Horwood et al., 2006); and (2) the mitogen-activated protein kinase (MAPK) cascade, which is responsible both for the induction of several genes required for neuronal and synapse growth, maintenance, and repair processes, as well as serves as a modulator of hippocampal synaptic plasticity that underlies learning and memory (Akter et al., 2011). These observations suggest that the impairment in insulin signaling in the brain may play a role in the development of neurological disorders.

MOLECULAR MECHANISMS CONTRIBUTING TO COMPLICATIONS IN TYPE II DIABETES

Long-term presence of hyperglycemia in the body results in type II diabetes complications. The pathophysiology of microvascular complications in type II diabetes involves major biochemical pathways leading to secondary complications in various tissues due to the insulin-independent uptake of high levels of glucose. Pathophysiological complications in patients with type II diabetes are not only caused by insulin resistance, increased synthesis of high levels of sorbitol, induction of oxidative and nitrosative stress, and alterations in hormonal responses but also due to the depletion of endogenous antioxidant, enhanced lipid peroxidation, and other metabolic changes (Fig. 5.3) (Brownlee, 2005). At the molecular level, all aforementioned processes and pathways involve high intracellular

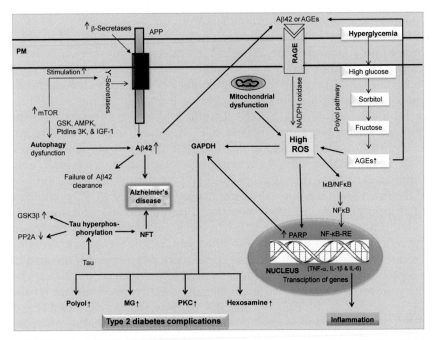

FIGURE 5.3　Hypothetical diagram showing the contribution of advanced glycated end products in complications caused by type II diabetes. *AGEs*, advanced glycation end products; *AMPK*, AMP-activated protein kinase; *GAPDH*, glyceraldehyde-3-phosphate dehydrogenase; *GSH*, glutathione; *IL-1β*, interleukin-1β; *IL-6*, interleukin-6; *IκB*, inhibitory subunit of NF-κB; *MG*, methylglyoxal; *mTOR*, mechanistic target of rapamycin; *NADP oxidase*, nicotinamide adenine dinucleotide phosphate oxidase; *NF-κB*, nuclear factor-κB; *NF-κB-RE*, nuclear factor-κB response element; *PARP*, poly(ADP-ribose)polymerase; *PKC*, protein kinase C; *PM*, plasma membrane; *RAGE*, receptor for advanced glycation end products; *ROS*, reactive oxygen species; *TNF-α*, tumor necrosis factor-α.

glucose levels and its downstream metabolic metabolism. Hyperglycemia leads to the generation of high levels of superoxides, which inhibits glyceraldehyde 3-phosphate dehydrogenase (GAPDH) activity in vivo by modifying the enzyme with polymers of ADP-ribose (Du et al., 2003). Specific inhibitors of poly(ADP-ribose) polymerase (PARP) prevent the inhibition of GAPDH. Under normal conditions, PARP resides in the nucleus in an inactive form. In type II diabetes increased production of ROS causes DNA damage through the activation of PARP in the nucleus. PARP hydrolyzes NAD^+ into nicotinic acid and ADP-ribose. PARP then proceeds to make polymers of ADP-ribose, which accumulate on GAPDH and other nuclear proteins. GAPDH is commonly thought to reside exclusively in the cytosol. However, it normally shuttles in and out of the nucleus, where it plays a critical role in DNA repair (Sawa et al., 1997; Du et al., 2003). In type II diabetes, hyperglycemia and insulin resistance promote the conversion of glucose into fructose, which is then metabolized into advanced glycation end products (AGEs) like methylglyoxal (Fig. 5.3) (Beisswenger et al., 2003; Thornalley, 1993). AGEs are formed via Maillard reaction between carbohydrates and proteins. In the clinical setting, AGEs and receptor of AGEs (RAGE) have been identified in neurons and hippocampus (Wang et al., 2009). It has been suggested that diabetes complications are caused by interactions between AGEs and RAGE via oxidative stress (Brownlee, 2005; Giacco and Brownlee, 2010). Under normal conditions, glucose is metabolized to glyceraldehyde-3-phosphate (G3P) via glycolysis. G3P is transformed into 1,3-diphosphoglycerate by GAPDH, which is upregulated. G3P is further metabolized to form pyruvate. During insulin resistance, abnormalities in insulin function may result in downregulation of GAPDH, slowing glucose metabolism and increasing its metabolism via the polyol pathway (Fig. 5.3) (Thornalley, 1993; Alexander et al., 1988). The polyol pathway involves aldoketo reductase, an enzyme that uses a wide variety of carbonyl compounds as substrates and reduces them into sugar alcohols (polyols) through the participation of nicotinamide adenine dinucleotide phosphate (NADPH). For example, glucose is metabolized into sorbitol by the enzyme aldose reductase. Sorbitol is then oxidized to fructose by the enzyme sorbitol dehydrogenase with NAD^+ as a cofactor (Giacco and Brownlee, 2010). In type II diabetes, mitochondrial dysfunction leads to superoxide overproduction not only in endothelial cells in blood vessels but also in the myocardium. The increase in superoxide production activates four major pathways contributing to diabetic complications through increase in nonenzymic glycosylation of proteins and lipids, which not only interfere with their normal function by disrupting molecular conformation and through alterations in enzymatic activity but also by reducing degradative capacity and interfering with receptor recognition. The interactions of glycosylated proteins with their receptor also led to (1) the induction of oxidative stress and proinflammatory responses, (2) activation of protein kinase C (PKC) with subsequent alteration in growth factor expression, (3) shunting of excess intracellular glucose into the hexosamine pathway resulting in O-linked glycosylation of various enzymes with perturbations in normal enzyme function, and (4) activation of RAGE by AGEs, which initiates a vicious cycle eliciting more oxidative stress and inducing an increase in inflammation through the induction of cytokine secretion by several cell types including monocytes and adipocytes (Brownlee, 2001; Dias and Griffiths, 2014; Yamagishi, 2011) and subsequently involvement of vascular inflammation (Toma et al., 2009) and thrombosis (Takenaka et al., 2006), thereby inducing potential vascular damage (Kook et al., 2012; Wan et al., 2014). Furthermore, abnormal mitochondrial dysfunction and insulin resistance in type II diabetes also contribute to the overproduction of ROS (Fig. 5.4), which inactivates endothelial nitric oxide synthase (eNOS) and prostacyclin synthase, thereby impairing the vascular tone (Mooradian, 1997).

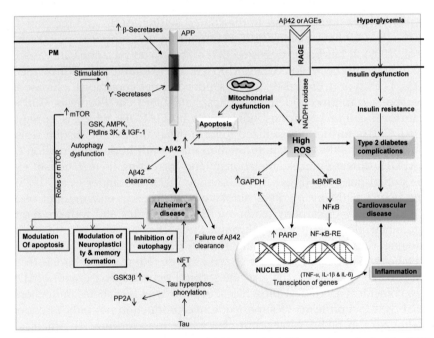

FIGURE 5.4   Interactions between amyloid beta and RAGE receptor and roles of mTOR in the brain. *AGEs*, advanced glycation end products; *AMPK*, AMP-activated protein kinase; *APP*, amyloid precursor protein; *Aβ*, amyloid β; *GSH*, glutathione; *GSK-3*, glycogen synthase kinase 3; *IGF-1*, insulin-like growth factor 1; *IL-1β*, interleukin-1β; *IL-6*, interleukin-6; *IκB*, inhibitory subunit of NF-κB; *mTOR*, mechanistic target of rapamycin; *NF-κB*, nuclear factor-κB; *NF-κB-RE*, nuclear factor-κB response element; *PM*, plasma membrane; *PtdIns 3K*, phosphoinositide 3-kinase; *RAGE*, receptor for advanced glycation end products; *ROS*, reactive oxygen species; *TNF-α*, tumor necrosis factor-α.

These pathways tightly link hyperglycemia-mediated oxidant stress with vascular damage in type II diabetes. Type II diabetes also has a strong genetic component. However, only a handful of genes have been identified. These genes include genes for calpain 10, potassium inward-rectifier 6.2, peroxisome proliferator–activated receptor gamma, IRS-1, KCNJ11 gene, TCF7L2 gene, ABCCC8 gene, and gene for adiponectin (Stumvoll et al., 2005; Dedoussis et al., 2007). As stated earlier, type II diabetes is also accompanied by the insulin resistance. The molecular mechanisms involved in insulin resistance are not fully understood. However, consumption of high-calorie diet (Western diet) and elevations in levels of saturated fatty acids as well as phospholipid-derived (diacylglycerol and isoprostanes), and sphingolipid-derived (ceramide) lipid mediators in the brain may contribute to insulin resistance (Holland and Summers, 2008; Farooqui, 2013).

Activation of AMP kinase, a central regulator of cellular energy status (Quintero et al., 2006), leads to accelerated atherosclerotic disease modulating blood flow to arteries that supply blood to the heart, brain, and lower extremities. Diabetes-mediated lower extremity arterial disease in conjunction with neuropathy accounts for 60% of all nontraumatic amputations in the United States (CDC Fact Sheet, 2007). Diabetes and impaired glucose tolerance increase cardiovascular disease risk by three- to eightfold (CDC Fact Sheet, 2007). Management of type II diabetes includes not only diet and exercise but also combinations of antihyperglycemic drug treatment with lipid-lowering, antihypertensive, and antiplatelet therapy. About 90%–95% of all North American cases of diabetes are type II diabetes and about 20% of the population older than 65 years has type II diabetes (Zimmet et al., 2001). Type II diabetes is the leading cause of premature deaths. Improper management of type II diabetes not only can result in nonneural diseases (heart disease, kidney failure, MetS, slow healing of wounds, leg and foot amputations, and arterial diseases) but also leads to neurological diseases (stroke, AD, depression, blindness, and neuropathy) (Farooqui, 2013). Collective evidence suggests that type II diabetes has significant impact on multiple systems of many tissues including the nervous system.

Cellular metabolism in type II diabetes is tightly linked with mechanistic target of rapamycin (mTOR), a 289-kDa enzyme with a kinase domain and a FKBP12 binding domain (Maiese, 2014; Johnson et al., 2015). This protein kinase is ubiquitously expressed throughout the body including the brain, vascular system, and immune system, which are crucial for maintaining neuronal health. It plays a key role in regulating protein synthesis and cell growth by phosphorylating downstream proteins such as p70 S6 kinase and eukaryotic initiation factor eIF4E-binding protein (Wullschleger et al., 2006). mTOR is a negative regulator of autophagy induction (Wullschleger et al., 2006). By simultaneously regulating protein synthesis and degradation, mTOR modulates and controls protein homeostasis, a process that is not only altered in type II diabetes but also in AD and other proteinopathies (Lee et al., 2013). mTOR has been shown to have important roles in neurons. It has not only been implicated in axon pathfinding and regeneration, but also in dendrite arborization and spine morphology (Jaworski and Sheng, 2006). Moreover, mTOR not only regulates synaptic plasticity (Hoeffer and Klann, 2010; Santini et al., 2014; Caccamo et al., 2015, 2010) but also is necessary for memory consolidation (Tang et al., 2002; Parsons et al., 2006). It is also suggested that mTOR hyperactivity is detrimental and causes cognitive deficits in animals and people (Ehninger et al., 2008; Puighermanal et al., 2009; Ehninger, 2013). A major upstream regulator of mTOR is tuberous sclerosis complex 2 (TSC2), which integrates signals from many other signaling molecules, including Akt (PKB) and GSK-3 (Wullschleger et al., 2006). This kinase is also associated with increase in lifespan and health span in several genetically different species (Lamming et al., 2013). mTOR also blocks neuronal cell apoptosis through the epidermal growth factor receptor (Kimura et al., 2013). This process protects against cognitive loss during type II diabetes through increases in the expression of acetylcholinesterase (Liu et al., 2015) and promotes glucose homeostasis (Miao et al., 2013; Malla et al., 2015). mTOR supports the aforementioned processes by coordinating and interacting with the upstream signal components, including insulin, growth factors, AMP-activated protein kinase (AMPK), PtdIns-3 K/ Akt, and GSK-3 (Gouras, 2013; O'Neill, 2013). mTOR is known to form two complexes, namely, mTORC1 and mTORC2 (Wullschleger et al., 2006). The rapamycin-sensitive mTORC1 contains the following proteins: raptor, GβL (also known as mLST8) and the proline-rich Akt substrate of 40 kDa (PRAS40). The mTORC2 complex contains rictor, mammalian stress-activated protein kinase (SAPK)-interacting protein 1, Protor-1, and GβL5 (Wullschleger et al., 2006).

TYPE II DIABETES AS RISK FACTOR FOR ALZHEIMER'S DISEASE

AD is a metabolic disease characterized by impairment in brain glucose utilization. In AD, amyloid precursor protein (APP) processing and amyloid β (Aβ) production produce mitochondrial activity defect and increase oxidative stresses. These processes impair key players of the glucose metabolic pathway (Farooqui, 2013; Butterfield et al., 2014). The binding of Aβ with RAGE may contribute to the pathogenesis of AD not only in humans but also in transgenic AD animal models (Farooqui, 2015). RAGE is a multiligand transmembrane receptor of the immunoglobulin superfamily. RAGE not only binds with AGEs, but also Aβ, amphoterin,

S100b/calgranulins, and other proinflammatory ligands (Bucciarelli et al., 2002; Schmidt et al., 2000). These molecules are often present in the altered tissue environments associated with type II diabetes. The binding of Aβ oligomer with RAGE induces AD-like pathology, oxidative stress (Fig. 5.5) (Farooqui, 2010), synapse deterioration and loss (De Felice et al., 2009; Lacor et al., 2007), and inhibition of synaptic plasticity. Interactions of RAGE with its ligands induces different signaling pathways making the RAGE-mediated cellular signaling extremely complex. Thus the activation of RAGE results in activation of not only ERK1/2 (p44/p42), p38 and SAPK/c-Jun N-terminal kinase (JNK) MAPK but also increased activities of rho-GTPases, PtdIns-3-kinase, JAK/STAT, and various isoforms of PKC. These enzymes and pathways play an important role in RAGE-mediated cellular responses (Bierhaus et al., 2005; Nitti et al., 2005).

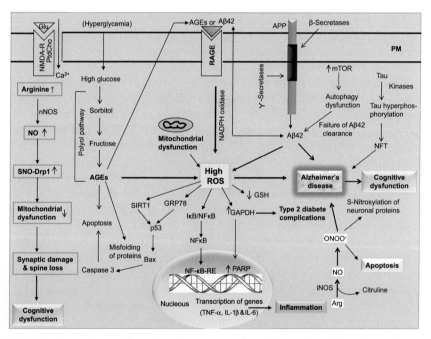

FIGURE 5.5 **Interactions among hyperglycemia, amyloid β, and NMDA receptor in Alzheimer's disease.** *AGEs,* advanced glycation end products; *APP,* amyloid precursor protein; *Arg,* arginine; *Aβ,* amyloid β; *Drp1,* dynamin-related protein 1; *eNOS,* endothelial nitric oxide synthase; *GADPH,* glyceraldehyde-3-phosphate dehydrogenase; *GSH,* glutathione; *IL-1β,* interleukin-1β; *IL-6,* interleukin-6; *IκB,* inhibitory subunit of NF-κB; *NADP oxidase,* nicotinamide adenine dinucleotide phosphate oxidase; *NFT,* neurofibrillary tangle; *NF-κB,* nuclear factor-κB; *NF-κB-RE,* nuclear factor-κB response element; *NO,* nitric oxide; *ONOO⁻,* peroxynitrite; *PARP,* poly(ADP-ribose)polymerase; *PM,* plasma membrane; *RAGE,* receptor for advanced glycation end products) *ROS,* reactive oxygen species; *SNO-Drp1,* nitrosylated dynamin-related protein; *TNF-α,* tumor necrosis factor-α.

RAGE-dependent signaling pathway activation directly induces ROS production mainly through NADPH oxidase activation. Moreover, it is important to underline that RAGE signaling leads to the activation of the transcription factor nuclear factor (NF)-κB, which in turn induces RAGE expression, making a positive loop that enhances cell response (Bierhaus et al., 2005). However, other transcription factors such as SP-1, AP-2, and NF-IL6 have been shown to regulate RAGE expression (Bierhaus et al., 2005). RAGE signaling is also accompanied by AGE-mediated upregulation of GRP78, p53, and caspase 3 along with decrease in the bcl-2/bax ratio. ER-stress also stimulates the expression of p53 (Lin et al., 2012) along with expression of GRP78 inducing the cell death pathway (Zhou et al., 2014). Moreover, Aβ has been shown to regulate the transcription of p53 (Liu et al., 2013). It is also reported that Aβ also induces stress-mediated apoptosis in the endoplasmic reticulum (ER) during AD progression (Costa et al., 2012). These reports support the view that generation of AGEs results in the production of more ROS and Aβ as well as the downstream ER stress-mediated apoptosis pathway. In addition, a number of reports have suggested that Sirt1 (a group of histone deacetylating enzymes, which are activated by NAD^+) and ER stress are antagonistic (Ghemrawi et al., 2013; Li et al., 2011). Both type II diabetes and AD are also accompanied by the presence of AGE-modified proteins (Takeuchi and Yamagishi, 2008). Like Aβ peptide, these modified proteins act as ligands for the RAGE. The interactions between RAGE and its ligands trigger the activation of a key cell signaling pathway, such as $p21^{ras}$ and transcription factor NF-κB (Takeuchi and Yamagishi, 2008). The extent to which this happens in vivo is still unclear, but increased RAGE expression along with increased amounts of AGE-modified proteins may be associated with vascular complications in AD as well as type II diabetes. Blocking interactions between Aβ and RAGE may impair the activation of microglial cells and astrocytes, resulting in reduction in the production of proinflammatory mediators (Ramasamy et al., 2009). RAGE also plays an important role not only in the clearance of Aβ but also contributes to apolipoprotein E (APOE)-mediated cellular processing and signaling (Bu, 2009). These observations suggest that in many age-related diseases, the accumulation of AGEs is a significant contributing factor in degenerative diseases, especially in retinopathy, cardiovascular diseases, complications of type II diabetes, AD, and dementia. This elevation in levels of AGE in early stage of aforementioned pathological conditions may contribute to a significantly higher risk for the development of muscle stiffness/hypertonia (defined as paratonia) compared to those with dementia without type II diabetes (Gasser and Forbes, 2008; Hobbelen et al., 2011). In AD, AGE also interacts with hyperphosphorylated Tau protein leading to increase in oxidative stress and chronic inflammation, accelerating the pathogenesis of AD (Angeloni et al., 2014). Moreover, accumulation of Aβ or Tau glycation results in increased aggregation and subsequent formation of senile plaques or neurofibrillary tangles (Loske et al., 2000) supporting the view that AGE modification is an important risk factor for AD and other neurodegenerative diseases (Grillo and Colombatto, 2008). Although increase in the accumulation of AGEs in brain, as seen in aging, diabetes, or AD, speeds up oxidative damage to neurons contributing to synaptic dysfunction and cognitive decline, its underlying mechanisms are not well understood. During insulin resistance, one develops reduced sensitivity to insulin resulting in hyperinsulinemia, and this impairment in insulin signaling is closely associated with the pathogenesis of AD. In type II diabetes, insulin increases the levels of Aβ in plasma in patients with AD, and these effects may contribute to the risk of developing AD in patients with type II diabetes. The accumulation of Aβ oligomers contribute to neuronal insulin resistance in the AD brain not only by blocking the insulin signaling network through insulin/Akt pathway (Townsend et al., 2007) but also by dissociating IRs after binding at the dendrites of synaptic sites (Zhao et al., 2008). The finding that insulin and Aβ are hydrolyzed by the similar enzymes has attracted considerable attention of many investigators on the role of insulin signaling in Aβ clearance. Increase in insulin levels frequently seen in insulin resistance may compete for these enzymes and thus contribute to Aβ accumulation. Indeed, insulin signaling has been shown to regulate expression of metalloproteases such as insulin degrading enzyme (IDE) (Qiu and Folstein, 2006), and influence aspects of Aβ metabolism and catabolism (Gasparini et al., 2001). In addition, intracellular Aβ oligomer interferes with IR signaling by interacting with phosphoinositide-dependent kinase and Akt (Lee et al., 2009). Impaired insulin signaling cannot efficiently block GSK-3β. Therefore, the activated GSK-3β triggers APP γ-secretase activity and increases Tau phosphorylation (Hooper et al., 2008), simultaneously aggravating the APP processing and Tau phosphorylation in AD brain. As stated earlier, Aβ oligomers are small, diffusible aggregates that can specifically attach with synapses in hippocampal and cortical neurons and act as pathogenic ligands for the loss of synapse in AD (De Felice et al., 2009).

Furthermore, mTOR signaling is also closely associated with the pathogenesis of type II diabetes and AD. The activation of mTOR enhances Aβ generation and deposition through the activation of regulating β- and γ-secretase (Spilman et al., 2010; Son et al., 2012; Chen et al., 2009). In addition, mTOR interacts with several key signaling pathways and regulate Aβ generation or Aβ clearance, including PtdIns 3-K/Akt (Damjanac et al., 2008; Bhaskar et al., 2009), GSK-3 (Maiese et al., 2012), AMPK (Cai et al., 2012a), and insulin/IGF-1 (Pei and Hugon, 2008). As stated above, mTOR not only contributes to synaptic plasticity but also facilitates memory formation in the hippocampus and the inhibition of mTOR impairs memory consolidation (Slipczuk et al., 2009). An increase in phosphorylation of mTOR by GSK-3β and Tau protein by p70 ribosome S6 kinase (p70S6K) have been reported to occur in AD brain, suggesting that phosphorylation of mTOR and hyperphosphorylation of Tau may promote AD progression (An et al., 2003; Griffin et al., 2005). Furthermore, inhibition of mTOR has been reported to increase autophagy in murine models of AD. This process not only improves memory but also contributes to the decrease in Aβ levels (Spilman et al., 2010). It is interesting to note that loss of mTOR signaling impairs long-term potentiation (LTP) and decreases synaptic plasticity in models of AD (d'Abramo et al., 2006). In addition, Aβ, which is toxic to neuronal cells, inhibits the activation of mTOR and p70S6K in neuroblastoma cells and in lymphocytes of patients with AD (Dal Col and Dolcetti, 2008) supporting the view that the activation of mTOR and p70S6K may prevent neurodegeneration following Aβ exposure in microglia (Zemke et al., 2007). Other studies have indicated that blockade of mTOR activity may lead to neuronal atrophy and apoptosis in AD (Prada et al., 2007).

mTOR also inhibits autophagy. This process decreases the Aβ clearance by autophagy/lysosome system. This system is responsible for the clearance of abnormal proteins (Pei and Hugon, 2008). It is proposed that chronic deterioration of the autophagy/lysosome pathway is an important factor in the failure of Aβ clearance from the AD brain (Spilman et al., 2010; Bhaskar et al., 2009). Furthermore, inhibition of mTOR, activity by mTOR inhibitor retards autophagy. This process is closely associated with high levels of Aβ (Spilman et al., 2010). In contrast, upregulation of mTOR signaling by lifestyle choices may at least partially determine the pathogenesis of AD in patients and animal models (Caccamo et al., 2010; Oddo, 2012). It is also reported that diabetes increases the risk of AD by an mTOR-dependent mechanism (Ma et al., 2013; Orr et al., 2014). In contrast, caloric restrictions produce reduction in mTOR signaling (Speakman and Mitchell, 2011). Converging evidence suggests that reduction in mTOR levels and signaling results in beneficial effects on AD-like pathology in animal models of AD such as Tg2576 mice and APP/mTOR$^{+/-}$ mice. These mice perform significantly better than APP mice and as well as CTL mice in a spatial learning and memory task. Collective evidence suggests that there is a link between mTOR and AD. mTOR is not only critical for long-lasting forms of synaptic plasticity and long-term memory formation. These processes involve messenger RNA translation and control and de novo protein synthesis. In addition, inhibition of the mTOR pathway modulates aging, a well-established risk factor for AD. Furthermore, autophagy has been reported to contribute to neurodegeneration in AD, and the well-characterized mTOR inhibitor, rapamycin, induces autophagy. Furthermore, investigations on single nucleotide polymorphisms (SNPs) associated with type II diabetes and AD have indicated 927 SNPs associated with both AD and type II diabetes with p-value≤0.01. This overlap significantly larger than random chance (overlapping p-value of

6.93E-28) suggesting common pathogenic mechanisms underlying the development of both AD and type II diabetes. These genes control immune responses, cell signaling, and neuronal plasticity. Abnormalities in expression of these genes are known to contribute to both type II diabetes and AD pathogenesis supporting the view that type II diabetes subjects have greater chances of developing AD (Hao et al., 2015). This suggestion is supported by neuroimaging studies. Using magnetic resonance imaging, it is shown that older patients with type II diabetes have a moderately increased risk for developing hippocampal atrophy and that the severity of lesions parallels the progression of type II diabetes. Moreover, patients with type II diabetes also have a high risk of developing MCI (Ferreira et al., 2010; Janson et al., 2004). Conversely, patients with AD exhibit significantly increased prevalence of type II diabetes and impaired fasting glucose as compared to control subjects, which over 80% of patients with AD with comorbid type II diabetes and impaired fasting glucose (Janson et al., 2004; Li and Holscher, 2007). Moreover, using an ex vivo stimulation protocol with near-physiological doses of insulin leads to onset of prominent brain insulin resistance in patients with AD undergoing postmortem even in the absence of diabetes (Talbot et al., 2012). Thus these results suggest that type II diabetes is closely correlated with AD in its epidemiology.

METABOLIC SYNDROME AND ITS EFFECTS ON THE BRAIN

The MetS is a complex pathological condition that is defined by a clustering of interconnected factors such as abdominal obesity, dyslipidemia, hypertension, and insulin resistance. These factors not only increase the risk of coronary heart disease and other forms of cardiovascular diseases (CVD) but also that of type II diabetes. At the molecular level, MetS is accompanied by hyperglycemia, elevation in triglycerides, high-density lipoprotein cholesterol, impaired mitochondrial oxidative phosphorylation, and mitochondrial biogenesis, dampened insulin metabolic signaling, and endothelial dysfunction. Other abnormalities such as chronic proinflammatory and prothrombotic states, nonalcoholic fatty liver disease, and sleep apnea can also be added to the entity of the syndrome, making its definition even more complex. The aforementioned factors predispose the individual to increased risk of developing stroke, AD, depression, and dementia (Fig. 5.6) (Komulainen et al., 2007; Ren et al., 2010; Farooqui et al., 2012; Muoio and Newgard, 2008).

As stated previously, the transport of glucose into most cells is controlled by insulin. The excess of glucose promotes hyperinsulinemia and insulin resistance, which is characterized by the failure of peripheral tissues to appropriately regulate glucose homeostasis in response to insulin

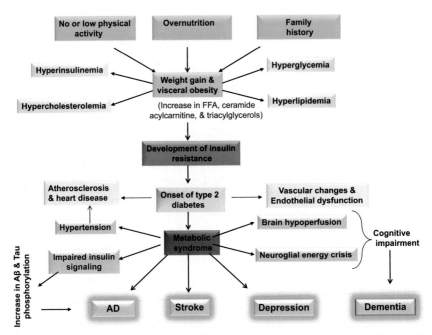

FIGURE 5.6 Factors contributing to the pathogenesis of metabolic syndrome. *AD*, Alzheimer's disease; *FFA*, free fatty acids.

(Fig. 5.6) (Parillo and Riccardi, 2004; Hwu et al., 2009). The increase in expression of inflammatory adipokines and cytokines and deranged substrate handling by adipose tissue produce most features of the MetS, such as insulin resistance, obesity (accumulation of excess body fat, which manifests as increased weight or waist circumference), type II diabetes, hypertension, hyperlipidemia, and their compounded combined effects. Thus, long-term consumption of high-calorie diet may induce difficulties in utilization and disposal of glucose, eliciting inflammation and ultimately developing MetS. Patients with MetS also have autonomic nervous system dysfunction, which is characterized by predominance of the sympathetic nervous system in many organs, including heart, kidneys, vasculature, adipose tissue, and muscles.

METABOLIC SYNDROME AS A RISK FACTOR FOR ALZHEIMER'S DISEASE

As stated earlier, insulin regulates glucose uptake and utilization by cells, and free fatty acid levels in peripheral blood and therefore plays the central role in regulating energy metabolism in the body. Insulin has also been implicated in the regulation of autonomic outflow and release of neurotrophic factors (Banks et al., 2012; Fernandez and Torres-Alemán, 2012). Patients with type II diabetes and MetS due to reduction in cerebral perfusion show insulin resistance (Brundel et al., 2012; Rusinek et al., 2015). Although neural cells also produce small amount of insulin (Banks, 2004), most insulin action in the brain is probably due to the circulating peripheral insulin (Banks, 2004). Insulin is transported across the BBB by three mechanisms: extracellular pathway, saturable transmembrane diffusion, and via the choroid plexus. Permeability of the BBB to insulin is variable among brain regions. It is shown that insulin crosses the BBB 2–8 times faster in the olfactory bulb, the most IR-enriched region in the whole brain (Banks et al., 1999, 2012). Peripheral insulin exerts its effects via the IR and insulin-like growth factor receptors, which are located in the olfactory bulb, cerebral cortex, hippocampus, cerebellum and hypothalamus. In the brain, levels of insulin can reach 10- to 100-fold higher than in plasma, especially in hippocampus, hypothalamus, cortex, olfactory bulb, substantia nigra, and pituitary (van der Heide et al., 2006). IRs are abundantly expressed in neurons and less abundantly in glia (Frolich et al., 1998). Insulin performs a wide range of functions in the brain and peripheral tissues. Thus in peripheral tissues insulin is important for cell growth and survival. In the brain, insulin, IGFs and their receptors regulate dendritic sprouting, neuronal stem cell activation, cell growth, repair, synaptic maintenance, and neuroprotection (Fig. 5.7) (Craft and Watson, 2004; Van Dam and Aleman, 2004; Kleinridders et al., 2014). Insulin not

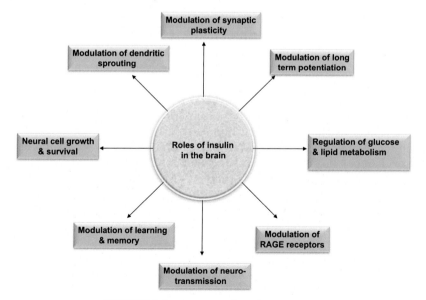

FIGURE 5.7   Functions of insulin in the brain.

only regulates glucose and lipid metabolism in the brain but also modulates neurotransmission and synaptic activities (Zhao and Alkon, 2001; Zhao et al., 2004) such as LTP (Nisticò et al., 2011), as well as promotes long-term depression (LTD) (Labouèbe et al., 2013). Insulin enhances cognitive abilities via activation of IRs in the hippocampal region of brain. Insulin also stimulates translocation of GLUT4 to hippocampal plasma membranes thereby enhancing glucose uptake in a time-dependent manner (Ren et al., 2014). Glucose utilization during neuronal activity is similar in both peripheral tissue and hippocampal region. Few studies have indicated that insulin is stored in synaptic vesicles at nerve endings in rat brain and is released on depolarization conditions (Blázquez et al., 2014). Insulin also potentiates the brain transport of molecules such as leptin (Kastin and Akerstrom, 2001) and amino acids (Tagliamonte et al., 1976). Due to structural and functional similarity, insulin and IGF-1 interact with both IR and IGF-1R (Conejo and Lorenzo, 2001; Duarte et al., 2012). Once bound to their respective receptors, insulin or IGF-1 mediate autophosphorylation of tyrosine residue, triggering its intrinsic tyrosine activity and phosphorylating IRS docking protein at tyrosine residue (Duarte et al., 2012). The phosphorylation of intracellular substrates leads to the recruitment and activation of multiple proteins and the initiation of several signaling cascades, including PtdIns 3K/Akt, and the MAPK signaling pathways (Fig. 5.8) (van der Heide et al., 2006). The activation of PtdIns 3K is followed by recruitment of downstream signaling proteins, such as serine (Ser)/threonine (Thr) kinase Akt, which is translocated from plasma membrane to the cytosol and nucleus. These processes lead to the phosphorylation of target proteins (GSK-3β) (Kim and Feldman, 2012). This enzyme occurs in two distinct forms: an active form (Ser9 dephosphorylated) that is mostly found in nuclei, mitochondria, and membrane lipid rafts and the cytosolic inactive form (Cole et al., 2007). The PtdIns 3K pathway is the major pathway involved in glucose transport.

The insulin/IGF-1-mediated activation of Akt, PKC, or cAMP-dependent protein kinase results in the inactivation of GSK-3β. This process triggers multiple neurochemical effects such as synthesis of proteins associated with neuronal glucose metabolism, induction of antiapoptotic mechanisms, and antioxidant defenses (van der Heide et al., 2006). The overexpression of constitutively active GSK-3β not only promotes neural cell death but also prevents apoptotic cell death (Duarte et al., 2012). Other known targets of PtdIns 3K/Akt signaling are FOXO3, NF-κB, and cAMP response element–binding (CREB) protein. Thus Akt also phosphorylates and inhibits FOXO3, preventing the disruption of mitochondrial membrane potential and cytochrome c release. This process promotes neuronal survival (Cole et al., 2007; van der Heide et al., 2006). NF-κB phosphorylation by Akt protects against oxidative stress and apoptosis by increasing Cu/Zn superoxide dismutase (SOD) expression and manganese SOD levels (Duarte et al., 2012; Cole et al., 2007).

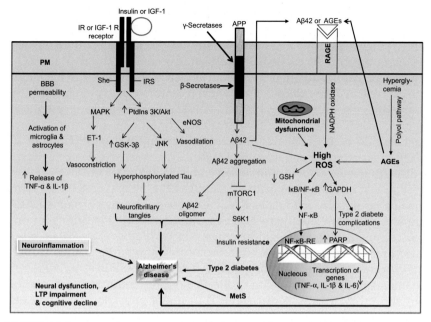

FIGURE 5.8 **Interactions among insulin and insulin-like growth factor receptor, amyloid β metabolism, and RAGE receptor.** *AGEs*, advanced glycation end products; *AMP*, adenosine monophosphate; *APP*, amyloid precursor protein; *Aβ*, amyloid β; *BBB*, blood–brain barrier; *eNOS*, endothelial nitric oxide synthase; *ET-1*, endothelin-1; *GADPH*, glyceraldehyde--phosphate dehydrogenase; *GSH*, glutathione; *GSK-3 β*, glycogen synthase kinase-3 β; *IGF-1*, insulin-like growth factor 1; *IL-1 β*, interleukin-1β; *IL-6*, interleukin-6; *IRS-1*, insulin receptor substrate-1; *IκB*, inhibitory subunit of NF-κB; *JNK*, c-Jun N-terminal kinase; *LTP*, long-term potentiation; *MetS*, metabolic syndrome; *mTOR*, mechanistic target of rapamycin; *NADP oxidase*, nicotinamide adenine dinucleotide phosphate oxidase; *NF-κB*, nuclear factor-κB; *NF-κB-RE*, nuclear factor-κB response element; *PARP*, poly(ADP-ribose)polymerase; *PKC*, protein kinase C; *PM*, plasma membrane; *RAGE*, receptor for advanced glycation end products; *ROS*, reactive oxygen species; *S6K1*, S6 kinase 1; *She and IRS*, docking protein; *TNF-α*, tumor necrosis factor-α.

Moreover, the CREB targeting of Akt is not only involved in the stimulation of neuronal glucose metabolism but also contribute to the enhancement of mitochondrial membrane potential, ATP levels, NADPH redox state, and hexokinase activity (Heras-Sandoval et al., 2011). Similarly, the activation of MAPK pathway promotes the expression of genes involved in cell and synapse growth as well as in cell repair and maintenance (Kim and Feldman, 2012), whereas Akt signaling is implicated in cell proliferation, cell growth, and protein synthesis (Brazil and Hemmings, 2001; Tremblay and Giguère, 2008) supporting the view that insulin signaling in the brain is associated with neuronal survival, neurotransmission, and modulation of synaptic activities (Zhao and Alkon, 2001). In addition, insulin is also involved in the regulation of synaptic plasticity and modulation of LTP (Nisticò et al., 2013), as well as in promotion of LTD (Labouèbe et al., 2013). These processes are involved in learning and memory. Insulin also potentiates the brain transport of molecules such as leptin (Kastin and Akerstrom, 2001) and amino acids (Tagliamonte et al., 1976). In streptozotocin-treated mice, insulin increases cerebral microvessels expression of occludin, claudin-5, and ZO-1 (Sun et al., 2015). Although there is no evidence on the effect of insulin on the expression of low-density lipoprotein receptor–related protein 1 (LRP-1) at the BBB, insulin regulates translocation and uptake of LRP-1 receptor in hepatic cells (Laatsch et al., 2009). Interestingly, LRP-1 expression is downregulated in brain capillaries of streptozotocin-injected mice (Hong et al., 2009) and cerebrospinal fluid–soluble LRP1 is increased in patients with type I diabetes treated with insulin for several years (Ouwens et al., 2014), suggesting that insulin might increase central LRP-1, at least on the long term. Finally, insulin also modulates RAGE expression. Levels of soluble RAGE are inversely correlated with plasma insulin concentration during an oral glucose tolerance test in healthy human subjects (Forbes et al., 2014). In isolated brain microvessels from streptozotocin-injected mice, insulin reduces the concentration of RAGE compared to diabetic mice (Sun et al., 2015).

Interestingly, the activation of the IR not only induces vasodilatation but also promotes vasoconstriction. Thus under physiological conditions there is a balance of both processes to regulate the immediate metabolic requirements of various tissues. In AD, accumulation of Aβ oligomers leads to increased TNF-α levels and activation of JNK resulting in inhibitory serine phosphorylation of IRS-1 (Fig. 5.8). Insulin promotes Aβ accumulation by limiting Aβ degradation via direct competition with IDE, a 110-kDa thiol zinc-metalloendopeptidase that cleaves small proteins of diverse sequence, many of which share a propensity to form β-pleated sheet–rich amyloid fibrils such as Aβ, insulin, glucagon, amylin, atrial natriuretic factor, and calcitonin (Bennett et al., 2000; Kurochkin, 2001). Increase in IDE activity results in more degradation of insulin than the normal physiological conditions (Lin et al., 1992). It is also reported that GLUT 4 is present in a subset of hippocampal region neurons, which are sensitive to insulin (Ren et al., 2014). Uptake of the glucose through these receptors may play an important role in the cognition and hippocampal-based learning and memory (Ren et al., 2014). IDE inhibitors have been used as a good therapeutic agent to stop learning and memory events at this level (Tang, 2016). Uncontrolled IDE activity may also downregulate IGF-1R protein expression (Ding et al., 1992) modulating synaptic transfer and neuroplasticity. Its downregulation may not only lead to Aβ generation but also to neurodegeneration (Bedse et al., 2015).

Aberrant phosphorylation of IRS results in an imbalance in homeostatic regulation of vascular function. This decrease in production of NO may result in decrease in cerebral blood flow and increase in pro-inflammatory cytokines (TNF-α, IL-1β, and IL-6) and ROS production resulting an increase in chronic inflammation and oxidative stress.

Both these processes are closely interlinked in the brain, and it is difficult to establish the temporal sequence of their relationship. Onset of chronic inflammation and oxidative stress not only supports the neurochemical and pathophysiological processes that contribute to type II diabetes and MetS but also is closely associated with the pathogenesis of AD.

AD is accompanied by excitotoxicity (Farooqui, 2010; Ong et al., 2013), onset of oxidative and nitrosative stress (Akhtar et al., 2016), and induction of neuroinflammation (Farooqui, 2010). These processes lead to the aberrant dysregulation of protein activity (Nakamura et al., 2013). In patients with type II diabetes and MetS the onset of oxidative and nitrosative stress also results in nitrosylation of specific protein with thiol targets. For example, glucose and $A\beta$ increase S-nitrosylation of IDE, which is closely associated with degradation of insulin as well as $A\beta$ peptide. In neurons, the formation of nitrosylated IDE inhibits degradation of both $A\beta$ and insulin (Neant-Fery et al., 2008; Ralat et al., 2009) supporting the view that this inhibitory effect may not only contribute to increase in $A\beta$ levels but also hyperinsulinemia along with an increase in Tau phosphorylation resulting in an imbalance in insulin-regulated Tau kinases and phosphatase. Chronic hyperinsulinemia may also exacerbate inflammatory responses and increase markers of oxidative stress (Farooqui, 2015). These observations are consistent with the concept that long-term consumption of high-calorie diet may promote alterations in insulin and IRs leading to the initiation of AD (Bero et al., 2011) along with increase in insulin levels, which correlates with insulin resistance in type II diabetes and MetS. These finding support the view that brain insulin resistance with compromised insulin signaling may be an important risk factor for human AD (Talbot et al., 2012).

Similarly, elevated glucose and $A\beta$ levels facilitate S-nitrosylation of the mitochondrial fission protein Drp1, which contributes to excessive mitochondrial fragmentation, energy compromise, and consequent synaptic dysfunction resulting in synaptic loss (Cho et al., 2009; Barsoum et al., 2006). Excitotoxicity increases in flux of Ca^{2+}, stimulates phospholipase A_2 (PLA_2) and eNOS activities leading to the generation of arachidonic acid (ARA) from membrane phospholipids and synthesis of NO from arginine. Increased synthesis of NO is also induced by hyperglycemia and accumulation of $A\beta$ (Fig. 5.5). These processes promote aberrant nitrosylation, which is attenuated by the NMDAR antagonist memantine in in vitro models (Akhtar et al., 2016). Moreover, in vivo studies in the db/db mouse model of type II diabetes have indicated that memantine can not only ameliorate dendritic damage but also induce synaptic spine loss (Akhtar et al., 2016). Importantly, in mouse model of AD, hyperglycemia modulates LTP, a process that is important for synaptic plasticity and learning and memory.

In similar mouse models of AD, sucrose feeding–mediated hyperglycemia induces deficits in spatial learning and memory (Cao et al., 2007; Soares et al., 2013) and is blocked by memantine. These studies are also supported by human phase 3 clinical trials of memantine indicating that memantine transiently stabilizes or improves behavioral deficits in AD (Reisberg et al., 2003). Collectively, these studies indicate that increased levels of glucose in type II diabetes and MetS may support accumulation of Aβ not only through the induction of oxidative stress but also via aberrant nitrosylation of two proteins, Drp1 and IDE, which can directly contribute to the pathogenesis and progression of AD.

Insulin resistance is modulated by both genetic and acquired factors. In MetS, insulin resistance is accompanied by hyperinsulinemia. These alterations in insulin signaling are closely associated with the pathogenesis of AD. As stated earlier, insulin increases the levels of Aβ in plasma in AD and MetS by inhibiting the insulin network not only through the involvement of insulin/Akt pathway (Townsend et al., 2007) but also by dissociating IRs after binding at the dendrites of synaptic sites (Zhao et al., 2008). As in type II diabetes, both MetS and AD are accompanied by the accumulation of AGE-modified proteins (Takeuchi and Yamagishi, 2008). These AGE-modified proteins like Aβ act as ligands for the RAGE leading to the activation of NF-κB (Takeuchi and Yamagishi, 2008). The extent of occurrence of this process in in vivo setting is still unclear, but increase in RAGE expression along with increase in amounts of AGE-modified proteins strongly suggests that vascular complications in MetS and AD may be caused by the accumulation of AGE-modified proteins. Inhibition of binding between Aβ and RAGE is known to impair the activation of microglia and reduce the production of proinflammatory mediators (Ramasamy et al., 2009). RAGE not only plays an important role in the clearance of Aβ but also contributes to APOE-mediated cellular processing and signaling (Bu, 2009). In addition to AGEs and Aβ, RAGE recognizes other ligands including serum amyloid A, S100 protein, and high-mobility group box1. These molecules are often present in the altered tissue environments associated with type II diabetes. The binding of Aβ oligomer with RAGE induces AD-like pathology, oxidative stress (Farooqui, 2010), synapse deterioration and loss (De Felice et al., 2009; Lacor et al., 2007), and inhibition of synaptic plasticity (Fig. 5.8) (Lambert et al., 1998). Thus converging evidence suggests that reduction in glucose utilization and deficient energy metabolism are early features of type II diabetes, MetS, and AD. These observations suggest that impaired insulin signaling may be characteristic features of metabolic and neurodegenerative diseases. In brain, the insulin/IGF signaling is important for neuronal growth, synaptic maintenance, and neuroprotection (Van Dam and Aleman, 2004). It is now proposed that impairments in brain insulin/IGF signaling are associated with increased accumulation of Aβ, phosphorylated Tau, reactive oxygen/nitrogen species, as well as proinflammatory and proapoptotic molecules (de la Monte, 2009; Farooqui, 2013; de la Monte and Tong, 2013, 2014). It has been suggested that AD may represent a metabolic disease of the brain associated with brain insulin and IGF-I resistance (de la Monte and Tong, 2013; Farooqui, 2013, 2015), which is defined by the reduction of insulin capacity to activate glucose utilization, either by insulin deficiency or by impairment in its secretion and/or utilization.

LIPID MEDIATORS IN TYPE II DIABETES, METABOLIC SYNDROME, AND ALZHEIMER'S DISEASE

As stated in Chapter 3, enzymatically derived lipid mediators of ARA include prostaglandins, leukotrienes, thromboxane, and lipoxins. These lipid mediators produce proinflammatory, antiinflammatory, prothrombotic, and proaggregatory properties. They are associated with neurotransmitter release, learning and memory, differentiation, modulation of cell cycle, membrane repair, and neurodegeneration (Farooqui, 2011). The corresponding lipid mediators of docosahexaenoic acid (DHA) metabolism are resolvins, neuroprotectins, and maresins (Farooqui, 2011). These lipid mediators produce antioxidant, anti-inflammatory, and antiapoptotic effects in the brain. The nonenzymic lipid mediators of ARA are 4-HNEs, isoprostanes, isoketal, isofuran, acrolein, and malondialdehyde. In contrast, nonenzymic lipid mediators are 4-hydroxyhexanal, neuroprostane, neuroketal, and neurofuran.

Very little is known about 4-HNE levels in visceral tissues of patients with MetS. However, it is reported that 4-HNE not only damages pancreatic beta cells but also can impair the ability of muscle and liver cells to respond to insulin (Mattson, 2009). In addition, 4-HNE also promote atherosclerosis by modifying lipoproteins and may mediate cardiac cell damage by impairing metabolic enzymes (Mattson, 2009). The levels of F_2-isoprostanes (IsoP), total hydroxyeicosatetraenoic (HETE) acid, cholesterol oxidation products (such as 24- and 27-hydroxycholesterol), 8-hydroxy-2'-deoxyguanosine (8-OHdG), and allantoin (a product of oxidative damage to uric acid) are not effected in MetS (Tsai et al., 2009; Seet et al., 2010). However, in type II diabetes levels of urinary F_2-isoprostanes are increased only in those individuals who have type II diabetes for 7 years indicating that oxidative damage in this condition is possibly a late event (Helmersson et al., 2004). Studies on the relationship between concentrations of plasma F_2-isoprostanes and isofurans, in obesity indicate that plasma F_2-isoprostanes and isofurans are elevated in obesity (Alkazemi et al., 2012). Collectively these studies indicate that levels of enzymatic and nonenzymatic lipid mediators of arachidonic and DHA metabolism are markedly increased in diabetes, MetS, and AD. The sphingolipid-derived lipid mediators include ceramide, ceramide-1-phosphate, sphingosine, and sphingosine-1-phosphate. Levels of sphingolipid-derived lipid mediators are increased in type II diabetes, MetS, and AD (Othman et al., 2012). Like ceramide, GM_3 ganglioside also contributes to the development of insulin resistance. This sphingolipid not only inhibits tyrosine phosphorylation of IRs and IRS-1 but also retards the activity of PLA_2 (Kabayama et al., 2005). It is proposed that GM_3 induces TNF-α-mediated insulin resistance and causes enhancement of tyrosine phosphorylation of IRs (Kabayama et al., 2005; Yamashita et al., 2003). Improved phosphorylation of IRs is also observed in GM_3 synthase knockout mice (Zhao et al., 2007). These studies indicate that complex sphingolipids play an important role in insulin resistance.

Elevated levels of oxy/hydroxysterols have been observed in patients with type II diabetes, MetS, hypercholesterolemia, and advanced carotid atherosclerosis. These individuals have increased risk for CVD (Murakami et al., 2001). Oxy/hydroxysterols not only regulate cholesterol metabolism (Krook, 2006) but also act as important modulators of insulin by acting as agonists of liver X receptors (Murakami et al., 2001). Oxy/hydroxysterols mediate proinflammatory, proapoptotic, and profibrogenic effects in various types of cell cultures (Sottero et al., 2009). Increase in levels of oxy/hydroxysterols in obesity and the MetS are linked with enhanced inflammatory stress, as indicated by the increase in C-reactive protein, a pentameric acute-phase protein that has emerged as the golden marker for inflammation (Knopp and Paramsothy, 2006). In the brain, cholesterol is mostly synthesized by de novo synthesis in astrocytes and plasma lipoproteins cannot cross the BBB. The BBB is permeable to oxy/hydroxysterols. Thus, excess cholesterol in the brain may under normal physiological conditions be excreted in the blood. The major oxysterol involved in this excretion mechanism are 24-hydroxycholesterol and 27-hydroxycholesterol (Bjorkhem, 2006). Oxy/hydroxylsterols promote neurodegeneration in various types of neuronal cultures (Farooqui, 2011). Thus in SH-SY5Y human neuroblastoma cells, 27-hydroxycholesterol enhances Aβ42 production by upregulating APP and BACE1 protein levels (Prasanthi et al., 2009). In the brain, high cholesterol levels not only decrease the activity of IDE but also decrease LRP-1 levels in rabbit hippocampus (Prasanthi et al., 2008) supporting the view that cholesterol-enriched diets and synthesis of oxy/hydroxysterols induce AD-like pathology by altering leptin signaling (Marwarha et al., 2010). Converging evidence from aforementioned studies suggest that in type II diabetes, MetS, and AD increase in degradation of membrane glycerophospholipids, sphingolipids, and cholesterol in visceral tissues and brain results in marked elevation in levels of lipid mediators, which not only disrupts the signaling networks but also results in the loss of communication among various receptors on neural and nonneural cells. These processes threaten the integrity of neural cell lipid homeostasis and facilitate cell death through the induction of inflammation and

TABLE 5.1   Levels of Lipid Mediators in Diabetes, MetS, and Alzheimer's Disease

Lipid Mediators	Type II Diabetes	Metabolic Syndrome	Alzheimer's Disease	References
Prostaglandins	Increased	Increased	Increased	Farooqui, (2010, 2011, 2013)
Leukotrienes	Increased	Increased	Increased	Farooqui, (2010, 2011, 2013)
Thromboxanes	Increased	Increased	Increased	Farooqui, (2010, 2011, 2013)
Isoprostanes	Increased	Increased	Increased	Farooqui, (2010, 2011, 2013)
Ceramide	Increased	Increased	Increased	Farooqui, (2010, 2011, 2013)
Sphingosine-1-phosphate	Increased	Increased	Increased	Farooqui, (2010, 2011, 2013)
Oxy/hydroxycholesterol	Increased	Increased	Increased	Farooqui, (2010, 2011, 2013)

oxidative stress. Thus enhancement in levels of glycerophospholipid-, sphingolipid-, and cholesterol-derived lipid mediators may be a common process in the pathogenesis of type II diabetes, MetS, and AD (Table 5.1). Collective evidence suggests that onset of AD is brought about a complex interplay among several neurochemical pathways including changes in APP metabolism, phosphorylation of the Tau protein, oxidative stress, impaired energetics, mitochondrial dysfunction, onset of neuroinflammation, membrane lipid dysregulation, and neurotransmitter pathway disruption (Kaddurah-Daouk et al., 2013). Most of these changes can be directly linked to metabolic abnormalities supporting the view that metabolic dysfunction is an important factor in the pathogenesis AD (Cai et al., 2012b). For example, impaired cerebral glucose uptake occurs decades prior to the onset of cognitive dysfunction and is an invariant feature of AD (Chen and Zhong, 2013). The well-documented neurotoxicity associated with Aβ42 is thought to participate in impaired neuronal energetics through initiating a cascade of pathological events; interaction between Aβ42 and mitochondrial enzymes leads to increased release of ROS, affecting glycolysis, the tricarboxylic acid cycle, and mitochondrial respiratory–chain activity through the accumulation of deleterious intermediate metabolites in the mitochondria (Chen and Yan, 2006).

LONG-TERM CONSUMPTION OF HIGH-CALORIE DIET AS A RISK FACTOR FOR ALZHEIMER'S DISEASE

High-calorie diet (Western diet) is not only enriched in high amounts of processed macronutrients (fats, cholesterol, red meat proteins, and simple sugars) but also contains high amounts of salt (sodium chloride).

Consumption of high levels of salt contributes to hypertension, a condition characterized by a chronic increase in blood pressure. In addition, high-calorie diet is low in fiber and deficient in minerals. High-calorie diet provides about 50% of total daily calories from refined carbohydrates, 35% calories from fat and refined oils, and 15% calories from proteins of animal origin. In addition, high-calorie diet contains 20 times more ARA than DHA (Farooqui, 2014). In contrast, Paleolithic diet on which our forefathers lived and survived thousands of years contained unprocessed carbohydrates (40%), fats (21%), and proteins (39%) (Simopoulos, 2009, 2013; Bengmark, 2013). In general, long-term consumption of high-calorie diet upregulates the metabolism of human cells toward biosynthetic pathways not only resulting in production of proinflammatory molecules but also leading to a dysbiotic gut microbiota, alteration of intestinal immunity, and low-grade systemic inflammation (Riccio and Rossano, 2015). In this metabolic state, intake of energy exceeds energy expenditure producing obesity. This condition is accompanied by the induction of chronic oxidative stress and inflammation, which not only contribute to cardiovascular diseases, type II diabetes, and MetS (Farooqui et al., 2012) but also are important risk factors for neurotraumatic (stroke), neurodegenerative (AD), and neuropsychiatric (depression) diseases (Farooqui, 2013). As stated previously, long-term consumption of high-calorie diet during adolescence results in obesity and decrease in BDNF in hippocampus. In contrast, low-calorie diets based on the consumption of vegetables, fruit, legumes, fish, prebiotics, and probiotics act on nuclear receptors and enzymes that upregulate oxidative metabolism, downregulate the synthesis of proinflammatory molecules, and restore or maintain a healthy symbiotic gut microbiota (Riccio and Rossano, 2015). In addition, the consumption of low-calorie diet during adolescence increases hippocampal and prefrontal BDNF levels, proliferative cells and neuron numbers in dentate gyrus, thus positively affecting spatial memory in adulthood (Kaptan et al., 2015). Moreover, the consumption of low-calorie diet decreases serum glucose levels and level of lipid peroxidation in serum and hippocampus, suggesting the importance of nutrition in adolescence for cognitive function in adulthood (Kaptan et al., 2015).

CONCLUSION

Type II diabetes and MetS are metabolic diseases, which are characterized by hyperglycemia, dyslipidemia, hypertension, and insulin resistance along with alterations in glycerophospholipid-, sphingolipid-, and cholesterol-derived lipid mediators. These processes and changes in lipid mediators markedly effect brain function. Type II diabetes and MetS are major risk factors for AD, a debilitating neurodegenerative disease, which is characterized by the progressive loss of cholinergic neurons, accumulation of Aβ, and hyperphosphorylated Tau protein. These processes result in severe behavioral, motor, and cognitive impairments due to decrease in synaptic plasticity in patients with AD. Diet and lifestyle play an important role in the development of type II diabetes and MetS. The levels of simple sugars and fatty acid composition of diet plays a major role in the pathogenesis of type II diabetes and MetS. Typical Western diet is characterized by high levels of saturated and n-6 fatty acids and low levels of n-3 fatty acids. Long-term consumption of Western diet induces insulin resistance, whereas diet containing n-3 fatty acids and low in simple sugars suppresses insulin resistance. Accumulation of glycerophospholipid-, sphingolipid-, and cholesterol-derived lipid mediators in the brain is responsible for insulin resistance and neurodegeneration in brains of patients with AD.

References

Akhtar, M.W., Sanz-Blasco, S., Dolatabadi, N., Parker, J., Chon, K., 2016. Elevated glucose and oligomeric β-amyloid disrupt synapses via a common pathway of aberrant protein S-nitrosylation. Nat. Commun. 7, 10242.

Akter, K., Lanza, E.A., Martin, S.A., Myronyuk, N., Rua, M., Raffa, R.B., 2011. Diabetes mellitus and Alzheimer's disease: shared pathology and treatment? Br. J. Clin. Pharmacol. 71, 365–376.

Alexander, M.C., Lomanto, M., Nasrin, N., Ramaika, C., 1988. Insulin stimulates glyceraldehyde-3-phosphate dehydrogenase gene expression through cis-acting DNA sequences. Proc. Natl. Acad. Sci. U.S.A. 85, 5092–5096.

Alkazemi, D.G., Egeland, G.M., Jackson Roberts, L., Kubow, S., 2012. Isoprostanes and isofurans as non-traditional risk factors for cardiovascular disease among Canadian Inuit. Free Radic. Res. June 20, 1258–1266.

American Diabetes Association, 2010. Diagnosis and classification of diabetes mellitus. Diabetes Care 33 (Suppl. 1), S62–S69.

An, W.L., Cowburn, R.F., Li, L., Braak, H., Alafuzoff, I., Iqbal, K., Iqbal, I.G., Winblad, B., Pei, J.J., 2003. Up-regulation of phosphorylated/activated p70 S6 kinase and its relationship to neurofibrillary pathology in Alzheimer's disease. Am. J. Pathol. 163, 591–607.

Angeloni, C., Maraldi, T., Vauzour, D., 2014. Redox signaling in degenerative diseases: from molecular mechanisms to health implications. Biomed. Res. Int. 2014, 245761.

Banks, W.A., Kastin, A.J., Pan, W., 1999. Uptake and degradation of blood-borne insulin by the olfactory bulb. Peptides 20, 373–378.

Banks, W.A., Owen, J.B., Erickson, M.A., 2012. Insulin in the brain: there and back again. Pharmacol. Ther. 136, 82–93.

Banks, W.A., 2004. The source of cerebral insulin. Eur. J. Pharmacol. 490, 5–12.

Barsoum, M.J., Yuan, H., Gerencser, A.A., Liot, G., Kushnareva, Y., 2006. Nitric oxide-induced mitochondrial fission is regulated by dynamin-related GTPases in neurons. EMBO J. 25, 3900–3911.

Bechmann, L.P., Hannivoort, R.A., Gerken, G., Hotamisligil, G.S., Trauner, M., Canbay, A., 2012. The interaction of hepatic lipid and glucose metabolism in liver diseases. J. Hepatol. 56, 952–964.

Bedse, G., Di Domenico, F., Serviddio, G., Cassano, T., 2015. Aberrant insulin signaling in Alzheimer's disease: current knowledge. Front. Neurosci. 9, 204.

Beisswenger, P.J., Howell, S.K., Smith, K., Szwergold, B.S., 2003. Glyceraldehyde-3-phosphate dehydrogenase activity as an independent modifier of methylglyoxal levels in diabetes. Biochim. Biophys. Acta 1637, 98–106.

Bengmark, S., 2013. Processed foods, dysbiosis, systemic inflammation, and poor health. Curr. Nutr. Food Sci. 9, 113–143.

Bennett, R.G., Duckworth, W.C., Hamel, F.G., 2000. Degradation of amylin by insulin-degrading enzyme. J. Biol. Chem. 275, 36621–36625.

Bero, A.W., Yan, P., Roh, J.H., Cirrito, J.R., Stewart, F.R., Raichle, M.E., et al., 2011. Neuronal activity regulates the regional vulnerability to amyloid-β deposition. Nat. Neurosci. 14, 750–756.

Bhaskar, K., Miller, M., Chludzinski, A., Herrup, K., Zagorski, M., Lamb, B.T., 2009. The PI3K-Akt-mTOR pathway regulates Abeta oligomer induced neuronal cell cycle events. Mol. Neurodegener. 4, 14.

Bierhaus, A., Humpert, P.M., Morcos, M., Wendt, T., Chavakis, T., et al., 2005. Understanding RAGE, the receptor for advanced glycation end products. J. Mol. Med. 83, 876–886.

Bjorkhem, I., 2006. Crossing the barrier: oxysterols as cholesterol transporters and metabolic modulators in the brain. J. Intern. Med. 260, 493–508.

Blázquez, E., Velázquez, E., Hurtado-Carneiro, V., Ruiz-Albusac, J.M., 2014. Insulin in the brain: its pathophysiological implications for states related with central insulin resistance, type2 diabetes and Alzheimer's disease. Front. Endocrinol. 5, 3389.

Brazil, D.P., Hemmings, B.A., 2001. Ten years of protein kinase B signalling: a hard Akt to follow. Trends Biochem. Sci. 26, 657–664.

Brownlee, M., 2001. Biochemistry and molecular cell biology of diabetic complications. Nature 414, 813–820.

Brownlee, M., 2005. The pathobiology of diabetic complications: a unifying mechanism. Diabetes 54, 1615–1625.

Brundel, M., van den Berg, E., Reijmer, Y.D., de Bresser, J., Kappelle, L.J., Biessels, G.J., 2012. Cerebral haemodynamics, cognition and brain volumes in patients with type 2 diabetes. J. Diabetes Complicat. 26, 205–209.

Bu, G., 2009. Apolipoprotein E and its receptors in Alzheimer's disease: pathways, pathogenesis and therapy. Nat. Rev. Neurosci. 10, 333–344.

Bucciarelli, L.G., Wendt, T., Rong, L., Lalla, E., Hofmann, M.A., Goova, M.T., et al., 2002. RAGE is a multiligand receptor of the immunoglobulin superfamily: implications for homeostasis and chronic disease. Cell Mol. Life Sci. 59, 1117–1128.

Butterfield, D.A., Di Domenico, F., Barone, E., 2014. Elevated risk of type 2 diabetes for development of Alzheimer disease: a key role for oxidative stress in brain. Biochim. Biophys. Acta 1842, 1693–1706.

Caccamo, A., Majumder, S., Richardson, A., Strong, R., Oddo, S., 2010. Molecular interplay between mammalian target of rapamycin (mTOR), amyloid-beta, and tau: effects on cognitive impairments. J. Biol. Chem. 285, 13107–13120.

Caccamo, A., Branca, C., Talboom, J.S., Shaw, D.M., Turner, D., Ma, L., et al., 2015. Genetic reduction of mammalian target of rapamycin ameliorates Alzheimer's disease-like cognitive and pathological deficits by restoring hippocampal gene expression signature. J. Neurosci. 35, 14042–14056.

Cai, Z., Zhao, B., Li, K., et al., 2012a. Mammalian target of rapamycin: a valid therapeutic target through the autophagy pathway for Alzheimer's disease? J. Neurosci. Res. 90, 1105–1118.

Cai, H., Cong, W.N., Ji, S., Rothman, S., Maudsley, S., et al., 2012b. Metabolic dysfunction in Alzheimer's disease and related neurodegenerative disorders. Curr. Alzheimer Res. 9, 5–17.

Campbell, C.T., Kolesar, J.E., Kaufman, B.A., 2012. Mitochondrial transcription factor A regulates mitochondrial transcription initiation, DNA packaging, and genome copy number. Biochim. Biophys. Acta 1819, 921–929.

Cao, D., Lu, H., Lewis, T.L., Li, L., 2007. Intake of sucrose-sweetened water induces insulin resistance and exacerbates memory deficits and amyloidosis in a transgenic mouse model of Alzheimer disease. J. Biol. Chem. 282, 36275–36282.

CDC National Diabetes Fact Sheet, 2007.

Centers for Medicare and Medicaid Services, 2013. National Health Expenditure Projections 2012–2022. www.cms.gov.

Chen, X., Yan, S.D., 2006. Mitochondrial abeta: a potential cause of metabolic dysfunction in Alzheimer's disease. IUBMB Life 58, 686–694.

Chen, Z., Zhong, C., 2013. Decoding Alzheimer's disease from perturbed cerebral glucose metabolism: implications for diagnostic and therapeutic strategies. Prog. Neurobiol. 108, 21–43.

Chen, T.J., Wang, D.C., Chen, S.S., 2009. Amyloid-beta interrupts the PI3K-Akt-mTOR signaling pathway that could be involved in brain-derived neurotrophic factor-induced Arc expression in rat cortical neurons. J. Neurosci. Res. 87, 2297–2307.

Cho, D.H., Nakamura, T., Fang, J., Cieplak, P., Godzik, A., Gu, Z., Lipton, S.A., 2009. S-nitrosylation of Drp1 mediates beta-amyloid-related mitochondrial fission and neuronal injury. Science 324, 102–105.

Cole, A.R., Astell, A., Green, C., Sutherland, C., 2007. Molecular connexions between dementia and diabetes. Neurosci. Behav. Rev. 31, 1046–1063.

Conejo, R., Lorenzo, M., 2001. Insulin signaling leading to proliferation, survival, and membrane ruffling in C2C12 myoblasts. J. Cell Physiol. 187, 96–108.

Costa, R.O., Ferreiro, E., Martins, I., Santana, I., Cardoso, S.M., et al., 2012. Amyloid beta-induced ER stress is enhanced under mitochondrial dysfunction conditions. Neurobiol. Aging 33, 824e5–824e16.

Craft, S., Watson, G.S., 2004. Insulin and neurodegenerative disease: shared and specific mechanisms. Lancet Neurol. 3, 169–178.

d'Abramo, C., Ricciarelli, R., Pronzato, M.A., Davies, P., 2006. Troglitazone, a peroxisome proliferator-activated receptor-gamma agonist, decreases tau phosphorylation in CHOtau4R cells. J. Neurochem. 98, 1068–1077.

Dal Col, J., Dolcetti, R., 2008. GSK-3beta inhibition: at the crossroad between Akt and mTOR constitutive activation to enhance cyclin D1 protein stability in mantle cell lymphoma. Cell Cycle 7, 2813–2816.

Damjanac, M., Rioux Bilan, A., Paccalin, M., et al., 2008. Dissociation of Akt/PKB and ribosomal S6 kinase signaling markers in a transgenic mouse model of Alzheimer's disease. Neurobiol. Dis. 29, 354–367.

De Felice, F.G., Vieira, M.N., Bomfim, T.R., Decker, H., Velasco, P.T., et al., 2009. Protection of synapses against Alzheimer's-linked toxins: insulin signaling prevents the pathogenic binding of Abeta oligomers. Proc. Natl. Acad. Sci. U.S.A. 106, 1971–1976.

de la Monte, S.M., Tong, M., 2013. Insulin resistance and metabolic failure underlie Alzheimer disease. In: Farooqui, T., Farooqui, A.A. (Eds.), Metabolic Syndrome and Neurological Disorders, Oxford UK. John Wiley & Sons, Inc., Oxford, UK, pp. 1–30.

de la Monte, S.M., Tong, M., 2014. Brain metabolic dysfunction at the core of Alzheimer's disease. Biochem. Pharmacol. 88, 548–559.

de la Monte, S.M., 2009. Insulin resistance and Alzheimer's disease. BMB Rep. 42, 475–481.

Dedoussis, C.V.Z., Kaliora, A.C., Panagiotakos, D.B., Spring 2007. Genes, diet and type 2 diabetes mellitus: a review. Rev. Diabet. Stud. 4, 13–24.

Dias, I.H., Griffiths, H.R., 2014. Oxidative stress in diabetes – circulating advanced glycation end products, lipid oxidation and vascular disease. Ann. Clin. Biochem. 51, 125–127.

Ding, L., Becker, A.B., Suzuki, A., Roth, R.A., 1992. Comparison of the enzymatic and biochemical properties of human insulin-degrading enzyme and Escherichia coli protease III. J. Biol. Chem. 267, 2414–2420.

Drew, B.G., Rye, K.A., Duffy, S.J., Barter, P., Kingwell, B.A., 2012. The emerging role of HDL in glucose metabolism. Nat. Rev. Endocrinol. 8, 237–245.

Du, X., Matsumura, T., Edelstein, D., Rossetti, L., Zsengeller, Z., Szabo, C., Brownlee, M., 2003. Inhibition of GAPDH activity by poly(ADP-ribose) polymerase activates three major pathways of hyperglycemic damage in endothelial cells. J. Clin. Invest. 112, 1049–1057.

Duarte, A.I., Moreira, P.I., Oliveira, C.R., 2012. Insulin in central nervous system: more than just a peripheral hormone. J. Aging Res. 384017.

Ehninger, D., Han, S., Shilyansky, C., Zhou, Y., Li, W., Kwiatkowski, D.J., et al., 2008. Reversal of learning deficits in a Tsc2± mouse model of tuberous sclerosis. Nat. Med. 14, 843–848.

Ehninger, D., 2013. From genes to cognition in tuberous sclerosis: implications for mTOR inhibitor-based treatment approaches. Neuropharmacology 68, 97–105.

Farooqui, A.A., Farooqui, T., Panza, F., Frisardi, V., 2012. Metabolic syndrome as a risk factor for neurological disorders. Cell Mol. Life Sci. 69, 741–762.

Farooqui, A.A., 2010. Neurochemical Aspects of Neurotraumatic and Neurodegenerative Diseases. Springer, New York.

Farooqui, A.A., 2011. Lipid Mediators and Their Metabolism in the Brain. Springer, New York.

Farooqui, A.A., 2013. Metabolic Syndrome: An Important Risk Factor for Stroke, Alzheimer, and Depression. Springer, New York.

Farooqui, A.A., 2014. Inflammation and Oxidative Stress in Neurological Disorders. Springer, New York.

Farooqui, A.A., 2015. High Calorie Diet and Human Brain. Springer International Publishing, Switzerland.

Fernandez, A.M., Torres-Alemán, I., 2012. The many faces of insulin-like peptide signaling in the brain. Nat. Rev. Neurosci. 13, 225–239.

Ferreira, I.L., Resende, R., Ferreiro, E., Rego, A.C., Pereira, C.F., 2010. Multiple defects in energy metabolism in Alzheimer's disease. Curr. Drug Targets 11, 1193–1206.

Forbes, J.M., Sourris, K.C., de Courten, M.P., Dougherty, S.L., Chand, V., Lyons, J.G., et al., 2014. Advanced glycation end products (AGEs) are cross-sectionally associated with insulin secretion in healthy subjects. Amino Acids 46, 321–326.

Freude, S., Schilbach, K., Schubert, M., 2009. The role of IGF-1 receptor and insulin receptor signaling for the pathogenesis of Alzheimer's disease: from model organisms to human disease. Curr. Alzheimer Res. 6, 213–223.

Frolich, L., Blum-Degen, D., Bernstein, H.G., Engelsberger, S., Humrich, J., et al., 1998. Brain insulin and insulin receptors in aging and sporadic Alzheimer's disease. J. Neural Transm. 105, 423–438.

Gasparini, L., Gouras, G.K., Wang, R., Gross, R.S., Beal, M.F., et al., 2001. Stimulation of beta-amyloid precursor protein trafficking by insulin reduces intraneuronal beta-amyloid and requires mitogen-activated protein kinase signaling. J. Neurosci. 21, 2561–2570.

Gasser, A., Forbes, J.M., 2008. Advanced glycation: implications in tissue damage and disease. Protein Pept. Lett. 15, 385–391.

Ghemrawi, R., Pooya, S., Lorentz, S., Gauchotte, G., Arnold, C., Gueant, J.L., et al., 2013. Decreased vitamin B12 availability induces ER stress through impaired SIRT1-deacetylation of HSF1. Cell Death Dis. 4, e553.

Giacco, F., Brownlee, M., 2010. Oxidative stress and diabetic complications. Circ. Res. 107, 1058–1070.

Gouras, G.K., 2013. mTOR: at the crossroads of aging, chaperones, and Alzheimer's disease. J. Neurochem. 124, 747–748.

Griffin, R.J., Moloney, A., Kelliher, M., Johnston, J.A., Ravid, R., Dockery, P., O'Connor, R., O'Neill, C., 2005. Activation of Akt/PKB, increased phosphorylation of Akt substrates and loss and altered distribution of Akt and PTEN are features of Alzheimer's disease pathology. J. Neurochem. 93, 105–117.

Grillo, M.A., Colombatto, S., 2008. Advanced glycation end-products (AGEs): involvement in aging and in neurodegenerative diseases. Amino Acids 35, 29–36.

Hao, K., Di Narzo, A.F., Ho, L., Luo, W., Li, S., et al., 2015. Shared genetic etiology underlying Alzheimer's disease and type 2 diabetes. Mol. Asp. Med. 43–44, 66–76.

Helmersson, J., Vessby, B., Larsson, A., Basu, S., 2004. Association of type 2 diabetes with cyclooxygenase-mediated inflammation and oxidative stress in an elderly population. Circulation 109, 1729–1734.

Heras-Sandoval, D., Avila-Muñoz, E., Arias, C., 2011. The phosphatidylinositol 3-kinase/ mTor pathway as a therapeutic target for brain aging and neurodegeneration. Pharmaceuticals 4, 1070–1087.

Hobbelen, J.S., Tan, F.E., Verhey, F.R., Koopmans, R.T., de Bie, R.A., 2011. Prevalence, incidence and risk factors of paratonia in patients with dementia: a one-year follow-up study. Int. Psychogeriatr. 23, 1051–1060.

Hoeffer, C.A., Klann, E., 2010. mTOR signaling: at the crossroads of plasticity, memory and disease. Trends Neurosci. 33, 67–75.

Holland, W.L., Summers, S.A., 2008. Sphingolipids, insulin resistance, and metabolic disease: new insights from in vivo manipulation of sphingolipid metabolism. Endocr. Rev. 29, 381–402.

Hong, H., Liu, L.P., Liao, J.M., Wang, T.S., Ye, F.Y., Wu, J., et al., 2009. Downregulation of LRP1 [correction of LPR1] at the blood–brain barrier in streptozotocin-induced diabetic mice. Neuropharmacology 56, 1054–1059.

Hooper, C., Killick, R., Lovestone, S., 2008. The GSK3 hypothesis of Alzheimer's disease. J. Neurochem. 104, 1433–1439.

Horwood, J.M., Dufour, F., Laroche, S., Davis, S., 2006. Signalling mechanisms mediated by the phosphoinositide 3-kinase/Akt cascade in synaptic plasticity and memory in the rat. Eur. J. Neurosci. 23, 3375–3384.

Hwu, C.M., Liou, T.L., Hsiao, L.C., Lin, M.W., 2009. Prehypertension is associated with insulin resistance. Q. J. Med. 102, 705–711.

Janson, T., Laedtke, J.E., Parisi, P., O'Brien, R.C., Petersen, P.C., et al., 2004. Butler increased risk of type 2 diabetes in Alzheimer disease. Diabetes 53, 474–481.

Jaworski, J., Sheng, M., 2006. The growing role of mTOR in neuronal development and plasticity. Mol. Neurobiol. 34, 205–219.

Johnson, S.C., Sangesland, M., Kaeberlein, M., Rabinovitch, P.S., 2015. Modulating mTOR in aging and health. Interdiscip. Top. Gerontol. 40, 107–127.

Kabayama, K., Sato, T., Kitamura, F., Uemura, S., Kang, B.W., et al., 2005. TNFalpha-induced insulin resistance in adipocytes as a membrane microdomain disorder: involvement of ganglioside GM3. Glycobiology 15, 21–29.

Kaddurah-Daouk, R., Zhu, H., Sharma, S., Bogdanov, M., Rozen, S.G., et al., 2013. Alterations in metabolic pathways and networks in Alzheimer's disease. Transl. Psychiatry 3, e244.

Kaptan, Z., Akgün-Dar, K., Kapucu, A., Dedeakayoğulları, H., Batu, Ş., et al., 2015. Long term consequences on spatial learning-memory of low-calorie diet during adolescence in female rats; hippocampal and prefrontal cortex BDNF level, expression of NeuN and cell proliferation in dentate gyrus. Brain Res. 1618, 194–204.

Kastin, A.J., Akerstrom, V., 2001. Glucose and insulin increase the transport of leptin through the blood-brain barrier in normal mice but not in streptozotocin-diabetic mice. Neuroendocrinology 2001 (73), 237–242.

Keller, J.N., Mattson, M.P., 1998. Roles of lipid peroxidation in modulation of cellular signaling pathways, cell dysfunction, and death in the nervous system. Rev. Neurosci. 9, 105–116.

Keller, J.N., Mark, R.J., Bruce, A.J., Blanc, E., Rothstein, J.D., et al., 1997. 4-Hydroxynonenal, an aldehydic product of membrane lipid peroxidation, impairs glutamate transport and mitochondrial function in synaptosomes. Neuroscience 80, 685–696.

Kim, B., Feldman, E.L., 2012. Insulin resistance in the nervous system. Trends Endocrinol. Metab. 23, 133–141.

Kimura, R., Okouchi, M., Kato, T., Imaeda, K., Okayama, N., Asai, K., Joh, T., 2013. Epidermal growth factor receptor transactivation is necessary for glucagon-like peptide-1 to protect PC12 cells from apoptosis. Neuroendocrinology 97, 300–308.

Kissebah, A.H., Vydelingum, N., Murray, R., Evans, D.J., Hartz, A.J., et al., 1982. Relation of body fat distribution to metabolic complications of obesity. J. Clin. Endocrinol. Metab. 54, 254–260.

Kleinridders, A., Ferris, H.A., Cai, W., Kahn, C.R., 2014. Insulin action in brain regulates systemic metabolism and brain function. Diabetes 63, 2232–2243.

Knopp, R.H., Paramsothy, P., 2006. Oxidized LDL and abdominal obesity: a key to understanding the metabolic syndrome. Am. J. Clin. Nutr. 83, 1–2.

Komulainen, P., Lakka, T.A., Kivipelto, M., Hassinen, M., Helkala, E.L., et al., 2007. Metabolic syndrome and cognitive function: a population-based follow-up study in elderly women. Dement. Geriatr. Cogn. Disord. 23, 29–34.

Kook, S.Y., Hong, H.S., Moon, M., Ha, C.M., Chang, S., et al., 2012. RAGE interaction disrupts tight junctions of the blood-brain barrier via Ca^{2+}-calcineurin signaling. J. Neurosci. 32, 8845–8854.

Krook, A., 2006. Can the liver X receptor work its magic in skeletal muscle too? Diabetologia 49, 819–821.

Kurochkin, I.V., 2001. Insulin-degrading enzyme: embarking on amyloid destruction. Trends Biochem. Sci. 26, 421–425.

Laatsch, A., Merkel, M., Talmud, P.J., Grewal, T., Beisiegel, U., et al., 2009. Insulin stimulates hepatic low density lipoprotein receptor-related protein 1 (LRP1) to increase postprandial lipoprotein clearance. Atherosclerosis 204, 105–111.

Labouèbe, G., Liu, S., Dias, C., Zou, H., Wong, J.C., et al., 2013. Insulin induces long-term depression of ventral tegmental area dopamine neurons via endocannabinoids. Nat. Neurosci. 16, 300–308.

Lacor, P.N., Buniel, M.C., Furlow, P.W., Clemente, A.S., Velasco, P.T., et al., 2007. Abeta oligomer-induced aberrations in synapse composition, shape, and density provide a molecular basis for loss of connectivity in Alzheimer's disease. J. Neurosci. 27, 796–807.

Lambert, M.P., Barlow, A.K., Chromy, B.A., Edwards, C., Freed, R., et al., 1998. Diffusible, nonfibrillar ligands derived from Abeta1-42 are potent central nervous system neurotoxins. Proc. Natl. Acad. Sci. U.S.A. 95, 6448–6453.

Lamming, D.W., Ye, L., Sabatini, D.M., Baur, J.A., 2013. Rapalogs and mTOR inhibitors as anti-aging therapeutics. J. Clin. Invest. 123, 980–989.

Lee, H.-K., Kumar, P., Fu, Q., Rosen, K.M., Querfurth, H.W., 2009. The insulin/Akt signaling pathway is targeted by intracellular β-amyloid. Mol. Biol. Cell 20, 1533–1544.

Lee, M.J., Lee, J.H., Rubinsztein, D.C., 2013. Tau degradation: the ubiquitin-proteasome system versus the autophagy-lysosome system. Prog. Neurobiol. 105, 49–59.

Leung, P.S., 2016. The potential protective action of vitamin D in hepatic insulin resistance and pancreatic islet dysfunction in type 2 diabetes mellitus. Nutrients 8 (3) pii: E147.

Li, L., Holscher, C., 2007. Common pathological processes in Alzheimer disease and type 2 diabetes: a review. Brain Res. Rev. 56, 384–402.

Li, Y., Xu, S., Giles, A., Nakamura, K., Lee, J.W., Hou, X., et al., 2011. Hepatic overexpression of SIRT1 in mice attenuates endoplasmic reticulum stress and insulin resistance in the liver. FASEB J. 25, 1664–1679.

Lin, T., Wang, D.E.L.I., Nagpal, M.L., Chang, W.E., Calkins, J.H., 1992. Down-regulation of Leydig cell insulin-like growth factor-I gene expression by interleukin-1. Endocrinology 130, 1217–1224.

Lin, W.C., Chuang, Y.C., Chang, Y.S., Lai, M.D., Teng, Y.N., Su, I.J., et al., 2012. Endoplasmic reticulum stress stimulates p53 expression through NF-kappaB activation. PLoS One 7, e39120.

Liu, T., Ren, D., Zhu, X., Yin, Z., Jin, G., Zhao, Z., et al., 2013. Transcriptional signaling pathways inversely regulated in Alzheimer's disease and glioblastoma multiform. Sci. Rep. 3, 3467.

Liu, Y.W., Zhang, L., Li, Y., Cheng, Y.Q., Zhu, X., Zhang, F., Yin, X.X., 2015. Activation of mTOR signaling mediates the increased expression of AChE in high glucose condition: in vitro and in vivo evidences. Mol. Neurobiol. 53.

Loske, C., Gerdemann, A., Schepl, W., Wycislo, M., Schinzel, R., et al., 2000. Transition metal-mediated glycoxidation accelerates cross-linking of beta-amyloid peptide. Eur. J. Biochem. 267, 4171–4178.

Ma, Y.Q., Wu, D.K., Liu, J.K., 2013. mTOR and tau phosphorylated proteins in the hippocampal tissue of rats with type 2 diabetes and Alzheimer's disease. Mol. Med. Rep. 7, 623–627.

Maiese, K., Chong, Z.Z., Wang, S., Shang, Y.C., 2012. Oxidant stress and signal transduction in the nervous system with the PI 3-K, Akt, and mTOR cascade. Int. J. Mol. Sci. 13, 13830–13866.

Maiese, K., 2014. Driving neural regeneration through the mammalian target of rapamycin. Neural Regen. Res. 9, 1413–1417.

Malla, R., Ashby Jr., C.R., Narayanan, N.K., Narayanan, B., Faridi, J.S., Tiwari, A.K., 2015. Proline-rich AKT substrate of 40-kDa (PRAS40) in the pathophysiology of cancer. Biochem. Biophys. Res. Commun. 463, 161–166.

Mark, R.J., Lovell, M.A., Markesbery, W.R., Uchida, K., Mattson, M.P., 1997. A role for 4-hydroxynonenal, an aldehydic product of lipid peroxidation, in disruption of ion homeostasis and neuronal death induced by amyloid β-peptide. J. Neurochem. 68, 255–264.

Marwarha, G., Dasari, B., Prasanthi, J.R., Schommer, J., Ghribi, O., 2010. Leptin reduces the accumulation of Abeta and phosphorylated tau induced by 27-hydroxycholesterol in rabbit organotypic slices. J. Alzheimer's Dis. 19, 1007–1019.

Mattson, M.P., 2009. Roles of the lipid peroxidation product 4-hydroxynonenal in obesity, the metabolic syndrome, and associated vascular and neurodegenerative disorders. Exp. Gerontol. 44, 625–633.

Miao, X.Y., Gu, Z.Y., Liu, P., Hu, Y., Li, L., Gong, Y.P., et al., 2013. The human glucagon-like peptide-1 analogue liraglutide regulates pancreatic beta-cell proliferation and apoptosis via an AMPK/mTOR/P70S6K signaling pathway. Peptides 39, 71–79.

Mooradian, A.D., 1997. Pathophysiology of central nervous system complications in diabetes mellitus. Clin. Neurosci. 4, 322–326.

Mooradian, A.D., 2009. Dyslipidemia in type 2 diabetes mellitus. Nat. Clin. Pract. Endocrinol. Metab. 5, 150–159.

Muoio, D.M., Newgard, C.B., 2008. Mechanisms of disease: molecular and metabolic mechanisms of insulin resistance and beta-cell failure in type 2 diabetes. Nat. Rev. Mol. Cell Biol. 9, 193–205.

Murakami, H., Tamasawa, N., Matsui, J., Yamato, K., JingZhi, G., et al., 2001. Plasma levels of soluble vascular adhesion molecule-1 and cholesterol oxidation product in type 2 diabetic patients with nephropathy. J. Atheroscler. Thromb. 8, 21–24.

Nakamura, T., Tu, S., Akhtar, M.W., Sunico, C.R., Okamoto, S., Lipton, S.A., 2013. Aberrant protein s-nitrosylation in neurodegenerative diseases. Neuron 78, 596–614.

Neant-Fery, M., Garcia-Ordoñez, R.D., Logan, T.P., Selkoe, D.J., Li, L., Reinstatler, L., et al., 2008. Molecular basis for the thiol sensitivity of insulin-degrading enzyme. Proc. Natl. Acad. Sci. U.S.A. 105, 9582–9587.

Nisticò, R., Dargan, S.L., Amici, M., Collingridge, G.L., Bortolotto, Z.A., 2011. Synergistic interactions between kainate and mGlu receptors regulate bouton Ca signalling and mossy fibre LTP. Sci. Rep. 1, 103.

Nisticò, R., Mango, D., Mandolesi, G., Piccinin, S., Berretta, N., et al., 2013. Inflammation subverts hippocampal synaptic plasticity in experimental multiple sclerosis. PLoS One 8, e54666.

Nitti, M., D'Abramo, C., Traverso, N., Verzola, D., Garibotto, G., et al., 2005. Central role of PKCδ in glycoxidation-dependent apoptosis of human neurons. Free Radic. Biol. Med. 38, 846–856.

O'Neill, C., 2013. PI3-kinase/Akt/mTOR signaling: impaired on/off switches in aging, cognitive decline and Alzheimer's disease. Exp. Gerontol. 48, 647–653.

Oddo, S., 2012. The role of mTOR signaling in Alzheimer disease. Front. Biosci. 4, 941–952.

Ong, W.Y., Tanaka, K., Dawe, G.S., Ittner, L.M., Farooqui, A.A., 2013. Slow excitotoxicity in Alzheimer's disease. J. Alzheimer's Dis. 35, 643–668.

Orr, M.E., Salinas, A., Buffenstein, R., Oddo, S., 2014. Mammalian target of rapamycin hyperactivity mediates the detrimental effects of a high sucrose diet on Alzheimer's disease pathology. Neurobiol. Aging 35, 1233–1242.

Othman, A., Rütti, M.F., Ernst, D., Saely, C.H., Rein, P., et al., 2012. Plasma deoxysphingolipids: a novel class of biomarkers for the metabolic syndrome? Diabetologia 55, 421–431.

Ouwens, D.M., van Duinkerken, E., Schoonenboom, S.N., Herzfeld de Wiza, D., Klein, M., et al., 2014. Cerebrospinal fluid levels of Alzheimer's disease biomarkers in middle-aged patients with type 1 diabetes. Diabetologia 57, 2208–2214.

Parillo, M., Riccardi, G., 2004. Diet composition and the risk of type 2 diabetes: epidemiological and clinical evidence. Br. J. Nutr. 92, 7–19.

Parsons, R.G., Gafford, G.M., Helmstetter, F.J., 2006. Translational control via the mammalian target of rapamycin pathway is critical for the formation and stability of long-term fear memory in amygdala neurons. J. Neurosci. 26, 12977–12983.

Pei, J.J., Hugon, J., 2008. mTOR-dependent signalling in Alzheimer's disease. J. Cell Mol. Med. 12 (6B), 2525–2532.

Plum, L., Schubert, M., Bruning, J.C., 2005. The role of insulin receptor signaling in the brain. Trends Endocrinol. Metab. 16, 59–65.

Prada, P.O., Hirabara, S.M., de Souza, C.T., et al., 2007. L-glutamine supplementation induces insulin resistance in adipose tissue and improves insulin signalling in liver and muscle of rats with diet-induced obesity. Diabetologia 50, 1949–1959.

Prasanthi, J.R.P., Schommer, E., Thomasson, S., Thompson, A., Ghribi, O., 2008. Regulation of beta-amyloid levels in the brain of cholesterol-fed rabbit, a model system for sporadic Alzheimer's disease. Mech. Ageing Dev. 129, 649–655.

Prasanthi, J.R., Huls, A., Thomasson, S., Thompson, A., Schommer, E., et al., 2009. Differential effects of 24-hydroxycholesterol and 27-hydroxycholesterol on beta-amyloid precursor protein levels and processing in human neuroblastoma SH-SY5Y cells. Mol. Neurodegener. 4, 1.

Puighermanal, E., Marsicano, G., Busquets-Garcia, A., Lutz, B., Maldonado, R., et al., 2009. Cannabinoid modulation of hippocampal long-term memory is mediated by mTOR signaling. Nat. Neurosci. 12, 1152–1158.

Qiu, W.Q., Folstein, M.F., 2006. Insulin, insulin-degrading enzyme and amyloid-beta peptide in Alzheimer's disease: review and hypothesis. Neurobiol. Aging 27, 190–198.

Quintero, M., Colombo, S.L., Godfrey, A., Moncada, S., 2006. Mitochondria as signaling organelles in the vascular endothelium. Proc. Natl. Acad. Sci. U.S.A. 103, 5379–5384.

Ralat, L.A., Ren, M., Schilling, A.B., Tang, W.J., 2009. Protective role of Cys-178 against the inactivation and oligomerization of human insulin-degrading enzyme by oxidation and nitrosylation. J. Biol. Chem. 284, 34005–34018.

Ramasamy, R., Yan, S.F., Schmidt, A.M., 2009. RAGE therapeutic target and biomarker of the inflammatory response—the evidence mounts. J. Leukoc. Biol. 86, 505–512.

Reisberg, B., Doody, R., Stöffler, A., Schmitt, F., Ferris, S., 2003. Memantine in moderate-to-severe Alzheimer's disease. N. Engl. J. Med. 348, 1333–1341.

Ren, J., Pulakat, L., Whaley-Connell, A., Sowers, J.R., 2010. Mitochondrial biogenesis in the metabolic syndrome and cardiovascular disease. J. Mol. Med. 88, 993–1001.

Ren, H., Yan, S., Zhang, B., Lu, T.Y., Arancio, O., et al., 2014. Glut4 expression defines an insulin-sensitive hypothalamic neuronal population. Mol. Metab. 3, 452–459.

Riccio, P., Rossano, R., 2015. Nutrition facts in multiple sclerosis. ASN Neuro 7 pii: 1759091414568185.

Rorsman, P., Braun, M., 2013. Regulation of insulin secretion in human pancreatic islets. Annu. Rev. Physiol. 75, 155–179.

Rusinek, H., Ha, J., Yau, P.L., Storey, P., Tirsi, A., Tsui, W.H., et al., 2015. Cerebral perfusion in insulin resistance and type 2 diabetes. J. Cereb. Blood Flow Metab. 35, 95–102.

Russell, A.P., Gastaldi, G., Bobbioni-Harsch, E., Arboit, P., Gobelet, C., et al., 2003. Lipid peroxidation in skeletal muscle of obese as compared to endurance-trained humans: a case of good vs. bad lipids? FEBS Lett. 551, 104–106.

Santini, E., Huynh, T.N., Klann, E., 2014. Mechanisms of translation control underlying long-lasting synaptic plasticity and the consolidation of long-term memory. Prog. Mol. Biol. Transl. Sci. 122, 131–167.

Sawa, A., Khan, A.A., Hester, L.D., Snyder, S.H., 1997. Glyceraldehyde-3-phosphate dehydrogenase: nuclear translocation participates in neuronal and nonneuronal cell death. Proc. Natl. Acad. Sci. U.S.A. 94, 11669–11674.

Schmidt, A.M., Yan, S.D., Yan, S.F., Stern, D.M., 2000. The biology of the receptor for advanced glycation end products and its ligands. Biochim. Biophys. Acta 1498, 99–111.

Seet, R.C., Lee, C.Y., Lim, E.C., Quek, A.M., Huang, S.H., et al., 2010. Markers of oxidative damage are not elevated in otherwise healthy individuals with the metabolic syndrome. Diabetes Care 33, 1140–1142.

Simopoulos, A.P., 2009. Evolutionary aspects of the dietary omega-6:omega-3 fatty acid ratio: medical implications. World Rev. Nutr. Diet. 100, 1–21.

Simopoulos, A.P., 2013. Dietary omega-3 fatty acid deficiency and high fructose intake in the development of metabolic syndrome, brain metabolic abnormalities, and non-alcoholic fatty liver disease. Nutrients 5, 2901–2923.

Slipczuk, L., Bekinschtein, P., Katche, C., Cammarota, M., Izquierdo, I., Medina, J.H., 2009. BDNF activates mTOR to regulate GluR1 expression required for memory formation. PLoS One. 4, e6007.

Soares, E., Prediger, R.D., Nunes, S., Castro, A.A., Viana, S.D., 2013. Spatial memory impairments in a prediabetic rat model. Neuroscience 250, 565–577.

Son, S.M., Song, H., Byun, J., et al., 2012. Altered APP processing in insulin-resistant conditions is mediated by autophagosome accumulation via the inhibition of mammalian target of rapamycin pathway. Diabetes 61, 3126–3138.

Sottero, B., Gamba, P., Gargiulo, S., Leonarduzzi, G., Poli, G., 2009. Cholesterol oxidation products and disease: an emerging topic of interest in medicinal chemistry. Curr. Med. Chem. 16, 685–705.

Speakman, J.R., Mitchell, S.E., 2011. Caloric restriction. Mol. Asp. Med. 32, 159–221.

Spilman, P., Podlutskaya, N., Hart, M.J., et al., 2010. Inhibition of mTOR by rapamycin abolishes cognitive deficits and reduces amyloid-beta levels in a mouse model of Alzheimer's disease. PLoS One. 5, e9979.

Stumvoll, M., Goldstein, B.J., van Haeften, T.W., 2005. Type 2 diabetes: principles of pathogenesis and therapy. Lancet 365, 1333–1346.

Sun, Y.N., Liu, L.B., Xue, Y.X., Wang, P., 2015. Effects of insulin combined with idebenone on blood-brain barrier permeability in diabetic rats. J. Neurosci. Res. 93, 666–677.

Tagliamonte, A., DeMontis, M.G., Olianas, M., Onali, P.L., Gessa, G.L., 1976. Possible role of insulin in the transport of tyrosine and tryptophan from blood to brain. Adv. Exp. Med. Biol. 69, 89–94.

Takenaka, K., Yamagishi, S., Matsui, T., Nakamura, K., Imaizumi, T., 2006. Role of advanced glycation end products (AGEs) in thrombogenic abnormalities in diabetes. Curr. Neurovasc. Res. 3, 73–77.

Takeuchi, M., Yamagishi, S., 2008. Possible involvement of advanced glycation end-products (AGEs) in the pathogenesis of Alzheimer's disease. Curr. Pharm. Des. 14, 973–978.

Talbot, K., Wang, H.Y., Kazi, H., Han, L.Y., Bakshi, K.P., Stucky, A., et al., 2012. Demonstrated brain insulin resistance in Alzheimer's disease patients is associated with IGF-1 resistance, IRS-1 dysregulation, and cognitive decline. J. Clin. Invest. 122, 1316–1338.

Tang, S.J., Reis, G., Kang, H., Gingras, A.C., Sonenberg, N., Schuman, E.M., 2002. A rapamy-cin-sensitive signaling pathway contributes to long-term synaptic plasticity in the hip-pocampus. Proc. Natl. Acad. Sci. U.S.A. 99, 467–472.

Tang, W.J., 2016. Targeting insulin-degrading enzyme to treat type 2 diabetes mellitus. Trends Endocrinol. Metab. 27, 24–34.

Thornalley, P.J., 1993. Modification of the glyoxalase system in disease processes and pros-pects for therapeutic strategies. Biochem. Soc. Trans. 21, 531–534.

Toma, L., Stancu, C.S., Botez, G.M., Sima, A.V., Simionescu, M., 2009. Irreversibly glycated LDL induce oxidative and inflammatory state in human endothelial cells; added effect of high glucose. Biochem. Biophys. Res. Commun. 390, 877–882.

Townsend, M., Mehta, T., Selkoe, D.J., 2007. Soluble Abeta inhibits specific signal transduc-tion cascades common to the insulin receptor pathway. J. Biol. Chem. 282, 33305–33312.

Tremblay, M.L., Giguère, V., 2008. Phosphatases at the heart of FoxO metabolic control. Cell Metab. 7, 101–103.

Tsai, I.J., Croft, K.D., Mori, T.A., Falck, J.R., Beilin, L.J., et al., 2009. 20-HETE and F2-isoprostanes in the metabolic syndrome: the effect of weight reduction. Free Radic. Biol. Med. 46, 263–270.

Van Dam, P.S., Aleman, A., 2004. Insulin-like growth factor-I, cognition and brain aging. Eur. J. Pharmacol. 490, 87–95.

van der Heide, L.P., Ramakers, G.M., Smidt, M.P., 2006. Insulin signaling in the central ner-vous system: learning to survive. Prog. Neurobiol. 79, 205–221.

Ventura-Clapier, R., Garnier, A., Veksler, V., 2008. Transcriptional control of mitochondrial biogenesis: the central role of PGC-1alpha. Cardiovasc. Res. 79, 208–217.

Vincent, H.K., Powers, S.K., Dirks, A.J., Scarpace, P.J., 2001. Mechanism for obesity-induced increase in myocardial lipid peroxidation. Int. J. Obes. Relat. Metab. Disord. 25, 378–388.

Wan, W., Chen, H., Li, Y., 2014. The potential mechanisms of AÎ²-receptor for advanced gly-cation end-products interaction disrupting tight junctions of the blood-brain barrier in Alzheimer's disease. Int. J. Neurosci. 124, 75–81.

Wang, M.Y., Ross-Cisneros, F.N., Aggarwal, D., Liang, C.Y., Sadun, A.A., 2009. Receptor for advanced glycation end products is upregulated in optic neuropathy of Alzheimer's disease. Acta Neuropathol. 118, 381–389.

Wang, Q., Jin, T., 2009. The role of insulin signaling in the development of beta-cell dysfunc-tion and diabetes. Islets 1, 95–101.

Wullschleger, S., Loewith, R., Hall, M.N., 2006. TOR signaling in growth and metabo-lism. Cell 124, 471–484.

Xing, Z., Gauldie, J., Cox, G., Baumann, H., Jordana, M., et al., 1998. IL-6 is an antiinflamma-tory cytokine required for controlling local or systemic acute inflammatory responses. J. Clin. Invest. 101, 311–320.

Yamagishi, S., 2011. Role of advanced glycation end products (AGEs) and receptor for AGEs (RAGE) in vascular damage in diabetes. Exp. Gerontol. 46, 217–224.

Yamashita, T., Hashiramoto, A., Haluzik, M., Mizukami, H., Beck, S., et al., 2003. Enhanced insulin sensitivity in mice lacking ganglioside GM3. Proc. Natl. Acad. Sci. U.S.A. 100, 3445–3449.

Zemke, D., Azhar, S., Majid, A., 2007. The mTOR pathway as a potential target for the devel-opment of therapies against neurological disease. Drug News Perspect. 20, 495–499.

Zhao, W.Q., Alkon, D.L., 2001. Role of insulin and insulin receptor in learning and memory. Mol. Cell Endocrinol. 177, 125–134.

Zhao, W.Q., Chen, H., Quon, M.J., Alkon, D.L., 2004. Insulin and the insulin receptor in experimental models of learning and memory. Eur. J. Pharmacol. 490, 71–81.

Zhao, H., Przybylska, M., Wu, I.H., Zhang, J., Siegel, C., et al., 2007. Inhibiting glycosphin-golipid synthesis improves glycemic control and insulin sensitivity in animal models of type 2 diabetes. Diabetes 56, 1210–1218.

Zhao, W.Q., De Felice, F.G., Fernandez, S., Chen, H., Lambert, M.P., et al., 2008. Amyloid beta oligomers induce impairment of neuronal insulin receptors. FASEB J. 22, 246–260.

Zhou, S., Yin, X., Zheng, Y., Miao, X., Feng, W., Cai, J., et al., 2014. Metallothionein prevents intermittent hypoxia-induced cardiac endoplasmic reticulum stress and cell death likely via activation of Akt signaling pathway in mice. Toxicol. Lett. 227, 113–123.

Zimmet, P., Alberti, K.G., Shaw, J., 2001. Global and societal implications of the diabetes epidemic. Nature 414, 782–787.

Contribution of Neuroinflammation in the Pathogenesis of Alzheimer's Disease

INTRODUCTION

Alzheimer's disease (AD) is characterized by extracellular accumulation of β-amyloid (Aβ) peptides in senile plaques and intraneuronal aggregates of hyperphosphorylated Tau protein in neurofibrillary tangles. In addition, it is also becoming increasingly evident that microglial cell–mediated neuroinflammatory processes along with the presence of reactive astrocytes directly contribute to the pathogenesis of AD (Farooqui, 2014). Participation of neuroinflammation in the pathogenesis of AD is not only supported by the generation of high levels of phospholipid-derived lipid mediators and increased expression and release of proinflammatory cytokines and chemokines (a heterogeneous group of proteins with molecular weights ranging from 8 to 40 kDa) but also by the identification of several AD risk genes that are specifically expressed by microglia within the central nervous system (CNS), including *CD33*, *CR1*, *HLA-DR*, and *TREM2* (Naj et al., 2011; Guerreiro et al., 2013; Jonsson et al., 2013; Heppner et al., 2015). In the brain tissue, activated microglial cells and astrocytes are major sources of neuroinflammation, a complex cellular and molecular defense mechanism that involves an innate immune response in the brain against harmful and irritable stimuli such as stress, injury, infection, and metabolic toxic waste, which contains high levels of reactive oxygen species (ROS) and nitric oxide (NO) (Tansey et al., 2007). Neuroinflammation isolates the damaged brain tissue from uninjured area, destroys affected cells, and repairs the extracellular matrix (ECM) (Correale and Villa, 2004). Neuroinflammation is mediated by oligomeric protein complexes known as inflammasomes. These protein complexes reside in cytoplasm and act as platforms for the induction of neuroinflammation. The basic components of inflammasomes are (1) a cytosolic danger sensor from the NOD-like receptor (NLR) family such as NALP3 (NACHT, LRR, and PYD domains-containing protein 3) also known as NLRP3 (NOD-like receptor family, pyrin domain–containing 3) and NALP1 or ice protease-activating factor (IPAF), (2) the proteolytic effector caspase-1, and (3) apoptosis-associated speck-like protein containing a CARD domain (ASC), an adaptor protein needed to recruit NLR and to stabilize caspase-1/NLR complexes (Minkiewicz et al., 2013; Schroder and Tschopp, 2010; Di Virgilio, 2013). Caspase-1 activity of inflammasomes acts as sensors of microbial components and sterile danger signals. The NLRP3 inflammasomes have been implicated in several chronic inflammatory diseases as they can sense inflammatory crystals and aggregated proteins, including Aβ (Halle et al., 2008). The activation of NLRP3 inflammasomes has been reported to not only contribute the pathology of AD but also may initiate AD-like pathology in certain mouse models. Inflammatory mediators, which are generated in NLRP3 inflammasome during their activation, contribute to synaptic dysfunction, cognitive impairment, and the restriction of beneficial microglial clearance functions. This key role of the NLRP3 inflammasome in Aβ-mediated inflammatory responses suggests that drugs that inhibit the activation of the NLRP3 inflammasomes, or inflammasome-derived cytokines, and chemokines may effectively interfere with the progression of AD. This is consistent with the hypothesis that Aβ-mediated activation of the NLRP3 inflammasome may enhance AD progression by inducing harmful chronic inflammatory responses in the brain. Crossing of NLRP3-deficient mice with a mouse model of AD that overexpresses mutant human amyloid precursor protein (APP) together with mutant presenilin 1 (PSEN1) results in the offspring that show rescue of spatial memory, reduction in caspase-1 activation, and increase in Aβ clearance (Heneka et al., 2013; Lee and Landreth, 2013), supporting the view that microglial activation may exacerbate AD-like pathology. Activation of proinflammatory caspase-1 and caspase-5 cleaves the precursor forms of proinflammatory cytokines interleukin (IL)-1β, IL-18, and IL-33 into their active forms. These proinflammatory cytokines have been shown to promote a variety of innate immune processes associated with infection, inflammation, and autoimmunity, and thereby play an instrumental role in instigating neuroinflammation during old age as well as in neurodegenerative diseases. The released cytokines not only stimulate phospholipid-hydrolyzing enzymes [phospholipase A_2 (PLA_2)] and arachidonic acid–oxidizing enzymes [cyclooxygenase (COX)-2 and lipooxygenase-5] but also promote the generation of eicosanoids, lysophosphatidylcholine (lyso-PtdCho), and platelet-activating factor (PAF). These lipid mediators facilitate the induction of neuroinflammation. Furthermore, lyso-PtdCho aids the activation of microglial cells through P2X7R signaling regulation (Chakraborty et al., 2010; Takenouchi et al., 2007). Neuroinflammation not only initiates the activation of microglial cells and astrocytes (Minghetti et al., 2005) but also promotes the reestablishment of cellular and metabolic homeostasis in the brain after injury-mediated disequilibrium of normal physiology. Microglial cells also respond to proinflammatory signals released from other nonneuronal cells such as mast cells. These immunity-related cells, while resident in the CNS, are capable of migrating across the blood–spinal cord barrier and blood–brain barrier (BBB) in situations where the barrier is compromised as a result of CNS pathology (Skaper et al., 2012). Mast cells not only produce a variety of mediators including biogenic amines (histamine, serotonin) and cytokines [IL-1 to IL-6, leukemia inhibitory factor, tumor necrosis factor (TNF)-α, interferon (IFN)-γ, transforming growth factor (TGF)-β, granulocyte-macrophage colony-stimulating factor] but also release enzymes (acid hydrolases, chymase, phospholipases, rat mast cell protease I and II, trypase), lipid metabolites (eicosanoids and PAF), ATP, neuropeptides (vasoactive intestinal peptide), growth factors [nerve growth factor (NGF)], NO, and heparin (Johnson and Krenger, 1992). It is becoming increasingly evident that in the brain inflammatory processes are markedly increased during normal aging (Lynch, 1998; Gemma and Bickford, 2007). However, the intensity and levels of neuroinflammation and inflammatory mediators are markedly increased in neurotraumatic (stroke), neurodegenerative (AD), and neuropsychiatric (depression) diseases. Two types of neuroinflammation, namely, acute inflammation and chronic inflammation, have been reported to occur in the brain.

ACUTE NEUROINFLAMMATION

Acute neuroinflammation develops rapidly with the experience of pain, whereas chronic inflammation develops slowly without any pain. Acute neuroinflammation is accompanied by generation of high levels of proinflammatory eicosanoids [prostaglandins (PGs), leukotrienes (LTs), and thromboxanes (TXs)] and rapid activation of microglia and astrocytes along with acute upregulation of proinflammatory cytokines (TNF-α and IL-1β) and adhesion molecules [human macrophage chemoattractant protein-1 (MCP-1) also known as chemokine (CCL2)], soluble intracellular adhesion molecule-1 and E-selectin, and damage to the cerebrovascular units, which consist of glial cells, endothelial cells, pericytes, and neurons (Fig. 6.1) (Farooqui et al., 2007; Kamouchi et al., 2011; Farooqui, 2014). Brain tissue also fails to maintain the neuronal microenvironment due to the breakdown of BBB, a specialized barrier between blood and brain mainly consisting of specific endothelial cells, tight liner sheets formed by astrocytic end-feet and pericytes, and tight junctions. This process leads to the infiltration of various blood-borne immune cells into the ischemic brain from disrupted vessels. These infiltrating immune cells and the injured brain cells produce inflammatory mediators, which not only exaggerate brain edema or directly promote the death of brain cells in the penumbra but also facilitate the clearing of necrotic debris. After the demolition of necrotic debris has been completed, brain inflammation subsides. The vascular deposition of Aβ induces the expression of adhesion molecules and alters the expression of tight junction proteins, potentially facilitating the transmigration of circulating leukocytes (Heppner et al., 2015). The resolution of acute inflammation is an active, complex, and dynamic process involving several distinct metabolic and cellular mechanisms and mediators. Inflammation is normally terminated by resolution, with the purpose of promoting the healing and return to homeostasis. Resolution results in reduced numbers of immune cells at the site of insult by decreased infiltration and apoptosis, and clearance of apoptotic cells and debris by increased phagocytic activity (Rossi et al., 2006; Uller et al., 2006; Serhan, 2010; Farooqui, 2011). The clearance of the inflammatory site is mediated in part via the nonphlogistic recruitment of monocytes, which, as macrophages, participate in the phagocytosis of apoptotic

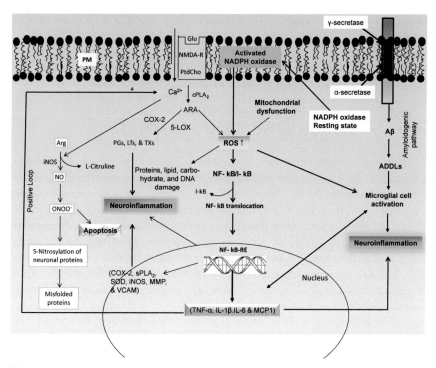

FIGURE 6.1 Processes contributing to neuroinflammation in the brain. *5-LOX*, 5-lipoxygenase; *Aβ*, β-amyloid; *ADDL*, Aβ-derived diffusible ligand; *ARA*, arachidonic acid; *Arg*, arginine; *COX-2*, cyclooxygenase-2; *cPLA₂*, cytosolic phospholipase A₂; *Glu*, glutamate; *IκB*, inhibitory subunit of NF-κB; *IL-1β*, interleukin-1β; *IL-6*, interleukin-6; *iNOS*, inducible nitric oxide synthase; *LTs*, leukotrienes; *lyso-PtdCho*, lyso-phosphatidylcholine; *MCP-1*, monocyte chemoattractant protein-1; *MMP*, matrix metalloproteinase; *NF-κB*, nuclear factor-κB; *NF-κB-RE*, nuclear factor-κB-response element; *NMDA*, N-methyl-D-aspartate receptor (NMDA-R); *NO*, nitric oxide; *ONOO⁻*, peroxynitrite; *PGs*, prostaglandins; *PM*, plasma membrane; *PtdCho*, phosphatidylcholine; *ROS*, reactive oxygen species; *SOD*, superoxide dismutase; *sPLA₂*, secretory phospholipase A₂; *TNF-α*, tumor necrosis factor-α; *TXs*, thromboxanes; *VCAM-1*, vascular cell adhesion molecule 1.

cells (Persidsky et al., 1999; Schwab et al., 2007). At the molecular level, the resolution of acute inflammation is orchestrated by lipid mediators called lipoxins, resolvins, neuroprotectins, and maresins (Serhan, 2009; Farooqui, 2011). Lipoxins are derived from arachidonic acid, whereas D series resolvins, neuroprotectins, and maresins are derived from docosahexaenoic acid. These lipid mediators not only suppress the expression of proinflammatory cytokines and chemokines but also inhibit the release of proinflammatory PGs, LTs, and TXs. Lipoxins, resolvins, and protectins decrease vascular permeability and retard polymorphonuclear leukocytes recruitment, while promoting recruitment of monocytes, neutrophils, and stimulating efferocytosis (Phillis et al., 2006; Serhan et al., 2008). In addition to the recruitment of neutrophils and monocytes, the brain also possesses native cells capable of organizing and expressing glial fibrillary acidic protein (GFAP) (Szmydynger-Chodobska et al., 2012; Schiff et al., 2012). Acute inflammation occurs in stroke, traumatic brain injury, chronic traumatic encephalitis, and spinal cord injury (Farooqui, 2010).

CHRONIC NEUROINFLAMMATION

Chronic inflammation develops slowly below the threshold of pain perception. As a result, the immune system continues to attack brain tissue at the cellular level. Chronic inflammation lingers for years causing continued insult to the brain tissue reaching the threshold of detection (Wood, 1998) and initiating the pathogenesis of chronic diseases such as AD (Cooper et al., 2000). A characteristic feature of chronically inflamed brain is the presence of an increased number of monocytes, microglia cells, and astrocytes in the CNS (Akiyama et al., 2000). Chronic inflammation disrupts hormonal signaling networks in the brain. In brain, chronic neuroinflammation is supported by long-term activation of microglia and astrocytes along with sustained release of low levels of inflammatory mediators leading to increase in oxidative and nitrosative stress (Tansey et al., 2007). The sustained release of inflammatory mediators (PGs, LTs, TXs, cytokines, and chemokines, monocyte chemoattractant proteins, complement factors, proteases, protease inhibitors, pentraxins) not only alters the inflammatory cycle and activates additional microglial cells but also promotes proliferation, leading to further release of inflammatory factors. In nonneural tissues sustained inflammation due to persistent activation of inflammatory cells or defects in the resolution program results in a fibrogenic response, which involves the overall remodeling of tissue structure and eventual organ failure. Fibrogenic response is supported not only by sustained inflammation and cell proliferation but also by growth factor responses. Owing to the chronic and sustained nature of the inflammation, there is often compromise of the BBB, which increases infiltration of peripheral macrophages into the brain parenchyma to further perpetuate the inflammation (Rivest, 2015). Chronic inflammation induces deficits in long-term potentiation (LTP), the major neuronal substrate for learning and memory, in middle-aged but not in young rats (Liu et al., 2012). Rather than serving a protective role as in acute neuroinflammation, chronic neuroinflammation is most often detrimental and damaging to the brain tissue. Thus whether neuroinflammation has beneficial or harmful outcomes in the brain may depend critically on the duration of the inflammatory response. Prolonged chronic inflammatory state has detrimental health effects and predisposes to a wide variety of chronic diseases, especially those that are more prevalent with advanced age, such as cardiovascular diseases, diabetes, and neurodegenerative diseases (Farooqui et al., 2007). Chronic inflammation is also a strong predictor of both disability and mortality in the elderly—even in the absence of clinical disease (Penninx et al., 2004; Farooqui, 2013). Although significant information is available on the generation of proinflammatory mediators, little is known about the internal and external factors that modulate the dynamic aspects of acute and chronic neuroinflammation. Depending on its timing and magnitude in brain tissue, neuroinflammation serves multiple purposes. As stated earlier, neuroinflammation is not only involved in protection of uninjured neurons and removal of degenerating neuronal debris but also contributes in assisting repair and recovery processes (Farooqui, 2010, 2011). In AD, it is proposed that the intractable nature of the Aβ plaques and tangles contribute to a chronic inflammatory reaction to clear these debris (Cai et al., 2014). Thus senile plaques contain dystrophic neurites, activated microglia, and reactive astrocytes (Akiyama et al., 2000). Aggregated amyloid fibrils and inflammatory mediators secreted by microglial and astrocytic cells contribute to neuronal dystrophy (Findeis, 2007). Chronically activated glia can mediate neurodegeneration in adjacent neurons by releasing highly toxic products such as ROS, NO, glutamate, proteolytic enzymes, and complementary factors (Halliday et al., 2000). Inflammatory mediators and a number of stress conditions promote the production of APP and the amyloidogenic processing of APP to generate Aβ peptide production and deposition. Under these circumstances, production of soluble APP fraction has a neuroprotective effect (Atwood et al., 2003). On the other hand, production and deposition of Aβ not only upregulates the expression of proinflammatory cytokines in glia cells in a vicious cycle (Lindberg et al., 2005) but also activates the complement cascade (Aisen, 1997) and induces the inflammatory enzyme systems such as iNOS and COX-2. Thus converging evidence suggests that all these factors can contribute to neuronal dysfunction and cell death, either alone or in concert (Abbas et al., 2002). It is not yet known whether the aforementioned events precede disease states or are a direct consequence of the damage that occurs in pathology of neurodegenerative diseases. For example, Aβ plaques have been shown to induce proinflammatory effects in animal models of AD (Halliday et al., 2000; Tuppo and Arias, 2005) supporting the view that neuroinflammatory events initiate or even aid in the progression of AD (Heneka and O'Banion, 2007; Bales et al., 2000).

ACTIVATION OF MICROGLIAL CELLS IN THE BRAIN

Glial cells consist of astrocytes, microglial cells, and oligodendrocytes. They perform a variety of functions in the brain, ranging from immune defense against external and endogenous hazardous stimuli, regulation of synaptic formation, calcium homeostasis, and metabolic support for neurons (Aguzzi et al., 2013; Farooqui, 2014; Ries and Sastre, 2016). Dysregulation of their metabolism and functions can contribute to the development and pathogenesis of not only AD and other neurodegenerative diseases but also stroke, traumatic brain injury, and spinal cord injury. One of the most important functions of glial cells in AD is the regulation of Aβ levels in the brain through Aβ clearance and degradation (Farooqui, 2014). The mechanisms of Aβ degradation by glial cells involves the production of proteases, including neprilysin, the insulin-degrading enzyme, plasminogen activators, angiotensin-converting enzyme, and matrix metalloproteinases. Other mediators that are released by glial cells and promote Aβ clearance include extracellular chaperones and receptors/transporters. Finally, astrocytes and microglia have an essential role in phagocytosing Aβ, in many cases via a number of receptors that are expressed on their surface (Yoon and Jo, 2012; Farooqui, 2014; Ries and Sastre, 2016).

Microglial cells, which originate from bone marrow–derived myeloid cells, are the brain's immune cells. They account for 10%–12% of the total glial cell population in the brain. They belong to the monocytic–macrophage lineage and orchestrate innate immune responses; however, they are distinct from other tissue macrophages due to their relatively quiescent phenotype and tight regulation by the brain microenvironment. Microglial cells are predominately found in the gray matter, with especially high concentrations in the hippocampus, hypothalamus, basal ganglia, and substantia nigra (Block et al., 2007). The total number and density of microglial cells increase significantly with age initially in various regions of the brain such as hippocampus (Mouton et al., 2002), visual and auditory cortices (Tremblay et al., 2010, 2012), and the retina (Damani et al., 2011). Under normal physiological conditions, microglial cells have a small, somewhat elongated cell body with long and fine processes. The activation of microglial cells not only results in hypertrophy in their cell bodies but also produces their ramification and wide distribution throughout the brain. Activated microglial cells can separate presynaptic axon terminals from postsynaptic neuronal perikarya or dendrites in a process known as synaptic stripping (Wake et al., 2009). The activation of microglial cells also results in proliferation and migration to the site of the lesion, where they develop abilities to phagocytize and ability to release proinflammatory cytokines (TNF-α and IL-1β). Microglial cells and neurons also produce antiinflammatory cytokines (IL-4, Il-10, IL-13, and TGF-β). Activated microglial cells constantly scan and survey the surrounding brain microenvironment and synapses for intruders/stressors, which may disrupt the structure and function of neuronal circuits and interfere with brain homeostasis (Wake et al., 2009; Kettenmann et al., 2011). Activation of microglial cells is crucial for maintaining the homeostasis of the brain in health and diseases (Lindsey et al., 1979; Mosher and Wyss-Coray, 2014). Thus microglial cells perform several beneficial

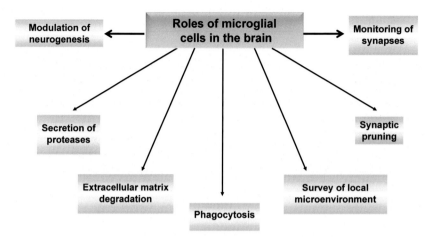

FIGURE 6.2   Roles of microglial cells in the brain.

functions, such as neuroprotection (Liang et al., 2010), clearance of cellular debris and toxic substances (Richard et al., 2008), guidance of stem cell migration in neuronal repair, and neurogenesis (Fig. 6.2) (Ziv et al., 2006). Microglial cells also contribute to the elimination of synapses, but not entire neurons, in response to an inflammatory stimulus (bacterial infection) in the cerebral cortex as a neuron-supportive mechanism (Trapp et al., 2007). Like macrophages, microglial cells can be identified by their expression of the marker CD11b (also known as complement receptor 3). Furthermore, the expression of the ionized calcium-binding adapter molecule 1 is restricted to microglial cells and is an excellent marker for the analysis of microglial ramifications. Neuroprotective effects of microglial cells are associated with the expression of YM1, a secretory protein related to neuroregeneration (Hung et al., 2002). Under pathological conditions, when neurons are under severe stress in aging or undergoing neurodegeneration in neurological disorders, microglial cells become activated via the release of ATP, neurotransmitters, growth factors, or cytokines, and via ion changes in the local environment (Hanisch and Kettenmann, 2007; Lucin and Wyss-Coray, 2009). A variety of other factors also activate microglial cells (Table 6.1) (Hanisch and Kettenmann, 2007). Neural injury is accompanied by migration of neighboring microglial cells toward the damaged site within 30 min of injury. It is suggested that chemotactic migration of microglial cells to the injury site is supported by astrocytes, which are supporting cells for neurons in the brain (Haynes et al., 2006). Following ischemic injury microglial cells not only make contact with neuronal synapses (Wake et al., 2009) but also express high levels of proinflammatory cytokines including TNF-α, IL-α (IL-1α), IL-1β, and IL-6, as well as cytokine receptors, which contribute to acute neuroinflammation. Under these

TABLE 6.1   Factors Contributing to the Activation of Microglial Cells

Factors	References
Blood clotting factors	Hanisch and Kettenmann (2007), Lucin and Wyss-Coray (2009)
RNA and DNA released by necrotic cells	Hanisch and Kettenmann (2007), Lucin and Wyss-Coray (2009)
Externalization of PtdSer	Hanisch and Kettenmann (2007), Lucin and Wyss-Coray (2009)
Immunoglobulin antigen complexes	Hanisch and Kettenmann (2007), Lucin and Wyss-Coray (2009)
Opsonizing complement	Hanisch and Kettenmann (2007), Lucin and Wyss-Coray (2009)
Aggregated proteins (such as β-amyloid)	Hanisch and Kettenmann (2007), Lucin and Wyss-Coray (2009)
Pathogen-related structures	Hanisch and Kettenmann (2007), Lucin and Wyss-Coray (2009)
Neuronally derived protein fractalkine	Hanisch and Kettenmann (2007), Lucin and Wyss-Coray (2009)
Neuronally derived CD200	Hanisch and Kettenmann (2007), Lucin and Wyss-Coray (2009)
Neuronally derived proteins TREM2	Hanisch and Kettenmann (2007), Lucin and Wyss-Coray (2009), and Hsieh et al., (2009)
Neuronally-derived semaphorin	Hanisch and Kettenmann (2007), Lucin and Wyss-Coray (2009), and Majed et al. (2006)
CD22 and CD45	Hanisch and Kettenmann (2007), Mott et al. (2004)

PtdSer, phosphatidylserine; TREM, triggering receptor expressed on myeloid cells-2.

conditions, microglial cells dramatically change their morphology and adopt an ameboid appearance in the activated state. Activated microglial cells not only express receptors for neurotransmitters such as ATP, adenosine, glutamate, γ-aminobutyric acid (GABA), acetylcholine, dopamine, and adrenaline (Lee, 2013) but also secrete Fc receptor (CD11b, CD11c, CD14), major histocompatibility complex (MHC) molecules, Toll-like receptors (TLRs), receptors for advanced glycation end products (RAGE), scavenger receptors, and cytokine/chemokine receptors (Rock et al., 2004), as well as a variety of immune system modulators including complement proteins, adhesion molecules, colony-stimulating factor-1, tumor and growth factors (TGF-α and TGF-β), MCP-1, macrophage inflammatory peptide-1α (MIP-1α), and triggering receptor expressed on myeloid cells-2 (TREM2), all of which contribute to the control and tight regulation of local immune responses (Minghetti et al., 2005; Galimberti et al., 2006). The TREM2 gene encodes five exons that code for a 693-bp DNA, located on chromosome 6p21.1, which is translated into 230 amino acids (Golde et al., 2013). TREM2 is one of the highest expressed cell surface receptors on microglia and is >300-fold enriched in microglia versus astrocytes (Hickman et al., 2013). In the human brain, TREM2 is found at high concentrations in white matter, the hippocampus, and the neocortex, but at very low concentrations in the cerebellum. In microglial cells, TREM2 is also localized in the Golgi complex. Phagocytic receptor recycling modulates microglial cell surface expression of TREM2, a process that results in the generation and release of soluble TREM2 (Kleinberger et al., 2014). The regulation of TREM2 is mediated by a disintegrin and metalloproteinase domain-containing protein 10 (ADAM10). This is not followed by intramembranous cleavage of the remaining C-terminal transmembrane domain by intracellular domain of γ-secretase (Kleinberger et al., 2014; Wunderlich et al., 2013). Inability to fully process the TREM2 C-terminal domain by blocking γ-secretase activity leads to the accumulation of membrane-bound nonfunctional TREM2 fragments lacking the extracellular domain inappropriately coupled to DAP12, thereby decreasing the availability of DAP12 to couple with functional TREM2 molecules (Wunderlich et al., 2013). Thus TREM2 transduces its intracellular signaling through DAP12. However, the natural ligands of TREM2 remain elusive. The binding of TREM2 with DAP12 induces downstream signaling in microglial cells.

Microglia play a key role in the immune response in the CNS and are the resident innate immune cells responsible for the early control of infections.

The identification of coding variants in *TREM2* that increase the risk for AD supports the role of the immune response and inflammation in AD pathogenesis. However, the mechanism by which these variants increase risk for AD remains unknown. Coding variants in other genes in the *TREM*-gene family may also modify AD disease risk as is the case for a common protective variant in *TREML2* (Benitez et al., 2014). In addition to innate immunity, activated microglial cells also play other beneficial roles, such as neuroprotection mediated by release of neurotrophic factors (Liang et al., 2010), maintenance of CNS homeostasis by clearance of cellular debris and toxic substances (Richard et al., 2008), and guidance of stem cell migration in neuronal repair and neurogenesis (Fig. 6.2) (Ziv et al., 2006).

Activation of microglial cells also results in interaction between neuronal chemokine called fractalkine (CX3CL1) with fractalkine receptor CX3CR1 (Hanisch and Kettenmann, 2007; Ransohoff and Perry, 2009). CX3CL1 is a 373-amino-acid membrane glycoprotein with an extended mucinlike stalk and a chemokine domain on top. It is expressed on hippocampal and cortical neuronal cell surface, whereas CX3CR1 receptors are exclusively expressed by microglia (Fig. 6.3) (Cardona et al., 2006). CX3CL1

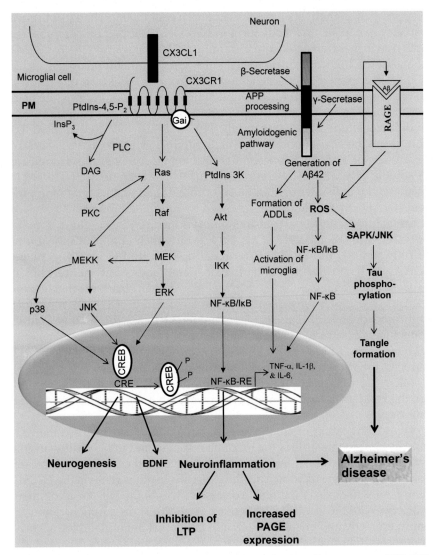

FIGURE 6.3   The binding of fractalkine (CX3CL1) with fractalkine receptor (CX3CR1) and stimulation of signal transduction mechanism associated with CX3CL1 and CX3CR1 receptor. *Aβ*, β-amyloid; *ADDL*, Aβ-derived diffusible ligand; *Akt/PKB*, serine/threonine kinase; *APP*, amyloid precursor protein; *CRE*, cAMP response elements; *CREB*, cAMP response element–binding protein; *CXCL1*, fractalkine; *DAG*, diacylglycerol; *ERKs*, extracellular-signal-regulated kinases; *IKK*, IκB kinase; *IL-1β*, interleukin-1β; *IL-6*, interleukin-6; *InsP₃*, inositol 1,4,5-trisphosphate; MAPK/ERK pathway (Ras-Raf-MEK-ERK pathway); *NF-κB*, nuclear factor-κB; *NF-κB-RE*, nuclear factor-κB-response element; *PKC*, protein kinase C; *PM*, plasma membrane; *PtdIns 3K*, phosphatidylinositol 3-kinase; *PtdIns 4,5-P₂*, phosphatidylinositol 4,5-bisphosphate; *ROS*, reactive oxygen species; *TNF-α*, tumor necrosis factor-α.

exists in secreted as well as membrane-bound form, which can be cleaved by cathepsin S, ADAM-10, and ADAM-17. The soluble form of CX3CL1 can serve as a signaling molecule mediating neuronal/microglial interactions. The interactions between CX3CL1 and CX3CR1 occur through a Gαi-coupled seven transmembrane cognate receptors. The interactions between CX3CL1 and CX3CR1 result in the activation of several intracellular signaling pathways involving phospholipase C (PLC), phosphatidylinositol 3-kinase (PtdIns 3K), and extracellular-signal-regulated kinase (ERK), and the recruitment of transcription factors [nuclear factor (NF-κB), cAMP response element–binding protein (CREB)] (Sheridan and Murphy, 2013; Wolf et al., 2013). Involvement of CX3CR1 in microbial call activation is a double-edged sword—it can confer neuroprotection or cause neurodestruction, depending on the pathophysiologic conditions (Zhang et al., 2012). Microglial cells also produce a number of neuroprotective substances in response to neural injury, such as antiinflammatory cytokines and neurotrophic factors, including NGF, TGF-β, IL-10, and IL-1 receptor antagonist (IL-1ra) (Farooqui, 2014). IL-1ra plays a major role in counteracting the biological effects of IL-1β, owing to its ability to bind to IL-1 receptor without initiating signal transduction (Allan et al., 2005; Farooqui, 2014). Furthermore, both TGF-β1 and IL-10 inhibit macrophage/microglia activation by downregulating the expression of molecules associated with antigen presentation and production of proinflammatory cytokines, chemokines, NO, and oxygen free radicals (Allan et al., 2005; Farooqui, 2014). These observations suggest that microglial cells play a rather active and dynamic role in maintaining neural homeostasis in a healthy brain and can detect and respond efficiently to CNS injury.

CONTRIBUTION OF MICROGLIAL CELL IN THE PATHOGENESIS OF ALZHEIMER'S DISEASE

Normal aging is accompanied by increase in oxidative stress and neuroinflammation. These processes are supported by the onset of lipid peroxidation and the ability of microglial cells to express proinflammatory cytokines (TNF-α, IL-1α, IL-1β, and IL-6) and chemokines [MHC II and complement receptor 3 (CD11b)] and their receptors (Table 6.2). In vivo microinjections of fibrillary Aβ in the cortex of aged rhesus monkeys induce neurodegeneration, Tau phosphorylation, and microglial cell proliferation, but not in those of young monkeys, suggesting that Aβ neurotoxicity is a pathological response of the aging brain (Geula et al., 1998). In this context, microglia-upregulated production of IL-1β is possibly implicated in age-associated cognitive impairments (Maher et al., 2006). Aging process also results in a decrease in antiinflammatory cytokines, such as IL-10 (Ye and Johnson, 2001). The expression of proinflammatory cytokines and

TABLE 6.2   Cytokines, Chemokines, and Their Receptors Found on Microglial Cells

Cytokine/Chemokine	Receptor	Effect
TNF-α	TNF-R1/TNF-R2	Proinflammatory
IFN-γ	IFN-γR	Proinflammatory
IL1α/IL-β	IL-1R1, IL-1R2	Proinflammatory
IL-2	IL-2Rα/β	Proinflammatory
IL-4	IL-4R	Antiinflammatory
IL-10	IL10R	Antiinflammatory
IL-13	IL-13R	Antiinflammatory
IL-15	IL-15R	Proinflammatory
IL-17	IL-17R	Proinflammatory
IL-18	IL-18R	Proinflammatory
CCL2		Proapoptotic
MCP-1		Antiinflammatory
CXCL10		Proapoptotic
CX3CL-1	CX3CR1	Proinflammatory
CXCL8	CXCR2	Proinflammatory
CXCL12	CXCR4	Proinflammatory
MIPs		

IFN, interferon; *IL*, interleukin; *MCP-1*, monocyte chemoattractant protein-1; *MIPs*, macrophage inflammatory proteins; *TNF*, tumor necrosis factor.
Summarized from Allaman et al., 2010; Sofroniew and Vinters, 2010; Sheridan and Murphy, 2013; Wolf et al., 2013; Cameron and Landreth, 2010.

chemokines is coupled with the activation of microglial cells and decrease in regulatory status. These processes may also promote the environment, which not only hinders the repair capabilities of microglial cells to downregulate neuroinflammation but also may promote recovery or tissue remodeling in the brain (Farooqui, 2014). In addition to the aforementioned changes, brain aging also reduces neuronal plasticity. Thus major differences in microglial biology between young and old brain occur when the immune system is challenged and microglia are activated. Converging evidence suggests that increase in inflammatory markers in the aged brain sets the stage for increased or exaggerated immune responses, which may cause cognitive and memory impairments. These impairments are linked with increases in IL-6 and IL-1β because high levels of IL-6 in the brain may inhibit memory formation and learning causing neurodegeneration (von Bernhardi et al., 2015a) and exacerbate sickness behavior (Godbout and Johnson, 2004). These effects are similar to that of IL-1β as high IL-1β concentrations in the hippocampus are associated with impaired memory (Rachal Pugh et al., 2001; Barrientos et al., 2004).

In AD, microglial cells not only play an important role in maintaining the brain's extracellular environment but also promote the clearance of aggregated proteins such as Aβ (Simard et al., 2006). As stated earlier, activated microglial cells perform many important functions such as presentation of antigens, phagocytosis of cellular debris, and secretion of proteases that promote microglial motility, as well as myelin and ECM degradation (Fig. 6.2) (Hanisch and Kettenmann, 2007; Lucin and Wyss-Coray, 2009). CX3CL1and CX3CR1 receptors play a unique role in the pathogenesis of AD. In several transgenic mouse models of AD, CX3CR1 deficiency ameliorated Aβ deposition by altering microglial activation and promoting microglial phagocytosis (Lee et al., 2010). On the other hand, CX3CR1 deficiency exacerbated microglial activation and increased Tau protein phosphorylation via neuronal p38 mitogen-activated protein kinase (MAPK) activation in the hyperphosphorylated Tau model of tauopathy (Bhaskar et al., 2010). While these studies indicate the importance of CX3CL1–CX3CR1 interactions in modulating AD-related pathologies, the detailed molecular mechanisms and signal transduction pathways underlying the divergent Aβ and Tau protein phenotypes, as well as the relative contribution of membrane-anchored versus soluble CX3CL1 entities, remain to be defined. In AD brain, activated microglia are found in the vicinity of extracellular deposits of Aβ and secrete proinflammatory cytokines (Cameron and Landreth, 2010). Chronic inflammatory state is closely associated with the progression of AD supporting the view that activated microglia not only exert neuroprotective effects but also may contribute to detrimental effects to the surviving neuronal tissue. It still remains unknown why and when, in the course of AD, microglial cells switch from being beneficial to becoming neurotoxic, but age-related disturbance of the physiological function and regulation of microglia (immunosenescence) has been suggested to play an important role in AD pathogenesis (Luo et al., 2010; Fuhrmann et al., 2010). Another important question that remains to be answered is whether the activation of microglial cells occurs as a consequence of extracellular Aβ deposition in AD or if it serves as a triggering factor for Aβ deposition in the initial stages of the disease. Studies on animal models have indicated that microglial cells can be activated by extracellular Aβ (Wirths et al., 2010). However, the presence of activated microglia has been observed even before the onset of Aβ deposition (Bradt et al., 1998; Heneka et al., 2005). Crossing CX3CL1 receptor–deficient mice with mice that overexpress the entire wild-type human Tau gene in the absence of mouse Tau (hTau mice) results in offspring with enhanced microglial activation and increased Tau phosphorylation and aggregation supporting the view that microglial CX3CL1 may play a neuroprotective as well as antiinflammatory roles in animals that model Tau-associated neurodegeneration (Bhaskar et al., 2010). It is also reported that CX3CL1 acts as a potent neuromodulator, which can evoke excitatory synaptic transmission and play a major role not only in memory formation and increase in synaptic plasticity but also in neuroprotection (Sheridan et al., 2014; Bertollini et al., 2006). Mechanisms associated with these processes have been explored. However, it is suggested that CX3CL1/CX3CR1 interactions may modulate cognitive function mainly by regulating LTP (Maggi et al., 2009), NO signaling, and production and secretion of brain-derived neurotrophic factor (BDNF) (Parkhurst et al., 2013). Converging evidence suggest that CX3CL1-mediated microglia–neuron interactions not only modulate synaptic pruning, neuronal survival, and neural cell precursors but also modulate synaptic transmission and plasticity by enhancing synapse and network maturation. These processes facilitate and promote the establishment of neuropathic pain circuits.

In vitro studies have indicated that Aβ activates NALP3-containing inflammasome, leading to neuroinflammation and brain damage (Halle et al., 2008). Moreover, the deficiency of NALP3 in the APP/PSEN1 AD mouse model produces a decrease in Aβ deposition and rescue from memory deficits (Heneka et al., 2013). Although in most studies involvement of microglial inflammasome in AD has been proposed, some studies have indicated that astrocytic inflammasome also play an important role in the pathogenesis of AD. Downregulation of astrocytic IPAF inflammasome reduces Aβ42 generation by primary neurons, and expressions of IPAF and ASC are significantly increased in a subgroup of patients with sporadic AD (Liu and Chan, 2014). Moreover, in cell culture system aggregated Aβ can directly induce neuroinflammation (Fig. 6.3) not only through the release of proinflammatory cytokines but also via binding with the RAGE (Han et al., 2011) and through the involvement of different scavenger receptors, such as TGF-β1 receptor, scavenger receptor A-1, MARCO, scavenger receptor B-1, and CD36 (von Bernhardi et al., 2015b; Yu and Ye, 2015). Compelling evidence supports the view that RAGE acts as inflammatory intermediary as well as oxidative stress inducer. RAGE consists of an extracellular domain (V-type followed by two C-type regions) and a single transmembrane domain followed by a short cytosolic tail, the latter mediating signal transduction. Interactions between Aβ and RAGE not only increase the intensity of oxidative stress but also stimulate levels of free NF-κB leading to the increase in expression of proinflammatory cytokines and chemokines. NF-κB is a redox-sensitive transcription factor that is located in the cytoplasm and responsible for the production and regulation of cytokine and chemokines. NF-κB is a dimer composed of members of the Rel family transcription factors: RelA (p65), RelB, c-Rel, p52, and p50. Under physiological conditions NF-κB remains inactive due to the presence of an inhibitor called IκB. High levels of ROS or the phosphorylation of IκB by IκB kinases promote the breakdown of NF-κB/IκB complex leading to the formation of free NF-κB, which migrates to the nucleus and induces the transcription of proinflammatory cytokines and chemokines (Jana et al., 2002; Nakajima et al., 2006). In relation to neuroinflammation, NF-κB activation has been shown to mediate cytokine expression and iNOS induction and NO production (Bhat et al., 2002) in microglia. These processes increase the production of Aβ and promote the formation of NFT, failure of synaptic transmission, and neuronal degeneration. As stated in Chapter 1, the steady-state level of Aβ depends on the balance between production and clearance. RAGE may play an important role in Aβ clearance. RAGE acts as an important transporter via regulating influx of circulating Aβ into brain, whereas the efflux of brain-derived Aβ into the circulation via BBB is implemented by LRP1 (Cai et al., 2016). Increasing the generation of Aβ RAGE may via enhancing β- and/or γ-secretase activities upregulate inflammatory response and oxidative stress in AD. Furthermore, RAGE may also induce the synaptic dysfunction and neuronal circuit dysfunction leading to cognitive dysfunction (Chen et al., 2014). RAGEs also interact with a broad repertoire of ligands, including products of nonenzymatic glycoxidation [advanced glycation end products (AGEs)], Aβ, the S100/calgranulin family of proinflammatory cytokinelike mediators, and high-mobility group box 1 nonhistone DNA-binding protein (amphoterin) (Chen et al., 2007). Under normal conditions, in mature healthy normal individuals, RAGE expression is very low in the brain. The accumulation of Aβ species in AD brain, or Tg models of β-amyloidosis, increases the expression of RAGE in affected cerebral vessels, neurons, and microglial cells (Lue et al., 2001). These studies support the view that production of AGE and induction of RAGE may be a trigger for the pathogenesis of Aβ and Tau hyperphosphorylation because both processes participate in the process of cognitive impairment (Cai et al., 2016).

In contrast to RAGE, TGF-β1 is a neurotrophic factor associated with the initiation and maintenance of neuronal differentiation and synaptic plasticity. The deficiency of TGF-β1 signaling contributes to Aβ pathology and neurofibrillary tangle formation in AD animal models. The overexpression of TGF-β1 has been reported to reduce plaque formation and induce Aβ accumulation in mice having human AβAPP, and this effect of TGF-β1 may be related to increase in Aβ clearance by BV-2 microglia (Wyss-Coray et al., 2001; Caraci et al., 2011). Moreover, reduction in TGF-β1 signaling seems to contribute both to microglial activation and to ectopic cell-cycle reactivation in neurons. Both these events contribute to neurodegeneration in the AD brain. The interaction between neuron and microglial cells induces the release of costimulatory factors like CD14 and CD40 ligand, suggesting that there is a close pathological link between Aβ plaque pathology and the chronic activation of microglial cells along with the modulation of innate immune system (Baglio et al., 2013; Tan et al., 2002). At physiological levels, TGF-β1 dampens microglial activation (Brionne et al., 2003), but overexpression of TGF-β1 promotes neuroinflammation (Wyss-Coray et al., 2000), simultaneously accelerates brain vascular Aβ deposits, and reduces parenchymal Aβ deposits (Wyss-Coray et al., 2001), and elicits neuronal Aβ secretion (Tesseur et al., 2006). It is also reported that in AD, Aβ contributes to the stimulation of microglial cells due to the production of NO and ROS (Heppner et al., 2015). Microglial cells also possess a number of receptors, most prominently the TLRs, a class of pattern recognition receptors (PRRs) in the innate immune system, through which they are able to detect microbial and viral pathogens and tissue damage (Palm and Medzhitov, 2009). Among TLRs, TLR4 and TLR2 are not only essential against gram-negative and gram-positive bacteria but also relevant to AD (Fig. 6.4). Fibrillar Aβ has been shown to activate microglia

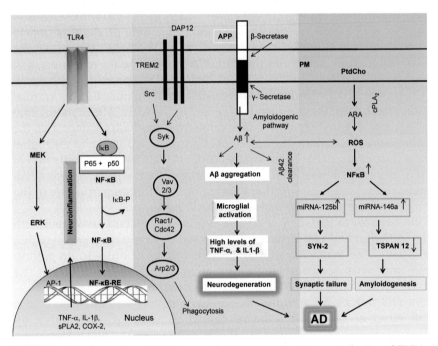

FIGURE 6.4 Generation of Aβ from amyloid precursor protein, contribution of TLR in neuroinflammation, and clearance of Aβ by microglial cells. *AP-1*, activator protein 1; *ARA*, arachidonic acid; *COX*, cyclooxygenase; *cPLA₂*, cytosolic phospholipase A₂; *DAP12*, DNAX-activating protein 12; *IL-1β*, interleukin 1β; *NF-κB*, nuclear factor-κB; *PM*, plasma membrane; *PtdCho*, phosphatidylcholine; *ROS*, reactive oxygen species; *sPLA₂*, secretory phospholipase A₂; *Syk*, spleen tyrosine kinase; *TLR*, Toll-like receptor; *TNF-α*, tumor necrosis factor-α; *TREM2*, triggering receptor expressed on myeloid cells-2.

via TLR receptor leading to the phagocytosis of fibrillar Aβ (Scholtzova et al., 2009; Reed-Geaghan et al., 2009). The significance of the aforementioned observation is that Aβ acts as an immune stimulus for neural cells through classical host defense mechanisms. By interacting with CD36, α6β1 integrin, CD47, and scavenger receptor A, TLRs bind to Aβ to initiate the activation of intracellular signaling pathways (Reed-Geaghan et al., 2009), which translate into a complex cellular response to Aβ. The recognition that TLRs participate in this receptor complex provides a mechanistic explanation for the ability of Aβ to elicit the production and secretion of proinflammatory (Th1-type) cytokines (Mantovani et al., 2004) and production of reactive oxygen and nitrogen radicals. Thus in AD, TLRs are necessary for recognition and clearance of the oligomeric Aβ species. At this time, it is difficult to establish the correct sequence of these events, so it is not clear whether activation of microglia and the associated inflammatory changes play a part in triggering neurodegenerative processes or whether glial cell activation is a response to the early changes associated with AD. The activation of microglia is beneficial for clearing toxic Aβ assemblies and secreting neurotrophic factors (Wyss-Coray, 2006). On the other hand, activated microglia can be synaptotoxic and neurotoxic by initiating and advancing the disease (Mandrekar-Colucci and Landreth, 2010). Converging evidence suggests that in AD, early microgliosis promotes Aβ clearance by protofibrillar Aβ phagocytosis, whereas chronic microgliosis promotes Aβ accumulation and subsequent neurodegeneration.

Activated microglial cells also secrete a variety of immune system modulators including complement proteins, adhesion molecules, colony-stimulating factor-1, tumor and growth factors (TGF-α and TGF-β), TREM2, MCP-1, and MIP-1α, all of which contribute to the control and tight regulation of local immune responses as well as Aβ-mediated neuroinflammation (Minghetti et al., 2005; Galimberti et al., 2006). As stated in Chapter 1, brain has a highly efficient system to clear Aβ and its oligomers. Microglial cell–mediated clearance and phagocytosis of Aβ is triggered by receptor expressed in myeloid/microglial cells-2 (TREM2) in concert with the membrane-spanning linker protein TYROBP (DAP12) (Fig. 6.4) (Malkki, 2015; Dempsey, 2015; Rivest, 2015; Bhattacharjee et al., 2014). TREM2 is a glycosylated 230-amino-acid signaling receptor of the immunoglobulin/lectinlike gene superfamily, which is encoded in mice on chromosome 17 and in humans on chromosome 6p21.1. TREM2 contains an extracellular domain, a transmembrane domain, and a short cytoplasmic tail (Bouchon et al., 2000). Oppositely charged intramembrane residues of TREM2 and DAP12 facilitate their interaction (Colonna, 2003). DAP12 is also a type I transmembrane protein with a cytoplasmic tail that harbors a single immunoreceptor tyrosine activation motif (Lanier et al., 1998). TREM2–DAP12 signaling not only plays an important role in the suppression of inflammatory cytokine production following receptor-mediated recognition of microbial or self-derived damage-associated motifs but also in the phagocytosis of a variety of microbial and endogenous ligands that facilitate debris clearance following injury or insult and induce a transcriptional profile reflective of enhanced myeloid cell proliferation and reduction of neural cell death. Induction of heterozygous missense mutations in TREM2 or DAP12 have been reported to contribute to progressive and presenile dementing diseases such as Nasu-Hakola syndrome, polycystic lipomembranous osteodysplasia with sclerosing leukoencephalopathy, sporadic amyotrophic lateral sclerosis (ALS), and sporadic AD (Nimmerjahn et al., 2005; Cady et al., 2014; Malkki, 2015; Sasaki et al., 2015a,b). In addition, studies on aluminum toxicity have indicated that this neurotoxin mediates its effects through NF-κB-mediated induction of microRNA-34a (miRNA-34a), which downregulates TREM2 and stimulates Aβ accumulation and aggregation in cultured microglial cells (Zhao et al., 2014). Studies on brain tissue from patients with sporadic AD have indicated that TREM2 downregulation is closely associated with the ability of microglial cells to phagocytize and remove Aβ from the brain tissue (Zhao et al., 2014; Hickman and El Khoury, 2014). Furthermore, knocking down of TREM2 has been shown to exacerbate age-related neuroinflammatory signaling and induce cognitive deficits in senescence-accelerated mouse prone-8 (SAMP8) mice (Jiang et al., 2014). TREM2 regulation is closely associated with microglial cell activation, Aβ clearance, and induction of neuroinflammation. Based on the aforementioned findings, it has been speculated that loss of function due to TREM2 mutations in familial forms of AD may have the same effects on deficiencies in phagocytosis as a downregulation of a fully functional TREM2 in sporadic AD and that modest TREM2 overexpression may result in enhancing the sensing, scavenging, phagocytizing, and removal of cellular debris in the aging brain, including neurotoxic and self-aggregating Aβ42 monomeric peptides (Zhao and Lukiw, 2015). Collective evidence suggests that during neuroinflammation, MCP-1 is a major chemokine that attracts more monocytes to the plaque to enhance the neuroinflammation in cerebrovascular systems. Interactions between CX3CL1–CX3CR1 modulate AD-related pathologies, regulation of TREM2 may be closely associated with microglial cell activation in AD, and TREM2 ligand on neurons help preserving the "immune-privileged" environment (Goldbaum and Richter-Landsberg, 2001), and neuronal semaphorins additionally assisting in immune regulation and repair by mediating oligodendrocyte precursor cell migration (Amor et al., 2014). It is also reported that diseased microglial cells are different from activated microglial cells in that they are incapacitated cells, whereas activated microglia are cells capable of responding to injury.

Thus in AD, microglial cells lose their intrinsic beneficial function during the course of the disease and may acquire a "toxic" phenotype over time. It is also suggested that Aβ may not be an appropriate trigger to induce phagocytosis and degradation by microglia in vivo.

CONTRIBUTION OF ASTROCYTE ACTIVATION IN THE PATHOGENESIS OF ALZHEIMER'S DISEASE

Astrocytes are complex, highly differentiated, and most abundant cells of the brain. They are intimately associated with the surrounding neurons and blood vessels, and their processes envelop all cellular components of the brain. It is estimated that astrocytes represent between 30% and 50% of human neural cells (Lent et al., 2012). They are functionally coupled to neurons, oligodendrocytes, and other astrocytes via both contact-dependent and non-contact-dependent pathways. Based on their morphology and localization in the brain astrocytes are classified into protoplasmic and fibrous types (Oberheim et al., 2012). Protoplasmic astrocytes are localized in the gray matter and are generally spongiform in nature. The processes of protoplasmic astrocytes spread radially from the cell body and have extensive fine branching that is distributed uniformly around the cell. In contrast, fibrous astrocytes are present in white matter. They have fewer, but longer, processes that extend along axon bundles providing structural support for axonal tracts (Sun et al., 2012). Astrocytes are traditionally viewed as supportive cells for neurons, which are responsible for CNS homeostasis and neuronal functions. Both types of astrocytes are in contact with blood vessels (Sofroniew and Vinters, 2010); however, fibrous astrocytes also send processes that contact axons at the nodes of Ranvier (Butt et al., 1994) while protoplasmic astrocytic foot processes ensheath neuronal synapses (Theodosis et al., 2008). Astrocytes contain many neurotransmitter receptors, and their activation leads to oscillations in internal Ca^{2+} (Koizumi, 2010). These oscillations induce the accumulation of arachidonic acid and the release of glutamate, GABA, glycine, D-serine, ATP, and NO. Ca^{2+} oscillations in astrocytic end-feet can control cerebral microcirculation through the arachidonic acid metabolites (PGE2 and epoxyeicosatrienoic acid) that induce arteriolar dilation and 20-Hydroxyeicosatetraenoic acid that induces arteriole constriction (Haydon and Carmignoto, 2006). Astrocytes also have processes in contact with both blood vessels and synapses. It is proposed that these contacts facilitate blood flow in relation to levels of synaptic activity, as demonstrated in the visual cortex where functional magnetic resonance imaging has detected changes in blood flow in response to visual stimuli (Wolf and Kirchhoff, 2008).

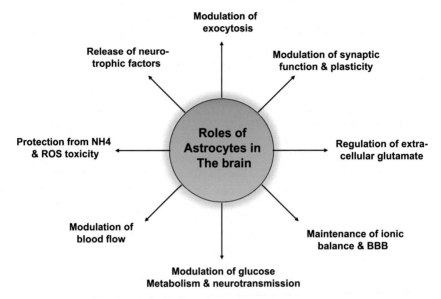

FIGURE 6.5   Roles of astrocytes in the brain. *BBB*, blood–brain barrier; *ROS*, reactive oxygen species.

Astrocytes perform multiple roles in the brain (Fig. 6.5). Thus astrocytes play different roles at different stages of the lifespan. During neurogenesis and early development, astrocytes provide a scaffold for the correct migration of neurons and growth cones. Astrocytes also provide guidance cues and may also be involved in neuronal proliferation. In the adult, astrocytes maintain neuronal homeostasis and synaptic plasticity. In addition, astrocytes contribute to the secretion of important neurotrophic factors, such as TGF-β, BDNF, NGF, and glial-derived neurotrophic factor (GDNF). They also participate in the transport of the vast majority of extracellular glutamate, and convert glucose into lactic acid, which is taken up by neurons and metabolized into pyruvate for energy metabolism (Allen and Barres, 2009). Major proportion of glutamate in the brain is cleared by glutamate transporters (GLTs) located on astrocytes. Among GLTs, GLT-1 accounts for up to 95% of glutamate uptake (Maragakis and Rothstein, 2001). Astrocytic glutamate transport plays a role in LTP induction and synaptic plasticity (Pita-Almenar et al., 2012a,b), and impaired glutamate clearance can be deleterious to neural cells via overactivation of glutamate receptors. This overactivation of glutamate receptor can lead to excitotoxicity, a process implicated in the pathogenesis of a number of neurodegenerative diseases (Maragakis and Rothstein, 2001; Farooqui et al., 2008). Astrocytes are also essential components of tripartite synapses (Ullian et al., 2001), which are responsible for bidirectional communication between pre- and postsynaptic neurons and integrated glial cells. Astrocytes are critical for basal synaptic transmission, LTP, and cortical circuit function (Pannasch and Rouach, 2013). Through the modulation of these processes, astrocytes play a crucial role not only in supporting the homeostatic environment in the brain but also maintain and promote the proper functioning of the neuronal network. In brain, astrocyte–neuron lactate shuttles couple synaptic plasticity and glucose metabolism. This coupling facilitates learning and memory (Yang et al., 2014). By forming connections to neuronal synapses as well as to each other through gap junctions, astrocytes can not only modulate neuronal activity and metabolic function but also have the ability to monitor synaptic activity, regulate synaptic plasticity, and synchronize neuronal networks. This observation supports the view that astrocytes contribute to cognitive processes (Henneberger et al., 2010; Lalo et al., 2011). Astrocytes contain chemokine CXCR4 receptors. Activation of the astrocytic CXCR4 by its natural ligand CXCL12 contributes to the release of the proinflammatory cytokine TNF-α and PGs leading to glutamate release. Another emerging role of the CXCR4–CXCL12 signaling axis in brain physiology comes from the observation that astrocytes promote the release of glutamate through a regulated exocytosis process, which occurs in a few hundred milliseconds (Calì and Bezzi, 2010). Astrocytes are electrically nonexcitable cells in which the onset of exocytosis depends on the Ca^{2+} release from the internal stores. This suggests that there is a close relationship between sites of Ca^{2+} release and fusion process that occurs during exocytosis (Calì and Bezzi, 2010). Converging evidence suggests that astrocytes contribute to the maintenance of glucose, ATP, and glutamate homeostasis, modulate synaptic formation and function, and are an essential component of the BBB.

Collective evidence suggests that the function of astrocytes is tightly integrated with the intracellular ionic signaling mediated by complex dynamics of cytosolic concentrations of free Ca^{2+} and Na^+. The ionic signaling of astroglial cells occurs not only through the ionotropic receptors at the plasmalemma and pumps solute carrier transporters but also at intracellular organelles (endoplasmic reticulum and mitochondria). The relative contribution of these molecular cascades/organelles can be plastically remodeled in development and under environmental stress (Verkhratsky et al., 2016).

Two distinct mechanisms contribute to the activation of astrocytes. First mechanism involves the downregulation of gap junction proteins restricting the overall syncytia of astrocytes leading to alterations in the morphology and number of astrocyte–neuronal connections and the BBB (Brand-Schieber et al., 2005). The second mechanism is associated with changes in astrocyte morphology due to immune regulation and inflammation under pathological conditions (Leonard, 2010). Astrocytes react to the neuronal damage by overexpressing the GFAP, vimentin, nestin, and ECM molecules, growth factors, and cytokines (IL-6, leukemia inhibitory factor, ciliary neurotrophic factor, TNF-α, INF-γ, IL-1, IL-10, TGF-β, fibroblast growth factor-2, etc.), releasing neurotransmitters such as glutamate and noradrenalin, ATP, ROS, NO, and molecules associated with systemic metabolic toxicity such as NH_4^+ (Allaman et al., 2010; Sofroniew and Vinters, 2010), and activating various intracellular signaling pathways, such as the MAPK, the NF-κB, and the Janus kinase/signal transducer and activator of transcription (JAK/STAT) pathways (Kang and Hébert, 2011). Among these pathways, the JAK/STAT3 pathway seems to be a central player in the induction of astrocyte reactivity, which is activated by a variety of cytokines and growth factors that signal through the gp130 receptor (Levy and Darnell, 2002). Activation of the JAK/STAT3 pathway in astrocytes has been reported not only in acute spinal cord injury (Herrmann et al., 2008) but also in patients and mouse models of ALS (http://www.jneurosci.org/content/35/6/2817.long; Shibata et al., 2010). This process is called astrogliosis (Hwang et al., 2016). Astrogliosis is supported by a series of cellular and molecular events, including the release of NO, ROS, proinflammatory cytokines (such as TNF-α, IL-1β, and IL-6) and prostanoids, all of which at high concentrations can produce deleterious effects on neuronal function. Converging evidence suggests that reactive astrogliosis is not merely a marker of neuropathological changes, but plays essential roles in orchestrating the injury response as well as in modulating neuroinflammation and repair in a manner that markedly impacts functional and clinical outcomes (Sofroniew, 2009; Sofroniew and Vinters, 2010).

Compared to the rapid microglial response, the astrocytic response usually occurs as a secondary event. The consequences of astrocyte activation in the aged brain not only include direct communication with neurons and microglial cells but also regulate the maintenance of BBB (Tichauer et al., 2007; von Bernhardi et al., 2015a; Abbott et al., 2006). As stated previously, astrocytes secrete a variety of signaling molecules such as the sonic hedgehog, retinoic acid, GDNF, and angiopoietin 1 (Lee et al., 2003; Zlokovic, 2008; Shearer et al., 2012). These factors enhance the BBB intactness not only via the activation of multiple receptors expressed in the endothelial cells but also through the increased expression of tight junction proteins (Igarashi et al., 1999; Acker et al., 2001). Astrocytes are important cellular component of the neurovascular unit, working hand in hand with the endothelial cells, neurons, microglia, and pericytes to maintain a state of tight regulation sensitive enough to detect neuronal metabolic and energy changes (Bell and Zlokovic, 2009). Neurovascular coupling and vascular tone are maintained by an astrocytic Ca^{2+} rise combined with PLA_2 activation, which leads to the release of arachidonic acid, later converted to vasoactive signals such as PGs and epoxyeicosatrienoic acids (Filosa et al., 2015). Furthermore, age-related changes in astrocytes can also affect BBB permeability, especially during the onset of neuroinflammation in neurodegenerative diseases (Zlokovic, 2008). Astrocytes interact with other glial cells (microglia and oligodendrocytes) through glia-to-glia signaling, which involves nonsynaptic communication by coupling of astrocytes. In this cell communication, cytoplasmic exchange of ions and small molecules among cells is accomplished through coupling of cells via cell-to-cell contacts, called gap junctions. Astrocytes also interact with neurons through neuron-to-glia signaling, which involves synaptic interactions (Vernadakis, 1996).

Activation of astrocytes by Aβ not only result in the release of inflammatory factors (TNF-α, IL-1 and IL-6, TGF-β, and IFN-γ) but also reduces synaptic and neuronal health in cell culture models of AD (Pekny et al., 2014; Correa-Cerro and Mandell, 2007; Garwood et al., 2011). In patients with AD and transgenic models of AD, the activation of astrocytes and onset of neuroinflammation are accompanied by increase in expression of high levels of calcineurin, a protein phosphatase, which mediates inflammatory responses. Importantly, astrocytes also mediate the effects of Aβ on Tau phosphorylation and cleavage, and this is linked with increased release of inflammatory mediators including IL-1β, TNF-α, and IL-6 (John et al., 2005; Furman et al., 2012). These studies are supported by positron emission tomography (PET). PET has indicated that astrocyte activation not only occurs in presymptomatic AD because reactive astrocytes surround Aβ plaques in autopsy brain tissue (Schöll et al., 2015; Rodriguez-Vieitez et al., 2015) but also in transgenic mice models of AD. Based on these studies, it is suggested that transgenic mice have widespread astroglial atrophy in the brain in earlier stages of AD prior to Aβ plaque deposition (Stalder et al., 2005; Yeh et al., 2011; Kulijewicz-Nawrot et al., 2012) and show presence of hypertrophic astrocytes surrounding amyloid plaques (Olabarria et al., 2010). In vitro studies on cell cultures systems have also indicated that the exposure of astrocytes to Aβ causes deleterious consequences in astrocytic morphology and metabolism. Thus Aβ treatment not only induces hypertrophy but also provokes alterations in calcium homeostasis, energetic modification, and degeneration of cocultured neurons (Scuderi et al., 2011). Like microglial cells, treatment of lipopolysaccharide (LPS)-primed murine astrocytes with Aβ42 treatment results in the release of IL-1β, depending on ASC expression, as ASC$^{-/-}$ astrocytes do not produce significant amount of IL-1β. Activation of inflammasome in astrocytes is a debatable issue due to the absence of expression of NALP3. In astrocytes, ASC and caspase-1 are colocalized with the GFAP, an astrocytic marker in a murine model of ALS supporting the view that Aβ-mediated activation of microglial cells as well as astrocytes results in production and secretion of IL-1β (Gustin et al., 2015; Johann et al., 2015). Lastly, Aβ-mediated insult not only produces increased oxidative and nitrosative stress (Frank-Cannon et al., 2009) but also results in the overexpression of a number of inflammation-related factors, such as IL-1α, IL-1β, and TNF-α (Li et al., 2003; Esposito et al., 2006). An increasing body of evidence supports the view that the generation of NO is one of the major effectors of neuronal cell death through mitochondrial depolarization (Solenski et al., 2003), while IL-1β enhances activation of caspase-3, an enzyme implicated in hippocampal neuronal apoptotic death in aged rats (Lynch and Lynch, 2002). Similarly, TNF-α contributes to the neuronal apoptosis mediated by Aβ, activating the caspase cascade, through interaction with its specific type 1 receptor (Li et al., 2004). Taken together, these results suggest that astrocytes play an important role in the progression of AD, probably due to their activation leading to the increase in the secretion of inflammatory mediators. Finally, by virtue of astrogliosis, astrocytes form the innate brain defense system, localizing the lesions and assisting in pathological remodeling of the affected circuitry (Sofroniew, 2009; Sofroniew and Vinters, 2010). Like microglial cells, astrocytes also possess the ability to internalize and metabolize Aβ in vivo. They express RAGE, which interacts with Aβ, and possess a complex apparatus able to take up Aβ (Wyss-Coray et al., 2003) supporting the view that astrocytes can take up Aβ from the extracellular space and channel it to lysosomes for degradation leading to the maintenance of Aβ homeostasis (Wyss-Coray et al., 2003; Koistinaho et al., 2004).

Astrocyte dysfunctions play a critical role not only in aging but also in AD (Fig. 6.6). It is now well established that astrocytes are essential in the control of cerebral homeostasis. However, the deposition of Aβ may

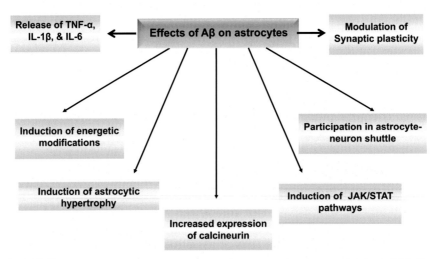

FIGURE 6.6 Effects of Aβ on neurochemical activities of astrocytes. *Aβ*, β-amyloid; *IL*, interleukin; *JAK/STAT*, Janus kinase/signal transducer and activator of transcription; *TNF-α*, tumor necrosis factor-α.

contribute to the modification of their physiological function (Verkhratsky et al., 2014, 2016). In general, astrocyte activation is a fundamentally protective response aimed at removing injurious stimuli. However, uncontrolled and prolonged activation of astrocytes goes beyond the physiological control and produces detrimental effects overriding the beneficial ones. Astrocytes contain many receptors, which modulate intracellular signaling cascades to respond quickly to changes in their environment (Buffo et al., 2010; Burda and Sofroniew, 2014). Thus astrocytes express many receptors, including PRRs, which not only detect abnormal signals in the extracellular space such as viral or bacterial molecules, serum proteins, aggregated proteins such as Aβ, and increased levels of cytokines, chemokines, and purines, but also "normal" signals from neighboring cells such as growth factors and neurotransmitters (Buffo et al., 2010; Burda and Sofroniew, 2014; Kigerl et al., 2014). As stated earlier, under pathological conditions in aged brain, astrocytes foster neuroinflammatory response accounting for the increased synthesis of different cytokines, chemokines, and TGF-β1. Astrocytes have also been reported to contribute to sustained inflammation in the brain through the upregulation of inflammatory pathways, which are modulated by TLR expression. A single injection of LPS in aged rats results in sustained astrocyte activation and prolonged increases in cytokine expression (Fu et al., 2014; Burda and Sofroniew, 2014). However, following activation, astrocytes release both pro- and antiinflammatory factors. After their release, proinflammatory signal molecules act, in an autocrine way, to self-perpetuate reactive gliosis in a paracrine way, to kill neighboring neurons expanding the neuropathological damage (Mrak and Griffin, 2001; Pekny et al., 2014). Collectively, these studies suggest that in AD, astrocytic Ca^{2+} signaling undergoes substantial reorganization due to an abnormal regulation of expression of Ca^{2+}-handling molecular cascades.

Converging evidence suggests that the regulation of inflammatory cytokine production during microgliosis and astrocytosis is under the control of intracellular signaling pathways, including the MAPK signaling pathway, the NF-κB signaling cascade, and peroxisome proliferator–activated receptor-γ.

Astrocytes are associated with the regulation of cerebral blood flow, water transport, and extracellular concentrations of ions, metabolites, and neurotransmitters (Anderson and Swanson, 2000). In addition, their processes contribute to the formation of BBB, directly contacting endothelial cells with vascular end-feet providing the structural and functional integrity to the BBB. The importance of astrocyte morphology in innate and adaptive immune responses in the brain has also been demonstrated (Dong and Benveniste, 2001). Thus astrocytes have the ability to express class II MHC antigens and costimulatory molecules (B7 and CD40), which are critical for antigen presentation and T-cell activation.

The functional role of astrocytes as immune effector cells involves antigen presentation and chemokine/cytokine production along with Th2 responses. These studies support the view that astrocytes function in innate and adaptive immune responses in the brain (Dong and Benveniste, 2001).

CROSSTALK AMONG NEURONS, ASTROCYTES, AND MICROGLIAL CELLS

It is well known that activation of neuronal and glial cells and interactions between neuron–glial cells are closely associated with brain development. Interactions between neurons and glial cells play important roles in neurogenesis, myelination, synaptogenesis, neuronal migration, phagocytosis, proliferation, differentiation, and neuronal–glial signaling (Alvarez-Maubecin et al., 2000; Gomes et al., 2001). Several soluble factors (neurotransmitters, hormones, and growth factors), which are secreted either by glial cells or neurons, contribute to the neuronal firing and intracellular signaling, which modulate neurotransmission and development of brain. The important roles played by microglial cells and astrocytes in normal and diseased brain functioning are growing, and a more complete picture of neuron–glia interactions is beginning to emerge. Thus prolonged activation of microglial cells and astrocytes, induction of neuroinflammation, and oxidative stress are closely related with the development and severity of neurodegenerative diseases such as AD, Parkinson's disease, ALS, and Huntington disease. Inflammatory and degenerating neural cells release ATP via pannexins (transmembrane protein channels) or connexin hemichannels (Junger, 2011), which connect the intracellular with the extracellular space. These channels have been implicated in the release of ATP from apoptotic cells (Chekeni et al., 2010; Eltzschig et al., 2008). In the extracellular compartment, ATP predominantly functions as a signaling molecule through the activation of purinergic P2 receptors. ATP acts as a neurotransmitter, since P2XR-mediated ATPergic transmission has been found in central synapses (Pankratov et al., 2002; Mori et al., 2001). ATP also controls neuroinflammation (Idzko et al., 2014), with multiple actions on microglia (Koizumi et al., 2013) leading to many metabolism-altering consequences in astrocytes and neurons. ATP also acts as a danger signal supporting the view that an abnormal and sustained elevation of extracellular ATP levels may cause neural cell dysfunction through the involvement of purine receptors, namely, P2X7R (ATP), P2Y1R (ADP), and $A_{2A}R$ (adenosine). In AD, activation and upregulation of P2X7R predominantly occurs in microglial cells around the Aβ plaques in mice (Parvathenani et al., 2003; Lee et al., 2011) and humans (McLarnon et al., 2006) and Aβ is known to not only stimulate the expression and release of IL-1β secretion but also promote the generation of ROS from microglial cell in a P2X7R-dependent manner (Sanz et al., 2009). These processes may lead to neuronal cell death. The release of ATP and other nucleotides from degenerating neurons under pathological conditions contributes to the "find-me" signal (Chekeni et al., 2010). This signal also activates microglial $P2Y_{12}$ receptors and mediates the extension of processes. In contrast, UDP functions as an "eat-me" signal and triggers microglial phagocytosis via $P2Y_6$ receptors (Inoue, 2007; Koizumi et al., 2007). In addition to the nucleotide-activated microglial cell function, microglial cells have been reported to release ATP (Imura et al., 2013). Microglial cell-derived ATP also activates the astrocytic $P2Y_1$ receptor and triggers ATP release. This process in turn regulates synaptic transmission (Pascual et al., 2012). The activation of $P2Y_1$ receptor on astrocytes not only contribute to neuroprotective effects against oxidative stress (Fujita et al., 2009) and traumatic brain injury (Talley Watts et al., 2013) but also produce neuroprotection against toxic substances (Noguchi et al., 2013). The activation of microglia and astrocytes occurs with spatially and temporarily distinct patterns. The activation of microglial cells occurs earlier than that of astrocytes supporting the onset of the crosstalk between microglia and astrocytes. The molecular mechanisms and physiological consequences of this crosstalk between microglial cells and astrocytes remain unknown. It is tempting to speculate that a better understanding of crosstalk between activated of microglia and astrocytes would be helpful to elucidate the role of glial cells in pathological conditions, which could accelerate the development of treatment for various diseases.

CX_3CL1 is one of the chemokines most abundantly expressed on neuronal surface, whereas the only known CX3CL1 receptor, CX3CR1, is expressed by microglial cells (Hanisch and Kettenman, 2007; Ransohoff and Perry, 2009). The CX_3CR1/CX_3CL1 axis plays a major role in neuron/microglia crosstalk and in neuroprotection under conditions of inflammation/injury (Hughes et al., 2002). Astrocytes also express CX3CL1 (Hatori et al., 2002), although at relatively lower levels than neurons. CX3CL1 has been reported to induce its own expression in a pertussis toxin–sensitive and G protein-dependent manner. CX3CL1 signaling is linked with cascade involving PLC, PtdIns 3K, and ERK, and the recruitment of transcription factors (NF-κB, CREB) (Sheridan and Murphy, 2013; Wolf et al., 2013). It is shown that CX3CL1 reduces neuronal death induced by glutamate through microglia-derived protective factors including adenosine (Limatola et al., 2005) and in rodent models of permanent ischemia, exogenous CX_3CL1 administration is neuroprotective (Cipriani et al., 2011). Alterations in CX3CL1 signaling have been reported to occur in a variety of neurological disorders such as ischemic brain injury, spinal cord injury, multiple sclerosis, and AD (Sheridan and

Murphy, 2013). Aging is accompanied not only by an increase in activation of microglial cells around Aβ plaques (Mandrekar-Colucci and Landreth, 2010) but also by a significant decrease in the expression of neuronal CX3CL1 levels (Lyons et al., 2009), which may contribute to excess microglial activation in the elderly subjects. Studies on the role of CX3CR1 signaling in a triple-transgenic mouse model of AD have indicated that this mouse model displays neuronal loss in layer III of the cortex between 4 and 6 months of age (Fuhrmann et al., 2010). Knocking down CX3CR1 in these mice retards neurodegeneration supporting the view that microglial cells play an important role in Aβ-induced neuronal death. Knocking down of CX3CR1 receptor in triple-transgenic mice has no effect on densities of microglial cells during this period. It is observed that the velocity of microglial cell migration to the site of neuronal damage was twofold greater than in healthy areas of the cortex in both transgenic mice and in CX3CR1$^{-/-}$ triple transgenics. The completion of neuronal elimination results in stopping the migration of microglial cells (Fuhrmann et al., 2010). These observations suggest that CX3CL1/CX3CR1 signaling in microglia during Aβ-induced neuronal stress is detrimental for neuronal survival.

Contribution of Neurons in Neuroinflammation

Neurons can also contribute to neuroinflammation through the generation of inflammatory molecules. Thus neurons can serve as a source of not only complement but also of COX-2-derived proinflammatory eicosanoids (Pavlov and Tracey, 2005; Davis and Laroche, 2003). They also express several cytokines (Murphy et al., 1999; Yermakova and O'Banion, 2000), and macrophage colony-stimulating factor (Yan et al., 1997), which play an important role in the onset of neuroinflammation. It is well known that the expression of COX-2 is driven by physiological synaptic activity (Yermakova and O'Banion, 2000) in neurons, which contribute to neuroinflammation and neurodestruction through the production of eicosanoids and PAF. Furthermore, under stress neurons also express iNOS, which degenerate in AD brains (Heneka et al., 2001; Lee et al., 1999). Compelling evidence suggests that long-term release of NO may be related to the chronic activation of iNOS and NO-dependent peroxynitrite formation (Smith et al., 1997). Glial- and neuronal-derived NO and peroxynitrite have been demonstrated to cause neuronal dysfunction and cell death in vitro and in vivo (Boje and Arora, 1992; Heneka et al., 1998).

Contribution of Complement System in Neuroinflammation

The complement system is a complex and tightly regulated attack system, which is designed to attack and destroy invaders and to assist in the phagocytosis of waste materials. The components of complement system perform four major functions: recognition, opsonization, inflammatory stimulation, and direct killing through the membrane attack complex (McGeer and McGeer, 2002). Proteins of complement system interact with cell surface receptors to promote a local inflammatory response, which is associated with the protection and healing of the injured or diseased brain. Complement activation induces inflammation and neural cell damage, but it is essential for eliminating cell debris and potentially toxic protein aggregates (Shen and Meri, 2003). The complement system consists of at least 30 fluid-phase and cell-membrane-associated proteins, which can be activated through different routes. The classical pathway (involving C1q, C1r, C1s, C4, C2, and C3 components) is activated primarily by the interaction of C1q with immune complexes (antibody antigen) (Jiang et al., 1994). The activation of classical pathway can also occur after interaction of C1q with nonimmune molecules such as DNA, RNA, C-reactive protein, serum amyloid P, bacterial LPSs, and some fungal and viral membranes. The initiation of the alternative pathway (involving C3, factor B, factor D, and properdin) does not require the presence of immune complexes and leads to the deposition of C3 fragments on target cells (Kohl, 2006; Bohlson et al., 2007). Many complement proteins and receptors can be synthesized locally in the brain (Gasque et al., 1995; Morgan and Gasque, 1996). The complement system activation has been observed in many neurological disorders, such as AD, multiple sclerosis, and stroke (Shen and Meri, 2003; Bonifati and Kishore, 2007). Both astrocytes and microglial cells express components of classical and alternative pathways. Astrocytes express all the components of classical and alternative pathways, such as C1–C9; regulatory factors B, D, H, and I; and several complement receptors, for example, C1qR, C3aR and C5aR, whereas microglial cells exhibit a narrower set of complement proteins, for example, C1q, C3, and receptors C1qR, CR3, and C5aR, which support the phagocytic uptake of targeted structures (Morgan and Gasque, 1996). Complement proteins (C1q, C4, C3, and Factor B) are also colocalized with fibrillar amyloid plaques and cerebral vascular amyloid in the cerebral cortex and hippocampus of patients with AD (Strohmeyer et al., 2000; Stoltzner et al., 2000). Astrocytic complement activation modulates Aβ dynamics in vitro and amyloid pathology in AD mouse models through microglial C3aR (Lian et al., 2016). It is reported that in primary microglial cultures, the activation of acute C3 or C3a promotes microglial phagocytosis. This phagocytosis can be attenuated by chronic C3/C3a treatment supporting the view that these interactions may play an important role in neuroinflammation. In addition, several studies have indicated that Aβ pathology and neuroinflammation in APP transgenic mice are worsened by hyperactivation of NF-κB in astrocytes. Furthermore, the treatment with the C3aR antagonist (C3aRA) also ameliorates plaque load and microgliosis suggesting that complement-dependent intercellular crosstalk between neurons and astrocytes may contribute to the overproduction of Aβ and induction of neuroinflammation in the brain (Lian et al., 2016). In addition to neuro-inflammatory responses, AD is often preceded by malfunctions in the cerebrovascular system, including accumulation of Aβ in cerebral blood vessels (cerebral amyloid angiopathy) and decrease in cerebral blood flow, which leads to hypoxia, breakdown of the BBB, and ultimately to brain atrophy and death (Bell et al., 2009; Kelleher and Soiza, 2013; Sagare et al., 2013). Studies in mice and humans indicate that Aβ peptides cause forceful constriction of cerebral blood vessels (Paris et al., 2003), suggesting that vascular factors play an early pathogenic role in AD. Collective evidence suggests that both positive and negative alterations have been reported to occur depending not only on neuron–glia signaling pathways but also on the presence of specific immune cells in specific regions of the brain (Orsini et al., 2014).

CONCLUSION

Neuroinflammation is a complex process, which not only involves a host of cellular and molecular changes and recruitment of peripheral immune cells but also induces some intracellular signaling pathways, which promote the release of inflammatory mediators in the brain. All these factors can contribute to synaptic dysfunction and neurodegeneration. Neuroinflammation is normally terminated by a process called resolution, with the purpose to promote the healing and return of damaged brain region to homeostasis. Resolution is accompanied not only by reduction in numbers of immune cells at the site of injury but also by decrease in apoptosis and increase in phagocytic activity. Neuroinflammation is an important feature of many neurodegenerative diseases including AD. Neuroinflammation is accompanied by the generation of high levels of inflammatory mediators and increased expression of proinflammatory cytokines and chemokines along with the activation of microglial cells and astrocytes. In AD, microgliosis is induced by Aβ, Aβ fibrils, and Aβ oligomers. Early microgliosis promotes Aβ clearance by protofibrillar Aβ phagocytosis, whereas chronic microgliosis promotes the release of several potentially cytotoxic substances, such as cytokines, NO, arachidonic acid–derived mediators, ROS, proteases, excitatory amino acids, and various neurotrophic factors. These mediators not only promote Aβ accumulation but also induce neuroinflammation leading to subsequent neurodegeneration. In response to Aβ oligomers, reactive astrocytes undergo astrocytosis, which is accompanied by the activation of NF-κB and upregulation of TNF-α, IL-1β, and COX-2 in the brain leading to neuroinflammation. The onset of astrocytosis and microgliosis in AD is also a sign that astrocytes and microglial cells are attempting to engulf and clear Aβ and its oligomers. In fact, microglia and astrocytes have been shown to contain Aβ in AD brain. This accumulation is due to the engulfment, and not due to endogenous production of Aβ by microglial cells and astrocytes. Brain also displays immune responses. Dendritic cells, microglia, and astrocytes are able to detect Aβ through TLR signaling. TLR2, in particular, plays a key role in triggering neuroinflammatory events that lead to the activation of signal-dependent transcription factors, driving expression of downstream genes associated with the inflammatory pathway.

References

Abbas, N., Bednar, I., Mix, E., et al., 2002. Up-regulation of the inflammatory cytokines IFN-γ and IL-12 and down-regulation of IL-4 in cerebral cortex regions of APPSWE transgenic mice. J. Neuroimmunol. 126, 50–57.

Abbott, N.J., Ronnback, L., Hansson, E., 2006. Astrocyte-endothelial interactions at the blood–brain barrier. Nat. Rev. Neurosci. 7, 41–53.

Acker, T., Beck, H., Plate, K.H., 2001. Cell type specific expression of vascular endothelial growth factor and angiopoietin-1 and -2 suggests an important role of astrocytes in cerebellar vascularization. Mech. Dev. 108, 45–57.

Aguzzi, A., Barres, B.A., Bennett, M.L., 2013. Microglia: scapegoat, saboteur, or something else? Science 339, 156–161.

Aisen, P.S., 1997. Inflammation and Alzheimer's disease: mechanisms and therapeutic strategies. Gerontology 43, 143–149.

Akiyama, H., Barger, S., Barnum, S., et al., 2000. Inflammation and Alzheimer's disease. Neurobiol. Aging 21, 383–421.

Allaman, I., Gavillet, M., Bélanger, M., Laroche, T., Viertl, D., Lashuel, H.A., et al., 2010. Amyloid-beta aggregates cause alterations of astrocytic metabolic phenotype: impact on neuronal viability. J. Neurosci. 30, 3326–3338.

Allan, S.M., Tyrrell, P.J., Rothwell, N.J., 2005. Interleukin-1 and neuronal injury. Nat. Rev. Immunol. 5, 629–640.

Allen, N.J., Barres, B.A., 2009. Neuroscience: glia – more than just brain glue. Nature 457, 675–677.

Alvarez-Maubecin, V., Garcia-Hernandez, F., Williams, J.T., Van Bockstaele, E.J., 2000. Functional coupling between neurons and glia. J. Neurosci. 20, 4091–4098.

Amor, S., Peferoen, L.A., Vogel, D.Y., Breur, M., van der Valk, P., Baker, D., van Noort, J.M., 2014. Inflammation in neurodegenerative diseases–an update. Immunology 142, 151–166.

Anderson, C.M., Swanson, R.A., 2000. Astrocyte glutamate transport: review of properties, regulation, and physiological functions. Glia 32, 1–14.

Atwood, C.S., Obrenovich, M.E., Liu, T., et al., 2003. Amyloid-β: a chameleon walking in two worlds: a review of the trophic and toxic properties of amyloid-β. Brain Res. Rev. 43, 1–16.

Baglio, F., Saresella, M., Preti, M.G., Cabinio, M., Griffanti, L., 2013. Neuroinflammation and brain functional disconnection in Alzheimer's disease. Front. Aging Neurosci. 5, 81.

Bales, K.R., Du, Y., Holtzman, D., Cordell, B., Paul, S.M., 2000. Neuroinflammation and Alzheimer's disease: critical roles for cytokine/Abetainduced glial activation, NF-kappaB, and apolipoprotein E. Neurobiol. Aging 21, 427–432 discussion 451–423.

Barrientos, R.M., Sprunger, D.B., Campeau, S., Watkins, L.R., Rudy, J.W., Maier, S.F., 2004. BDNF mRNA expression in rat hippocampus following contextual learning is blocked by intrahippocampal IL-1beta administration. J. Neuroimmunol. 155, 119–126.

Bell, R.D., Zlokovic, B.V., 2009. Neurovascular mechanisms and blood–brain barrier disorder in Alzheimer's disease. Acta Neuropathol. 118, 103–113.

Bell, R.D., Deane, R., Chow, N., Long, X., Sagare, A., et al., 2009. SRF and myocardin regulate LRP-mediated amyloid-beta clearance in brain vascular cells. Nat. Cell Biol. 11, 143–153.

Benitez, B.A., Jin, S.C., Guerreiro, R., Graham, R., Lord, J., et al., 2014. Missense variant in TREML2 protects against Alzheimer's disease. Neurobiol. Aging 35 1510.e19–1510.e26.

Bertollini, C., Ragozzino, D., Gross, C., Limatola, C., Eusebi, F., 2006. Fractalkine/CX3CL1 depresses central synaptic transmission in mouse hippocampal slices. Neuropharmacology 51, 816–821.

Bhaskar, K., Konerth, M., Kokiko-Cochran, O.N., Cardona, A., Ransohoff, R.M., Lamb, B.T., 2010. Regulation of tau pathology by the microglial fractalkine receptor. Neuron 68, 19–31.

Bhat, N.R., Feinstein, D.L., Shen, Q., Bhat, A.N., 2002. p38 MAPK-mediated transcriptional activation of inducible nitric-oxide synthase in glial cells: roles of nuclear factors, nuclear factor κB, cAMP response element-binding protein, CCAAT/enhance-binding protein-β, and activating transcription factor-2. J. Biol. Chem. 277, 29584–29592.

Bhattacharjee, S., Zhao, Y., Lukiw, W.J., 2014. Deficits in the miRNA-34a-regulated endogenous TREM2 phagocytosis sensor-receptor in Alzheimer's disease (AD); an update. Front. Aging Neurosci. 6, 116.

Block, M.L., Zecca, L., Hong, J.S., 2007. Microglia-mediated neurotoxicity: uncovering the molecular mechanisms. Nat. Rev. Neurosci. 8, 57–69.

Bohlson, S.S., Fraser, D.A., Tenner, A.J., 2007. Complement proteins C1q and MBL are pattern recognition molecules that signal immediate and long-term protective immune functions. Mol. Immunol. 44, 33–43.

Boje, K.M., Arora, P.K., 1992. Microglial-produced nitric oxide and reactive nitrogen oxides mediate neuronal cell death. Brain Res. 587, 250–256.

Bonifati, D.M., Kishore, U., 2007. Role of complement in neurodegeneration and neuroinflammation. Mol. Immunol. 44, 999–1010.

Bouchon, A., Dietrich, J., Colonna, M., 2000. Cutting edge: inflammatory responses can be triggered by TREM-1, a novel receptor expressed on neutrophils and monocytes. J. Immunol. 164, 4991–4995.

Bradt, B.M., Kolb, W.P., Cooper, N.R., 1998. Complement-dependent proinflammatory properties of the Alzheimer's disease β-peptide. J. Exp. Med. 188, 431–438.

Brand-Schieber, E., Werner, P., Iacobas, D.A., Iacobas, S., Beelitz, M., Lowery, S.L., et al., 2005. Connexin43, the major gap junction protein of astrocytes, is down-regulated in inflamed white matter in an animal model of multiple sclerosis. J. Neurosci. Res. 80, 798–808.

Brionne, T.C., Tesseur, I., Masliah, E., Wyss-Coray, T., 2003. Loss of TGF-β1 leads to increased neuronal cell death and microgliosis in mouse brain. Neuron 40, 1133–1145.

Buffo, A., Rolando, C., Ceruti, S., 2010. Astrocytes in the damaged brain: molecular and cellular insights into their reactive response and healing potential. Biochem. Pharmacol. 79, 77–89.

Burda, J.E., Sofroniew, M.V., 2014. Reactive gliosis and the multicellular response to CNS damage and disease. Neuron 81, 229–248.

Butt, A.M., Duncan, A., Berry, M., 1994. Astrocyte associations with nodes of Ranvier: ultrastructural analysis of HRP-filled astrocytes in the mouse optic nerve. J. Neurocytol. 23, 486–499.

Cady, J., Koval, E.D., Benitez, B.A., Zaidman, C., Jockel-Balsarotti, J., Allred, P., et al., 2014. TREM2 variant p.R47H as a risk factor for sporadic amyotrophic lateral sclerosis. JAMA Neurol. 71, 449–453.

Cai, Z., Hussain, M.D., Yan, L.J., 2014. Microglia, neuroinflammation, and beta-amyloid protein in Alzheimer's disease. Int. J. Neurosci. 124, 307–321.

Cai, Z., Liu, N., Wang, C., Qin, B., Zhou, Y., Xiao, M., et al., July 15, 2016. Role of RAGE in Alzheimer's disease. Cell. Mol. Neurobiol. 36, 483–495.

Calì, C., Bezzi, P., 2010. CXCR4-mediated glutamate exocytosis from astrocytes. J. Neuroimmunol. 224, 13–21.

Cameron, B., Landreth, G.E., 2010. Inflammation, microglia, and Alzheimer's disease. Neurobiol. Dis. 37, 503–509.

Caraci, F., Pistara, V., Corsaro, A., Tomasello, F., Giuffrida, M.L., et al., 2011. Neurotoxic properties of the anabolic androgenic steroids nandrolone and methandrostenolone in primary neuronal cultures. J. Neurosci. Res. 89, 592–600.

Cardona, A.E., Pioro, E.P., Sasse, M.E., Kostenko, V., Cardona, S.M., Dijkstra, I.M., 2006. Control of microglial neurotoxicity by the fractalkine receptor. Nat. Neurosci. 9, 917–924.

Chakraborty, S., Kaushik, D.K., Gupta, M., Basu, A., 2010. Inflammasome signaling at the heart of central nervous system pathology. J. Neurosci. Res. 88, 1615–1631.

Chekeni, F.B., Elliott, M.R., Sandilos, J.K., Walk, S.F., Kinchen, J.M., Lazarowski, E.R., et al., 2010. Pannexin 1 channels mediate 'find-me' signal release and membrane permeability during apoptosis. Nature 467, 863–867.

Chen, X., Walker, D.G., Schmidt, A.M., Arancio, O., Lue, L.F., Yan, S.D., 2007. RAGE: a potential target for Abeta-mediated cellular perturbation in Alzheimer's disease. Curr. Mol. Med. 7, 735–742.

Chen, C., Li, X.H., Tu, Y., Sun, H.T., Liang, H.Q., Cheng, S.X., et al., 2014. Aβ-AGE aggravates cognitive deficit in rats via RAGE pathway. Neuroscience 257, 1–10.

Cipriani, R., Villa, P., Chece, G., Lauro, C., Paladini, A., et al., 2011. CX3CL1 is neuroprotective in permanent focal cerebral ischemia in rodents. J. Neurosci. 31, 16327–16335.

Colonna, M., 2003. TREMs in the immune system and beyond. Nat. Rev. Immunol. 3, 445–453.

Cooper, N.R., Kalaria, R.N., McGeer, P.L., Rogers, J., 2000. Key issues in Alzheimer's disease inflammation. Neurobiol. Aging 21, 451–453.

Correa-Cerro, L.S., Mandell, J.W., 2007. Molecular mechanisms of astrogliosis: new approaches with mouse genetics. J. Neuropathol. Exp. Neurol. 66, 169–176.

Correale, J., Villa, A., 2004. The neuroprotective role of inflammation in nervous system injuries. J. Neurol. 251, 1304–1316.

Damani, M.R., Zhao, L., Fontainhas, A.M., Amaral, J., Fariss, R.N., Wong, W.T., 2011. Age-related alterations in the dynamic behavior of microglia. Aging Cell 10, 263–276.

Davis, S., Laroche, S., 2003. What can rodent models tell us about cognitive decline in Alzheimer's disease? Mol. Neurobiol. 27, 249–276.

Dempsey, L.A., 2015. TREM2 function in microglia. Nat. Immunol. 16, 447.

Di Virgilio, F., 2013. The therapeutic potential of modifying inflammasomes and NOD-like receptors. Pharmacol. Rev. 65, 872–905.

Dong, Y., Benveniste, E.N., 2001. Immune function of astrocytes. Glia 36, 180–190.

Eltzschig, H.K., Macmanus, C.F., Colgan, S.P., 2008. Neutrophils as sources of extracellular nucleotides: functional consequences at the vascular interface. Trends Cardiovasc. Med. 18, 103–107.

Esposito, G., De Filippis, D., Steardo, L., Scuderi, C., Savani, C., Cuomo, V., et al., 2006. CB1 receptor selective activation inhibits beta-amyloid-induced iNOS protein expression in C6 cells and subsequently blunts tau protein hyperphosphorylation in co-cultured neurons. Neurosci. Lett. 404, 342–346.

Farooqui, A.A., Horrocks, L.A., Farooqui, T., 2007. Modulation of inflammation in brain: a matter of fat. J. Neurochem. 101, 577–599.

Farooqui, A.A., Ong, W.Y., Horrocks, L.A., 2008. Neurochemical Aspects of Excitotoxicity. Springer, New York.

Farooqui, A.A., 2010. Neurochemical Aspects of Neurotraumatic and Neurodegeneration Disease. Springer, New York.

Farooqui, A.A., 2011. Lipid Mediators and Their Metabolism in the Brain. Springer, New York.

Farooqui, A.A., 2013. Metabolic Syndrome: An Important Risk Factor for Stroke, Alzheimer Disease, and Depression. Springer, New York.

Farooqui, A.A., 2014. Inflammation and Oxidative Stress in Neurological Disorders. Springer, New York.

Filosa, J.A., Morrison, H.W., Iddings, J.A., Du, W., Kim, K.J., 2015. Beyond neurovascular coupling, role of astrocytes in the regulation of vascular tone. Neuroscience 323, 96–109.

Findeis, M.A., 2007. The role of amyloid β peptide 42 in Alzheimer's disease. Pharmacol. Ther. 116, 266–286.

Frank-Cannon, T.C., Alto, L.T., McAlpine, F.E., Tansey, M.G., 2009. Does neuroinflammation fan the flame in neurodegenerative diseases? Mol. Neurodegener. 4, 47.

Fu, H.Q., Yang, T., Xiao, W., Fan, L., Wu, Y., 2014. Prolonged neuroinflammation after lipopolysaccharide exposure in aged rats. PLoS One 9, e106331.

Fuhrmann, M., Bittner, T., Jung, C.K., Burgold, S., Page, R.M., Mitteregger, G., et al., 2010. Microglial Cx3cr1 knockout prevents neuron loss in a mouse model of Alzheimer's disease. Nat. Neurosci. 13, 411–413.

Fujita, T., Tozaki-Saitoh, H., Inoue, K., 2009. P2Y1 receptor signaling enhances neuroprotection by astrocytes against oxidative stress via IL-6 release in hippocampal cultures. Glia 57, 244–257.

Furman, J.L., Sama, D.M., Gant, J.C., Beckett, T.L., Murphy, M.P., et al., 2012. Targeting astrocytes ameliorates neurologic changes in a mouse model of Alzheimer's disease. J. Neurosci. 32, 16129–16140.

Galimberti, D., Fenoglio, C., Lovati, C., Venturelli, E., Guidi, I., et al., 2006. Serum MCP-1 levels are increased in mild cognitive impairment and mild Alzheimer's disease. Neurobiol. Aging 27, 1763–1768.

Garwood, C.J., Pooler, A.M., Atherton, J., Hanger, D.P., Noble, W., 2011. Astrocytes play a key role in Aβ-induced tau phosphorylation and neurotoxicity in primary culture. Cell Death Dis. 2, e167.

Gasque, P., Fontaine, M., Morgan, B.P., 1995. Complement expression in human brain: biosynthesis of terminal pathway components and regulators in human glial cells and cell lines. J. Immunol. 154, 4726–4733.

Gemma, C., Bickford, P.C., 2007. Interleukin-1beta and caspase-1: players in the regulation of age-related cognitive dysfunction. Rev. Neurosci. 18, 137–148.

Geula, C., Wu, C.K., Saroff, D., Lorenzo, A., Yuan, M., 1998. Aging renders the brain vulnerable to amyloid beta-protein neurotoxicity. Nat. Med. 4, 827–831.

Godbout, J.P., Johnson, R.W., 2004. Interleukin-6 in the aging brain. J. Neuroimmunol. 147, 141–144.

Golde, T.E., Streit, W.J., Chakrabarty, P., 2013. Alzheimer's disease risk alleles in TREM2 illuminate innate immunity in Alzheimer's disease. Alzheimers Res. Ther. 5, 24.

Goldbaum, O., Richter-Landsberg, C., 2001. Stress proteins in oligodendrocytes: differential effects of heat shock and oxidative stress. J. Neurochem. 78, 1233–1242.

Gomes, F.C., Spohr, T.C., Martinez, R., Moura Neto, V., 2001. Cross-talk between neurons and glia: highlights on soluble factors. Braz. J. Med. Biol. Res. 34, 611–620.

Guerreiro, R., Wojtas, A., Bras, J., Carrasquillo, M., Rogaeva, E., et al., 2013. TREM2 variants in Alzheimer's disease. N. Engl. J. Med. 368, 117–127.

Gustin, A., Kirchmeyer, M., Koncina, E., Felten, P., Losciuto, S., Heurtaux, T., et al., 2015. NLRP3 inflammasome is expressed and functional in mouse brain microglia but not in astrocytes. PLoS One 10, e0130624.

Haydon, P.G., Carmignoto, G., 2006. Astrocyte control of synaptic transmission and neurovascular coupling. Physiol. Rev. 86, 1009–1031.

Halle, A., Hornung, V., Petzold, G.C., Stewart, C.R., Monks, B.G., Reinheckel, T., Fitzgerald, K.A., 2008. The NALP3 inflammasome is involved in the innate immune response to amyloid-beta. Nat. Immunol. 9, 857–865.

Halliday, G., Robinson, S.R., Shepherd, C., Kril, J., 2000. Alzheimer's disease and inflammation: a review of cellular and therapeutic mechanisms. Clin. Exp. Pharmacol. Physiol. 27, 1–8.

Han, S.H., Kim, Y.H., Mook-Jung, I., 2011. RAGE: the beneficial and deleterious effects by diverse mechanisms of actions. Mol. Cells 31, 91–97.

Hanisch, U.-K., Kettenmann, H., 2007. Microglia: active sensor and versatile effector cells in the normal and pathologic brain. Nat. Neurosci. 10, 1387–1394.

Hatori, K., Nagai, A., Heisel, R., Ryu, J.K., Kim, S.U., 2002. Fractalkine and fractalkine receptors in human neurons and glial cells. J. Neurosci. Res. 69, 418–426.

Haynes, S.E., Hollopeter, G., Yang, G., Kurpius, D., Dailey, M.E., Gan, W.B., Julius, D., 2006. The P2Y12 receptor regulates microglial activation by extracellular nucleotides. Nat. Neurosci. 9, 1512–1519.

Heneka, M.T., Löschmann, P.A., Gleichmann, M., et al., 1998. Induction of nitric oxide synthase and nitric oxide-mediated apoptosis in neuronal PC12 cells after stimulation with tumor necrosis factor-α/lipopolysaccharide. J. Neurochem. 71, 88–94.

Heneka, M.T., Wiesinger, H., Dumitrescu-Ozimek, L., Riederer, P., Feinstein, D.L., Klockgether, T., 2001. Neuronal and glial coexpression of argininosuccinate synthetase and inducible nitric oxide synthase in Alzheimer disease. J. Neuropathol. Exp. Neurol. 60, 906–916.

Heneka, M.T., Sastre, M., Dumitrescu-Ozimek, L., et al., 2005. Focal glial activation coincides with increased BACE1 activation and precedes amyloid plaque deposition in APP[V717I] transgenic mice. J. Neuroinflamm. 2, 22.

Heneka, M.T., O'Banion, M.K., 2007. Inflammatory processes in Alzheimer's disease. J. Neuroimmunol. 184, 69–91.

Heneka, M.T., Kummer, M.P., Stutz, A., Delekate, A., Schwartz, S., et al., 2013. NLRP3 is activated in Alzheimer's disease and contributes to pathology in APP/PS1 mice. Nature 493, 674–678.

Henneberger, C., Papouin, T., Oliet, S.H., Rusakov, D.A., 2010. Long-term potentiation depends on release of d-serine from astrocytes. Nature 463, 232–236.

Heppner, F.L., Ransohoff, R.M., Becher, B., 2015. Immune attack: the role of inflammation in Alzheimer disease. Nat. Rev. Neurosci. 16, 358–372.

Herrmann, J.E., Imura, T., Song, B., Qi, J., Ao, Y., Nguyen, T.K., Korsak, R.A., Takeda, K., Akira, S., Sofroniew, M.V., 2008. STAT3 is a critical regulator of astrogliosis and scar formation after spinal cord injury. J. Neurosci. 28, 7231–7243.

Hickman, S.E., Kingery, N.D., Ohsumi, T.K., Borowsky, M.L., Wang, L.C., et al., 2013. The microglial sensome revealed by direct RNA sequencing. Nat. Neurosci. 16, 1896–1905.

Hickman, S.E., El Khoury, J., 2014. TREM2 and the neuroimmunology of Alzheimer's disease. Biochem. Pharmacol. 88, 495–498.

Hsieh, C.L., Koike, M., Spusta, S., Niemi, E., Yenari, M., Nakamura, M.C., Seaman, W.E., 2009. A role for TREM2 ligands in the phagocytosis of apoptotic neuronal cells by microglia. J. Neurochem. 109, 1144–1156.

Hughes, P.M., Botham, M.S., Frentzel, S., Mir, A., Perry, V.H., 2002. Expression of fractalkine (CX3CL1) and its receptor, CX3CR1, during acute and chronic inflammation in the rodent CNS. Glia 37, 314–327.

Hung, S.-I., Chang, A.C., Kato, I., Chang, N.-C.A., 2002. Transient expression of Ym1, a heparin-binding lectin, during developmental hematopoiesis and inflammation. J. Leukoc. Biol. 72, 72–82.

Hwang, C.J., Park, M.H., Hwang, J.Y., Kim, J.H., Yun, N.Y., Oh, S.Y., et al., February 17, 2016. CCR5 deficiency. Oncotarget. http://dx.doi.org/10.18632/oncotarget.7453 (Epub ahead of print).

Idzko, M., Ferrari, D., Eltzschig, H.K., 2014. Nucleotide signalling during inflammation. Nature 509, 310–317.

Igarashi, Y., Utsumi, H., Chiba, H., Yamada-Sasamori, Y., Tobioka, H., et al., 1999. Glial cell line-derived neurotrophic factor induces barrier function of endothelial cells forming the blood–brain barrier. Biochem. Biophys. Res. Commun. 261, 108–112.

Imura, Y., Morizawa, Y., Komatsu, R., Shibata, K., Shinozaki, Y., Kasai, H., et al., 2013. Microglia release ATP by exocytosis. Glia 61, 1320–1330.

Inoue, K., 2007. UDP facilitates microglial phagocytosis through P2Y6 receptors. Cell Adh. Migr. 1, 131–132.

Jana, M., Dasgupta, S., Liu, X., Pahan, K., 2002. Regulation of tumor necrosis factor-α expression by CD40 ligation in BV-2 microglial cells. J. Neurochem. 80, 197–206.

Jiang, H., Burdick, D., Glabe, C.G., Cotman, C.W., Tenner, A.J., 1994. β-amyloid activates complement by binding to a specific region of the collagen-like domain of the C1q A chain. J. Immunol. 152, 5050–5059.

Jiang, T., Yu, J.T., Zhu, X.C., Tan, M.S., Gu, L.Z., Zhang, Y.D., Tan, L., 2014. Triggering receptor expressed on myeloid cells 2 knockdown exacerbates aging-related neuroinflammation and cognitive deficiency in senescence-accelerated mouse prone 8 mice. Neurobiol. Aging 35, 1243–1251.

Johann, S., Heitzer, M., Kanagaratnam, M., Goswami, A., Rizo, T., Weis, J., et al., 2015. NLRP3 inflammasome is expressed by astrocytes in the SOD1 mouse model of ALS and in human sporadic ALS patients. Glia 63, 2260–2273.

John, G.R., Lee, S.C., Song, X., Rivieccio, M., Brosnan, C.F., 2005. IL-1-regulated responses in astrocytes: relevance to injury and recovery. Glia 49, 161–176.

Johnson, D., Krenger, W., 1992. Interactions of mast cells with the nervous system—recent advances. Neurochem. Res. 17, 939–951.

Jonsson, T., Stefansson, H., Steinberg, S., Jonsdottir, I., Jonsson, P.V., Snaedal, J., et al., 2013. Variant of TREM2 associated with the risk of Alzheimer's disease. N. Engl. J. Med. 368, 107–116.

Junger, W.G., 2011. Immune cell regulation by autocrine purinergic signalling. Nat. Rev. Immunol. 11, 201–212.

Kamouchi, M., Ago, T., Kitazono, T., 2011. Brain pericytes: emerging concepts and functional roles in brain homeostasis. Cell. Mol. Neurobiol. 31, 175–193.

Kang, W., Hébert, J.M., 2011. Signaling pathways in reactive astrocytes, a genetic perspective. Mol. Neurobiol. 43, 147–154.

Kelleher, R.J., Soiza, R.L., 2013. Evidence of endothelial dysfunction in the development of Alzheimer's disease: is Alzheimer's a vascular disorder? Am. J. Cardiovasc. Dis. 3, 197–226.

Kettenmann, H., Hanisch, U.K., Noda, M., Verkhratsky, A., 2011. Physiology of microglia. Physiol. Rev. 91, 461–553.

Kigerl, K.A., De Rivero Vaccari, J.P., Dietrich, W.D., Popovich, P.G., Keane, R.W., 2014. Pattern recognition receptors and central nervous system repair. Exp. Neurol. 258, 5–16.

Kleinberger, G., Yamanishi, Y., Suárez-Calvet, M., et al., 2014. TREM2 mutations implicated in neurodegeneration impair cell surface transport and phagocytosis. Sci. Transl. Med. 6, 243ra286.

Kohl, J., 2006. The role of complement in danger sensing and transmission. Immunol. Res. 34, 157–176.

Koistinaho, M., Lin, S., Wu, X., Esterman, M., Koger, D., Hanson, J., et al., 2004. Apolipoprotein E promotes astrocyte colocalization and degradation of deposited amyloid-beta peptides. Nat. Med. 10, 719–726.

Koizumi, S., 2010. Synchronization of Ca^{2+} oscillations: involvement of ATP release in astrocytes. FEBS J. 277, 286–292.

Koizumi, S., Shigemoto-Mogami, Y., Nasu-Tada, K., Shinozaki, Y., Ohsawa, K., Tsuda, M., et al., 2007. UDP acting at P2Y6 receptors is a mediator of microglial phagocytosis. Nature 446, 1091–1095.

Koizumi, H., Koshiya, N., Chia, J.X., Cao, F., Nugent, J., et al., 2013. Structural-functional properties of identified excitatory and inhibitory interneurons within pre-Botzinger complex respiratory microcircuits. J. Neurosci. 33, 2994–3009.

Kulijewicz-Nawrot, M., Verkhratsky, A., Chvátal, A., Syková, E., Rodríguez, J.J., 2012. Astrocytic cytoskeletal atrophy in the medial prefrontal cortex of a triple transgenic mouse model of Alzheimer's disease. J. Anat. 221, 252–262.

Lalo, U., Pankratov, Y., Parpura, V., Verkhratsky, A., 2011. Ionotropic receptors in neuronal-astroglial signalling: what is the role of 'excitable' molecules in non-excitable cells. Biochim. Biophys. Acta 1813, 992–1002.

Lanier, L.L., Corliss, B.C., Wu, J., Leong, C., Phillips, J.H., 1998. Immunoreceptor DAP12 bearing a tyrosine-based activation motif is involved in activating NK cells. Nature 391, 703–707.

Lee, M., 2013. Neurotransmitters and microglial-mediated neuroinflammation. Curr. Protein Pept. Sci. 14, 21–32.

Lee, S.C., Zhao, M.L., Hirano, A., Dickson, D.W., 1999. Inducible nitric oxide synthase immunoreactivity in the Alzheimer disease hippocampus: association with Hirano bodies, neurofibrillary tangles, and senile plaques. J. Neuropathol. Exp. Neurol. 58, 1163–1169.

Lee, S.W., Kim, W.J., Choi, Y.K., Song, H.S., Son, M.J., et al., 2003. SSeCKS regulates angiogenesis and tight junction formation in blood–brain barrier. Nat. Med. 9, 900–906.

Lee, S., Varvel, N.H., Konerth, M.E., Xu, G., Cardona, A.E., Ransohoff, R.M., Lamb, B.T., 2010. CX3CR1 deficiency alters microglial activation and reduces beta-amyloid deposition in two Alzheimer's disease mouse models. Am. J. Pathol. 177, 2549–2562.

Lee, H.G., Won, S.M., Gwag, B.J., Lee, Y.B., 2011. Microglial P2X7 receptor expression is accompanied by neuronal damage in the cerebral cortex of the APPswe/PS1dE9 mouse model of Alzheimer's disease. Exp. Mol. Med. 43, 7–14.

Lee, C.Y.D., Landreth, G.E., 2013. The role of microglia in amyloid clearance from the AD brain. J. Neural Transm. 117, 949–960.

Lent, R., Azevedo, F.A., Andrade-Moraes, C.H., Pinto, A.V., 2012. How many neurons do you have? Some dogmas of quantitative neuroscience under revision. Eur. J. Neurosci. 35, 1–9.

Leonard, B.E., 2010. The concept of depression as a dysfunction of the immune system. Curr. Immunol. Rev. 6, 205–212.

Levy, D.E., Darnell Jr., J.E., 2002. Stats: transcriptional control and biological impact. Nat. Rev. Mol. Cell Biol. 3, 651–662.

Li, Y., Liu, L., Barger, S.W., Griffin, W.S., 2003. Interleukin-1 mediates pathological effects of microglia on tau phosphorylation and on synaptophysin synthesis in cortical neurons through a p38-MAPK pathway. J. Neurosci. 23, 1605–1611.

Li, R., Yang, L., Lindholm, K., Konishi, Y., Yue, X., Hampel, H., et al., 2004. Tumor necrosis factor death receptor signaling cascade is required for amyloid-beta protein-induced neuron death. J. Neurosci. 24, 1760–1771.

Lian, H., Litvinchuk, A., Chiang, A.C., Aithmitti, N., Jankowsky, J.L., et al., 2016. Astrocyte-microglia cross talk through complement activation modulates amyloid pathology in mouse models of Alzheimer's disease. J. Neurosci. 36, 577–589.

Liang, J., Takeuchi, H., Jin, S., Noda, M., Li, H., Doi, Y., et al., 2010. Glutamate induces neurotrophic factor production from microglia via protein kinase C pathway. Brain Res. 1322, 8–23.

Limatola, C., Lauro, C., Catalano, M., Ciotti, M.T., Bertollini, C., Di Angelantonio, S., Ragozzino, D., Eusebi, F., 2005. Chemokine CX3CL1 protects rat hippocampal neurons against glutamate-mediated excitotoxicity. J. Neuroimmunol. 166, 19–28.

Lindberg, C., Hjorth, E., Post, C., Winblad, B., Schultzberg, M., 2005. Cytokine production by a human microglial cell line: effects of βamyloid and α-melanocyte-stimulating hormone. Neurotox. Res. 8, 267–276.

Lindsey, J.D., Landfield, P.W., Lynch, G., 1979. Early onset and topographical distribution of hypertrophied astrocytes in hippocampus of aging rats: a quantitative study. J. Gerontol. 34, 661–671.

Liu, X., Wu, Z., Hayashi, Y., Nakanishi, H., 2012. Age-dependent neuroinflammatory responses and deficits in long-term potentiation in the hippocampus during systemic inflammation. Neuroscience 216, 133–142.

Liu, L., Chan, C., 2014. IPAF inflammasome is involved in interleukin-1beta production from astrocytes, induced by palmitate; implications for Alzheimer's Disease. Neurobiol. Aging 35, 309–321.

Lucin, K.M., Wyss-Coray, T., 2009. Immune activation in brain aging and neurodegeneration: too much or too little? Neuron 64, 110–122.

Lue, L.F., Walker, D.G., Brachova, L., Beach, T.G., Rogers, J., Schmidt, A.M., et al., 2001. Involvement of microglial receptor for advanced glycation endproducts (RAGE) in Alzheimer's disease: identification of a cellular activation mechanism. Exp. Neurol. 171, 29–45.

Luo, X.G., Ding, J.Q., Chen, S.D., 2010. Microglia in the aging brain: relevance to neurodegeneration. Mol. Neurodegener. 5, 12.

Lynch, M.A., 1998. Age-related impairment in long-term potentiation in hippocampus: a role for the cytokine, interleukin-1 beta? Prog. Neurobiol. 56, 571–589.

Lynch, A.M., Lynch, M.A., 2002. The age-related increase in IL-1 type I receptor in rat hippocampus is coupled with an increase in caspase-3 activation. Eur. J. Neurosci. 15, 1779–1788.

Lyons, A., Lynch, A.M., Downer, E.J., Hanley, R., O'Sullivan, J.B., Smith, A., Lynch, M.A., 2009. Fractalkine-induced activation of the phosphatidylinositol-3 kinase pathway attenuates microglial activation in vivo and in vitro. J. Neurochem. 110, 1547–1556.

Maggi, L., Trettel, F., Scianni, M., Bertollini, C., Eusebi, F., et al., 2009. LTP impairment by fractalkine/CX3CL1 in mouse hippocampus is mediated through the activity of adenosine receptor type 3 (A3R). J. Neuroimmunol. 215, 36–42.

Maher, F.O., Clarke, R.M., Kelly, A., Nally, R.E., Lynch, M.A., 2006. Interaction between interferon gamma and insulin-like growth factor-1 in hippocampus impacts on the ability of rats to sustain long-term potentiation. J. Neurochem. 96, 1560–1571.

Majed, H.H., Chandran, S., Niclou, S.P., Nicholas, R.S., Wilkins, A., Wing, M.G., Rhodes, K.E., Spillantini, M.G., Compston, A., 2006. A novel role for Sema3A in neuroprotection from injury mediated by activated microglia. J. Neurosci. 26, 1730–1738.

Malkki, H., 2015. Alzheimer disease: the involvement of TREM2 R47H variant in Alzheimer disease confirmed, but mechanisms remain elusive. Nat. Rev. Neurol. 11, 307.

Mandrekar-Colucci, S., Landreth, G.E., 2010. Microglia and inflammation in Alzheimer's disease. CNS Neurol. Disord. Drug Targets 9, 156–167.

Mantovani, A., Sica, A., Sozzani, S., Allavena, P., Vecchi, A., Locati, M., 2004. The chemokine system in diverse forms of macrophage activation and polarization. Trends Immunol. 25, 677–686.

Maragakis, N.J., Rothstein, J.D., 2001. Glutamate transporters in neurologic disease. Arch. Neurol. 58, 365–370.

McGeer, P.L., McGeer, E.G., 2002. The possible role of complement activation in Alzheimer disease. Trends Mol. Med. 8, 519–523.

McLarnon, J.G., Ryu, J.K., Walker, D.G., Choi, H.B., 2006. Upregulated expression of purinergic P2X$_7$ receptor in Alzheimer disease and amyloid-beta peptide-treated microglia and in peptide-injected rat hippocampus. J. Neuropathol. Exp. Neurol. 65, 1090–1097.

Minghetti, L., Ajmone-Cat, M.A., De Berardinis, M.A., De Simone, R., 2005. Microglial activation in chronic neurodegenerative diseases: roles of apoptotic neurons and chronic stimulation. Brain Res. Rev. 48, 251–256.

Minkiewicz, J., de Rivero Vaccari, J.P., Keane, R.W., 2013. Human astrocytes express a novel NLRP2 inflammasome. Glia 61, 1113–1121.

Morgan, B.P., Gasque, P., 1996. Expression of complement in the brain: role in health and disease. Immunol. Today 17, 461–466.

Mori, M., Heuss, C., Gähwiler, B.H., Gerber, U., 2001. Fast synaptic transmission mediated by P2X receptors in CA3 pyramidal cells of rat hippocampal slice cultures. J. Physiol. 535, 115–123.

Mosher, K.I., Wyss-Coray, T., 2014. Microglial dysfunction in brain aging and Alzheimer's disease. Biochem. Pharmacol. 88, 594–604.

Mott, R.T., Ait-Ghezala, G., Town, T., Mori, T., Vendrame, M., Zeng, J., Ehrhart, J., Mullan, M., Tan, J., 2004. Neuronal expression of CD22: novel mechanism for inhibiting microglial proinflammatory cytokine production. Glia 46, 369–379.

Mouton, P.R., Long, J.M., Lei, D.L., Howard, V., Jucker, M., et al., 2002. Age and gender effects on microglia and astrocyte numbers in brains of mice. Brain Res. 956, 30–35.

Mrak, R.E., Griffin, W.S., 2001. The role of activated astrocytes and of the neurotrophic cytokine S100B in the pathogenesis of Alzheimer's disease. Neurobiol. Aging 22, 915–922.

Murphy, P.G., Borthwick, L.S., Johnston, R.S., Kuchel, G., Richardson, P.M., 1999. Nature of the retrograde signal from injured nerves that induces interleukin-6 mRNA in neurons. J. Neurosci. 19, 3791–3800.

Naj, A.C., Jun, G., Beecham, G.W., Wang, L.S., Vardarajan, B.N., Buros, J., et al., 2011. Common variants at MS4A4/MS4A6E, CD2AP, CD33 and EPHA1 are associated with late-onset Alzheimer's disease. Nat. Genet. 43, 436–441.

Nakajima, K., Matsushita, Y., Tohyama, Y., Kohsaka, S., Kurihara, T., 2006. Differential suppression of endotoxin-inducible inflammatory cytokines by nuclear factor kappa B (NFκB) inhibitor in rat microglia. Neurosci. Lett. 401, 199–202.

Nimmerjahn, A., Kirchhoff, F., Helmchen, F., 2005. Resting microglial cells are highly dynamic surveillants of brain parenchyma in vivo. Science 308, 1314–1318.

Noguchi, Y., Shinozaki, Y., Fujishita, K., Shibata, K., Imura, Y., Morizawa, Y., et al., 2013. Astrocytes protect neurons against methylmercury via ATP/P2Y(1) receptor-mediated pathways in astrocytes. PLoS One 8, e57898.

Oberheim, N.A., Goldman, S.A., Nedergaard, M., 2012. Heterogeneity of astrocytic form and function. Methods Mol. Biol. 814, 23–45.

Olabarria, M., Noristani, H.N., Verkhratsky, A., Rodríguez, J.J., 2010. Concomitant astroglial atrophy and astrogliosis in a triple transgenic animal model of Alzheimer's disease. Glia 58, 831–838.

Orsini, F., De Blasio, D., Zangari, R., Zanier, E.R., De Simoni, M.G., 2014. Versatility of the complement system in neuroinflammation, neurodegeneration and brain homeostasis. Front. Cell. Neurosci. 8, 380.

Palm, N.W., Medzhitov, R., 2009. Pattern recognition receptors and control of adaptive immunity. Immunol. Rev. 227, 221–233.

Pankratov, Y., Lalo, U., Krishtal, O., Verkhratsky, A., 2002. Ionotropic P2X purinoreceptors mediate synaptic transmission in rat pyramidal neurones of layer II/III of somato-sensory cortex. J. Physiol. 542, 529–536.

Pannasch, U., Rouach, N., 2013. Emerging role for astroglial networks in information processing: from synapse to behavior. Trends Neurosci. 36, 405–417.

Paris, D., Humphrey, J., Quadros, A., Patel, N., Crescentini, R., Crawford, F., Mullan, M., 2003. Vasoactive effects of A beta in isolated human cerebrovessels and in a transgenic mouse model of Alzheimer's disease: role of inflammation. Neurol. Res. 25, 642–651.

Parkhurst, C.N., Yang, G., Ninan, I., Savas, J.N., Yates 3rd, J.R., et al., 2013. Microglia promote learning-dependent synapse formation through brain-derived neurotrophic factor. Cell 155, 1596–1609.

Parvathenani, L.K., Tertyshnikova, S., Greco, C.R., Roberts, S.B., Robertson, B., Posmantur, R., 2003. P2X$_7$ mediates superoxide production in primary microglia and is up-regulated in a transgenic mouse model of Alzheimer's disease. J. Biol. Chem. 278, 13309–13317.

Pascual, O., Ben Achour, S., Rostaing, P., Triller, A., Bessis, A., 2012. Microglia activation triggers astrocyte-mediated modulation of excitatory neurotransmission. Proc. Natl. Acad. Sci. U.S.A. 109, E197–E205.

Pavlov, V.A., Tracey, K.J., 2005. The cholinergic anti-inflammatory pathway. Brain Behav. Immun. 19, 493–499.

Pekny, M., Wilhelmsson, U., Pekna, M., 2014. The dual role of astrocyte activation and reactive gliosis. Neurosci. Lett. 565, 30–38.

Penninx, B.W., Kritchevsky, S.B., Newman, A.B., Nicklas, B.J., Simonsick, E.M., Rubin, S., et al., 2004. Inflammatory markers and incident mobility limitation in the elderly. J. Am. Geriatr. Soc. 52, 1105–1113.

Persidsky, Y., Ghorpade, A., Rasmussen, J., Limoges, J., Liu, X.J., Stins, M., et al., 1999. Microglial and astrocyte chemokines regulate monocyte migration through the blood–brain barrier in human immunodeficiency virus-1 encephalitis. Am. J. Pathol. 155, 1599–1611.

Phillis, J.W., Horrocks, L.A., Farooqui, A.A., 2006. Cyclooxygenases, lipoxygenases, and epoxygenases in CNS: their role and involvement in neurological disorders. Brain Res. Rev. 52, 201–243.

Pita-Almenar, J.D., Ranganathan, G.N., Koester, H.J., 2012a. Impact of cortical plasticity on patterns of suprathreshold activity in the cerebral cortex. J. Neurophysiol. 107, 850–858.

Pita-Almenar, J.D., Zou, S., Colbert, C.M., Eskin, A., 2012b. Relationship between increase in astrocytic GLT-1 glutamate transport and late-LTP. Learn. Mem. 19, 615–626.

Rachal Pugh, C., Fleshner, M., Watkins, L.R., Maier, S.F., Rudy, J.W., 2001. The immune system and memory consolidation: a role for the cytokine IL-1beta. Neurosci. Biobehav. Rev. 25, 29–41.

Ransohoff, R.M., Perry, V.H., 2009. Microglial physiology: unique stimuli, specialized responses. Annu. Rev. Immunol. 27, 119–145.

Reed-Geaghan, E.G., Savage, J.C., Hise, A.G., Landreth, G.E., 2009. CD14 and toll-like receptors 2 and 4 are required for fibrillar A{beta}-stimulated microglial activation. J. Neurosci. 29, 11982–11992.

Richard, K.L., Filali, M., Prefontaine, P., Rivest, S., 2008. Toll-like receptor 2 acts as a natural innate immune receptor to clear amyloid beta 1–42 and delay the cognitive decline in a mouse model of Alzheimer's disease. J. Neurosci. 28, 5784–5793.

Ries, M., Sastre, M., 2016. Mechanisms of Aβ clearance and degradation by glial cells. Front. Aging Neurosci. 8, 160.

Rivest, S., 2015. TREM2 enables amyloid β clearance by microglia. Cell Res. 25, 535–536.

Rock, R.B., Gekker, G., Hu, S., Sheng, W.S., Cheeran, M., Lokensgard, J.R., et al., 2004. Role of microglia in central nervous system infections. Clin. Microbiol. Rev. 17, 942–964.

Rodriguez-Vieitez, E., Ni, R., Gulyás, B., Tóth, M., Häggkvist, J., et al., 2015. Astrocytosis precedes amyloid plaque deposition in Alzheimer APPswe transgenic mouse brain: a correlative positron emission tomography and in vitro imaging study. Eur. J. Nucl. Med. Mol. Imaging 42, 1119–1132.

Rossi, A.G., Sawatzky, D.A., Walker, A., Ward, C., Sheldrake, T.A., Riley, N.A., et al., 2006. Cyclin-dependent kinase inhibitors enhance the resolution of inflammation by promoting inflammatory cell apoptosis. Nat. Med. 12, 1056–1064.

Sagare, A.P., Bell, R.D., Zlokovic, B.V., 2013. Neurovascular defects and faulty amyloid-beta vascular clearance in Alzheimer's disease. J. Alzheimers Dis. 33 (Suppl. 1), S87–S100.

Sanz, J.M., Chiozzi, P., Ferrari, D., Colaianna, M., Idzko, M., Falzoni, S., et al., 2009. Activation of microglia by amyloid {beta} requires P2X7 receptor expression. J. Immunol. 182, 4378–4385.

Sasaki, A., Kakita, A., Yoshida, K., Konno, T., Ikeuchi, T., Hayashi, S., Matsuo, H., Shioda, K., 2015a. Variable expression of microglial DAP12 and TREM2 genes in Nasu-Hakola disease. Neurogenetics 71, 449–453.

Sasaki, A., Kakita, A., Yoshida, K., Konno, T., Ikeuchi, T., et al., 2015b. Variable expression of microglial DAP12 and TREM2 genes in Nasu-Hakola disease. Neurogenetics 16, 265–276.

Schiff, L., Hadker, N., Weiser, S., Rausch, C., 2012. A literature review of the feasibility of glial fibrillary acidic protein as a biomarker for stroke and traumatic brain injury. Mol. Diagn. Ther. 16, 79–92.

Scholtzova, H., Kascsak, R.J., Bates, K.A., Boutajangout, A., Kerr, D.J., Meeker, H.C., et al., 2009. Induction of toll-like receptor 9 signaling as a method for ameliorating Alzheimer's disease-related pathology. J. Neurosci. 29, 1846–1854.

Schroder, K., Tschopp, J., 2010. The inflammasomes. Cell 140, 821–832.

Schöll, M., Carter, S.F., Westman, E., Rodriguez-Vieitez, E., Almkvist, O., Thordardottir, S., et al., November 10, 2015. Early astrocytosis in autosomal dominant Alzheimer's disease measured in vivo by multi-tracer positron emission tomography. Sci. Rep. 5, 16404.

Schwab, J.M., Chiang, N., Arita, M., Serhan, C.N., 2007. Resolvin E1 and protectin D1 activate inflammation-resolution programmes. Nature 447, 869–874.

Scuderi, C., Esposito, G., Blasio, A., Valenza, M., Arietti, P., Steardo Jr., L., et al., 2011. Palmitoylethanolamide counteracts reactive astrogliosis induced by β-amyloid peptide. J. Cell. Mol. Med. 15, 2664–2674.

Serhan, C.N., Chiang, N., Van Dyke, T.E., 2008. Resolving inflammation: dual anti-inflammatory and pro-resolution lipid mediators. Nat. Rev. Immunol. 8, 349–361.

Serhan, C.N., 2009. Systems approach to inflammation resolution: identification of novel anti-inflammatory and pro-resolving mediators. J. Thromb. Haemost. 7 (Suppl. 1), 44–48.

Serhan, C.N., 2010. Novel lipid mediators and resolution mechanisms in acute inflammation: to resolve or not? Am. J. Pathol. 177, 1576–1591.

Shen, Y., Meri, S., 2003. Yin and Yang: complement activation and regulation in Alzheimer's disease. Prog. Neurobiol. 70, 463–472.

Sheridan, G.K., Murphy, K.J., 2013. Neuron-glia crosstalk in health and disease: fractalkine and CX3CR1 take centre stage. Open Biol. 3, 130181.

Sheridan, G.K., Wdowicz, A., Pickering, M., et al., 2014. CX3CL1 is up-regulated in the rat hippocampus during memory-associated synaptic plasticity. Front. Cell. Neurosci. 8, 233.

Shearer, K.D., Fragoso, Y.D., Clagett-Dame, M., McCaffery, P.J., 2012. Astrocytes as a regulated source of retinoic acid for the brain. Glia 60, 1964–1976.

Shibata, N., Yamamoto, T., Hiroi, A., Omi, Y., Kato, Y., Kobayashi, M., 2010. Activation of STAT3 and inhibitory effects of pioglitazone on STAT3 activity in a mouse model of SOD1-mutated amyotrophic lateral sclerosis. Neuropathology 30, 353–360.

Simard, A.R., Soulet, D., Gowing, G., Julien, J.P., Rivest, S., 2006. Bone marrow-derived microglia play a critical role in restricting senile plaque formation in Alzheimer's disease. Neuron 49, 489–502.

Skaper, S.D., Giusti, P., Facci, L., 2012. Microglia and mast cells: two tracks on the road to neuroinflammation. FASEB J. 26, 3103–3117.

Smith, M.A., Richey Harris, P.L., Sayre, L.M., Beckman, J.S., Perry, G., 1997. Widespread peroxynitrite-mediated damage in Alzheimer's disease. J. Neurosci. 17, 2653–2657.

Sofroniew, M.V., 2009. Molecular dissection of reactive astrogliosis and glial scar formation. Trends Neurosci. 32, 638–647.

Sofroniew, M.V., Vinters, H.V., 2010. Astrocytes: biology and pathology. Acta Neuropathol. 119, 7–35.

Solenski, N.J., Kostecki, V.K., Dovey, S., Periasamy, A., 2003. Nitric-oxide-induced depolarization of neuronal mitochondria: implications for neuronal cell death. Mol. Cell. Neurosci. 24, 1151–1169.

Stalder, A.K., Ermini, F., Bondolfi, L., Krenger, W., Burbach, G.J., Deller, T., Coomaraswamy, J., Staufenbiel, M., Landmann, R., Jucker, M., 2005. Invasion of hematopoietic cells into the brain of amyloid precursor protein transgenic mice. J. Neurosci. 25, 11125–11132.

Stoltzner, S.E., Grenfell, T.J., Mori, C., Wisniewski, K.E., Wisniewski, T.M., et al., 2000. Temporal accrual of complement proteins in amyloid plaques in Down's syndrome with Alzheimer's disease. Am. J. Pathol. 156, 489–499.

Strohmeyer, R., Shen, Y., Rogers, J., 2000. Detection of complement alternative pathway mRNA and proteins in the Alzheimer's disease brain. Brain Res. Mol. Brain Res. 81, 7–18.

Sun, J.D., Liu, Y., Yuan, Y.H., Li, J., Chen, N.H., 2012. Gap junction dysfunction in the prefrontal cortex induces depressive-like behaviors in rats. Neuropsychopharmacology 37, 1305–1320.

Szmydynger-Chodobska, J., Strazielle, N., Gandy, J.R., Keefe, T.H., Zink, B.J., Ghersi-Egea, J.F., Chodobski, A., 2012. Posttraumatic invasion of monocytes across the blood-cerebrospinal fluid barrier. J. Cereb. Blood Flow Metab. 32, 93–104.

Takenouchi, T., Sato, M., Kitani, H., 2007. Lysophosphatidylcholine potentiates Ca^{2+} influx, pore formation and p44/42 MAP kinase phosphorylation mediated by P2X7 receptor activation in mouse microglial cells. J. Neurochem. 102, 1518–1532.

Talley Watts, L., Sprague, S., Zheng, W., Garling, R.J., Jimenez, D., Digicaylioglu, M., et al., 2013. Purinergic 2Y1 receptor stimulation decreases cerebral edema and reactive gliosis in a traumatic brain injury model. J. Neurotrauma 30, 55–66.

Tan, D.X., Reiter, R.J., Manchester, L.C., Yan, M.T., El-Sawi, M., et al., 2002. Chemical and physical properties and potential mechanisms: melatonin as a broad-spectrum antioxidant and free radical scavenger. Curr. Top. Med. Chem. 2, 181–198.

Tansey, M.G., McCoy, M.K., Frank-Cannon, T.C., 2007. Neuroinflammatory mechanisms in Parkinson's disease: potential environmental triggers, pathways, and targets for early therapeutic intervention. Exp. Neurol. 208, 1–25.

Tesseur, I., Zou, K., Esposito, L., Bard, F., Berber, E., et al., 2006. Deficiency in neuronal TGF-β signaling promotes neurodegeneration and Alzheimer's pathology. J. Clin. Invest. 116, 3060–3069.

Theodosis, D.T., Poulain, D.A., Oliet, S.H., 2008. Activity-dependent structural and functional plasticity of astrocyte-neuron interactions. Physiol. Rev. 88, 983–1008.

Tichauer, J., Saud, K., von Bernhardi, R., 2007. Modulation by astrocytes of microglial cell-mediated neuroinflammation: effect on the activation of microglial signaling pathways. Neuroimmunomodulation. 14, 168–174.

Trapp, B.D., Wujek, J.R., Criste, G.A., Jalabi, W., Yin, X., et al., 2007. Evidence for synaptic stripping by cortical microglia. Glia 55, 360–368.

Tremblay, M.E., Lowery, R.L., Majewska, A.K., 2010. Microglial interactions with synapses are modulated by visual experience. PLoS Biol. 8, e1000527.

Tremblay, M.E., Zettel, M.L., Ison, J.R., Allen, P.D., Majewska, A.K., 2012. Effects of aging and sensory loss on glial cells in mouse visual and auditory cortices. Glia 60, 541–558.

Tuppo, E.E., Arias, H.R., 2005. The role of inflammation in Alzheimer's disease. Int. J. Biochem. Cell Biol. 37, 289–305.

Uller, L., Persson, C.G., Erjefalt, J.S., 2006. Resolution of airway disease: removal of inflammatory cells through apoptosis, egression or both? Trends Pharmacol. Sci. 27, 461–466.

Ullian, E.M., Sapperstein, S.K., Christopherson, K.S., Barres, B.A., 2001. Control of synapse number by glia. Science 291, 657–661.

Vernadakis, A., 1996. Glia-neuron intercommunications and synaptic plasticity. Prog. Neurobiol. 49, 185–214.

Verkhratsky, A., Parpura, V., Pekna, M., Pekny, M., Sofroniew, M., 2014. Glia in the pathogenesis of neurodegenerative diseases. Biochem. Soc. Trans. 42, 1291–1301.

Verkhratsky, A., Rodríguez-Arellano, J.J., Parpura, V., Zorec, R., 2016. Astroglial calcium signalling in Alzheimer's disease. Biochem. Biophys. Res. Commun. pii: S0006-291X(16)31345-6.

von Bernhardi, R., Eugenín-von Bernhardi, L., Eugenín, J., 2015a. Microglial cell dysregulation in brain aging and neurodegeneration. Front. Aging Neurosci. 7, 124.

von Bernhardi, R., Cornejo, F., Parada, G.E., Eugenín, J., 2015b. Role of TGFβ signaling in the pathogenesis of Alzheimer's disease. Front. Cell. Neurosci. 9, 426.

Wake, H., Moorhouse, A.J., Jinno, S., Kohsaka, S., Nabekura, J., 2009. Resting microglia directly monitor the functional state of synapses in vivo and determine the fate of ischemic terminals. J. Neurosci. 29, 3974–3980.

Wirths, O., Breyhan, H., Marcello, A., Cotel, M.C., Brück, W., Bayer, T.A., 2010. Inflammatory changes are tightly associated with neurodegeneration in the brain and spinal cord of the APP/PS1KI mouse model of Alzheimer's disease. Neurobiol. Aging 31, 747–757.

Wolf, F., Kirchhoff, F., 2008. Neuroscience. Imaging astrocyte activity. Science 320, 1597–1599.

Wolf, Y., Yona, S., Kim, K.-W., Jung, S., 2013. Microglia, seen from the CX3CR1 angle. Front. Cell. Neurosci. 7, 26.

Wood, P.L., 1998. Neuroinflammation: Mechanisms and Management. Humana Press, Totowa, New Jersey.

Wunderlich, P., Glebov, K., Kemmerling, N., Tien, N.T., Neumann, H., Walter, J., 2013. Sequential proteolytic processing of the triggering receptor expressed on myeloid cells-2 (TREM2) protein by ectodomain shedding and gamma-secretase-dependent intramembranous cleavage. J. Biol. Chem. 288, 33027–33036.

Wyss-Coray, T., Lin, C., Sanan, D.A., Mucke, L., Masliah, E., 2000. Chronic overproduction of transforming growth factor-β1 by astrocytes promotes Alzheimer's disease–like microvascular degeneration in transgenic mice. Am. J. Pathol. 156, 139–150.

Wyss-Coray, T., Lin, C., Yan, F., Yu, G.Q., Rohde, M., McConlogue, L., et al., 2001. TGF-β1 promotes microglial amyloid-β clearance and reduces plaque burden in transgenic mice. Nat. Med. 7, 612–618.

Wyss-Coray, T., Loike, J.D., Brionne, T.C., Lu, E., Anankov, R., Yan, F., et al., 2003. Adult mouse astrocytes degrade amyloid-beta in vitro and in situ. Nat. Med. 9, 453–457.

Wyss-Coray, T., 2006. Inflammation in Alzheimer disease: driving force, bystander or beneficial response? Nat. Med. 12, 1005–1015.

Yan, S.D., Zhu, H., Fu, J., et al., 1997. Amyloid-β peptide-receptor for advanced glycation endproduct interaction elicits neuronal expression of macrophage-colony stimulating factor: a proinflammatory pathway in Alzheimer disease. Proc. Natl. Acad. Sci. U.S.A. 94, 5296–5301.

Yang, J., Ruchti, E., Petit, J.M., Jourdain, P., Grenningloh, G., Allaman, I., Magistretti, P.J., 2014. Lactate promotes plasticity gene expression by potentiating NMDA signaling in neurons. Proc. Natl. Acad. Sci. U.S.A. 111, 12228–12233.

Ye, S.M., Johnson, R.W., 2001. An age-related decline in interleukin-10 may contribute to the increased expression of interleukin-6 in brain of aged mice. Neuroimmunomodulation. 9, 183–192.

Yeh, C.Y., Vadhwana, B., Verkhratsky, A., Rodríguez, J.J., 2011. Early astrocytic atrophy in the entorhinal cortex of a triple transgenic animal model of Alzheimer's disease. ASN Neuro 3, 271–279.

Yermakova, A., O'Banion, M.K., 2000. Cyclooxygenases in the central nervous system: implications for treatment of neurological disorders. Curr. Pharm. Des. 6, 1755–1776.

Yoon, S.S., Jo, S.A., 2012. Mechanisms of amyloid-β peptide clearance: potential therapeutic targets for Alzheimer's disease. Biomol. Ther. (Seoul). 20, 245–255.

Yu, Y., Ye, R.D., 2015. Microglial Aβ receptors in Alzheimer's disease. Cell. Mol. Neurobiol. 35, 71–83.

Zhang, M., Xu, G., Liu, W., Ni, Y., Zhou, W., 2012. Role of fractalkine/CX3CR1 interaction in light-induced photoreceptor degeneration through regulating retinal microglial activation and migration. PLoS One 7, e35446.

Zhao, Y., Bhattacharjee, S., Jones, B.M., Hill, J., Dua, P., Lukiw, W.J., 2014. Regulation of neurotropic signaling by the inducible, NF-kB-sensitive miRNA-125b in Alzheimer's disease (AD) and in primary human neuronal-glial (HNG) cells. Mol. Neurobiol. 50, 97–106.

Zhao, Y., Lukiw, W.J., 2015. Microbiome-generated amyloid and potential impact on amyloidogenesis in Alzheimer's disease (AD). J. Nat. Sci. 1, e138.

Ziv, Y., Ron, N., Butovsky, O., Landa, G., Sudai, E., Greenberg, N., et al., 2006. Immune cells contribute to the maintenance of neurogenesis and spatial learning abilities in adulthood. Nat. Neurosci. 9, 268–275.

Zlokovic, B.V., 2008. The blood–brain barrier in health and chronic neurodegenerative disorders. Neuron 57, 178–201.

Biomarkers for Alzheimer's Disease

INTRODUCTION

Alzheimer's disease (AD) is an irreversible age-related progressive neurodegenerative disease, which is accompanied by many different pathophysiological features like impairment of cognitive domains, a characteristic pathological cortical and hippocampal atrophy, histological feature of senile plaques composed of amyloid deposits and neurofibrillary tangles (NFTs) consisting of intraneuronal Tau fibrillary tangles, and a resultant decrease in neurons along with and neuropsychiatric disturbances. AD is also accompanied by biochemical changes like abnormalities of cholesterol metabolism, onset of neuroinflammation, and induction of oxidative stress and lysosomal dysfunction. Clinical diagnosis of AD remains difficult in initial stages. Currently, the average life expectancy has increased dramatically, but the healthy life span has not increased with the same pace. The increase of life expectancy has resulted in increased occurrence of age-related neurodegenerative diseases including AD, which has developed into a major health-care problem. Thus AD has become the most common cause of dementia in the elderly. It is estimated that in 2050, approximately 80 million people will suffer from AD worldwide. The pathological hallmark of AD is the extracellular accumulation of amyloid plaques, which are composed of β-amyloid (Aβ), a 4-kDa peptide, which has a tendency to aggregate. Aβ is derived from the cleavage of the amyloid precursor protein (APP). Another hallmark of AD is intraneuronal accumulation of NFTs, which are made up of hyperphosphorylated Tau protein (Liu et al., 2013). As stated in Chapter 1, in the healthy brain, APP plays a role in modulation of neurogenesis, Ca^{2+} homeostasis, neurotransmission, and the establishment of synaptic network (Löffler and Huber, 1992). Accumulation of Aβ plaques contributes to the disruption of cellular activities and communication in the brain, leading to oxidative stress, neurotoxic neuroinflammation, and neuronal death (Fig. 7.1). In healthy neurons, Tau is an integral component of microtubules, which are the internal support structures associated with the transport of nutrients, vesicles, mitochondria, and chromosomes from the cell body to the ends of the axon and backward. In AD, however, Tau becomes hyperphosphorylated. This hyperphosphorylation allows Tau to bind together and form tangled threads, which deposit as NFTs (Braak et al., 1994). The severity of NFT accumulation correlates well with neuronal loss and dementia in patients with AD (Farooqui, 2010). Although mutations in certain genes cause familial AD (FAD), more than 90% of patients have sporadic AD, suggesting that aging is the biggest risk factor for AD. Aβ accumulation in amyloid plaques leads to oxidative stress and chronic neuroinflammation in the brain, thereby contributing to disease progression and poor functional outcomes (Rubio-Perez and Morillas-Ruiz, 2012). A reduction in suppressor cell function in the periphery has also been observed in patients with AD, as manifested by loss of balance in immune cell populations and decreased interleukin-10 production in the blood (Guerreiro et al., 2007). Therefore activation of immunoregulatory mechanisms during the progression of AD may contribute to reestablishment of immune homeostasis. The prevalence of AD is expected to increase dramatically as the population around the

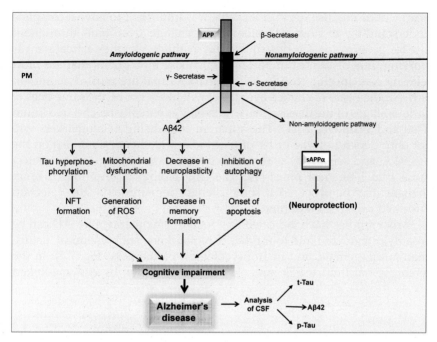

FIGURE 7.1 **APP processing pathway stimulating Aβ plaque and NFT deposition in Alzheimer's disease.** *APP*, amyloid precursor protein; *Aβ42*, β-amyloid; *NFTs*, neurofibrillary tangles; *PM*, plasma membrane; *p-Tau*, phosphorylated Tau; *ROS*, reactive oxygen species; *t-Tau*, total Tau.

globe continues to age (Kim and Hwang, 2016). Early diagnosis of AD, however, is essential not only for the development of therapeutics but also for comprehensive management of the disease. It is proposed that AD can be treated most effectively in preclinical stages, before cognitive functions become impaired and neurons and synapses are damaged irreversibly (Golde et al., 2011; Kim and Hwang, 2016). Currently, one of the major handicaps toward identifying patients with AD is the difficulty in early and definitive diagnosis of this disease. However, a probable diagnosis of AD can be made with a confidence of >90%, based on clinical criteria, including medical history, physical examination, laboratory tests, neuroimaging, and neuropsychological evaluation. To this end, it is important to establish reliable biomarkers to diagnose and monitor AD progression. The main problem in developing an ideal biomarker for AD has been the slow understanding of the pathogenesis of AD, unavailability of histopathological and biochemical diagnosis during patient's lifetime, occurrence of large overlap with other types of dementia [dementia with Lewy bodies (DLB) and vascular dementia (VaD)] and their influence on the different biomarkers, and lack of information on progression and treatment of the disease. Furthermore, it is important to have a complete understanding of how AD biomarkers change over time throughout disease progression. To determine and evaluate whether a treatment is working, it is of utmost importance to choose biomarkers that are most relevant and specific to AD. The biomarkers that are most dynamic (in terms of their rate of change over time) will likely not be the same ones at various stages of the disease, since each may eventually reach a maximum (Fjell and Walhovd, 2011). The apparent sequencing of biomarkers will not only depend on the order of true biological changes but also on the precision and sensitivity with which the assessment methods can detect those true biological processes. The true biological ordering of brain changes may be obscured if some changes remain below the detection threshold for a period of time.

Earlier studies have indicated that tentative biomarkers for AD can be broadly categorized into four types of biomarkers: neurochemical, neuroanatomical, genetic, and neuropsycological biomarkers (Fig. 7.2). In the cerebrospinal fluid (CSF), Aβ analysis and detection by enzyme-linked

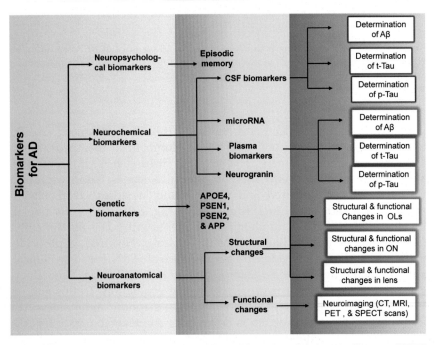

FIGURE 7.2 **Classification of potential biomarkers for Alzheimer's disease.** *APOE4,* apolipoprotein E4; *APP,* amyloid precursor protein; *Aβ42,* β-amyloid; *CSF,* cerebrospinal fluid; *OL,* olfactory lobes; *ON,* optic nerve; *PET,* positron emission tomography; *PSEN1 and PSEN2,* presenilin proteins; *p-Tau,* phosphorylated Tau; *SPECT,* single photon emission computed tomography; *t-Tau,* total Tau.

immunosorbent assays (ELISA) and cerebral Pittsburgh compound B positron emission tomography (PiB-PET) examination reflects abnormal amyloid metabolism in brains of patients with AD (McKhann et al., 2011; Rosenmann, 2012). Magnetic resonance imaging (MRI) at increasing field strength and resolution detects and provides the evolution of distinct types of structural and functional abnormality pattern throughout early to late AD stages. Anatomical or volumetric MRI is the most widely used technique, which gives information on local and global measures of atrophy. The use of advanced image analysis techniques provides information not only on the hippocampus but also on alterations in fiber tract and neural network disintegration that may substantially contribute to early detection and the mapping of AD progression. The use of molecular in vivo amyloid imaging agents (the fifth validated biomarker), such as the Pittsburgh compound B, and markers of neurodegeneration, such as fluoro-2-deoxy-D-glucose (FDG) (as the sixth validated biomarker), support the detection of early AD pathological processes and associated neurodegeneration. Information on AD biomarkers can be combined to increase accuracy or risk. AD is accompanied by the activation of microglia, which leads to increase in expression of mitochondrial protein called 18-kDa translocator protein (TSPO). This protein can be imaged using (R)-[11C] PK11195 (Yokokura et al., 2011; Schuitemaker et al., 2013). The imaging of TSPO has been confirmed, but this increase has not been detected in individual patients and is much weaker than the signal on amyloid PET (Wiley et al., 2009). In addition, (R)-[11C]PK11195 has not been able to separate patients with clinically stable prodromal AD from those who progressed to dementia, and there is no correlation with cognitive function. The discovery of several new TSPO ligands (Venneti et al., 2013), has resulted in development and identification of new drug targets for the treatment of AD (Chua et al., 2014). In particular, studies using [11C] PBR28 have shown a signal, which correlates with cognitive performance (Kreisl et al., 2013), providing a means for detecting changes early in the pathogenesis of AD. However, a major disadvantage of many new TSPO ligands is that due to genetic polymorphism (Owen et al., 2012), a subpopulation of subjects will not show binding. So there is a need for TSPO ligands that provide high signal, but are insensitive to this polymorphism.

The Alzheimer's Association's Research Roundtable held on May 2014 has indicated that some progress has been made on developing biomarkers for AD and these biomarkers have been useful for clinical trials, but there is a clear need to develop novel biomarkers that are minimally invasive and that more broadly characterize underlying pathogenic mechanisms, including neurodegeneration, neuroinflammation, and synaptic dysfunction. So major attempts have also been made in developing tools to enable noninvasive detection of amyloid plaques (e.g., through the skull) in living patients with AD and animal models (Nakada et al., 2008;

Ng et al., 2007). However, such noninvasive monitoring of Aβ plaques is still clinically challenging and is of limited resolution and availability (Klunk et al., 2005; Lockhart et al., 2007). Optical detection (neuroimaging) constitutes a powerful, high-resolution, and specific tool for in vivo imaging as reported using multiphoton microscopy to detect Aβ plaques in the mouse brain via an invasive cranial window (Meyer-Luehmann et al., 2008; Javaid et al., 2016). The Alzheimer's Disease Neuroimaging Initiative is a unique ongoing, longitudinal, multicenter study that is designed to develop all types (clinical, imaging, genetic, and biochemical) of biomarkers for the early detection and monitoring of AD (Weiner et al., 2012). Many articles have been published through this initiative. However, an ideal biomarker remains to be discovered, and in the light of limitations of β-amyloid hypothesis, more studies are required on early detection and monitoring of AD in large human population.

BIOMARKERS FOR ALZHEIMER'S DISEASE IN THE CEREBROSPINAL FLUID

CSF assays for Aβ detect soluble monomeric species of Aβ (Aβ38, Aβ40, and Aβ42). In AD, levels of CSF Aβ40 remain stable, whereas Aβ42 levels have consistently been reported to drop by <50% of normal age-matched control subjects (Zetterberg et al., 2010). This is due to its deposition in the senile plaques. Generation of Aβ42 is an early event in AD, hence measuring CSF Aβ42 is a very relevant strategy in prodromal AD to screen for early cases as well as for monitoring disease progression. However, the strategy to measure Aβ42 in the CSF only provides a supplementary test to support the diagnosis once cognitive dysfunction is apparent and gives little information on the progression of AD as this biomarker has already found a steady state of abnormality early in the disease progression (Hansson et al., 2007). Changes in Aβ42 levels have also been studied for other types of dementia, and it is reported that levels of Aβ42 are slightly decreased in frontotemporal lobar degeneration (FTLD), DLB, and VaD (Hansson et al., 2009). Levels of CSF Aβ42 have been shown to predict the rate of cognitive decline in patients with very mild dementia and predict AD in subjects with mild cognitive impairment (MCI) (Snider et al., 2009). With the development of amyloid PET imaging, the relationship between low CSF Aβ42 levels and amyloid plaque has been established in vivo in several studies (Fagan et al., 2006; Weigand et al., 2011; Tolboom et al., 2009). The quantitative detection of anatomical and structural features of AD can also be visualized by conventional radiography, ultrasonography, computed tomographic (CT) scanning (e.g., PET) or MRI, including functional MRI (fMRI) scans along with hypometabolism in AD with FDG-PET and amyloid deposition with amyloid-PET imaging.

Among the aforementioned techniques, functional imaging like PET, single photon emission computed tomography (SPECT), fMRI, and proton magnetic resonance spectroscopy provide information on detecting and characterizing the regional changes in cerebral blood flow, metabolism, and receptor binding sites that are associated with the pathogenesis of AD. These neuroimaging techniques along with molecular probes have been used to visualize the cellular activity and chemical processes involved in metabolism, oxygen distribution, or blood flow. Injection of imaging agent into the body results in distribution and accumulation of imaging agent at the site, where abnormal metabolic processes are taking place (Craig-Schapiro et al., 2009; Jiang et al., 2014). Collective evidence suggests that advancements in PET and SPECT technologies not only have applications in neurochemical imaging of neurotransmitters but can also be utilized as noninvasive tools for the diagnosis of neurological syndromes (AD) with high specificity when compared with previous procedures (Apurva et al., 2013). The key advantage of molecular imaging techniques is due to its ability to elucidate sophisticated biological phenomenon at the cellular and molecular level, linking investigations to specific pathologies (Weissleder and Mahmood, 2001). It must be stated that all biomarkers of AD may not reach abnormal levels or peak simultaneously but do so in an ordered manner because in AD neurochemical changes may develop decades before the earliest clinical symptoms. The CSF is largely produced by the choroid plexus. It is found in the ventricular system and subarachnoid spaces surrounding the brain and spinal cord. It is an ideal source of biomarkers for neurological diseases because CSF is in direct contact with the brain, and the molecular composition of CSF can reflect biochemical changes that occur in the brain following the onset of AD (Chintamaneni and Bhaskar, 2012). The use of amyloid imaging with N-methyl-[^{11}C]2-(4-methylaminophenyl)-6-hydroxybenzothiazole (Pittsburgh compound B, or [^{11}C]PIB) has also been proposed (Ewers et al., 2011; Sperling et al., 2011). This method detects AD pathology early in the course of disease (Fagan et al., 2006) and helps distinguishing AD from other types of dementia (Engler et al., 2008). [^{11}C]PIB, and analogs of PIB, detect amyloid plaques, mainly consisting of insoluble fibrils of Aβ. However, the load of insoluble Aβ does not correlate well with disease progression (Engler et al., 2006). The measurement of insulin resistance of the hippocampal/temporal lobe by PET with FDG integrated with CT (FDG-PET/CT) has become, besides amyloid-PET diagnostics (Perani et al., 2014), a highly specific and sensitive biomarker for AD, having predictive value even decades before the first clinical symptoms of AD manifest themselves (O'Brien et al., 2014).

As stated in Chapter 1, double Aβ42 is a better marker of disease status, and many therapeutic as well as diagnostic efforts are currently targeting soluble Aβ42 aggregates. Attempts are underway to develop an imaging agent that can visualize soluble Aβ42 aggregates. In addition to the aforementioned efforts, decreased glucose metabolism in temporal and parietal cortex is considered as a biomarker of synaptic dysfunction. Similarly, brain atrophy in the medial temporal lobes as well as paralimbic, temporal, and parietal cortex on structural MRI is a biomarker of AD-related neurodegeneration. Based on the above information, a biomarker model has been proposed. This model suggests abnormalities of biomarkers in AD occur in an ordered manner, which parallels the hypothetical pathophysiological sequence of AD and is particularly relevant to tracking the preclinical stages of AD (Sperling et al., 2011; Jack et al., 2010). There is now the consensus that only the combination of the CSF biomarkers mentioned earlier significantly increases the diagnostic validity for sporadic AD (Blennow, 2010; Marksteiner, 2007) (Table 7.1). The sensitivity and specificity of CSF Aβ42 for distinguishing AD from other dementias are estimated to be 73% and 67%, respectively. For total CSF Tau the sensitivity and specificity have been estimated to be 70%–75% and 74%–90% and for p-Tau, 79%–88% and 78%–83%, respectively, for separating AD from other dementias (Van Harten et al., 2011). These

TABLE 7.1   Biomarkers for Alzheimer's Disease in the CSF

Biomarker	Effect	References
Aβ	Decreased	Hoglund et al. (2008) and Zetterberg et al. (2010)
Total Tau	Increased	Hampel et al. (2004) and Thal et al. (2006)
Hyperphosphorylated Tau	Increased	Hampel et al. (2004) and Thal et al. (2006)
APOE4	Increased	Lautner et al. (2014) and Monge-Argilés et al. (2016)
TREM2	Increased	Heslegrave et al. (2016)
Neurogranin	Increased	Janelidze et al. (2015) and Tarawneh et al. (2016)
YKL-40	Increased	Janelidze et al. (2016)
VLP1	Increased	Tarawneh et al. (2015)
Resistin	Increased	Hu et al. (2010)
Thrombospondin-1	Increased	Hu et al. (2010)
Growth-associated protein	Increased	Sjögren et al. (2001)
F2-isoprostane	Increased	Quinn et al. (2004)
27-Hydroxycholesterol	Increased	Mateos et al. (2011)

APOE4, apolipoprotein 4; *Aβ*, β-amyloid; *CSF*, cerebrospinal fluid; *VLP1*, β-amyloid.

biomarkers can provide information about the neuronal activity in the brain. Low levels of biomarkers can also be detected by mass spectrometry (MS)-based proteomic technologies (Varghese et al., 2013; Kim and Hwang, 2016). Advanced MS-based proteomic technology has facilitated the generation of proteome and posttranslational modifications profiles in tissue and body fluid (plasma/serum and CSF) samples collected not only from patients with AD but also from patients with a broad spectrum of neurodegenerative diseases. Other biomarkers of AD include neurogranin (a postsynaptic neuronal protein), resistin (a marker for preferential microglial activation), thrombospondin-1 (a key molecule in astrocyte-induced neurogenesis), growth-associated protein (GAP43) (Table 7.1), cellular immune responses (autoantibodies), genetic traits [apolipoprotein E (APOE4)], histologic characteristics of diseased tissue, and proteins or RNA expressed in tissues (Hu et al., 2010; Sjögren et al., 2001; Janelidze et al., 2015; Tarawneh et al., 2016). It is also reported that soluble triggering receptor expressed on myeloid cells 2 (TREM2), a receptor glycoprotein of 230 amino acids of microglial origin, is increased in CSF of patients with AD. Soluble TREM2 can also be used as biomarker for AD (Heslegrave et al., 2016). Thus the use of biomarkers for AD is becoming increasingly intrinsic for medical practice (Varghese et al., 2013; Kim and Hwang, 2016). It is also becoming increasingly evident that insulin resistance contributes not only to the production of Aβ and its accumulation (Ho et al., 2004; Stöhr et al., 2013), and induction of Tau pathology (Špolcová et al., 2014), but also to impairment in synaptic transmission (Nisticò et al., 2012) and induction of neurodegeneration. Insulin resistance also enhances β- and γ-secretase activity inducing Aβ production (Ho et al., 2004). One study has indicated the utility of evaluating indices of insulin resistance and their consequences, that is, mitochondrial dysfunction, oxidative stress, neuroinflammation, and reduction in neuronal plasticity combined with p-Tau and Aβ in CSF-based multiplex assays (Lee et al., 2013). In the brain, insulin resistance can contribute to Aβ and Tau pathology by means of oxidative stress and inflammation. In turn, Aβ accumulation can enhance insulin resistance through Aβ-mediated inflammation and oxidative stress. Additionally, hyperinsulinemia and hyperglycemia caused by insulin resistance may also increase the formation of neuropathologic changes in AD brain (Matsuzaki et al., 2010), and preclinical and clinical studies support the view that insulin injections can be beneficial for the treatment of AD. Thus the insulin level may also be a novel biomarker in AD. The mentioned biomarkers can be used not only to assess the progression of AD in all stages of AD (early, middle, and advanced stages) pathogenesis but also to guide clinical treatment. It is proposed that the inclusion of metabolic biomarker (insulin and insulin resistance) in Aβ42, total Tau (t-Tau), and phosphorylated Tau (p-Tau) in CSF-based panels may provide more information about the status and progression of neurodegeneration, as well as aid in predicting progression from early- to late-stage AD, suggesting that the magnitude of biomarkers can be followed constantly by monitoring the progression of AD. Accumulating evidence from both genetic at-risk individuals and clinically normal older cohorts suggests that the pathophysiological process of AD begins one to two decades before the emergence of the clinical manifestation of dementia (Holtzman et al., 2011).

β-AMYLOID AS A BIOMARKER FOR ALZHEIMER'S DISEASE

As stated in Chapter 1, AD is characterized by the accumulation of extracellular Aβ plaque depositions and intraneuronal accumulation of hyperphosphorylated Tau in the form of NFTs. Aβ is cleaved from the large APP by β and γ-secretases (amyloidogenic pathways). This pathway produces a 42-amino-acid peptide (Aβ 1–42). This peptide aggregates in the brain under certain conditions (e.g., acidosis, metals). One of the most commonly used methods for determining the concentration of Aβ42 in the CSF is ELISA. Initially, ELISAs for the detection of Aβ42 are developed for monomeric proteins and may be ill-suited for detecting Aβ aggregates. Therefore, investigators have used ELISAs for the measurement of Aβ42 aggregates. In these assays Aβ42 aggregates are disaggregated with trifluoroacetic acid and then total Aβ42 concentration is determined in the body fluids. Analysis of CSF Aβ42 by ELISA shows a highly significant reduction in patients with AD compared to controls, with a cutoff of <500 pg/mL. The decrease in Aβ42 levels in the CSF is due to reduction in clearance of Aβ42 from the brain to the blood/CSF, as well as enhanced aggregation and deposition of senile plaques in the brain. The main drawbacks of Aβ42 oligomer–based ELISAs in body fluids are the very low levels of Aβ42 oligomers and the high background of monomeric Aβ in the CSF (Rosén et al., 2013). To overcome these drawbacks, investigators have designed surface-based fluorescence intensity distribution analysis (sFIDA) (Birkmann et al., 2007; Funke et al., 2010; Bannach et al., 2012). The methodology of sFIDA resembles a conventional sandwich ELISA. In sFIDA, all Aβ42 species are immobilized on a functionalized glass surface via Aβ-specific capture antibodies. After immobilization, Aβ42 aggregates are multiply loaded by at least two detection antibodies, each of them labeled with a different "fluorochrome" (Kühbach et al., 2016). Because capture and detection antibodies recognize the same or an overlapping epitope on Aβ42, Aβ42 monomers cannot bind any detection antibodies while bound to the capture antibody (Kühbach et al., 2016). In contrast to a classical ELISA, the result of the measurement is not a single readout for the whole sample. Instead, the surface is imaged by high-resolution fluorescence microscopy, such as dual-color total internal reflection fluorescence microscopy (Kühbach et al., 2016). Only those pixels that show signal intensities above the background noise in both channels are counted. Thus the number of colocalized pixels above the background noise is expected to correlate with the concentration of Aβ42 oligomers in the sample (Kühbach et al., 2016). In addition to aggregation and oligomerization, Aβ42 fragments also fibrillize into cross-β-sheets, forming the insoluble plaques that constitute the main neuropathological finding in AD. The primary function of senile plaques in AD brain is to serve as large reservoirs for soluble amyloid (the amount of insoluble fibrillar Aβ42 is 100-fold greater than the amount of soluble Aβ42 in the brain) (Mathis et al., 2007). Senile plaques may contribute to changes in the amount of circulating amyloid. Array tomography has revealed that plaques are not entirely benign species. They are surrounded by a ring of dystrophic and disfigured neurons (Mucke and Selkoe, 2012), implying that they exert local neurodegenerative effects (Bezprozvanny, 2009). Plaque burden, however, correlates poorly with disease severity (Savva et al., 2009), and it is now widely accepted that Aβ42's primary role in the pathogenesis of AD is to trigger abnormal signal processes, such as production of modulation of protein kinases, protein phosphatases, phospholipases, cyclooxygenases/lipoxygenases, cholesterol hydroxylases, and nitric oxide synthases. Upregulation of these enzymes results in production of high levels of glycerophospholipid-, sphingolipid-, and cholesterol-derived lipid mediators. These mediators contribute to oxidative stress and neuroinflammation in AD (Karran et al., 2011; Farooqui, 2010, 2011).

Determination and clinical relevance of Aβ42 in CSF and plasma are hampered by (1) the intrinsic nature of Aβ42, including its aggregation and adsorption properties; (2) the complexity and heterogeneity of Aβ42 isoforms, including modifications or different conformational forms; (3) the presence of confounding factors; (4) low concentrations of Aβ42 in biological fluids; (5) high variability in outcomes of each assay between study centers; and (6) the absence of a reference method or reference materials (relative quantitative assays) (Vanderstichele et al., 2012a,b; Andreasson et al., 2012). It is also reported that the combined determination and evaluation of Aβ37, Aβ38, Aβ39, Aβ40, and Aβ42 levels may increase the sensitivity and specificity in predicting progression of AD (Hoglund et al., 2008). In addition, some investigators have suggested that the determination of Aβ42/Aβ40 ratio in CSF can improve AD diagnosis (Hansson et al., 2007). To support this suggestion, it is proposed that the Aβ42/Aβ40 ratio normalizes individuals according to their Aβ production level, so that low CSF Aβ42 can be more easily detected in "high Aβ42 producers," and vice versa (Lewczuk et al., 2015). It is suggested that the CSF Aβ42/Aβ40 ratio is also valuable in the clinical setting (Dumurgier et al., 2015). Other studies have indicated that in CSF

Aβ42/Aβ40 and Aβ42/Aβ38 ratios may better reflect AD-type pathology compared with CSF Aβ42 alone. In a study with three different cohorts with 1182 individuals the determination of CSF Aβ42/Aβ40 and Aβ42/Aβ38 ratios by using three different immunoassays shows improved agreement with amyloid (^{18}F-flutemetamol) PET imaging compared with CSF Aβ42. Similarly, studies on determination of CSF Aβ42, the Aβ42/Aβ40, and Aβ42/Aβ38 ratios and determination of hippocampal volume, lateral ventricular volume, and white matter lesions by MRI indicate that Aβ42/Aβ40 and Aβ42/Aβ38 ratios are more selective biomarkers for differentiating AD from other neurodegenerative dementias (Janelidze et al., 2016). However, other investigators have not reported such changes (Barabash et al., 2007).

TAU AND PHOSPHORYLATED TAU AS BIOMARKERS FOR ALZHEIMER'S DISEASE

Tau belongs to the family of microtubule-associated proteins and binds to, stabilizes, and promotes the assembly of microtubules. Tau also contribute to signaling pathways and cytoskeletal organization (Iqbal et al., 2005). Tau is mainly expressed in the central and peripheral nervous system and most abundant in axons. It exists in six isoforms in the adult human brain, which vary in size and have either three or four microtubule-binding domains. The six forms each show functional differences (Spillantini and Goedert, 2013). The ratio between Tau containing three and four domains is 1:1 in normal human brain, but this ratio is altered in the different tauopathies such as FTLD, progressive supranuclear palsy (PSP), corticobasal degeneration, and prion diseases (Schoonenboom et al., 2012). Tau can be measured in the CSF and blood. So far, little is known about Tau levels in blood, and most studies have failed due to the low abundance of the protein in blood (Zetterberg et al., 2013). It is reported that there is no correlation between CSF Tau levels and plasma Tau indicating that the clearance of Tau is differently regulated (Zetterberg et al., 2013). Studies on the determination of Tau in the blood and CSF of healthy subjects indicate that Tau protein concentration ranges between <10 and >100 pg/mL and the ratio CSF:serum Tau is 10:1 (Reiber, 2003). An ultrasensitive immunoassay for plasma Tau detection has been introduced (Randall et al., 2013). Several ELISAs have been developed for detecting t-Tau and p-Tau. In AD, CSF levels of p-Tau increase to approximately 2–3 times of CSF levels in age-matched normal subjects. Commonly used assays measure p-Tau at either residue 181 or 231, both of which increase to similar levels in AD (Hampel et al., 2004; Thal et al., 2006). Autopsy studies reveal that CSF p-Tau correlates with NFT burden in AD (Buerger et al., 2006a). Because levels of p-Tau are thought to reflect both NFT load and phosphorylation state, elevations in p-Tau are generally thought to be a more specific finding in AD than elevations in CSF t-Tau (Hampel et al., 2004). This evaluation about t-Tau and p-Tau allows a better correlation with the later stages of synaptic dysfunction and early neurodegeneration in AD (Lavados et al., 2005). Dissociations between high t-Tau and normal p-Tau levels have been reported in several dementing diseases including Creutzfeldt–Jakob disease (CJD) (Buerger et al., 2006b). As stated earlier, in patients with AD, CSF levels of Tau are increased threefold compared to CSF from age-matched normal subjects (Blennow, 2004). Increases in CSF Tau are associated with both NFT burden and Braak staging (Tapiola et al., 2009). Elevations in CSF t-Tau levels are not specific for AD because transient elevations in t-Tau levels have been reported in patients with stroke (Hesse et al., 2001) and traumatic brain injury (Franz et al., 2003). These observations suggest that increase in CSF t-Tau levels are reflective of neurodegeneration in neurotraumatic and neurodegenerative diseases. The highest levels of CSF t-Tau are found in CJD, a prion disease characterized by accelerated neurodegeneration (Otto et al., 2002). Therefore, t-Tau alone may be not sensitive enough to differentiate patients with AD from those with other diseases, especially CJD and VaD. The sensitivity and specificity for diagnosis of AD can be improved by combining the two CSF markers (t-tau and Aβ42 levels). However, differentiating AD from other primary degenerative dementias is still an important problem.

APOLIPOPROTEIN E IN ALZHEIMER'S DISEASE

APOE is a protein involved in the transport of cholesterols and other lipids in the plasma and brain. It is abundantly found in the brain and liver (Mahley, 1988). In the brain, APOE is mainly synthesized in astrocytes, but low levels of APOE are present in some neurons, activated microglia, oligodendrocytes, and ependymal layer cells. In neurons, APOE is induced under neuronal stress and damaged conditions. APOE has been detected in cortical and hippocampal neurons (Huang et al., 2004). As stated in Chapter 2, in the normal brain, APOE is associated with the maintenance and repair of neurons and is involved in cholesterol homeostasis (Leduc et al., 2010). APOE occurs in polymorphic forms (APOE ε2, APOE ε3, and APOE ε4). These isoforms differ by single amino acid substitutions at positions 112 and 158 (Hatters et al., 2006). The APOE ε4 allele is a risk factor for late-onset FAD and sporadic AD (Bertram and Tanzi, 2008). Around 10%–15% of the general population has the ε4 allele, whereas the prevalence is 40%–65% in patients with AD. The homozygosity of APOE ε4 results in a 50%–90% risk of developing AD by the age of 85years, whereas individuals with one copy have a risk of 45%. For individuals with no APOE ε4 alleles the risk is about 20% (Harris et al., 2004). APOE is colocalized with amyloid plaques and NFTs in AD brain (Huang et al., 2004). Several mechanisms have been proposed for the role of APOE ε4 in the pathology of AD including regulation of the deposition and clearance of Aβ and amyloid plaques, regulation of phosphorylation and assembly of tau into NFTs, dysfunction of the neuronal signaling pathways, induction of Aβ-regulated lysosomal leakage, increased atherosclerosis and vascular inflammation in AD, and apoptosis in neurons (Huang et al., 2004; Tibolla et al., 2010). However, its exact role in the AD pathology still remains unclear (Huang et al., 2004). The ε4 allele of APOE is an important risk for the development of AD. The CSF APOE levels have been determined by several studies: some have found decreased levels in CSF of patients with AD, whereas other studies have shown an increase (Hesse et al., 2000). Increased CSF levels of APOE were also detected in DLB and patients with Parkinson's disease (PD) (Vijayaraghavan et al., 2014).

OTHER METABOLIC BIOMARKERS OF ALZHEIMER'S DISEASE IN THE CEREBROSPINAL FLUID AND BLOOD

It is becoming increasingly evident that analysis of routine CSF and blood samples has given some promising results on the platform for early AD detection and monitoring. For example, AD-linked changes in phospholipid-derived metabolites, ceramides, and sphingomyelins. As stated in Chapter 3, these metabolites play an important role in amyloidogenesis and inflammatory stress related to neuronal apoptosis in AD. Several clinical studies suggest that these biomarkers may have reproducibility in assessing early-stage AD (Han et al., 2011; Mielke et al., 2011; Farooqui, 2011). Reduction of hypercholesterolemia is an area of interest in AD clinical trials and is another possible approach to AD detection and monitoring, as elevated cholesterol blood level markers have been shown to be positively correlated to amyloid plaques in the brain (Lesser et al., 2009, 2011; Farooqui, 2011). Changes in the cholesterol metabolism–related gene expression in mononuclear cells from patients with AD can have potential diagnostic implications, and blood samples stained with Oil Red O, a fat-soluble lysochrome dye used to identify neutral lipids, may help to distinguish healthy elderly patients from patients with AD (Mandas et al., 2012). Beside the aforementioned biomarkers, CSF visinin-like protein-1 (VILIP-1), a calcium-mediated neuronal injury biomarker, has been used as a novel biomarker for AD. It is reported that the levels of CSF VILIP-1 level are significantly increased in patients with AD compared with both normal controls (Luo et al., 2013). Furthermore, CSF VILIP-1 levels positively correlate with t-Tau and p-Tau 181P within each group and with α-synuclein in the AD and control groups. Based on these observations, it is proposed that CSF VILIP-1 can be a diagnostic marker for AD (Luo et al., 2013).

OLFACTORY DYSFUNCTION AS A DIAGNOSTIC TOOL FOR ALZHEIMER'S DISEASE

It is becoming increasingly evident that the early diagnosis of dementia in several neurodegenerative diseases can be made by using olfactory sensory dysfunction in combination with neuropsychological measurements (Murphy et al., 2003; Zou et al., 2016). Thus 100% olfactory dysfunction occurs in AD, 90% in PD, 96% in FTLD, and 15% in VaD (Duff et al., 2002; Doty, 2012; Pardini et al., 2009). Patients with AD often have a reduced ability to detect, discriminate, and identify odors (Mesholam et al., 1998). Olfactory dysfunction occurs prior to the development of neuropathology and other cognitive dysfunctions. There are many common causes of olfactory dysfunction including traumatic brain injury, endocrine dysfunction, and inflammatory sinusitis along with normal aging (Doty, 2012). For example, over half of people aged 60 years and older have some problem with smell (Westervelt et al., 2007).

Entorhinal cortex (ERC) is the gateway for olfactory input to the hippocampus. Transmission of olfactory information to the hippocampus is sequential process. Thus, olfactory information is projected from the olfactory receptors to olfactory bulb to the primary olfactory cortex to the ERC from where through perforant pathway information is transferred to the hippocampal neurons. Olfactory dysfunction is modulated not only by hypoperfusion, and hypometabolism, but also by impaired synaptic transmission, and variable atrophy in olfaction-related regions (Daulatzai, 2015). One of the earliest pathological changes in AD includes the olfactory dysfunction and the atrophy in ERC and hippocampus. Very little is known about the neurochemical mechanisms associated with olfactory dysfunction in patients with AD. However, studies on molecular mechanisms of AD-related olfactory sensory loss in APP transgenic mouse models have indicated that olfaction involves processing stages spanning from sensory neuron input to the olfactory bulb, decoding, and plasticity in the piriform cortex, and ultimately downstream neurons in the hippocampus (Wachowiak and Shipley, 2006; Wilson and Stevenson, 2006).

Studies in patients with AD have indicated that pathological changes of Tau protein in the olfactory bulb and olfactory projection area are associated with pathological changes in Tau protein levels (Attems and Jellinger, 2006; Sun et al., 2012). More studies performed on olfactory dysfunction tests in 166 participants at baseline and performed on brain autopsy in 77 patients with AD indicate that the density of NFTs is the main pathological factor that affected olfactory function, especially in the ERC, hippocampus CA1 region, and subiculum. No significant association has been reported between NFTs, senile plaque deposition, and olfactory function in other brain regions (Wilson et al., 2007). These are supported by olfactory function tests performed with PiB-PET on 19 healthy volunteers, 24 patients with amnestic MCI, and 20 patients with AD, and results indicate that AD-related olfactory dysfunction is not directly related to Aβ burden but caused by pathological changes in Tau protein (Bahar-Fuchs et al., 2010). It is well known that olfactory nervous system is rich in acetylcholine, glutamate, γ-aminobutyric acid, and other neurotransmitters. A reduction in acetylcholine is not only associated with memory impairment and cognitive dysfunctions but also contributes to olfactory dysfunction in AD (Beach et al., 2000). Olfactory dysfunction in patients with AD can be reversed or improved by treatment with the cholinesterase inhibitor, donepezil, supporting the view that functional changes in olfactory recognition can be used to predict therapeutic effects in patients with AD (Velayudhan and Lovestone, 2009). However, there are limitations in olfactory tests not only due to the complexity of the brain areas but also due to the olfactory fatigue that occurs in the behavioral olfactory tests in patients with AD. Different patterns of olfactory dysfunction have been observed in patients with AD at early stages and even MCI, but the causes and mechanisms of the olfactory dysfunction remain unclear. Converging evidence suggests that olfactory dysfunction can be used as a clinical marker for severity and progression of AD. However, many questions remain unanswered. For example, it is unclear how to identify and differentiate age-related olfactory changes and olfactory dysfunction caused by AD. It is also unclear during which stage of AD (initial vs. late) pathological changes contribute to olfactory structure dysfunction. Moreover, the validity and clinical relevance in predicting AD using a combination of assessments including olfactory dysfunction and other biomarkers of AD remain unclear (Chen and Zhong, 2013).

VISUAL DYSFUNCTION AS A DIAGNOSTIC TOOL FOR ALZHEIMER'S DISEASE

During embryo development, the eyes and brain have a similar origin. The eyes are derived from the anterior neural tube, an area that later gives rise to the forebrain. Ocular development occurs through specification of the eye field postneural induction (Chow and Lang, 2001). This process involves specific transcription factors, which are also conserved in brain development. One such factor, a "master regulator" gene of the development of the eye field, Pax6, plays an essential role in neural development. When expressed ectopically, Pax6 can induce ocular formation in other parts of the body (Gehring, 1996), whereas its impairment or knockout disrupts neurogenesis in the cortex (Tuoc et al., 2009). In visual system, retinal neurons are comparable in many ways to their counterparts in the brain. Retinal neurons have dendrites, a soma, and an axon, which are the essential neuronal features (London et al., 2013). They are stained by many typical neuronal markers (Ma and Wang, 2006). It is reported that neurodegeneration in the brain of patients with AD can extend into the visual system (eyes). Changes in visual symptoms have been well documented in AD, and there is significant evidence to illustrate that ocular pathology occurs as part of the disorder (Sadun and Bassi, 1990; Paquet et al., 2007). Accumulation of Aβ in neurons of the retinal ganglion cell layer, inner nuclear layer, and outer nuclear layer has been detected in the retinas of APP/Presenilin-1 (PSEN-1) transgenic mice and TG2576 transgenic mice (Lu et al., 2010; Williams et al., 2013). Studies on time course of pathological changes in the retina and brain as well as cognitive dysfunction in APP/PSEN-1 transgenic mice indicate that (1) abnormalities in expression of β-secretase [β-site APP cleaving enzyme 1 (BACE1)] in the retina can be detected in the AD transgenic mice from 3 months of age and (2) BACE1 plaques in the cortex and cognitive impairment appear in AD transgenic mice from 6 months of age. These results suggest that abnormal expression of BACE1 in the retina is an early pathological change in AD transgenic mice (Li et al., 2016). The abnormal expression of BACE1 in the retina may be an early pathological change in APP/PSEN-1 transgenic mice, and it is likely that BACE1 activity in retina may have potential as a biomarker for the early diagnosis of AD in humans (Li et al., 2016). These observations provide an opportunity to use a minimally invasive approach to examine the pathological features in the brain—through the transparent medium of the eye. Tests such as optical coherence tomography can be used to detect decrease in retinal nerve fiber layer thickness (Yoneda et al., 2005; Ning et al., 2008; Koronyo-Hamaoui et al., 2011). Several studies have reported abnormalities in the optic nerve head, a reduction in the number of optic nerve fibers, and a reduction in the thickness of the parapapillary and macular retinal nerve fiber layers (Danesh-Meyer et al., 2006; Iseri et al., 2006). In addition to these anatomic changes, there are changes in the pattern of electroretinograms associated with AD (Parisi et al., 2001). Furthermore, studies on retinal blood flow have demonstrated a significant narrowing of the retinal veins and decreased retinal blood flow (Iseri et al., 2006; Berisha et al., 2007) supporting the view that the retina is affected in AD. Furthermore, the aggregation of Aβ can also be detected in ocular structures such as lenses (Melov et al., 2005). The conformational alteration in lens proteins may induce cataract formation (Surguchev and Surguchov, 2010), and it is suggested that there may be a possible relationship between AD and a specific subtype of age-related cataract called supranuclear cataract (Goldstein et al., 2003) and the existence of a pathway leading to AD-linked pathology in the brain and lens (Jun et al., 2012).

Patients with AD have ocular motility dysfunctions and poor visual attention. They are often unable to focus on a fixed object, due to their difficulty in suppressing reflexive saccades. Studies on saccadic eye movements have indicated that patients with AD have slower reaction times as compared to the control group (Crawford et al., 2015). Other studies have also indicated that patients with AD show abnormal hypometric saccades as compared to controls (Garbutt et al., 2008). In particular, antisaccades require movement of the eye in the opposite direction of a visual stimulus and suppression of the reflexive response. fMRI studies have indicated that patients with AD exhibit significantly more antisaccade errors and demonstrate a decrease in activity in all oculomotor regions of the brain as compared to age-matched control subjects (Wright and MacAskill, 2012). Furthermore, the convergence angle has also been noted to be smaller and more irregular in patients with AD (Uomori et al., 1993). Very little is known of the cellular mechanisms underlying the loss of visual function. However, histologic studies have shown that retinal ganglion cells undergo extensive neurodegeneration in AD (Blanks et al., 1991). Several mechanisms have been proposed to explain retinal neurodegeneration on the basis of inflammatory events, amyloid misfolding, and amyloid angiopathy (Berisha et al., 2007).

MICRORNAs AS BIOMARKERS FOR ALZHEIMER'S DISEASE

MicroRNAs (miRNAs) are small, noncoding RNA fragments. miRNAs are approximately 22 nucleotides long and are associated not only with posttranscriptional regulation of gene expression (Guan et al., 2010; Gallego et al., 2012; Cheng et al., 2013) but also play important roles in cell proliferation, differentiation, and apoptosis. In mammalian cells, miRNAs work through base pairing with complementary sequences within messenger RNA (mRNA) molecules, usually resulting in gene silencing via translational repression. These small nucleotide sequences of miRNA have several features making them favorable candidates as biomarkers, including function in multiple tissues, stability in bodily fluids (plasma, serum, urine, saliva, milk, and CSF), role in pathogenesis of disease, and the ability to be detected early in the disease course (Siegel et al., 2012; Galimberti et al., 2014). After processing by endonucleases (Cheng et al., 2013), the resultant single-stranded miRNAs combine with other macromolecules to form RNA-induced silencing complexes (RISCs). These RISCs target complementary mRNA strands for degradation, altering cellular function (Cheng et al., 2013; Freischmidt et al., 2013).

miRNAs are found in high abundance within the nervous system, where they often replicate a brain-specific expression pattern and are usually found to be coexpressed with their targets. In the brain, miRNAs are key regulators of different biological functions including synaptic plasticity and neurogenesis, where they channelize the cellular physiology toward neuronal differentiation. Also, they can indirectly influence neurogenesis by regulating the proliferation and self-renewal of neural stem cells (Grasso et al., 2014). miRNAs are aberrantly expressed in patients with AD (Zhao et al., 2015; Grasso et al., 2014; Hu et al., 2016). These miRNAs may act as biomarkers for AD before clinical symptoms and functional decline become apparent (Bekris et al., 2013; Hu et al., 2016). Downregulation of miR-15a/b, mR-16, miR-29a/b, miR-195, miR-103, and miR-107 have been observed in the brain of patients with AD compared to control subjects. These miRNAs target the BACE1. This enzyme is involved in the formation of amyloid plaques (Hébert et al., 2008; Lei et al., 2015). Studies on miRNA expression in hippocampal tissue of patients with AD have indicated an upregulation of specific proinflammatory miRNAs including miR-9, miR-125b, miR-146a, and miR-155. These miRNAs are induced by nuclear factor-κB, thus indicating a possible role of these miRNAs in neuronal inflammation in AD (Lukiw, 2007; Sethi and Lukiw, 2009). On the other hand, others have found a significant downregulation of miR-9 in the human hippocampus of patients with AD (Cogswell et al., 2008). Collective evidence thus suggests that most of miRNAs are downregulated; by contrast, some specific miRNAs are significantly upregulated, such as miR-29, miR-339, and miR-214 (Bekris et al., 2013; Hu et al., 2016). A subset of miRNAs can also be detected in CSF. Therefore, these miRNAs can be used as potential diagnostic tools for AD, provided that their levels can be quantified in volumes that are suitable for diagnostic purposes (<1 mL) (Sala Frigerio et al., 2013; Kiko et al., 2014; Müller et al., 2014). As stated earlier, marked alterations are observed in the miRNAs from hippocampus; however, only a few miRNAs (miR-16 and miR-146a, miR-27a, miR-29a, miR-29b, and miR-125b) are detectable in the CSF, and among these miRNAs, miR-146a is influenced by blood contamination (Müller et al., 2014). Many studies have been performed on miRNA determination in plasma and CSF samples, but the results have not always been consistent and no consensus has yet been reached. The exploration of the dysregulation of miRNAs in the pathological process of AD including APP metabolism, Tau pathology, neuroinflammation, and apoptosis indicate that miRNAs serve as critical regulators in the aforementioned processes and the dysregulation of miRNAs may influence gene transcription and protein expression and induce the spread of neurodegeneration (Faghihi et al., 2010). In addition to the CSF biomarkers mentioned earlier, it is also reported that routine analysis of blood and CSF samples shows promising results through the estimation of ceramides and sphingomyelins and their lipid mediators because they play roles in amyloidogenesis and neuroinflammatory stress. Several clinical studies suggest that these biomarkers may have reproducibility in assessing early-stage AD (Han et al., 2011; Mielke et al., 2011). Reduction of hypercholesterolemia is not only another important area in clinical trials but also another possible approach to detect and monitor pathogenesis of AD because hypercholesterolemia is positively correlated with the accumulation of senile plaques in the brain (Lesser et al., 2009, 2011; Matsubara et al., 2013).

Based on these findings, these days patients with AD are clinically diagnosed and monitored by using following tests (Knight et al., 2016):

1. *Neuropsychological tests*: These tests provide information on the pattern of deficits in the early stages of AD, but changes are nonspecific and an individual's baseline can be hard to ascertain.
2. *Structural imaging tests*: CT or MRI can diagnose groups of individuals with MCI, which can progress into AD (Risacher et al., 2009), but at present these tests do not identify human subjects with 100% accuracy.
3. *PET and SPECT* (O'Brien et al., 2014; Weiner et al., 2015): Aβ PET can not only identify patients with AD from healthy controls with over 95% accuracy but also diagnose MCI subjects who can progress to have AD (Vandenberghe et al., 2013). Although not in routine clinical use, there are also PET ligands that bind to p-Tau (FDDNP molecule (Small et al., 2006) and ^{18}F AV1451) that have been shown to track progressive Tau accumulation in aging and AD (Schöll et al., 2016).
4. *Lumbar puncture*: Decrease in Aβ levels and increase in Tau in CSF can predict patients with AD with 95% sensitivity and 87% specificity after 4–6 years' follow-up (Hansson et al., 2006), but lumbar puncture is an invasive test. However, beside Aβ and Tau levels, CSF can be used to determine levels of APOE, VILIP-1, cholesterol, as well as phospholipid-, sphingolipid-, and cholesterol-derived lipid mediators.

LIMITATIONS OF BIOMARKER ASSAY SYSTEMS

Due to the multifactorial nature of the pathogenesis of AD, it is unlikely that a single biomarker can be used for the clinical diagnosis of AD. However, the list of biomarkers shown in Table 7.1 offers the appropriate sensitivity, specificity, and positive and negative predictive values. An important limitation of fluid biomarkers is the lack of anatomical precision in the measurements. Another limitation is the lack of assay standardization. Different assay procedures give different absolute concentrations for Aβ42 and Tau at different stages of disease. The main challenges for Aβ oligomer–based diagnostics in CSF and other body fluids are the very low concentrations of Aβ oligomers and the high background of monomeric Aβ (Rosén et al., 2013). It should also be mentioned that lumbar puncture is an invasive procedure and may not be practically favorable for conducting large-scale studies on AD. A lot of research is currently ongoing to achieve better way of obtaining CSF and assay standardization through the establishment of reference measurement procedures for Aβ42 and Tau to validate these in multicenter settings (Carrillo et al., 2013; Rosén et al., 2013). Although AD biomarkers (Aβ42 and Tau) in CSF t-Tau have been extensively studied, there are no large, double-blind multicenter studies presently available for the diagnosis of AD. What does steady-state concentrations of Aβ and Tau among healthy individuals reflect; is it a low-grade neurodegeneration or perhaps normal neuroaxonal plasticity? It is also not known why some Tau pathologies, for example, PSP, typically have normal levels of Tau and why CSF levels of Tau remain largely unaltered with progression of AD. An explanation to the latter may be that the levels of Tau depend more on the rate of neurodegeneration than the state, but questions like these warrant translational studies spanning from models to human disease.

CONCLUSION

AD is the most common form of dementia in the growing aging population today, with prevalence expected to rise over the next 40 years. Clinically, patients exhibit a progressive decline in cognition, memory, social functioning, as well as visuospatial and olfactory dysfunction due to deposition of Aβ protein and intracellular hyperphosphorylated Tau protein. These pathological hallmarks of AD are measured through CSF analysis or neuroimaging (MRI, PET, and SPECT) or are diagnosed at postmortem. Among these neuroimaging measures, hippocampal atrophy is measured by MRI, amyloid uptake is measured by PiB-PET, and decrease in 18F-FDG uptake is measured by PET (FDG-PET). These days, the measurement of insulin resistance of the hippocampal/temporal lobe by PET with 18F-FDG integrated with CT (FDG-PET/CT) has become, besides amyloid-PET diagnostics, a highly specific and sensitive biomarker for AD, having predictive value even decades before the first clinical symptoms of AD manifest themselves.

CSF analysis provides information about levels of Aβ42, t-Tau, and p-Tau. SPECT scanning has been used to detect regional reduction in cerebral blood flow thought to be present in patients with AD. In addition to the sources mentioned previously, neuropathological progression of AD can also be detected in the olfactory as well as the visual systems both in human and animal models of AD. Thus olfactory and visual systems offer a transparent medium to cerebral pathology and have potentiated the development of ocular and olfactory biomarkers for the pathogenesis of AD. The use of noninvasive screening, such as retinal imaging and visual and olfactory testing, may enable earlier diagnosis in the clinical setting using minimizing invasive and less expensive testing. Visual and olfactory testing can not only improve disease management and quality of life for patients with AD but also promote monitoring of drug treatment.

References

Andreasson, U., Vanmechelen, E., Shaw, L., Zetterberg, H., Vanderstichele, H., 2012. Analytical aspects of molecular Alzheimer's disease biomarkers. Biomark. Med. 6, 377–389.

Apurva, P.A., Bipin, P.M., Kirti, P.M., 2013. Role of PET scan in clinical practice. Gujarat Med. J. 68, 19–22.

Attems, J., Jellinger, K.A., 2006. Olfactory tau pathology in Alzheimer disease and mild cognitive impairment. Clin. Neuropathol. 25, 265–271.

Bahar-Fuchs, A., Chételat, G., Villemagne, V.L., et al., 2010. Olfactory deficits and amyloid-β burden in Alzheimer's disease, mild cognitive impairment, and healthy aging: a PiB PET study. J. Alzheimer's Dis. 22, 1081–1087.

Bannach, O., Birkmann, E., Reinartz, E., Jaeger, K.E., Langeveld, J.P., et al., 2012. Detection of prion protein particles in blood plasma of scrapie infected sheep. PLoS One 7, e36620.

Barabash, A., Marcos, A., Ancin, I., Vazquez-Alvarez, B., de Ugarte, C., et al., 2007. APOE, ACT and CHRNA7 genes in the conversion from amnestic mild cognitive impairment to Alzheimer's disease. Neurobiol. Aging 30, 1254–1264.

Berisha, F., Feke, G.T., Trempe, C.L., McMeel, J.W., Schepens, C.L., 2007. Retinal abnormalities in early Alzheimer's disease. Invest. Ophthalmol. Vis. Sci. 48, 2285–2289.

Beach, T.G., Kuo, Y.M., Spiegel, K., Emmerling, M.R., Sue, L.I., et al., 2000. The cholinergic deficit coincides with Abeta deposition at the earliest histopathologic stages of Alzheimer disease. J. Neuropathol. Exp. Neurol. 59, 308–313.

Bekris, L.M., Lutz, F., Montine, T.J., Yu, C.E., Tsuang, D., et al., 2013. MicroRNA in Alzheimer's disease: an exploratory study in brain, cerebrospinal fluid and plasma. Biomarkers 18, 455–466.

Bertram, L., Tanzi, R.E., 2008. Thirty years of Alzheimer's disease genetics: the implications of systematic meta-analyses. Nat. Rev. Neurosci. 9, 768–778.

Bezprozvanny, I., 2009. Amyloid goes global. Sci. Signal. 2, pe16.

Birkmann, E., Henke, F., Weinmann, N., Dumpitak, C., Groschup, M., et al., 2007. Counting of single prion particles bound to a capture-antibody surface (surface-FIDA). Vet. Microbiol. 123, 294–304.

Blanks, J.C., Torigoe, Y., Hinton, D.R., Blanks, R.H., 1991. Retinal degeneration in the macula of patients with Alzheimer's disease. Ann. N. Y. Acad. Sci. 640, 44–46.

Blennow, K., 2004. Cerebrospinal fluid protein biomarkers for Alzheimer's disease. NeuroRx. 1, 213–225.

Blennow, K., 2010. Cerebrospinal fluid and plasma biomarkers in Alzheimer disease. Nat. Rev. Neurol. 6, 131–144.

Braak, H., Braak, E., Strothjohann, M., 1994. Abnormally phosphorylated tau protein related to the formation of neurofibrillary tangles and neuropil threads in the cerebral cortex of sheep and goat. Neurosci. Lett. 171, 1–4.

Buerger, K., Ewers, M., Pirttila, T., Zinkowski, R., Alafuzoff, I., Teipel, S.J., et al., 2006a. CSF phosphorylated tau protein correlates with neocortical neurofibrillary pathology in Alzheimer's disease. Brain 129, 3035–3041.

Buerger, K., Otto, M., Teipel, S.J., Zinkowski, R., Blennow, K., et al., 2006b. Dissociation between CSF total tau and tau protein phosphorylated at threonine 231 in Creutzfeldt-Jakob disease. Neurobiol. Aging 27, 10–15.

Carrillo, M.C., Blennow, K., Soares, H., Lewczuk, P., Mattsson, N., et al., 2013. Global standardization measurement of cerebral spinal fluid for Alzheimer's disease: an update from the Alzheimer's Association Global Biomarkers Consortium. Alzheimer's Dement. J. Alzheimer's Assoc. 9, 137–140.

Chen, Z., Zhong, C., 2013. Decoding Alzheimer's disease from perturbed cerebral glucose metabolism: implications for diagnostic and therapeutic strategies. Prog. Neurobiol. 108, 21–43.

Cheng, L., Quek, C.Y., Sun, X., Bellingham, S.A., Hill, A.F., 2013. The detection of microRNA associated with Alzheimer's disease in biological fluids using next-generation sequencing technologies. Front. Genet. 4, 150.

Chintamaneni, M., Bhaskar, M., 2012. Biomarkers in Alzheimer's disease: a review. ISRN Pharmacol. 2012, 984786.

Chua, S.W., Kassiou, M., Ittner, L.M., 2014. The translocator protein as a drug target in Alzheimer's disease. Expert Rev. Neurother. 14, 439–448.

Chow, R.L., Lang, R.A., 2001. Early eye development in vertebrates. Annu. Rev. Cell Dev. Biol. 17, 255–296.

Cogswell, J.P., Ward, J., Taylor, I.A., Waters, M., Shi, Y., et al., 2008. Identification of miRNA changes in Alzheimer's disease brain and CSF yields putative biomarkers and insights into disease pathways. J. Alzheimer's Dis. 14, 27–41.

Craig-Schapiro, R., Fagan, A.M., Holtzman, D.M., 2009. Biomarkers of Alzheimer's disease. Neurobiol. Dis. 35, 128–140.

Crawford, T.J., Devereaux, A., Higham, S., Kelly, C., 2015. The disengagement of visual attention in Alzheimer's disease: a longitudinal eye-tracking study. Front. Aging Neurosci. 7, 118.

Danesh-Meyer, H.V., Birch, H., Ku, J.Y.-F., Carroll, S., Gamble, G., 2006. Reduction of optic nerve fibers in patients with Alzheimer disease identified by laser imaging. Neurology 67, 1852–1854.

Daulatzai, M.A., 2015. Olfactory dysfunction: its early temporal relationship and neural correlates in the pathogenesis of Alzheimer's disease. J. Neural Transm. (Vienna) 122, 1475–1497.

Doty, R.L., 2012. Olfactory dysfunction in Parkinson disease. Nat. Rev. Neurol. 8, 329–339.

Duff, K., McCaffrey, R.J., Solomon, G.S., 2002. The pocket smell test: successfully discriminating probable Alzheimer's dementia from vascular dementia and major depression. J. Neuropsychiatry Clin. Neurosci. 14, 197–201.

Dumurgier, J., Schraen, S., Gabelle, A., Vercruysse, O., Bombois, S., et al., 2015. Cerebrospinal fluid amyloid-beta 42/40 ratio in clinical setting of memory centers: a multicentric study. Alzheimer's Res. Ther. 7, 30.

Engler, H., Forsberg, A., Almkvist, O., Blomquist, G., Larsson, E., et al., 2006. Two-year follow-up of amyloid deposition in patients with Alzheimer's disease. Brain 129, 2856–2866.

Engler, H., Santillo, A.F., Wang, S.X., Lindau, M., Savitcheva, I., et al., 2008. In vivo amyloid imaging with PET in frontotemporal dementia. Eur. J. Nucl. Med. Mol. Imaging 35, 100–106.

Ewers, M., Sperling, R.A., Klunk, W.E., Weiner, M.W., Hampel, H., 2011. Neuroimaging markers for the prediction and early diagnosis of Alzheimer's disease dementia. Trends Neurosci. 34, 430–442.

Fagan, A.M., Mintun, M.A., Mach, R.H., Lee, S.Y., Dence, C.S., Shah, A.R., et al., 2006. Inverse relation between in vivo amyloid imaging load and cerebrospinal fluid Abeta42 in humans. Ann. Neurol. 59, 512–519.

Faghihi, M.A., Zhang, M., Huang, J., Modarresi, F., Van Der Brug, M.P., et al., 2010. Evidence for natural antisense transcript-mediated inhibition of microRNA function. Genome Biol. 11, R56.

Farooqui, A.A., 2010. Neurochemical Aspects of Neurotraumatic and Neurodegenerative Diseases. Springer, New York.

Farooqui, A.A., 2011. Lipid Mediators and Their Metabolism in the Brain. Springer, New York.

Fjell, A.M., Walhovd, K.B., 2011. New tools for the study of Alzheimer's disease: what are biomarkers and morphometric markers teaching us? Neuroscientist 17, 592–605.

Franz, G., Beer, R., Kampfl, A., Engelhardt, K., Schmutzhard, E., et al., 2003. Amyloid beta 1-42 and tau in cerebrospinal fluid after severe traumatic brain injury. Neurology 60, 1457–1461.

Freischmidt, A., Müller, K., Ludolph, A.C., Weishaupt, J.H., 2013. Systemic dysregulation of TDP-43 binding microRNAs in amyotrophic lateral sclerosis. Acta Neuropathol. Commun. 1, 42.

Funke, S.A., Wang, L., Birkmann, E., Willbold, D., 2010. Single-particle detection system for Aβ aggregates: adaptation of surface-fluorescence intensity distribution analysis to laser scanning microscopy. Rejuvenation Res. 13, 206–209.

Galimberti, D., Villa, C., Fenoglio, C., Serpente, M., Ghezzi, L., et al., 2014. Circulating miRNAs as potential biomarkers in Alzheimer's disease. J. Alzheimer's Dis. 42, 1261–1267.

Gallego, J.A., Gordon, M.L., Claycomb, K., Bhatt, M., Lencz, T., et al., 2012. In vivo microRNA detection and quantitation in cerebrospinal fluid. J. Mol. Neurosci. 47, 243–248.

Garbutt, S., Matlin, A., Hellmuth, J., Schenk, A.K., Johnson, J.K., et al., 2008. Oculomotor function in frontotemporal lobar degeneration, related disorders and Alzheimer's disease. Brain 131, 1268–1281.

Gehring, W.J., 1996. The master control gene for morphogenesis and evolution of the eye. Genes Cells 1, 11–15.

Golde, T.E., Schneider, L.S., Koo, E.H., 2011. Anti-aβ therapeutics in Alzheimer's disease: the need for a paradigm shift. Neuron 69, 203–213.

Goldstein, L.E., Muffat, J.A., Cherny, R.A., et al., 2003. Cytosolic β-amyloid deposition and supranuclear cataracts in lenses from people with Alzheimer's disease. Lancet 361, 1258–1265.

Grasso, M., Piscopo, P., Confaloni, A., Denti, M.A., 2014. Circulating miRNAs as biomarkers for neurodegenerative disorders. Molecules 19, 6891–6910.

Guan, Y., Mizoguchi, M., Yoshimoto, K., Hata, N., Shono, T., et al., 2010. MiRNA-196 is upregulated in glioblastoma but not in anaplastic astrocytoma and has prognostic significance. Clin. Cancer Res. 16, 4289–4297.

Guerreiro, R.J., Santana, I., Bras, J.M., Santiago, B., Paiva, A., Oliveira, C., et al., 2007. Peripheral inflammatory cytokines as biomarkers in Alzheimer's disease and mild cognitive impairment. Neurodegener. Dis. 4, 406–412.

Hampel, H., Buerger, K., Zinkowski, R., Teipel, S.J., Goernitz, A., et al., 2004. Measurement of phosphorylated tau epitopes in the differential diagnosis of Alzheimer disease: a comparative cerebrospinal fluid study. Arch. Gen. Psychiatry 61, 95–102.

Han, X., Rozen, S., Boyle, S.H., Hellegers, C., Cheng, H., et al., 2011. Metabolomics in early Alzheimer's disease: identification of altered plasma sphingolipidome using shotgun lipidomics. PLoS One 6, e21643.

Hansson, O., Buchhave, P., Zetterberg, H., Blennow, K., Minthon, L., et al., 2009. Combined rCBF and CSF biomarkers predict progression from mild cognitive impairment to Alzheimer's disease. Neurobiol. Aging 30, 165–173.

Hansson, O., Zetterberg, H., Buchhave, P., Andreasson, U., Londos, E., et al., 2007. Prediction of Alzheimer's disease using the CSF Abeta42/Abeta40 ratio in patients with mild cognitive impairment. Dement. Geriatr. Cogn. Disord. 23, 316–320.

Hansson, O., Zetterberg, H., Buchhave, P., Londos, E., Blennow, K., Minthon, L., 2006. Association between CSF biomarkers and incipient Alzheimer's disease in patients with mild cognitive impairment: a follow-up study. Lancet Neurol. 5, 228–234.

Harris, F.M., Tesseur, I., Brecht, W.J., Xu, Q., Mullendorff, K., Chang, S.J., et al., 2004. Astroglial regulation of apolipoprotein E expression in neuronal cells – implications for Alzheimer's disease. J. Biol. Chem. 279, 3862–3868.

Hatters, D.M., Peters-Libeu, C.A., Weisgraber, K.H., 2006. Apolipoprotein E structure: insights into function. Trends Biochem. Sci. 31, 445–454.

Hébert, S.S., Horré, K., Nicolaï, L., Papadopoulou, A.S., Mandemakers, W., et al., 2008. Loss of microRNA cluster miR-29a/b-1 in sporadic Alzheimer's disease correlates with increased BACE1/beta-secretase expression. Proc. Natl. Acad. Sci. U.S.A. 105, 6415–6420.

Heslegrave, A., Heywood, W., Paterson, R., Magdalinou, N., Svensson, J., et al., 2016. Mol. Neurodegener. 11, 3.

Hesse, C., Larsson, H., Fredman, P., Minthon, L., Andreasen, N., et al., 2000. Measurement of apolipoprotein E (apoE) in cerebrospinal fluid. Neurochem. Res. 25, 511–517.

Hesse, C., Rosengren, L., Andreasen, N., Davidsson, P., Vanderstichele, H., Vanmechelen, E., et al., 2001. Transient increase in total tau but not phospho-tau in human cerebrospinal fluid after acute stroke. Neurosci. Lett. 297, 187–190.

Ho, L., Qin, W., Pompl, P.N., Xiang, Z., Wang, J., Zhao, Z., et al., 2004. Diet-induced insulin resistance promotes amyloidosis in a transgenic mouse model of Alzheimer's disease. FASEB J. 18, 902–904.

Hoglund, K., Hansson, O., Buchhave, P., Zetterberg, H., Lewczuk, P., et al., 2008. Prediction of Alzheimer's disease using a cerebrospinal fluid pattern of C-terminally truncated beta-amyloid peptides. Neuro-degener. Dis. 5, 268–276.

Holtzman, D.M., Morris, J.C., Goate, A.M., 2011. Alzheimer's disease: the challenge of the second century. Sci. Transl. Med. 3, 77sr1.

Hu, W.T., Chen-Plotkin, A., Arnold, S.E., Grossman, M., Clark, C.M., et al., 2010. Novel CSF biomarkers for Alzheimer's disease and mild cognitive impairment. Acta Neuropathol. 119, 669–678.

Hu, Y.B., Li, C.B., Song, N., Zou, Y., Chen, S.D., et al., 2016. Diagnostic value of microRNA for Alzheimer's disease: a systematic review and meta-analysis. Front. Aging Neurosci. 8, 13.

Huang, Y., Weisgraber, K., Mucke, L., Mahley, R., Apolipoprotein, E., 2004. J. Mol. Neurosci. 23, 189–204.

Iseri, P.K., Altinas, O., Tokay, T., Yuksel, N., 2006. Relationship between cognitive impairment and retinal morphological and visual functional abnormalities in Alzheimer disease. J. Neuroophthalmol. 26, 18–24.

Iqbal, K., Del, A., Alonso, C., Chen, S., Chohan, M.O., et al., 2005. Tau pathology in Alzheimer disease and other tauopathies. Biochim. Biophys. Acta 1739, 198–210.

Jack Jr., C.R., Knopman, D.S., Jagust, W.J., Shaw, L.M., Aisen, P.S., et al., 2010. Hypothetical model of dynamic biomarkers of the Alzheimer's pathological cascade. Lancet Neurol. 9, 119–128.

Janelidze, S., Hertze, J., Zetterberg, H., Landqvist Waldö, M., Santillo, A., et al., 2015. Cerebrospinal fluid neurogranin and YKL-40 as biomarkers of Alzheimer's disease. Ann. Clin. Transl. Neurol. 3, 12–20.

Janelidze, S., Zetterberg, H., Mattsson, N., Palmqvist, S., Swedish BioFINDER Study Group, Hansson, O., 2016. CSF Aβ42/Aβ40 and Aβ42/Aβ38 ratios: better diagnostic markers of Alzheimer disease. Ann. Clin. Transl. Neurol. 3, 154–165.

Javaid, F.Z., Brenton, J., Guo, L., Cordeiro, M.F., 2016. Visual and ocular manifestations of Alzheimer's disease and their use as biomarkers for diagnosis and progression. Front. Neurol. 7, 55.

Jiang, L., Tu, Y., Shi, H., Cheng, Z., 2014. PET probes beyond ^{18}F-FDG. J. Biomed. Res. 28, 435–446.

Jun, G., Moncaster, J.A., Koutras, C., et al., 2012. δ-catenin is genetically and biologically associated with cortical cataract and future Alzheimer-related structural and functional brain changes. PLoS One 7, e43728.

Karran, E., Mercken, M., De Strooper, B., 2011. The amyloid cascade hypothesis for Alzheimer's disease: an appraisal for the development of therapeutics. Nat. Rev. Drug Discov. 10, 698–712.

Kiko, T., Nakagawa, K., Tsuduki, T., Furukawa, K., Arai, H., et al., 2014. MicroRNAs in plasma and cerebrospinal fluid as potential markers for Alzheimer's disease. J. Alzheimer's Dis. 39, 253–259.

Kim, M., Hwang, D., 2016. Network-based protein biomarker discovery platforms. Genomics Inform. 14, 2–11.

Klunk, W.E., Lopresti, B.J., Ikonomovic, M.D., Lefterov, I.M., Koldamova, R.P., et al., 2005. Binding of the positron emission tomography tracer Pittsburgh compound-B reflects the amount of amyloid-β in Alzheimer's disease brain but not in transgenic mouse brain. J. Neurosci. 25, 10598–10606.

Knight, M.J., McCann, B., Kauppinen, R.A., Coulthard, E.J., 2016. Magnetic resonance imaging to detect early molecular and cellular changes in Alzheimer's disease. Front. Aging Neurosci. 16 (8), 139.

Koronyo-Hamaoui, M., Koronyo, Y., Ljubimov, A.V., Miller, C.A., Ko, M.K., et al., 2011. Identification of amyloid plaques in retinas from Alzheimer's patients and noninvasive in vivo optical imaging of retinal plaques in a mouse model. NeuroImage 54 (Suppl. 1), S204–S217.

Kreisl, W.C., Lyoo, C.H., McGwier, M., Snow, J., Jenko, K.J., et al., 2013. In vivo radioligand binding to translocator protein correlates with severity of Alzheimer's disease. Brain 17, 17.

Kühbach, K., Hülsemann, M., Herrmann, Y., Kravchenko, K., Kulawik, A., et al., 2016. Application of an amyloid beta oligomer standard in the sFIDA assay. Front. Neurosci. 10, 8.

Lavados, M., Farias, G., Rothhammer, F., Guillon, M., Mujica, M.C., et al., 2005. ApoE alleles and tau markers in patients with different levels of cognitive impairment. Arch. Med. Res. 36, 474–479.

Lautner, R., Palmqvist, S., Mattsson, N., Andreasson, U., Wallin, A., et al., 2014. Apolipoprotein E genotype and the diagnostic accuracy of cerebrospinal fluid biomarkers for Alzheimer disease. JAMA Psychiatry 71, 1183–1191.

Leduc, V., Jasmin-Bélanger, S., Poirier, J., 2010. APOE and cholesterol homeostasis in Alzheimer's disease. Trends Mol. Med. 16, 469–477.

Lee, S., Tong, M., Hang, S., Deochand, C., de la Monte, S., et al., 2013. CSF and brain indices of insulin resistance, oxidative stress and neuro-inflammation in early versus late Alzheimer's disease. J. Alzheimer's Dis. Parkinsonism 3, 128.

Lei, X., Lei, L., Zhang, Z., Zhang, Z., Cheng, Y., 2015. Downregulated miR-29c correlates with increased BACE1 expression in sporadic Alzheimer's disease. Int. J. Clin. Exp. Pathol. 8, 1565–1574.

Lesser, G.T., Haroutunian, V., Purohit, D.P., Schnaider Beeri, M., et al., 2009. Serum lipids are related to Alzheimer's pathology in nursing home residents. Dement. Geriatr. Cogn. Disord. 27, 42–49.

Lesser, G.T., Beeri, M.S., Schmeidler, J., Purohit, D.P., Haroutunian, V., 2011. Cholesterol and LDL relate to neuritic plaques and to APOE4 presence but not to neurofibrillary tangles. Curr. Alzheimer Res. 8, 303–312.

Lewczuk, P., Lelental, N., Spitzer, P., Maler, J.M., Kornhuber, J., 2015. Amyloid-beta 42/40 cerebrospinal fluid concentration ratio in the diagnostics of Alzheimer's disease: validation of two novel assays. J. Alzheimer's Dis. 43, 183–191.

Li, L., Luo, J., Chen, D., Tong, J., Zeng, L., et al., 2016. BACE1 in the retina: a sensitive biomarker for monitoring early pathological changes in Alzheimer's disease. Neural Regen. Res. 11, 447–453.

Liu, Y.H., Zeng, F., Wang, Y.R., Zhou, H.D., Giunta, B., et al., 2013. Immunity and Alzheimer's disease: immunological perspectives on the development of novel therapies. Drug Discov. Today 18, 1212–1220.

Lockhart, A., Lamb, J.R., Osredkar, T., Sue, L.I., Joyce, J.N., et al., 2007. PIB is a non-specific imaging marker of amyloid-beta (Aβ) peptide-related cerebral amyloidosis. Brain 130, 2607–2615.

Löffler, J., Huber, G., 1992. β-Amyloid precursor protein isoforms in various rat brain regions and during brain development. J. Neurochem. 59, 1316–1324.

London, A., Benhar, I., Schwartz, M., 2013. The retina as a window to the brain – from eye research to CNS disorders. Nat. Rev. Neurol. 9, 44–53.

Lu, Y., Li, Z., Zhang, X., Ming, B., Jia, J., et al., 2010. Retinal nerve fiber layer structure abnormalities in early Alzheimer's disease: evidence in optical coherence tomography. Neurosci. Lett. 480, 69–72.

Lukiw, W.J., 2007. Micro-RNA speciation in fetal, adult and Alzheimer's disease hippocampus. NeuroReport 18, 297–300.

Luo, X., Hou, L., Shi, H., Zhong, X., Zhang, Y., et al., 2013. CSF levels of the neuronal injury biomarker visinin-like protein-1 in Alzheimer's disease and dementia with Lewy bodies. J. Neurochem. 127, 681–690.

Ma, W., Wang, S.-Z., 2006. The final fates of neurogenin2-expressing cells include all major neuron types in the mouse retina. Mol. Cell Neurosci. 31, 463–469.

Mahley, R.W., 1988. Apolipoprotein E: cholesterol transport protein with expanding role in cell biology. Science 240, 622–630.

Mandas, A., Abete, C., Putzu, P.F., la Colla, P., Dessì, S., et al., 2012. Changes in cholesterol metabolism-related gene expression in peripheral blood mononuclear cells from Alzheimer patients. Lipids Health Dis. 11, 39.

Marksteiner, J., 2007. Cerebrospinal fluid biomarkers for diagnosis of Alzheimer's disease: beta-amyloid(1-42), tau, phospho-tau-181 and total protein. Drugs Today 43, 423–431.

Mateos, L., Ismail, M.A., Gil-Bea, F.J., Leoni, V., Winblad, B., et al., 2011. Upregulation of brain renin angiotensin system by 27-hydroxycholesterol in Alzheimer's disease. J. Alzheimer's Dis. 24, 669–679.

Mathis, C.A., Lopresti, B.J., Klunk, W.E., 2007. Impact of amyloid imaging on drug development in Alzheimer's disease. Nucl. Med. Biol. 34, 809–822.

Matsubara, E., Takamura, A., Okamoto, Y., et al., 2013. Disease modifying therapies for Alzheimer's disease targeting Abeta oligomers: implications for therapeutic mechanisms. Biomed. Res. Int. 2013, 984041.

Matsuzaki, T., Sasaki, K., Tanizaki, Y., Hata, J., Fujimi, K., et al., 2010. Insulin resistance is associated with the pathology of Alzheimer disease: the Hisayama study. Neurology 75, 764–770.

McKhann, G., Knopman, D., Chertkow, H., Hyman, B., Jack, C.J., et al., 2011. The diagnosis of dementia due to Alzheimer's disease: recommendations from the National Institute on Aging-Alzheimer's Association workgroups on diagnostic guidelines for Alzheimer's disease. Alzheimer's Dement. 7, 263–269.

Melov, S., Wolf, N., Strozyk, D., Doctrow, S.R., Bush, A.I., 2005. Mice transgenic for Alzheimer disease β-amyloid develop lens cataracts that are rescued by antioxidant treatment. Free Radic. Biol. Med. 38, 258–261.

Mesholam, R.I., Moberg, P.J., Mahr, R.N., et al., 1998. Olfaction in neurodegenerative disease: a meta-analysis of olfactory functioning in Alzheimer's and Parkinson's diseases. Arch. Neurol. 55, 84–90.

Meyer-Luehmann, M., Spires-Jones, T.L., Prada, C., Garcia-Alloza, M., de Calignonz, A., et al., 2008. Rapid appearance and local toxicity of amyloid-β plaques in a mouse model of Alzheimer's disease. Nature 451, 720–724.

Mielke, M.M., Haughey, N.J., Bandaru, V.V., Weinberg, D.D., Darby, E., Zaidi, N., et al., 2011. Plasma sphingomyelins are associated with cognitive progression in Alzheimer's disease. J. Alzheimer's Dis. 27, 259–269.

Monge-Argilés, J.A., Gasparini-Berenguer, R., Gutierrez-Agulló, M., Muñoz-Ruiz, C., Sánchez-Payá, J., et al., 2016. Influence of APOE genotype on Alzheimer's disease CSF biomarkers in a Spanish population. Biomed. Res. Int. 2016, 1390620.

Mucke, L., Selkoe, D.J., 2012. Neurotoxicity of amyloid beta-protein: synaptic and network dysfunction. Cold Spring Harb. Perspect. Med. 2, a006338.

Müller, M., Kuiperij, H.B., Claassen, J.A., Kusters, B., Verbeek, M.M., 2014. MicroRNAs in Alzheimer's disease: differential expression in hippocampus and cell-free cerebrospinal fluid. Neurobiol. Aging 35, 152–158.

Murphy, C., Jernigan, T.L., Fennema-Notestine, C., 2003. Left hippocampal volume loss in Alzheimer's disease is reflected in performance on odor identification: a structural MRI study. J. Int. Neuropsychol. Soc. 9, 459–471.

Nakada, T., Matsuzawa, H., Igarashi, H., Fujii, Y., Kwee, I.L., 2008. In vivo visualization of senile-plaque-like pathology in Alzheimer's disease patients by MR microscopy on a 7T system. J. Neuroimaging 18, 125–129.

Ng, S., Villemagne, V.L., Berlangieri, S., Lee, S.T., Cherk, M., et al., 2007. Visual assessment versus quantitative assessment of ^{11}C-PIB PET and ^{18}F-FDG PET for detection of Alzheimer's disease. J. Nucl. Med. 48, 547–552.

Ning, A., Cui, J., To, E., Ashe, K.H., Matsubara, J., 2008. Amyloid-β deposits lead to retinal degeneration in a mouse model of Alzheimer disease. Invest. Ophthalmol. Vis. Sci. 49, 5136–5143.

Nisticò, R., Cavallucci, V., Piccinin, S., Macrì, S., Pignatelli, M., et al., 2012. Insulin receptor β-subunit haploinsufficiency impairs hippocampal late-phase LTP and recognition memory. Neuromolecular Med. 14, 262–269.

O'Brien, J.T., Firbank, M.J., Davison, C., Barnett, N., Bamford, C., et al., 2014. 18F-FDG PET and perfusion SPECT in the diagnosis of Alzheimer and Lewy body dementias. J. Nucl. Med. 55, 1959–1965.

Otto, M., Wiltfang, J., Cepek, L., Neumann, M., Mollenhauer, B., et al., 2002. Tau protein and 14-3-3 protein in the differential diagnosis of Creutzfeldt-Jakob disease. Neurology 58, 192–197.

Owen, D.R., Yeo, A.J., Gunn, R.N., Song, K., Wadsworth, G., et al., 2012. An 18-kDa translocator protein (TSPO) polymorphism explains differences in binding affinity of the PET radioligand PBR28. J. Cereb. Blood Flow. Metab. 32, 1–5.

Paquet, C., Boissonnot, M., Roger, F., Dighiero, P., Gil, R., et al., 2007. Abnormal retinal thickness in patients with mild cognitive impairment and Alzheimer's disease. Neurosci. Lett. 420, 97–99.

Pardini, M., Huey, E.D., Cavanagh, A.L., Grafman, J., 2009. Olfactory function in corticobasal syndrome and frontotemporal dementia. Arch. Neurol. 66, 92–96.

Parisi, V., Restuccia, R., Fattapposta, F., Mina, C., Bucci, M.G., et al., 2001. Morphological and functional retinal impairment in Alzheimer's disease patients. Clin. Neurophysiol. 112, 1860–1867.

Perani, D., Schillaci, O., Padovani, A., Nobili, F.M., Iaccarino, L., et al., 2014. A survey of FDG- and amyloid-PET imaging in dementia and GRADE analysis. Biomed. Res. Int. 2014, 785039.

Quinn, J.F., Montine, K.S., Moore, M., Morrow, J.D., Kaye, J.A., et al., 2004. Suppression of longitudinal increase in CSF F2-isoprostanes in Alzheimer's disease. J. Alzheimer's Dis. 6, 93–97.

Randall, J., Mörtberg, E., Provuncher, G.K., Fournier, D.R., Duffy, D.C., et al., 2013. Tau proteins in serum predict neurological outcome after hypoxic brain injury from cardiac arrest: results of a pilot study. Resuscitation 84, 351–356.

Reiber, H., 2003. Proteins in cerebrospinal fluid and blood: barriers, CSF flow rate and source-related dynamics. Restor. Neurol. Neurosci. 21, 79–96.

Risacher, S.L., Saykin, A.J., West, J.D., Shen, L., Firpi, H.A., et al., 2009. Baseline MRI predictors of conversion from MCI to probable AD in the ADNI cohort. Curr. Alzheimer Res. 6, 347–361.

Rosenmann, H., 2012. CSF biomarkers for amyloid and tau pathology in Alzheimer's disease. J. Mol. Neurosci. 47, 1–14.

Rosén, C., Hansson, O., Blennow, K., Zetterberg, H., 2013. Fluid biomarkers in Alzheimer's disease – current concepts. Mol. Neurodegener. 8, 20.

Rubio-Perez, J.M., Morillas-Ruiz, J.M., 2012. A review: inflammatory process in Alzheimer's disease, role of cytokines. Sci. World J. 756357.

Sadun, A.A., Bassi, C.J., 1990. Optic nerve damage in Alzheimer's disease. Ophthalmology 97, 9–17.

Savva, G.M., Wharton, S.B., Ince, P.G., Forster, G., Matthews, F.E., Brayne, C., 2009. Age, neuropathology, and dementia. N. Engl. J. Med. 360, 2302–2309.

Sala Frigerio, C., Lau, P., Salta, E., Tournoy, J., Bossers, K., et al., 2013. Reduced expression of hsa-miR-27a-3p in CSF of patients with Alzheimer disease. Neurology 81, 2103–2106.

Schöll, M., Samuel Lockhart, N., Daniel Schonhaut, R., James O'Neil, P., Janabi, M., et al., 2016. PET imaging of Tau deposition in the aging human brain. Neuron 89, 971–982.

Sethi, P., Lukiw, W.J., 2009. Micro-RNA abundance and stability in human brain: specific alterations in Alzheimer's disease temporal lobe neocortex. Neurosci. Lett. 459, 100–104.

Schoonenboom, N.S.M., Reesink, F.E., Verwey, N.A., Kester, M.I., Teunissen, C.E., et al., 2012. Cerebrospinal fluid markers for differential dementia diagnosis in a large memory clinic cohort. Neurology 78, 47–54.

Schuitemaker, A., Kropholler, M.A., Boellaard, R., van der Flier, W.M., Kloet, R.W., et al., 2013. Microglial activation in Alzheimer's disease: an (R)-[(1)(1)C]PK11195 positron emission tomography study. Neurobiol. Aging 34, 128–136.

Siegel, S.R., Mackenzie, J., Chaplin, G., Jablonski, N.G., Griffiths, L., 2012. Circulating microRNAs involved in multiple sclerosis. Mol. Biol. Rep. 39, 6219–6225.

Sjögren, M., Davidsson, P., Gottfries, J., Vanderstichele, H., Edman, A., et al., 2001. The cerebrospinal fluid levels of tau, growth-associated protein-43 and soluble amyloid precursor protein correlate in Alzheimer's disease, reflecting a common pathophysiological process. Dement. Geriatr. Cogn. Disord. 12, 257–264.

Small, G.W., Kepe, V., Ercoli, L.M., Siddarth, P., Bookheimer, S.Y., et al., 2006. PET of brain amyloid and Tau in mild cognitive impairment. N. Engl. J. Med. 355, 2652–2663.

Snider, B.J., Fagan, A.M., Roe, C., Shah, A.R., Grant, E.A., et al., 2009. Cerebrospinal fluid biomarkers and rate of cognitive decline in very mild dementia of the Alzheimer type. Arch. Neurol. 66, 638–645.

Sperling, R.A., Aisen, P.S., Beckett, L.A., Bennett, D.A., Craft, S., Fagan, A.M., et al., 2011. Toward defining the preclinical stages of Alzheimer's disease: recommendations from the National Institute on Aging-Alzheimer's Association workgroups on diagnostic guidelines for Alzheimer's disease. Alzheimer's Dement. 7, 280–292.

Spillantini, M.G., Goedert, M., 2013. Tau pathology and neurodegeneration. Lancet Neurol. 12, 609–622.

Špolcová, A., Mikulášková, B., Kršková, K., Gajdošechová, L., Zórad, Š., et al., 2014. Deficient hippocampal insulin signaling and augmented Tau phosphorylation is related to obesity- and age-induced peripheral insulin resistance: a study in Zucker rats. BMC Neurosci. 15, 111.

Stöhr, O., Schilbach, K., Moll, L., Hettich, M.M., Freude, S., et al., 2013. Insulin receptor signaling mediates APP processing and beta-amyloid accumulation without altering survival in a transgenic mouse model of Alzheimer's disease. Age (Dordr.) 35, 83–101.

Sun, G.H., Raji, C.A., Maceachern, M.P., Burke, J.F., 2012. Olfactory identification testing as a predictor of the development of Alzheimer's dementia: a systematic review. Laryngoscope 122, 1455–1462.

Surguchev, A., Surguchov, A., 2010. Conformational diseases: looking into the eyes. Brain Res. Bull. 81, 12–24.

Tapiola, T., Alafuzoff, I., Herukka, S.K., Parkkinen, L., Hartikainen, P., et al., 2009. Cerebrospinal fluid {beta}-amyloid 42 and tau proteins as biomarkers of Alzheimer-type pathologic changes in the brain. Arch. Neurol. 66, 382–389.

Tarawneh, R., Head, D., Allison, S., Buckles, V., Fagan, A.M., Ladenson, J.H., Morris, J.C., Holtzman, D.M., 2015. Cerebrospinal fluid markers of neurodegeneration and rates of brain atrophy in early Alzheimer disease. JAMA Neurol. 72, 656–665.

Tarawneh, R., D'Angelo, G., Crimmins, D., Herries, E., Griest, T., et al., 2016. Diagnostic and prognostic utility of the synaptic marker neurogranin in Alzheimer disease. JAMA Neurol. 73, 561–571.

Thal, L.J., Kantarci, K., Reiman, E.M., Klunk, W.E., Weiner, M.W., et al., 2006. The role of biomarkers in clinical trials for Alzheimer disease. Alzheimer Dis. Assoc. Disord. 20, 6–15.

Tibolla, G., Norata, G.D., Meda, C., Arnaboldi, L., Uboldi, P., et al., 2010. Increased atherosclerosis and vascular inflammation in APP transgenic mice with apolipoprotein E deficiency. Atherosclerosis 210, 78–87.

Tolboom, N., van der Flier, W.M., Yaqub, M., Boellaard, R., Verwey, N.A., et al., 2009. Relationship of cerebrospinal fluid markers to 11C-PiB and 18F-FDDNP binding. J. Nucl. Med. 50, 1464–1470.

Tuoc, T.C., Radyushkin, K., Tonchev, A.B., Piñon, M.C., Ashery-Padan, R., et al., 2009. Selective cortical layering abnormalities and behavioral deficits in cortex-specific Pax6 knock-out mice. J. Neurosci. 29, 8335–8349.

Uomori, K., Murakami, S., Yamada, M., Fujii, M., Yoshimatsu, H., et al., 1993. Analysis of gaze shift in-depth in Alzheimers-disease patients. IEICE Trans. Inf. Syst. E76D, 963–973.

Varghese, T., Sheelakumari, R., James, J.S., Mathuranath, P., 2013. A review of neuroimaging biomarkers of Alzheimer's disease. Neurol. Asia 18, 239–248.

Velayudhan, L., Lovestone, S., 2009. Smell identification test as a treatment response marker in patients with Alzheimer disease receiving donepezil. J. Clin. Psychopharmacol. 29, 387–390.

Vanderstichele, H., Bibl, M., Engelborghs, S., Le Bastard, N., Lewczuk, P., et al., 2012a. Standardization of preanalytical aspects of cerebrospinal fluid biomarker testing for Alzheimer's disease diagnosis: a consensus paper from the Alzheimer's Biomarkers Standardization Initiative. Alzheimer's Dement. 8, 65–73.

Van Harten, A.C., Kester, M.I., Visser, P.J., Blankenstein, M.A., Pijnenburg, Y.A.L., et al., 2011. Tau and p-tau as CSF biomarkers in dementia: a meta-analysis. Clin. Chem. Lab. Med. 49, 353–366.

Vanderstichele, H., Stoops, E., Vanmechelen, E., Jeromin, A., 2012b. Potential sources of interference on Abeta immunoassays in biological samples. Alzheimer Res. Ther. 4, 39.

Vandenberghe, R., Adamczuk, K., Dupont, P., Laere, K.V., Chételat, G., 2013. Amyloid PET in clinical practice: its place in the multidimensional space of Alzheimer's disease. Neuroimage Clin. 2, 497–511.

Venneti, S., Lopresti, B.J., Wiley, C.A., 2013. Molecular imaging of microglia/macrophages in the brain. Glia 61, 10–23.

Vijayaraghavan, S., Maetzler, W., Reimold, M., Lithner, C.U., Liepelt-Scarfone, I., Berg, D., et al., 2014. High apolipoprotein E in cerebrospinal fluid of patients with Lewy body disorders is associated with dementia. Alzheimer's Dement. 10, 530–540.

Wachowiak, M., Shipley, M.T., 2006. Coding and synaptic processing of sensory information in the glomerular layer of the olfactory bulb. Semin. Cell Dev. Biol. 17, 411–423.

Weigand, S.D., Vemuri, P., Wiste, H.J., Senjem, M.L., Pankratz, V.S., et al., 2011. Transforming cerebrospinal fluid Abeta42 measures into calculated Pittsburgh compound B units of brain Abeta amyloid. Alzheimer's Dement. 7, 133–141.

Weiner, M.W., Veitch, D.P., Aisen, P.S., Beckett, L.A., Cairns, N.J., et al., 2012. The Alzheimer's Disease Neuroimaging Initiative: a review of papers published since its inception. Alzheimer's Dement. 8, S1–S68.

Weiner, M.W., Veitch, D.P., Aisen, P.S., Beckett, L.A., Cairns, N.J., et al., 2015. 2014 Update of the Alzheimer's Disease Neuroimaging Initiative: a review of papers published since its inception. Alzheimer's Dement. 11, e1–e120.

Weissleder, R., Mahmood, U., 2001. Molecular imaging. Radiology 219, 316–333.

Westervelt, H.J., Carvalho, J., Duff, K., 2007. Presentation of Alzheimer's disease in patients with and without olfactory deficits. Arch. Clin. Neuropsychol. 22, 117–122.

Wiley, C.A., Lopresti, B.J., Venneti, S., Price, J., Klunk, W.E., et al., 2009. Carbon 11-labeled Pittsburgh compound B and carbon 11-labeled (R)-PK11195 positron emission tomographic imaging in Alzheimer disease. Arch. Neurol. 66, 60–67.

Williams, P.A., Thirgood, R.A., Oliphant, H., Frizzati, A., Littlewood, E., et al., 2013. Retinal ganglion cell dendritic degeneration in a mouse model of Alzheimer's disease. Neurobiol. Aging 34, 1799–1806.

Wilson, D.A., Stevenson, R.J., 2006. Learning to Smell Olfactory Perception from Neurobiology to Behavior. The Johns Hopkins University Press.

Wilson, R.S., Arnold, S.E., Schneider, J.A., Tang, Y., Bennett, D.A., 2007. The relationship between cerebral Alzheimer's disease pathology and odour identification in old age. J. Neurol. Neurosurg. Psychiatry 78, 30–35.

Wright, S., MacAskill, M., 2012. Antisaccades in Alzheimer's disease using fMRI. Alzheimer's Dement. 8, 51.

Yokokura, M., Mori, N., Yagi, S., Yoshikawa, E., Kikuchi, M., et al., 2011. In vivo changes in microglial activation and amyloid deposits in brain regions with hypometabolism in Alzheimer's disease. Eur. J. Nucl. Med. Mol. Imaging 38, 343–351.

Yoneda, S., Hara, H., Hirata, A., Fukushima, M., Inomata, Y., Tanihara, H., 2005. Vitreous fluid levels of β-amyloid (1–42) and tau in patients with retinal diseases. Jpn. J. Ophthalmol. 49, 106–108.

Zetterberg, H., Blennow, K., Hanse, E., 2010. Amyloid beta and APP as biomarkers for Alzheimer's disease. Exp. Gerontol. 45, 23–29.

Zetterberg, H., Wilson, D., Andreasson, U., Minthon, L., Blennow, K., et al., 2013. Plasma tau levels in Alzheimer's disease. Alzheimer's Res. Ther. 5, 9.

Zhao, Y., Pogue, A.I., Lukiw, W.J., 2015. MicroRNA (miRNA) signaling in the human CNS in sporadic Alzheimer's disease (AD)-novel and unique pathological features. Int. J. Mol. Sci. 16, 30105–30116.

Zou, Y.M., Lu, D., Liu, L.P., Zhang, H.H., Zhou, Y.Y., 2016. Olfactory dysfunction in Alzheimer's disease. Neuropsychiatr. Dis. Treat. 12, 869–875.

Potential Treatments for Alzheimer's Disease

INTRODUCTION

Alzheimer's disease (AD) is an irreversible, progressive, and multifactorial neurodegenerative disorder associated with deterioration of memory that hampers the affected person's day-to-day life. AD is featured by the accumulation of β-amyloid (Aβ) containing senile plaques and microtubule-associated protein (Tau) containing neurofibrillary tangles (NFTs). The pathogenesis of AD includes a long preclinical phase, lasting a decade or more (Sperling et al., 2011; Bateman et al., 2012; Villemagne et al., 2013). During this preclinical phase, patients with AD are cognitively intact (i.e., they score within the normal range on neuropsychiatric tests), but the start of neurological disease can be observed through cerebrospinal fluid (CSF) testing or brain imaging analyses. In this preclinical phase (the dementia phase), before the onset of irreversible neurodegeneration an effective treatment of AD can be performed. As stated in earlier chapters, AD is not only accompanied by blood–brain barrier (BBB) disruption, oxidative stress, mitochondrial impairment, neuroinflammation, hypometabolism, and aberrant cell cycle reentry but also is related to Aβ peptide accumulation and Tau hyperphosphorylation as well as a decrease in acetylcholine levels and a reduction of cerebral blood flow (Mondragon-Rodriguez et al., 2010; Farooqui, 2016). Consumption of Western diet and resulting obesity are major risk factors for the onset of AD, because Western diet induces adipokine dysregulation, which consists of the release of the proinflammatory adipokines and decrease in antiinflammatory adipokines along with increase in levels of proinflammatory lipid mediators and proinflammatory cytokines (Farooqui, 2016). Causes of AD are not known, and controversy persists over which abnormalities initiate the pathogenesis of AD, what are the factors that contribute to the neurodegenerative process, and even whether some of these abnormalities represent neuroprotective mechanisms in the brain (Mondragon-Rodriguez et al., 2010; Shinohara et al., 2014). It is estimated that currently 47 million patients with AD exist worldwide and that number is expected to grow to more than 130 million by 2050 as a result of life expectancy increase over the coming decades (Thies and Bleiler, 2012, 2013). Epidemiological data project that while the prevalence of AD is 5% after the age of 65 years and 25% after the age of 85 years, delaying the onset of the clinical phase of the disease by just 1 year reduces its prevalence by 25%, and a 5-year delay in onset would decrease the prevalence in the population by 50% after 5 years of application of preventive measures (Brookmeyer and Gray, 2000). In 2015, the World Alzheimer Report has estimated that the current annual societal and economic cost of dementia is US $818 billion worldwide and that this amount is expected to rise to 1 trillion by 2018. This study also reported that the cost associated with dementia has increased by 35% since 2010. As stated in Chapters 1 and 2, hypercholesterolemia, type 2 diabetes, and MetS are important risk factors for the development and onset of AD. Thus, the treatment of AD should target multiple pathological processes including oxidative and metabolic stress and neuroinflammation. Increasing evidence supports the view that vascular and neurodegenerative pathologies often coexist with metabolic dysfunction (insulin resistance, inflammation, and oxidative stress) along with neurovascular dysfunction (BBB disruption and decrease in blood flow. These factors play a critical role in the development or progression of AD (Iadecola, 2013; Farooqui, 2013, 2015).

Key to treat AD is the complete understanding of the processes and mechanisms that trigger neurodegenerative process. The neuropharmacological treatments for AD are divided into two categories: symptomatic treatments such as acetylcholinesterase (AChE) inhibitors and N-methyl-D-aspartate (NMDA) receptor (NMDAR) antagonists and etiology-based treatments such as secretase inhibitors, amyloid binders, and Tau therapies. Strategies for prevention or retarding AD through nonpharmacological treatments involve not only lifestyle such as healthy diet but also regular exercise, optimal sleep, mental challenges, and socialization as well as caloric restriction. The most commonly used therapeutic agents are inhibitors of the AChE, which mainly relieves cognitive symptoms and produces weak disease-modifying effects on AD. Magnetic resonance imaging (MRI) and positron emission tomographic brain scanning studies have indicated that early signs of AD pathology in patients appear ~4–17 years before the onset of dementia (Villemagne et al., 2013). In addition to the AChE inhibitors, other therapeutic approaches that are commonly available at present include antioxidant and antiinflammatory strategies, statin therapy, memantine and nitromemantine therapy, gene therapy, immunization with vaccines, insulin therapy, and stem cell therapy.

CHOLINERGIC STRATEGIES FOR THE TREATMENT OF ALZHEIMER'S DISEASE

According to cholinergic hypothesis, AD is caused by reduction in the synthesis of acetylcholine cholinergic neurons of basal forebrain and loss of cholinergic neurotransmission in the cerebral cortex. The decrease in acetylcholine significantly contributes to the deterioration in cognitive function seen in patients with AD (Birks, 2006). AChE hydrolyzes acetylcholine, a neurotransmitter that plays an important role in learning, remembering, and thinking. The inhibition of AChE by AChE inhibitors reduces the breakdown of acetylcholine and increases the availability of acetylcholine at the synapse resulting in restoration of cognition and memory function. AChE inhibitors have poor oral bioavailability, brain penetration ability, and pharmacokinetic parameters. To overcome these disadvantages, new generation of AChE inhibitors, such as donepezil, galantamine, and rivastigmine, have been synthesized and used for the treatment of AD in animal models and human patients (Fig. 8.1) (Birks, 2006).

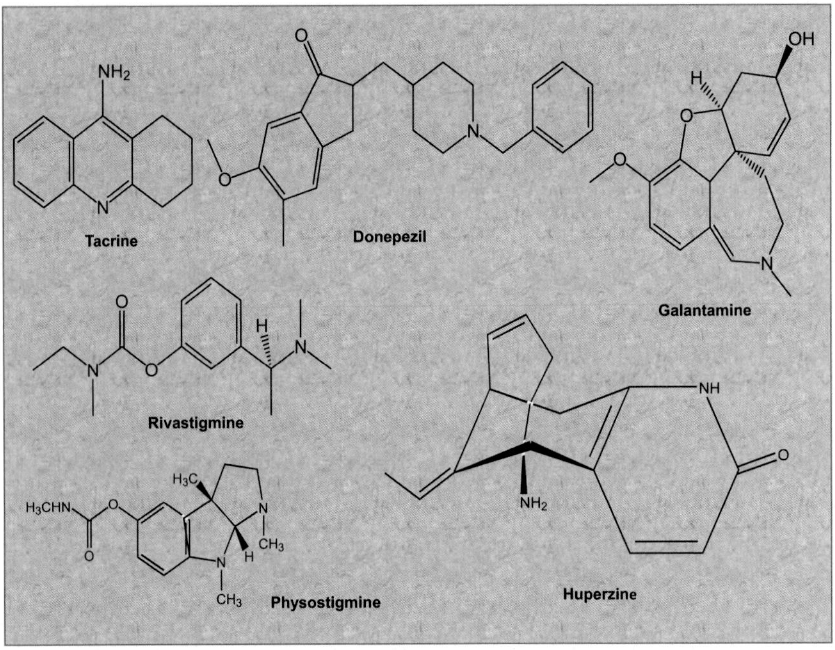

FIGURE 8.1 Chemical structures of acetylcholinesterase inhibitors (tacrine, donepezil, galantamine, rivastigmine, and physostigmine) and huperzine.

Donepezil, galantamine, and rivastigmine are currently approved for use in mild to moderate AD, with rivastigmine also available as a transdermal patch. These inhibitors not only reduce AChE activity but also retard processing and deposition of Aβ (Muñoz-Torrero, 2008). In addition, these inhibitors also increase the cerebral blood flow in patients with AD both after acute and fairly short period of treatment (Nordberg, 1999). Most AChE inhibitors produce only modest effects in delaying the progression of mild to moderate AD. The common side effects associated with cholinesterase inhibitors include nausea, vomiting, diarrhea, abdominal pain, loss of appetite, muscle cramps, insomnia, and nightmares (Alva and Cummings, 2008).

Huperzine A (HupA) is a plant-based alkaloid and potent, selective, and well-tolerated inhibitor of AChE (Fig. 8.1) (Damar et al., 2016). It is derived from a Chinese herb called *Huperzia serrata*. HupA can cross the BBB, and its potency is similar or superior to other AChE inhibitors (physostigmine, galantamine, donepezil, and tacrine) (Wang and Tang, 2005). HupA also displays good pharmacokinetics with a rapid absorption and a wide distribution in the body at a low to moderate rate of elimination. In addition to anti-AChE activity, HupA produces neuroprotective effects through brain iron regulation. HupA treatment not only reduces insoluble and soluble $A\beta$ levels and ameliorates $A\beta$ plaques formation but also modulates hyperphosphorylation of Tau in the cortex and hippocampus of APPswe/PSEN1dE9 transgenic AD mice (Wang et al., 2012). In addition, HupA reduces $A\beta$ oligomers and amyloid precursor protein (APP) levels and increases disintegrin A and metalloprotease domain 10 ["a disintegrin and metalloproteinase" 10 (ADAM10)] expression in HupA-treated AD mice. However, these beneficial effects of HupA are largely abolished by feeding the animals with a high-iron diet. It is also reported that HupA decreases iron content in the brain and plays an important role in reducing the expression of transferrin receptor 1 as well as the transferrin-bound iron uptake in cultured neurons. Recent studies have also indicated that the loss of dendritic spine density and synaptotagmin levels in the brain of APPswe/presenilin-1 (PSEN1) transgenic mice can be significantly ameliorated by chronic HupA treatment supporting the view that HupA-mediated neuroprotection is associated with reduction in $A\beta$ plaque burden and oligomeric $A\beta$ levels in the cortex and hippocampus of APPswe/PSEN1dE9 transgenic mice (Wang et al., 2012). It is also demonstrated that the amelioration effect of HupA on $A\beta$ deposits may be mediated, at least in part, by regulation of the compromised expression of a disintegrin and metalloprotease 10 (ADAM10) and excessive membrane trafficking of β-site APP cleavage enzyme 1 (BACE1) in these transgenic mice. In addition, extracellular signal–regulated kinases 1/2 (Erk1/2) phosphorylation may also be partially involved in the effect of HupA on APP processing. In conclusion, our work for the first time demonstrates the neuroprotective effect of HupA on synaptic deficits in APPswe/PSEN1dE9 transgenic mice and further clarifies the potential pharmacological targets for this protective effect, in which modulation of nonamyloidogenic and amyloidogenic APP processing pathways may be both involved. These findings may provide adequate evidence for the clinical and experimental benefits gained from HupA treatment. These observations support the view that reduction of iron levels in the brain is a novel mechanism of HupA action in the treatment of AD (Huang et al., 2014).

Based on the aforementioned information, HupA has been used for the treatment of AD in human clinical trials in China and the United States. So far, Chinese clinical studies have shown an improvement in the memory of patients with AD (Wang et al., 2009; Zhang et al., 2002). The phase 4 clinical trials in China have demonstrated that HupA can significantly improve memory deficits not only in elderly people with benign senescent forgetfulness but also in patients with AD and vascular dementia, with minimal peripheral cholinergic side effects and no unexpected toxicity (Zhang et al., 2008; Wang et al., 2009). In the United States, a multicenter (29 centers in 17 states), double-blind, placebo-controlled phase 2 clinical trial has shown cognitive improvement in patients with mild to moderate AD (ClinicalTrials.gov; NCT00083590; Rafii et al., 2011).

MEMANTINE FOR THE TREATMENT OF ALZHEIMER'S DISEASE

Memantine (*Namenda* or *Ebixa*) is a derivative of amantadine and noncompetitive NMDAR antagonist (Fig. 8.2). Memantine binds to NMDARs

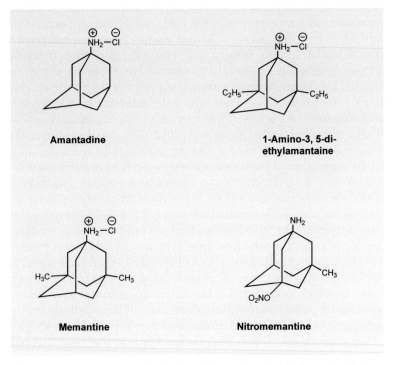

FIGURE 8.2 Chemical structures of amantadine, 1-amino-3,5-diethylamantaine, memantine, and nitromemantine.

with a low micromolar IC_{50} value. Furthermore, it exhibits poor selectivity among NMDAR subtypes, with less than micromolar IC_{50} values for NR2A, NR2B, NR2C, and NR2D receptors expressed in *Xenopus* oocytes (Parsons et al., 1999). Memantine exhibits neuroprotective activities against

Aβ toxicity (Hu et al., 2007), Tau phosphorylation (Song et al., 2008), neuro-inflammation (Willard et al., 2000), and oxidative stress (Figueiredo et al., 2013; Liu et al., 2013). It has been used for the treatment of not only AD, but also Parkinson's disease (PD), epilepsy, attention-deficit/hyperactivity disorder, vascular dementia, and fibromyalgia. Memantine acts as a neuroprotective agent by decreasing glutamate excitotoxicity. Memantine also increases levels of brain-derived neurotrophic factor (BDNF), a growth factor that modulates synaptic plasticity in rats (Picada et al., 2011). In addition, memantine not only blocks Kv1.3 potassium channels and inhibits CD3-antibody-induced and alloantigen-induced proliferation but also suppresses chemokine-mediated migration of peripheral blood T cells of healthy donors. Furthermore, memantine treatment results in a profound depletion of peripheral blood memory CD45RO + CD4+ T cells supporting the view that standard doses of memantine markedly reduce T-cell responses in treated patients through blockade of Kv1.3 channels. This may normalize deviant immunopathology in AD and contribute to the beneficial effects of memantine, but may also account for the enhanced infection rate (Lowinus et al., 2016). In the Morris water maze, memantine improves acquisition performance, improves spatial accuracy, and increases durability of synaptic plasticity (Sahiner et al., 2011). The clinically approved dose of memantine for humans starts with 5 mg/day, increasing progressively over a period of several weeks to 20 mg/day. This progressive dose adjustment may contribute to the drug's lack of side effects (Reisberg et al., 2003). At higher doses (7.5–20 mg/kg, administered subcutaneously), memantine attenuates morphine-induced tolerance, physical dependence, and drug-seeking effects in animals (Ribeiro Do Couto et al., 2004). The administration of memantine to various transgenic AD mice has been reported to improve cognitive deficits, very often completely back to normal wild-type control levels. However, such great benefits of memantine in preclinical studies do not translate into clinical results of this drug, showing only marginal and transient efficacy in moderate to severe AD. In patients with AD, memantine produces statistically significant effect on cognition, behavior, and the ability to perform activities of daily living (McShane et al., 2006). A small reduction in agitation has been consistently observed. However, trials examining memantine have been limited by high dropout rates, and the benefits identified, although results were statistically significant to a small magnitude. A recent 2-year trial has provided further evidence that memantine does not modify disease progression and is ineffective in mild AD (Dysken et al., 2014). Collective evidence suggests that like the cholinesterase inhibitors, memantine provides symptomatic relief to some but has failed to provide universal benefit in AD. It produces side effects such as dizziness, anorexia, vomiting, and diarrhea (Alva and Cummings, 2008). Recently, investigators have used Namzaric [fixed-dose combination of memantine extended-release (ER)/donepezil 28/10 mg] for the treatment of patients with moderate to severe AD (Raina et al., 2008; Greig, 2015). This fixed-dose formulation is bioequivalent to coadministration of the individual drugs. In a 24-week, phase 3 trial in patients with moderate to severe AD, addition of memantine ER 28 mg once daily to stable AChE inhibitor therapy proves more effective than add-on placebo on measures of cognition, global clinical status, dementia behavior, and semantic processing ability. Namzaric is generally well tolerated in the phase 3 trial, with diarrhea, dizziness, and influenza occurring at least twice (Greig, 2015).

SECTRETASE INHIBITORS AND MODULATORS FOR THE TREATMENT OF ALZHEIMER'S DISEASE

As stated in Chapter 1, the processing of APP for Aβ generation requires the sequential cleavage of APP by β-secretase and γ-secretase (Holtzman et al., 2012; Selkoe et al., 2012). A large body of evidence supports the view that an increase in Aβ42 to Aβ40 ratio may modulate the structure of toxic species and that excessive Aβ40/42 peptides induce AD-relevant changes in neuronal structure and function (Palop and Mucke, 2010; Freir et al., 2011). Soluble Aβ oligomer contributes to Aβ toxicity in animal model and human patients (Holtzman et al., 2012; Shankar et al., 2007). Current therapeutic efforts on drug development are targeted toward elimination/reduction in the production of Aβ40 and Aβ42 (Freir et al., 2011; Coric et al., 2012). The γ-secretase complex is responsible for the final stage of amyloidogenesis, leading to the generation of Aβ40 and Aβ42. One approach involves the use of γ-secretase inhibitors (GSIs) to prevent production of all Aβ peptides (Wolfe, 2012; Tamayev and D'Adamio, 2012). Examples of substrate-based inhibitor of γ-secretase are the difluoroketone peptidomimetic compound called MW167 and DFK167 (Wolfe et al., 1998). These compounds inhibit serine or cysteine protease in its keto form or an aspartyl protease in its hydrated form. However, difluoroalcohol analogs of MW167 and DFK167 (Fig. 8.3) have been reported to inhibit γ-secretase activity (Wolfe et al., 1999). Other GSIs include LY450139 (semagacestat) from Eli Lilly. This inhibitor shows very little selectivity for APP compared to Notch protein (a protein responsible for regulating cell proliferation, development, differentiation, and cellular communication) raising concern that doses that effectively lower brain Aβ production may cause systemic toxicity owing to inhibition of Notch signaling. Phase 3 clinical trials of semagacestat have been discontinued due to detrimental impacts not only on both cognition and daily function but also due to increased incidences of skin cancer and gastrointestinal side effects (Fleisher et al., 2008; Coric et al., 2012; Doody et al., 2013; Quintero-Monzon et al., 2011; Crump et al., 2012). Increase in cognitive dysfunction

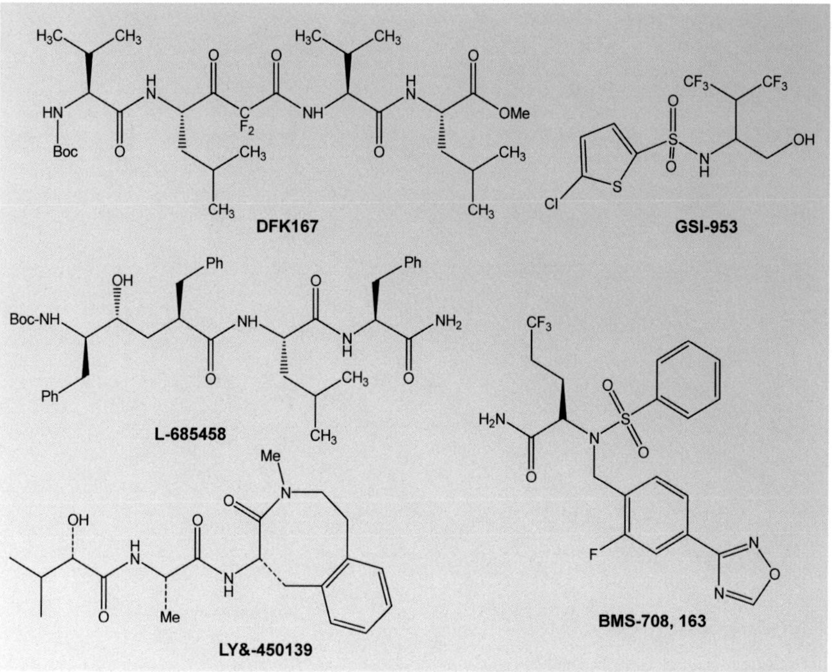

FIGURE 8.3 Chemical structures of transition-state analog inhibitors of γ-secretase (GSI-953 and L-685458) and DAPT and related analogs (LY-450139, GSI-953, and BMS-708).

following semagacestat treatment has been recently observed in the AD mouse model of Tg2576 as well as in wild-type mice (Mitani et al., 2012). An alternative approach for decreasing Aβ42 and Aβ40 levels in the brain is to enhance rather than inhibit the activity of γ-secretase via modulators of this enzyme complex. These small molecules are called as γ-secretase modulators (GSMs) (Fig. 8.4) (Kounnas et al., 2010; Wagner et al., 2014; Wolfe, 2012). GSMs are known to decrease levels of Aβ42 and Aβ40 while increasing the levels of shorter Aβ peptides, such as Aβ38, without affecting total Aβ levels (Kounnas et al., 2010; Mitani et al., 2012; Oehlrich et al., 2011). Because shorter Aβ peptides are regarded as nonpathogenic or less pathogenic (Holtzman et al., 2012; Selkoe et al., 2012; Wolfe, 2012), the use of GSMs has been suggested for decreasing Aβ levels in the brain.

5-Lipoxygenase (5-LOX) is a dioxygenase that catalyzes the formation of leukotrienes (LTs). 5-LOX catalyzes the first two steps in LT formation. Biosynthesis starts with dioxygenation of arachidonic acid (ARA) released by cytosolic phospholipase A_2 from cellular glycerophospholipids. This reaction results in two chemically unstable intermediates, 5(S)-hydroperoxy-6-trans-8,11,14-cis-eicosatetraenoic acid (5-HPETE) and LTA4. A noheme coordinated iron atom, located in the catalytic site

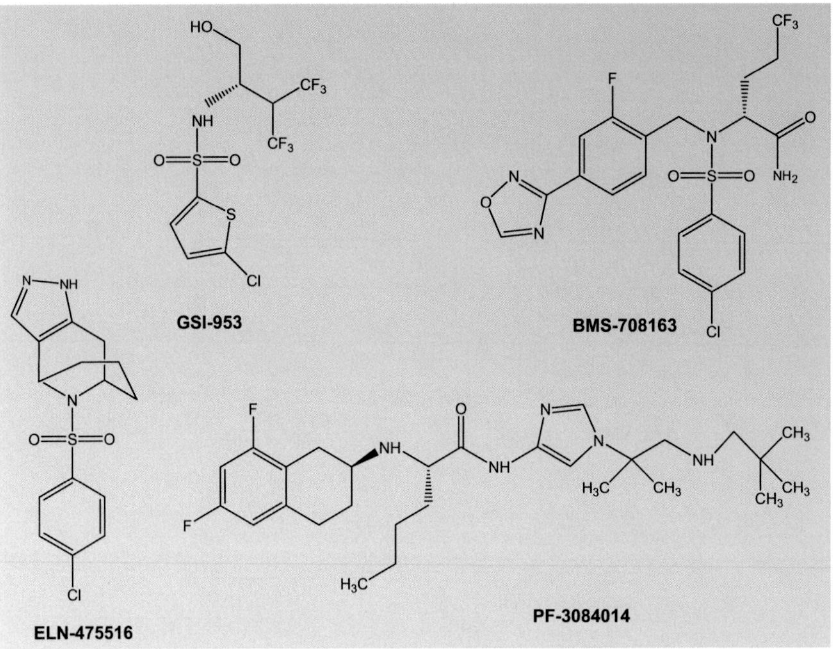

FIGURE 8.4   Chemical structures of notch-sparing γ-secretase modulators.

of 5-LOX, is involved in both chemical reactions. Depending on the cellular enzymes present, LTA4 can be either converted to LTB4 by LTA4 hydrolase or conjugated with glutathione by LTC4 synthase to generate LTC4. Further processing of LTC4 produces LTD4 and LTE4. 5-HPETE can be reduced by glutathione peroxidases to form the corresponding 5(S)-hydroxy-6-trans-8,11,14-cis-eicosatetraenoic acid (5-HETE) (Radmark et al., 2007). In resting cells, 5-LOX resides in either the nucleus or the cytosol, depending on the cell type. Upon activation, 5-LOX translocates to the nuclear membrane, where the 5-LOX–activating protein (FLAP) is thought to facilitate the transfer of glycerophospholipid-derived ARA to 5-LOX and to enhance the efficiency of conversion of 5-HPETE to LTA4 thereby triggering 5-LOX product formation (Abramovitz et al., 1993; Mancini et al., 1993). These lipid mediators contribute to oxidative stress and neuroinflammation in AD. 5-LOX also contributes the Aβ peptide formation not only by modulating γ-secretase activity but also by influencing other molecular pathologies such as synaptic integrity and cognitive dysfunctions (Fig. 8.5) (Chu and Praticò, 2011). 5-LOX also modulates γ-secretase messenger RNA (mRNA) level through the phosphorylation of cyclic adenosine response element–binding (CREB) protein. γ-Secretase not only produces Aβ peptides but also contribute to the processing of Notch, a single-pass transmembrane receptor protein critical for neuronal

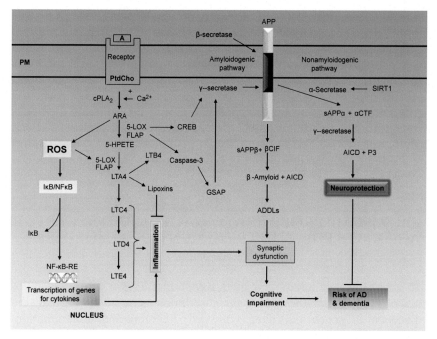

FIGURE 8.5   **Modulation of Aβ generation by 5-lipoxygenase and inhibition of inflammation by lipoxins.** *5-HPETE*, 5-hydroperoxyeicosatetraenoic acid; *5-LOX*, 5-lipoxygenase; *A*, agonist; *ADDL*, soluble Aβ oligomer; *APP*, amyloid precursor protein; *ARA*, arachidonic acid; *ASID*, substrate inhibitory domain; *cPLA$_2$*, cytosolic phospholipase A$_2$; *CREB*, cyclic adenosine response element–binding protein; *CTFα*, C-terminal fragments of APP reflecting α-secretase processing; *CTFβ*, C-terminal fragments of APP reflecting γ-secretase processing; *IκB*, inhibitory subunit of NF-κB; *LTA4, LTC4, LTD4, LTE4,* and *LTB4*, leukotrienes; *NF-κB*, nuclear factor-κB; *NF-κB-RE*, nuclear factor-κB response element; *PM*, plasma membrane; *PtdCho*, phosphatidylcholine; *ROS*, reactive oxygen species.

functioning and differentiation (Imbimbo and Giardina, 2011). As stated earlier, clinical trials on the use of GSIs in AD have to date been largely unsuccessful because they alter Notch signaling as well as Aβ production (Doody et al., 2013). With 5-LOX modulation, γ secretase-dependent Notch production is unperturbed, suggesting feasible future use of this class of drug. It is reported that FLAP, a protein that activates 5-LOX, also produces similar effect on Aβ production in FLAP knockouts (Chu and Praticò, 2012, 2013). Similar results have been obtained not only on neuronal cells in vitro but also in mouse models of AD-like amyloidosis (Giannopoulos et al., 2013, 2014). Recently, these results have also been reproduced in TgCRND8 animals, with reduction found in amyloid-associated angiopathy upon administration of MK886, an inhibitor of FLAP (Hawkes et al., 2014). Collective evidence suggests that LTs, which are generated by 5-LOX, can increase β- and γ-secretase-mediated generation of Aβ (Wang et al., 2013).

Other GSMs include a naturally occurring molecule called NIC5-15 or pinitol). It is a natural cyclic sugar alcohol (Pitt et al., 2013) that modulates γ-secretase and is reportedly capable of reducing Aβ production, while not affecting the substrate cleavage of Notch. No published information is available for this compound. However, it is proposed that the compound not only improves cognitive function and memory but also improves AD neuropathology. It is proposed that NIC5-15 may be a suitable therapeutic agent for the treatment of AD for two reasons: (1) it preserves Notch activity and (2) it is also potentially an insulin sensitizer. More studies are needed on pharmacokinetics of NIC5-15 in mice and human patients with AD.

PEROXISOME PROLIFERATOR-ACTIVATED RECEPTOR AGONISTS FOR THE TREATMENT OF ALZHEIMER'S DISEASE

Peroxisome proliferator-activated receptors (PPARs) are a group of nuclear receptor proteins that act as transcription factors. They regulate the expression of genes associated with cell differentiation, inflammation, development, and metabolism through the modulation of carbohydrate and lipid metabolism (Heneka et al., 2007). Three types of PPARs, namely, α, γ, and δ (β) have been identified in mammalian tissues. Upon activation, these receptors migrate to the nucleus, where they modulate a number of specific genes. PPARγ is a DNA-binding transcription factor whose transcriptional regulatory actions are activated after agonist binding (Kummer and Heneka, 2008). Thiazolidinediones (TZDs) are agonists for PPARγ (Fig. 8.6). They act by promoting PPARγ heterodimerization with the retinoid X receptor (RXR). PPARγ/RXR heterodimer is a transcription factor, which regulates the expression of genes involved in the lipid and glucose metabolism. In the brain, TZDs not only improve insulin sensitivity and reduce cytokine-dependent inflammation but also induce neurogenesis, gliogenesis, insulin sensitivity, and synaptic plasticity (Gold et al., 2010; Blalock et al., 2010; Rizvi et al., 2015). Examples of TZDs are rosiglitazone and pioglitazone. These drugs are used to treat type 2 diabetes, a metabolic disease characterized by chronic high glucose levels in the blood (hyperglycemia) caused by the inability of the body to produce insulin or the inability of the cells to respond to the insulin produced by the pancreas (Ginter and Simko, 2012). They act by regulating glucose homeostasis by increasing insulin sensitivity, reducing blood glucose levels, and improving lipid metabolism. Both compounds have also been studied as potential therapeutics for AD treatment, with reported improvements in mitochondrial oxidative metabolism (Blalock et al., 2010; Katsouri et al., 2011). Rosiglitazone and pioglitazone also induce the expression of PPARγ coactivator 1 alpha (PGC-1α), a

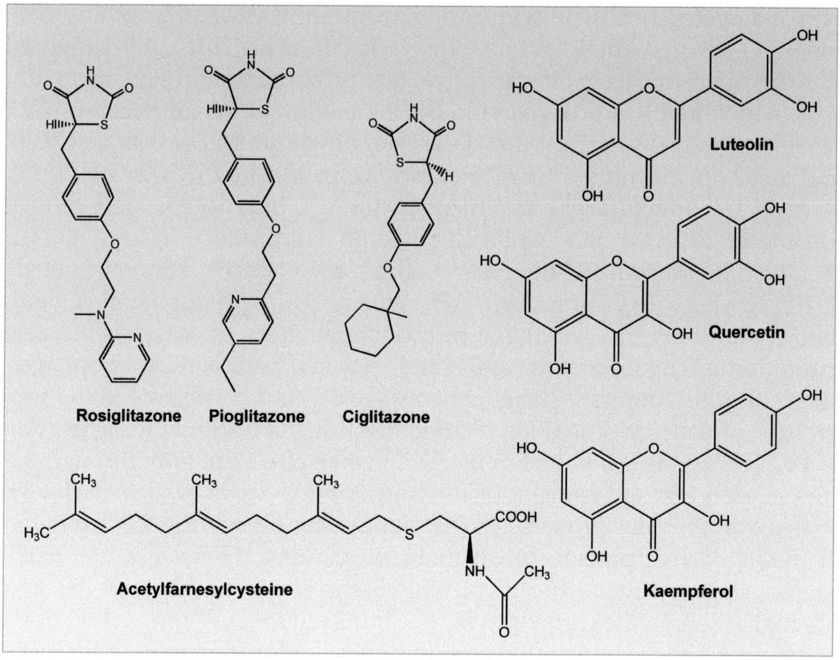

FIGURE 8.6 Chemical structures of thiazolidinediones, which are used for the treatment of type 2 diabetes and AD in animal and cell culture models.

molecule that plays multiple roles in mitochondrial biogenesis, energy metabolism, and mitochondrial antioxidants expression. Type 2 diabetes is a risk factor for the development of metabolic syndrome (MetS), a metabolic disease characterized by increase in abdominal fat, hyperglycemia, insulin resistance, high concentration of triglycerides, low levels of high-density lipoprotein (HDL), high blood pressure, and a generalized inflammatory state, and AD. It is suggested that the choice of drug treatment may influence the risk for patients with type 2 diabetes to develop AD (Imfeld et al., 2012; Farooqui, 2013). In fact, new findings suggest that medication with pioglitazone and rosiglitazone is associated with a lower risk of dementia in patients with type 2 diabetes, when these are compared to patients treated with metformin or insulin (Qin et al., 2009; Katsouri et al., 2011; Heneka et al., 2015). TZDs are PPARγ agonists used as antidiabetic drugs, whose mechanism of action induces a decrease in plasma free fatty acid concentrations and fasting hyperglycemia via an insulin-sensitizing effect (Desvergne et al., 2006). The endogenous agonists for PPARγ are polyunsaturated fatty acids (PUFAs) and 15-deoxy-(12,14)- prostaglandin J_2. TZDs are known to sensitize insulin with beneficial effects against neurodegeneration. However, patients with type 2 diabetes on TZDs treatment show cognitive decline (Heneka et al., 2015). In contrast, in a rat model of type 2 diabetes, treatment with rosiglitazone shows a significant improvement in spatial learning and memory task (Ma et al., 2015). It is proposed that this effect may be related to the regulation of the insulin signaling pathway, which involves a decrease in the expression of insulin receptor (IR), insulin receptor substrate-1 (IRS-1), serine/threonine kinase (Akt), p-CREB, and Bcl-2 in the hippocampal neurons from the rats (Ma et al., 2015). Furthermore, pioglitazone treatment of the A/T bitransgenic mice, which are known to accumulate senile plaques and overproduce Aβ and TGF-β1, results in the improvement of learning and memory (Papadopoulos et al., 2013). As stated in Chapter 6, in AD, cerebral neuroinflammation is mediated not only by the generation of proinflammatory eicosanoids but also through the activation of microglia and recruited peripheral macrophages, and both cell types may promote neurotoxicity and participate in the progression of neurodegeneration (Farooqui, 2013). Interestingly, the activation of PPARγ produces a potent antiinflammatory effect not only through the suppression and antagonism of the transcription factor nuclear factor-κB (NF-κB) or activator protein 1 (AP-1) but also by preventing the activation of microglia mediated by Aβ (Combs et al., 2001; Heneka et al., 2007). Moreover, pioglitazone treatment inhibits the activation of astrocytes and microglial cells in the cortex and hippocampus of the A/T mouse that overproduces Aβ and transforming growth factor-β1 (Papadopoulos et al., 2013). Similarly, injection of rosiglitazone into the brain of Wistar rats, which are treated with Aβ oligomers, prevents the increase in inflammatory cytokine expression, and this process may contribute to an improvement in cognitive decline and prevention of microglial cell activation (Xu et al., 2014). Similar effects occur in the AD transgenic mouse models J20 and APP/PSEN1 with previous oral administration with rosiglitazone (Escribano et al., 2010). Furthermore, the treatment with TZDs in animal models of AD decreases Aβ accumulation. For example, it is reported that oral treatment of APP transgenic mice with pioglitazone reduces the transcription and expression of BACE1 (β-secretase), an enzyme that contributes to the processing of APP protein (Heneka et al., 2005). However, other studies have reported that the amyloidogenic APP processing and Aβ production are not affected by treatment with rosiglitazone (Escribano et al., 2010) or pioglitazone (Mandrekar-Colucci et al., 2012), suggesting that the decrease in Aβ plaque produced by TZDs is related to amyloid degradation (Escribano et al., 2010; Mandrekar-Colucci et al., 2012). Collectively, these studies suggest that the inhibition of inflammatory response by TZDs involves the activation of PPARγ. These processes may contribute to the beneficial effect of TZDs for the treatment in AD. TZDs may also produce severe adverse effects in patients with type 2 diabetes. This has led to their withdrawal from the market or restricted clinical application. So, intense research efforts have recently been undertaken to discover selective PPARγ modulators (SPPARMs), compounds that improve glucose homeostasis but elicit reduced side effects due to partial PPARγ agonism based on selective receptor–cofactor interactions and target gene regulation (Balakumar and Kathuria, 2012; Schupp et al., 2005; Higgins and Depaoli, 2010). Using a gene expression-based screening approach, investigators have identified N-acetylfarnesylcysteine (AFC) (Fig. 8.6) as both a full and partial agonist depending on the PPARγ target gene (differential SPPARM) (Bhalla et al., 2011). AFC activated PPARγ as effectively as rosiglitazone with regard to Adrp, Angptl4, and AdipoQ, but is a partial agonist of aP2, a PPARγ target gene associated with increased adiposity (Bhalla et al., 2011). Induction of adipogenesis by AFC can be attenuated when compared with rosiglitazone. Reporter, ligand-binding assays, and dynamic modeling have indicated that AFC binds and activates PPARγ in a unique manner compared with other PPARγ ligands. Importantly, treatment of mice with AFC improves glucose tolerance similar to rosiglitazone, but AFC does not promote weight gain to the same extent. Finally, AFC promotes adipose tissue remodeling similar to rosiglitazone and induces enhanced antiinflammatory effects (Bhalla et al., 2011).

As stated in Chapter 1, sequential proteolysis of APP by β- and γ-secretase generates Aβ. Conversely, the α-secretase ADAM10 cleaves APP within the eventual Aβ sequence and precludes Aβ generation (Fig. 8.5). Therefore, upregulation of ADAM10 represents a plausible therapeutic strategy to combat overproduction of neurotoxic Aβ. PPARα is a transcription factor that regulates genes involved in fatty acid transport and catabolism. However, the hippocampus does not metabolize fat. However, recent studies have demonstrated that PPARα is constitutively expressed in the hippocampal neurons (Roy et al., 2013). The activation of PPARα induces the expression of ADAM10 and subsequent α-secretase–mediated proteolysis of APP. Furthermore, 5XFAD mice null for PPARα ($5X/\alpha^{-/-}$) have been reported to exhibit exacerbated brain Aβ load relative to traditional 5XFAD mice highlighting the view that PPARα plays an important role in reducing endogenous Aβ production by shifting APP processing toward the α-secretase pathway (Roy et al., 2013).

STATINS FOR THE TREATMENT OF ALZHEIMER'S DISEASE

Statins are inhibitors of cholesterol synthesis. They retard cholesterol synthesis by inhibiting 3-hydroxy-3-methylglutaryl-coenzyme A (HMG-CoA) reductase, an enzyme that not only initiates the syntheses of cholesterol and isoprenoid lipids but also is a rate-limiting step for the synthesis of cholesterol in the liver. Statins significantly reduce the risk for cardiovascular and cerebrovascular diseases (Endres, 2005). Cardioprotective and neuroprotective effects of statins in cardiovascular and cerebrovascular systems are due to their antiexcitotoxic, antioxidant, vasculoprotective, and antiinflammatory properties (Cordle and Landreth, 2005; Farooqui et al., 2007). In liver, sterol regulatory element-binding proteins sense changes in cholesterol level and increase the expression of low-density lipoprotein (LDL) receptor to uptake serum LDL cholesterol for compensating for the reduction in cellular cholesterol content (Goldstein and Brown, 2009). In addition, treatment with statin also promotes greater bioavailability of nitric oxide, improvement in endothelial cell function, increase in cerebral blood flow, and decrease in platelet aggregation. As stated in Chapter 3, cholesterol is a major component of neural membranes and is required for the formation of lipid rafts, which are the platforms for signal transduction processes, including isoprenoid-dependent assembly and activation of raftophilic β- and γ-secretases that facilitate the synthesis of Aβ42 fragments and its oligomers from APP. In brain, statins activate β-secretase and increase Aβ synthesis and accumulation in a transgenic mouse model of AD (Pedrini et al., 2005; Zimmermann et al., 2005). Similarly, statins decrease Aβ production, whereas cholesterol itself increases Aβ production in cell culture systems. Based on these observations, it is proposed that statins may produce preventive effects against the onset of AD in cell culture and animal model systems. Furthermore, statins may also protect neural cells not only through the inhibition of calpain and increase in generation and secretion of soluble form of APP via α-secretase stimulation but also via contribution of phosphatidylinosilol-3-kinase (PtdIns 3K) pathway and inhibition of ROCK signaling (Ma et al., 2009a, 2009b). Statins also suppress Aβ-induced inflammatory response through their ability to prevent the isoprenylation of members of the Rho family of small G-proteins, producing a functional inactivation of these G-proteins (Pedrini et al., 2005; Cordle et al., 2005). Treatment of microglial cells with statins results in perturbation of the cytoskeleton and morphological changes due to alteration in Rho family function. In transgenic and nontransgenic mice with AD-like pathology with a mixed genetic background, simvastatin enhances learning and memory (Mans et al., 2010). However, the molecular mechanism associated with enhancement of long-term potentiation (LTP) and memory formation by statins remains elusive. Prolonged in vitro simvastatin treatment (2–4 h) significantly increases the magnitude of LTP at CA3-CA1 synapses without altering basal synaptic transmission or the paired-pulse facilitation ratio in hippocampal slices from C57BL/6 mice supporting the view that simvastatin-mediated enhancement of hippocampal LTP may be a potential cellular mechanism underlying the beneficial effects of simvastatin on cognitive function (Mans et al., 2010, 2012). It should be noted that two recent randomized clinical studies enrolling 400 patients with mild to moderate AD have failed to show benefits of atorvastatin and simvastatin on cognitive function (Feldman et al., 2010; Sano et al., 2011). Although these two randomized studies have failed to produce the beneficial effects of statins in patients with AD, these studies were performed with several limitations including the duration of intervention (18-month period) and the selection of subjects such as the exclusion of patients with AD requiring treatment for dyslipidemia with lipid-lowering agents. On the other hand, several prospective studies have indicated that statins may prevent the onset of AD (Sparks et al., 2008; Haag et al., 2009; Bettermann et al., 2012). Follow-up period and cohort size of such supportive prospective studies have been relatively longer and larger (over 6 years of follow-up and more than 1000 subjects enrolled). A recent metaanalysis has indicated preventive effects of statins in AD (Wong et al., 2013). Many studies have indicated that statins produce preventive effects on AD only when they are given to relatively young subjects (Rockwood et al., 2002; Li et al., 2007, 2010). Thus the aforementioned studies indicate that the preventive and therapeutic effects of statins against AD should be started at a relatively young age. Such idea corresponds with current understanding of "anti-Aβ" therapies: once clinical symptoms develop, other mechanisms would be required to treat AD (Holmes et al., 2008; Burns, 2009). Therefore possible benefits of statins against AD can be obtained if statins are administered decades before the onset of the disease. A recent study has also indicated that inhibition of cholesterol biosynthesis, using AY9944, which blocks the final step of cholesterol biosynthesis, results in reduction of γ-secretase, an enzyme involved in the generation of Aβ peptides (Kim et al., 2016). Moreover, low cholesterol levels increase α-secretase activity on APP (Kojro et al., 2001), promoting neuroprotection by increasing the levels of αAPP fragments, which are involved in neurotrophic functions (Vingtdeux and Marambaud, 2012). It is tempting to speculate that large, multicenter, double-blind clinical trials are needed on the effects of various statins in patients with AD.

MEDITERRANEAN DIET AND ALZHEIMER'S DISEASE

The Mediterranean diet may not prevent AD, but it can delay the onset of AD for several years by slowing cognitive decline and prolonging longevity (Sofi et al., 2010, 2013; van de Rest et al., 2015). The Mediterranean diet also produces beneficial effects by improving the quality of life, which is translated into better psychological/physiological and metabolic profiles (Sofi et al., 2008, 2013). Components of the Mediterranean diet include vegetables, legumes, fruits, whole grains, fish, olive oil, fresh garlic, low levels of dairy products (cheese and yogurt), nuts, and modest intake of red wine (Willett et al., 1995). Fruits and vegetables are enriched in vitamins, carotenoids, flavonoids, fiber, potassium, magnesium, and other minerals. These components prevent age-related neurologic dysfunction because brain tissue contains low levels of endogenous antioxidants. Therefore, the brain is particularly vulnerable to free-radical damage, a process which is primarily associated with age-related neuronal decline (Farooqui, 2010). The consumption of Mediterranean diet not only retards heart disease, stroke, some cancers, hypertension, and osteoporosis, but also delays the onset of chronic neurological disorders such as AD, PD, and Huntington disease (Farooqui, 2010). Mediterranean diet also increases longevity and delays the onset of AD not only by inhibiting oxidative stress and neuroinflammation, normalizing mitochondrial dysfunction, improving endothelial function, inhibiting Aβ and Tau accumulation, and increasing cognitive function but also by maintaining the length of telomeres (Lopez-Miranda et al., 2007, 2012; Boccardi et al., 2016). Telomeres are long sequences of nucleotides at the ends of chromosomes, forming with specific proteins complex, "end caps" which preserve genome stability and lead a cell to correctly divide (Sfeir et al., 2005). Lots of important information has been published on various components of Mediterranean diet such as fish oil, olive oil, red wine pigment (resveratrol), and garlic. The following paragraph provides information on beneficial effects of components of Mediterranean diet on AD.

Olive oil contains phenolic compounds (tyrosol, hydroxytyrosols, oleocanthal, and oleuropein) and carotenes (Fig. 8.7). Consumption of olive oil results in the entry of tyrosol, hydroxytyrosols, oleocanthal, and oleuropein in the brain, where they produce neuroprotective effects. Studies on the effects of tyrosol and hydroxytyrosols against Aβ-induced toxicity in cultured neuroblastoma N2a cells have indicated that these phenols decrease neural cell death, but do not prevent the decrease of GSH induced by H_2O_2 or Aβ. Furthermore, Aβ-mediated increase in the nuclear translocation of the NF-κB subunits is attenuated by tyrosol and hydroxytyrosols (St-Laurent-Thibault et al., 2011). Oleocanthal (deacetoxy-ligstroside aglycon) has the ability to interact with Aβ and alters Aβ oligomer structure or function. It not only causes nonselective COX inhibition but also has antioxidant properties (Qosa et al., 2015; Beauchamp et al., 2005; Abuznait et al., 2013). Treatment of mice with oleocanthal for 4 weeks significantly decreases Aβ load in the hippocampal parenchyma and microvessels (Qosa et al., 2015; Abuznait et al., 2013). This reduction in Aβ is due to the increased cerebral clearance of Aβ across the BBB. Further mechanistic studies have demonstrated that oleocanthal not only increases the expression of Aβ clearance proteins (P-glycoprotein and LRP1) at the BBB but also activates the apolipoprotein E (APOE)-dependent amyloid clearance pathway in the mice brains (Qosa et al., 2015; Abuznait et al., 2013). The antiinflammatory effect of oleocanthal in the brains of TgSwDI mice also reduces astrocyte activation and decreases levels of interleukin (IL)-1β, a cytokine that produces neuroinflammation. Oleocanthal also inhibits the fibrillization of both Aβ40 and Aβ42 in vitro (Li et al., 2009). Oleocanthal also alters the oligomerization

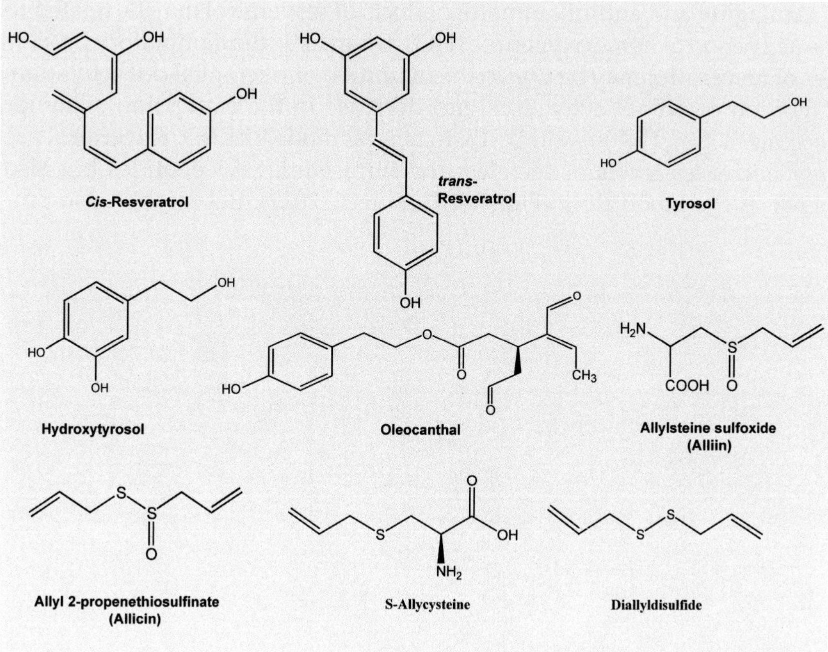

FIGURE 8.7 Chemical structures of compounds found in components of Mediterranean diet.

state of Aβ oligomers whilst protecting neurons from the synaptotoxicity effects of Aβ. Thus, oleocanthal protects neurons from Aβ-induced synaptic deterioration (Li et al., 2009). Similarly, in young/middle-aged TgCRND8 mice, diet supplementation with another oleuropein aglycone for 8 weeks improves animal behavior in two memory tests with respect to normally fed littermates, with scores reaching those displayed by age-matched wt mice (Grossi et al., 2013). Oleocanthal also inhibits COXs, LOX, and thromboxane synthases, the enzymes responsible for the generation of prostaglandins, LTs, and thromboxanes from ARA (de la Puerta et al., 1999; Petroni et al., 1995).

Resveratrol (3,4′,5-trihydroxystilbene) is a natural compound found in red grapes and wine. Red wine is an important component of Mediterranean diet (Fig. 8.7). Studies on the effect of resveratrol in animal models of AD have indicated that resveratrol produces beneficial effects not only due to antioxidant and free radical scavenging properties but also due to antiaging effects. Due to the presence of hydroxyl groups, resveratrol has the ability to transfer hydrogen atoms or electrons to the free radicals (Hussein, 2011) according to the following reaction.

$$\text{Resveratrol} + {}^{\bullet}\text{OH} \rightarrow \text{Resveratrol}^+ + \text{OH}^- \leftrightarrow \text{RV}(-\text{H})^{\bullet} + \text{H}_2\text{O}$$

Antiaging and antiinflammatory effects of resveratrol may be related to the activation of heme oxygenase (HO), whereas antiinflammatory properties of resveratrol may be due to the inhibition of microglial cell activation, inhibition of NF-κB activation, and decrease in the expression of tumor necrosis factor (TNF)-α and IL-1β. In animal models of AD, resveratrol not only decreases levels of secreted and intracellular Aβ peptides but also increases the blood flow (Fig. 8.8) (Pasinetti, 2012). Resveratrol also pro-

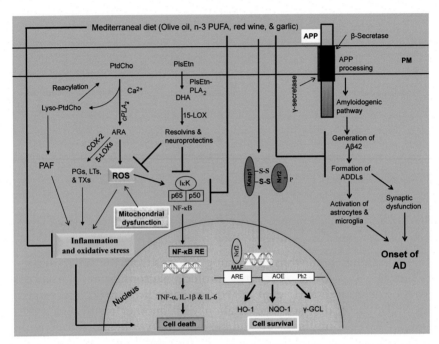

FIGURE 8.8 Signal transduction diagram showing the site of action of active components of Mediterranean diet. This diagram is based on in vitro and in vivo experimental data in cell culture and animal models. *15-LOX*, 15-lipoxygenase; *5-LOX*, 5-lipoxygenase; *ADDL*, soluble Aβ oligomer; *APP*, amyloid precursor protein; *ARA*, arachidonic acid; *ARE*, antioxidant response element; *Aβ*, β amyloid; *COX-2*, cyclooxygenase-2; *cPLA$_2$*, cytosolic phospholipase A$_2$; *HO-1*, hemeoxygenase; *Keap1*, Kelch like-ECH-associated protein 1; *LTs*, leukotrienes; *lyso-PtdCho*, lyso-phosphatidylcholine; *NF-κB*, nuclear factor-κB; *NF-κB-RE*, nuclear factor-κB response element; *NQO 1*, NAD(P)H quinone oxidoreductase 1; *Nrf2*, nuclear factor (erythroid-derived 2)-like 2; *PAF*, platelet-activating factor; *PGs*, prostaglandins; *PlsEtn*, ethanolamine plasmalogen; *plsEtn$_2$*, plasmalogen-selective phospholipase A$_2$; *PM*, plasma membrane; *PtdCho*, phosphatidylcholine; *ROS*, reactive oxygen species; *TXs*, thromboxanes; *γ-GCL*, glutamate cysteine ligase.

motes intracellular degradation of Aβ through a mechanism involving the proteasome activity (Marambaud et al., 2005). In the Tg2576 mouse model of AD, moderate consumption of Cabernet Sauvignon, a red wine enriched in resveratrol stimulates nonamyloidogenic α-secretase-mediated APP processing leading to neuroprotection and preventing the generation of amyloidogenic Aβ42-mediated cognitive deterioration (Wang et al., 2006). Resveratrol potently inhibits polymerization of the Aβ peptide (Riviere et al., 2007) by a mechanism that does not involve Aβ production because resveratrol has no effect on activity of β- and γ-secretases but stimulates indirectly the proteosomal degradation of Aβ peptides (Marambaud et al., 2005). The resveratrol-induced decrease of Aβ can be prevented by several selective proteasome inhibitors and by short interfering RNA (siRNA)-directed silencing of the proteasome subunit β5. Many resveratrol derivatives have been synthesized to treat AD. Among the synthesized compounds, 5d (E)-2-((4-(3,5-dimethoxystyryl) phenylamino) methyl)-4-(dimethylamino) phenol and 10d (E)-5-(4-(5-(dimethylamino)-2-hydroxybenzylamino) styryl)-benzene-1,3-diol not only significantly inhibit Aβ aggregation, have ability to chelate metals, and disintegrate highly structured Aβ fibrils but also prevent Cu^{2+}-induced Aβ aggregation and have antioxidant activity along with low neurotoxicity (Lu et al., 2012, 2013). Collective evidence suggests that resveratrol derivatives with improved bioavailability and neuroprotective effects can be used as novel multifunctional drugs in the treatment of AD.

Garlic (*Allium sativum*) is an important component of Mediterranean diet. It contains several organosulfur compounds, such as allicin, alliin, diallyl sulfide, diallyl disulfide, diallyl trisulfide, S-allyl-L-cysteine (SAC), dithiins, ajoene, methyl allyl disulfide, methyl allyl trisulfide, 2-vinyl-1,3-dithiin, and 3-vinyl-1,2-dithiin (Fig. 8.7) (Rybak et al., 2004). These are lipophilic thioesters derived from oxidized allicin, which are synthesized when garlic cloves are crushed. Garlic preparations induce their effects via different antioxidant mechanisms such as their ability to (1) scavenge reactive oxygen species (ROS) and reactive nitrogen species, (2) increase enzymic and nonenzymic antioxidants levels, and (3) inhibit some pro-oxidant enzymes (xanthine oxidase, cyclooxygenase, and NADPH oxidase). Neurochemical actions of garlic effects are induced through the modulation of transcription factors such as NF-κB and nuclear factor (erythroid-derived 2)-like 2 (Nrf2) (Fig. 8.8). Thus inhibition of NF-κB by garlic components retards the expression of proinflammatory cytokines (TNF-α, IL-1β, and IL-6) (Xiao et al., 2013). Organosulfur compounds from garlic not only promote migration of NF-κB to the nucleus but also inhibit 5-LOX, the enzyme that transforms ARA into LTs (Farooqui, 2014). Like NF-κB, garlic mediates the activation of Nrf2 pathway. Under physiological conditions, Keap1 and Nrf2 complex resides in the cytoplasm. Following oxidative stress or treatment with garlic constituents, NrF2 translocates from the cytoplasm into the nucleus where it forms heterodimers with other transcription factors such as Maf or c-Jun and interacts with antioxidant response elements (AREs). This ARE–Nrf2 binding leads to activation of the Keap1–Nrf2 pathway, which induces the expression of over 100 cytoprotective genes. These include (1) the cellular antioxidant and antiinflammatory defense enzymes such as NAD(P)H quinone oxyreductase (NQO1) and genes modulating levels of glutathione; (2) enzymes of glutathione biosynthesis, extracellular superoxide dismutase, and glutamate-6-phosphate dehydrogenase; and (c) pro- and antiinflammatory enzymes such as COX-2, inducible nitric oxide synthase (iNOS), and HO-1 (Fig. 8.8). In addition, Nrf2 activation also produces the expression of genes that regulate mitochondrial biogenesis and glycolytic metabolism, angiogenesis, cell survival, apoptosis, and other processes that interfere with cell survival (Suzuki et al., 2013; Suzuki and Yamamoto, 2015). Upon the recovery of cellular redox homeostasis, Keap1 travels into the nucleus to dissociate Nrf2 from the ARE. Subsequently, the Nrf2–Keap1 complex is transported out of the nucleus terminating the Nrf2/ARE signaling pathway (Suzuki and Yamamoto, 2015). Garlic components have the ability to induce antioxidant and antiinflammatory effects (Ho and Su, 2014; Xiao et al., 2013; Suzuki and Yamamoto, 2015). Most studies on the effects of garlic components in neurological disorders have been performed in animal and cell culture models. Thus, SAC is the most active component of aged garlic extract (AGE). SAC lowers Aβ levels and toxicity (Ray et al., 2011). Treatment of neuronal cultures with AGE and SAC protects against H_2O_2-mediated oxidative stress. AGE was also found to preserve presynaptic protein, synaptosomal associated protein of 25 kDa (SNAP25), from ROS-mediated insult. In vitro studies have also indicated that SAC has ability to block Aβ fibrillation through the destabilization of Aβ fibrils (Gupta and Rao, 2007). SAC reduces the brain levels of Aβ40 and Aβ42, lowers APP level, and downregulates BACE1 expressions resulting in lower Aβ accumulation and reduction in the levels of advanced glycation end products (carboxymethyllysine, pentosidine) (Tsai et al., 2011). In addition, SAC lowers aldose reductase activity and expression. SAC also has ability to protect neurons against the caspase-12-dependent neurotoxicity mediated by Aβ (Kosuge et al., 2003). Hypercholesterolemia is a risk factor for AD. In AD transgenic model (Tg2576), it is shown that SAC may act as HMG-CoA reductase inhibitor and induce its beneficial effect by lowering cholesterol levels (Chauhan, 2006).

Finally, another important component of Mediterranean diet is fish, which contains high levels of n-3 or ω-3 fatty acids [docosahaexenoic acid (DHA), and eicosapentaenoic acid (EPA)]. DHA is the most abundant n-3 fatty acid in the brain. DHA and EPA promote the formation of lipid rafts. Like other components of Mediterranean diet, these fatty acids produce beneficial effects (Farooqui, 2009) not only due to their effect on physicochemical properties of neural membrane (such as acyl chain order, permeability, fluidity, and synaptic connectivity) but also through the generation of n-3 fatty acid–derived docosanoids (resolvins, neuroprotectins, and maresins). These mediators produce antioxidant, antiinflammatory, and antiapoptotic effects and protect neuronal cells from neurodegeneration

(Farooqui, 2010). DHA- and EPA-enriched lipid rafts are not only platforms for signal transduction process but also absorb the oxidative stress induced by free radicals and increase cell membrane fluidity necessary for lipid raft creation and formation of effective synaptic contacts (Farooqui, 2009). Studies with aged rodents have consistently shown that n-3 fatty acid supplementation improves neurogenesis and synaptogenesis, executive functions, and learning abilities, while n-3 PUFA deficiency is associated with memory deficits and impaired hippocampal plasticity (Denis et al., 2013; Luchtman and Song, 2013; Maruszak et al., 2014). Epidemiological studies have indicated that sufficient DHA intake reduces the risk of developing AD and other neurodegenerative diseases (Morris et al., 2003; Schaefer et al., 2006; Farooqui, 2009). In addition, the consumption of fish and fish oil has been reported to reduce cognitive decline (Farooqui, 2009). A recent trial of fish oil has indicated positive effects of DHA supplementation on cognition in patients with very mild AD (Freund-Levi et al., 2006; Chiu et al., 2008). Similarly, investigations on three different transgenic models of AD indicate that animal models of AD are more vulnerable to DHA depletion than controls and that DHA exerts a beneficial effect against pathological signs of AD, including Aβ accumulation, cognitive impairment, synaptic marker loss, and hyperphosphorylation of Tau (Lim et al., 2005; Calon and Cole, 2007).

Based on studies on neurochemical effects of garlic components and some human trial studies on the effect of Mediterranean diet, it is proposed that this diet pattern is the healthiest dietary patterns in the world not only due to its relation with a low morbidity, mortality, and better quality of life but also due to its beneficial effects on lipoprotein levels, endothelium vasodilatation, hyperglycemia, insulin resistance, dyslipidemia, and obesity. Mediterranean diet also decreases the prevalence of the type 2 diabetes, metabolic syndrome, as well as cardiovascular and cerebrovascular diseases (Sofi et al., 2013; Estruch et al., 2013; Castro-Quezada et al., 2014).

Long-term randomized controlled trials promoting the consumption of Mediterranean diet may help in establishing whether improved adherence to this diet may help in preventing or delaying the onset of AD and dementia.

CURCUMIN AND THE TREATMENT OF ALZHEIMER'S DISEASE

Curcumin is a highly promising natural polyphenol with antioxidant and antiinflammatory properties (Fig. 8.9) (Gupta et al., 2012; Farooqui, 2016). Epidemiological studies have revealed that in India, where dietary curcumin is consumed daily in the form of curry compared with the United

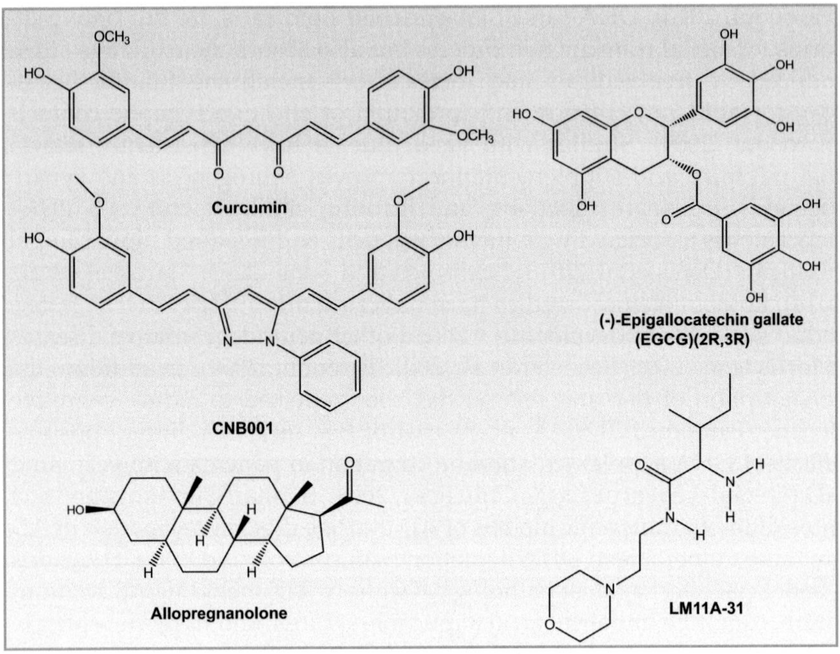

FIGURE 8.9 Chemical structures of phytochemical and other drugs used for the treatment of AD.

States, the morbidity rate attributed to AD for Indian elders (70–79 years old) is 4.4 times lower compared to the that of the same age group of Americans (Ganguli et al., 2000). The consumption of curcumin-enriched food by healthy elderly human subjects results in better cognitive performance than seniors who did not consume curcumin-enriched food (Ng et al., 2006). These observations suggest that curcumin can be used for the treatment of AD. Curcumin has the ability to cross BBB as well as all cell membranes and produces its intracellular effects (Farooqui, 2016). The antioxidant activity of curcumin may be due to the presence of phenolic and the methoxy group on the phenyl ring and the 1,3-diketone systems (Fig. 8.9). The degradation of curcumin in vivo produces smaller phenols like ferulic acid (*trans*-4-hydroxy-3-methoxycinnamic acid) (Ghosh et al., 2015), which is capable of producing neuroprotective effects. Recently, investigators have synthesized curcumin analogs (Fig. 8.9) (Priyadarsini, 2013). The replacement of the 1,3-dicarbonyl moiety in curcumin with isosteric isoxazoles and pyrazoles generated compounds that not only inhibit γ-secretase activity (Narlawar et al., 2007) but also prevent both Aβ and Tau aggregation (Narlawar et al., 2008). As stated in Chapter 1, elevated levels of redox-active Fe^{3+} and Cu^{2+} in AD generate ROS causing DNA damage in neural cell, by producing hydroxyl and superoxide radicals via

Fenton reaction. Curcumin binds Cu^{2+} and Fe^{3+} and form tight and inactive complexes, and can protect neural cell DNA against ROS and singlet oxygen–induced strand breaks, (Kim et al., 2005). In aged mice with advanced plaque deposits similar to those of AD, feeding of curcumin not only reduces the amount of plaque deposition in mice cortex and hippocampus but also decreases levels of proinflammatory cytokines (Begum et al., 2008; Yang et al., 2005) and increases levels of antiinflammatory cytokine (IL-4) in microglial cell cultures from curcumin-treated mice (Shytle et al., 2012). Antioxidant properties of curcumin are also promoted by the upregulation of PtdIns 3K/Akt/Nrf2 pathway. Curcumin-mediated upregulation of PtdIns 3K/Akt/Nrf2 pathway can also be blocked by the Nrf2 siRNA strongly indicating that curcumin mediates cytoprotection in rodents (Yin et al., 2012). Based on these studies, it is suggested that curcumin not only reduces oxidative damage and neuroinflammation but also reverses the amyloid pathology in an AD transgenic mouse (Yang et al., 2005; Park et al., 2008; Ye and Zhang, 2012; Yao and Xue, 2014). Another major curcumin-mediated defense against intraneuronal Aβ aggregate formation is the induction of heat shock proteins (HSPs) (Scapagnini et al., 2006; Yuan et al., 2008; Hu et al., 2015). Curcumin-mediated induction of HSPs may not only limit Aβ accumulation (Maiti et al., 2014), but also increase the expression of BDNF and ERK/P38 signaling pathways and degradation of protein kinase C delta (PKCδ) (Dong et al., 2012). Curcumin also inhibits COX-2, 5-LOX, phospholipases, transcription factors (AP-1 and NF-κB), and other enzymes involved in metabolizing the neural membrane phospholipids into proinflammatory eicosanoids, a group of lipid mediators, which contribute to neuroinflammation (Ong et al., 2015). Furthermore, curcumin produces a broad cytokine-suppressive antiinflammatory action. It downregulates the expression of iNOS, TNF-α, IL-1, IL-2, IL-6, IL-8, and IL-12. Curcumin blocks IL-6-mediated signaling via inhibition of IL-6-induced signal transducer and activator of transcription 3 (STAT3) phosphorylation and consequent STAT3 nuclear translocation and interferes with the first signaling steps downstream of the IL-6 receptor in microglial activation (Farooqui, 2016). Curcumin also modulates the phosphorylation of Tau by inhibiting glycogen synthase kinase 3β (GSK3β). The inhibition of GSK3β by curcumin may be one possible mechanism by which curcumin reduces Tau phosphorylation. Curcumin also induces phase II enzymes in astrocytes and HO-1 in neurons in vitro (Jiao et al., 2006). Curcumin upregulates neprilysin (NEP), an important Aβ-degrading enzyme, which is decreased with age and is inversely correlated with Aβ accumulation supporting the view that there may be a correlation between its activity and the late-onset AD (Farooqui, 2016). Twenty-five curcumin analogs are tested for their ability to upregulate NEP activity using a sensitive fluorescence-based Aβ digestion assay. Four compounds, dihydroxylated curcumin, monohydroxylated demethoxycurcumin, and mono- and dihydroxylated bisdemethoxycurcumin, increase NEP activity, while curcumin produces no effect on NEP activity (Chen et al., 2016). The ability of these polyhydroxycurcuminoids to upregulate NEP has also been confirmed by mRNA and protein expression levels in the cell and mouse models. Finally, treating feeding monohydroxylated demethoxycurcumin or dihydroxylated bisdemethoxycurcumin to $APP_{swe}/PSEN_1dE_9$ double transgenic mice not only upregulates NEP levels in the brain but also reduces Aβ accumulation in the hippocampus and cortex. Future studies on polyhydroxycurcuminoids may offer hope in the prevention of AD in more animal models and in patients with AD (Chen et al., 2016). Collective evidence suggests that curcumin fulfills the characteristics for an ideal neuroprotective agent with its low toxicity, affordability, and easy accessibility. Large, double-blind multicenter studies are needed in patients with AD to estimate the efficacy of curcumin in AD.

PHYTOCHEMICALS FOR THE TREATMENT OF ALZHEIMER'S DISEASE

Dietary phytochemicals are heterogeneous and nonnutrient compounds from a wide range of plant-derived foods. The consumption of dietary phytochemicals by humans results in health-promoting antioxidant and antiinflammatory effects not only in normal individuals but also in patients with acute and chronic neurological disorders (Farooqui, 2012). Beside antioxidant and antiinflammatory effects, phytochemicals not only stimulate the synthesis of adaptive and detoxification enzymes but also stimulate immune responses, and epigenetic effects in humans (Williams and Spencer, 2012; Farooqui, 2012). In addition, phytochemicals also stimulate the synthesis of adaptive enzymes involved in synaptic plasticity and neuronal repair (Eggler et al., 2008; Williams and Spencer, 2012). Epidemiological studies have demonstrated that incidences of neurological disorders among people living in Asia are lower than that among those in the Western world (Chandra et al., 2001; Chen et al., 2011a, 2011b). It is suggested that this may be due to the regular consumption of phytochemicals in the form of spices (turmeric, red pepper, black pepper, licorice, clove, ginger, garlic, coriander, and cinnamon). Many phytochemicals produce beneficial effects in cell culture and animal models of AD (Farooqui, 2012). Phytochemicals have poor bioavailability not only because of their poor absorption but also due to rapid excretion. The bioavailability of most phytochemicals in the peripheral tissues is higher than that in the brain due to BBB (Farooqui, 2012). However, small amounts of phytochemicals still cross the BBB and mediate antioxidant, antiinflammatory, and anticarcinogenic effects at low doses in the brain (Williams and Spencer, 2012; Farooqui, 2012). It is realized that phytochemicals are consumed in crude forms (roots, shoots, leaves, and fruits) and there are many questions concerning their specific medicinal effects and reproducibility. However, even in crude forms they produce beneficial effects in neurological disorders (Williams and Spencer, 2012; Farooqui, 2012). These days, investigators are not only focusing on identification of active ingredients of crude phytochemicals but also on evaluating their mechanism of action (Farooqui, 2012). Commonly used phytochemicals include polyphenols (green tea, curcumin, resveratrol, *Ginkgo biloba*), flavonoids, ginsenosides, sulfur compounds from garlic, and ginseng (from root of *Panax* species). The precise mechanism and site of action of phytochemicals in the signaling pathways remains elusive. However, evidence indicates that phytochemicals act on neural cells not only through the modulation of the activity of protein kinases [Mitogen activated kinase kinase kinase (MAPKKK), mitogen-activated protein kinase kinase (MAPKK), or mitogen-activated protein kinase (MAPK)] directly or indirectly but also through the preservation of Ca^{2+} homeostasis and modulation of transcription factors (Farooqui, 2012). Molecular mechanisms of action of resveratrol, sulfur compounds from garlic, and n-3 fatty acids have been discussed in the Mediterranean diet section. I will briefly mention some information on the effect of green tea and ginsenosides in animal models of AD.

Green tea (*Camellia sinensis*) contains four major catechins, namely, (–)-epicatechin (EC), (–) epicatechin-3-gallate, (–)-epigallocatechin, and (–)-epigallocatechin-3-gallate (EGCG) (Velayutham et al., 2008) (Fig. 8.9). The bioavailability of green tea catechins depends upon their structures. Catechin monomers can be easily absorbed through the gut barrier. In contrast, the large-molecular-weight catechins, such as EGCG, are poorly absorbed. In AD, EGCG produces beneficial effects by regulating APP processing under in vitro and in vivo conditions (Mandel et al., 2008; Smith et al., 2005, 2010). In neuronal cell cultures and mouse model of AD, EGCG enhances the nonamyloidogenic α-secretase pathway via PKC-dependent activation of α-secretase (Mandel et al., 2008; Smith et al., 2010), while EC reduces the formation of Aβ fibrils. EGCG modulates Aβ levels, either via translational inhibition of APP or by stimulating soluble APPα (sAPPα) generation and secretion. This process results in a significant reduction in cerebral Aβ levels and Aβ plaques. EGCG also inhibits the activation of ERK and nuclear transcription factor-κB in the Aβ-injected mouse brains and blocks Aβ-mediated apoptotic neuronal cell death in the brain. These studies suggest that EGCG may retard the development or progression of AD in cell culture and animal models. In APP/PSEN1 mice, EGCG (2 or 6 mg/kg/day) ameliorates the impaired learning and memory in 4 weeks not only by inhibiting TNF-α/ c-Jun-N-terminal kinase (JNK) signaling but also by increasing the phosphorylation of Akt and GSK3β in the hippocampus. In addition, ECGC consumption may alleviate AD-related cognitive deficits by effectively attenuating central insulin resistance (Jia et al., 2013).

Ginseng (*Panax ginseng*) belongs to the family Araliaceae. The bioactive constituents of ginseng root include more than 60 ginsenosides (Rg), which have poor bioavailability not only because of low membrane permeability and active biliary excretion but also due to biotransformation (Liu et al., 2009). Ginsenosides can cross the BBB and produce many neurochemical effects including modulation of ion channels and neurotransmitter receptors (NMDAR, nicotinic acetylcholine, and 5-hydroxytryptamine type 3 receptors), prevention of oxidative stress and neuroinflammation, and retardation of memory deficit (Liu et al., 2010; Chen et al., 2010).

It is reported that Siberian ginseng produces neuroprotective effects against Aβ-induced neurite atrophy. Gintonin, an important component of ginseng decreases Aβ42 and attenuates Aβ42-induced cytotoxicity in SH-SY5Y cells. Gintonin also rescues Aβ42-induced cognitive dysfunction in mice. Moreover, in a transgenic mouse AD model, long-term oral administration of gintonin attenuates amyloid plaque deposition as well as short- and long-term memory impairment (Hwang et al., 2012). Fermented ginseng also reduces the level of soluble Aβ42 in HeLa cells, which stably express the Swedish mutant form of APP, and decreases memory impairment in animal models of AD (Kim et al., 2013). Similarly, in aged transgenic AD mice (TgmAPP), which overexpress APP/Aβ, treatment with ginsenoside (Rg1) not only induces a significant reduction in cerebral Aβ levels but also mediates reversal of neuropathological changes and preservation of spatial learning and memory, as compared to vehicle-treated mice. Rg1 treatment inhibits the activity of γ-secretase in both TgmAPP mice and B103-APP cells, indicating the involvement of Rg1 in APP regulation pathway. Furthermore, administration of Rg1 enhances the protein kinase A (PKA)/CREB pathway activation in mice over-expressing amyloid precursor protein (mAPP) and in cultured cortical neurons exposed to Aβ or glutamate-mediated synaptic stress (Fang et al., 2012). *P. ginseng* also modulates Tau phosphorylation. Total ginsenoside extracts from stems and leaves of *P. ginseng* enhance the phosphatase activity of purified calcineurin. This may be useful in AD, since inhibition of calcineurin induces hyperphosphorylation of Tau at multiple sites (Tu et al., 2009). Accumulating evidence suggests that ginsenosides produce beneficial effects in cell culture and animal models of AD.

THERAPEUTIC POTENTIALS OF NEUROGENESIS FOR THE TREATMENT OF ALZHEIMER'S DISEASE

In the brain, neurogenesis has been reported to occur both in prenatal developmental as well as in adult brains. However, neurogenesis is much more active in the prenatal development than in the adult brain, and still occurs in adult brain (Mu and Gage, 2011). In most mammals, adult neurogenesis is restricted to the subgranular zone (SGZ) of the dentate gyrus in the hippocampus (Aimone et al., 2010) and subventricular zone (SVZ) of the lateral ventricle (Alvarez-Buylla et al., 2008).

These adult-born neurons function and integrate into the rest of the brain circuit. Both SGZ and SVZ neurogenesis play an important role not only in various forms of learning and memory but also in emotional behavior (Bath et al., 2008; Wang et al., 2010). The regenerative potential of the mammalian brain is sustained throughout the life span, whereas the magnitude of the proliferative efficacy of neural progenitors declines with age and diseases, such as AD (Hattiangady and Shetty, 2008; Lazarov et al., 2010). Age and AD have been reported to promote decline in hippocampal neurogenesis in multiple mouse models of AD along with a concomitant decline in cognitive function (Rodriguez et al., 2008, 2009). The impairment of neurogenesis also occurs early in AD and precedes plaque and tangle formation in patients with AD (Hamilton et al., 2010; Mu and Gage, 2011). Enhancement of neurogenesis is necessary to promote neuronal regeneration in AD animal models (MacMillan et al., 2011). However, many challenges remain in the regeneration of lost neural populations and regeneration and restoration of neural circuits necessary for cognitive function. Neural stem cells, despite having great potential for neuronal regeneration, have limited use in the clinic because of the challenge of their delivery to the brain (Fan et al., 2014; Taupin, 2009). An alternative approach that may be more amenable to clinical application in patients with AD may be the use of extracellular small molecules called morphogens. Morphogens stimulate neurogenesis in the brain. In adult neurogenesis, a number of morphogens play important role in critically establishing/regulating the stem cell niche, including Notch, sonic hedgehog, Wnts, and bone morphogenetic proteins (Liu and Song, 2016). In addition, growth factors [BDNF, insulinlike growth factor (IGF-1), and fibroblast growth factor-2] and neurotransmitters [γ-aminobutyric acid (GABA), dopamine, glutamate, and serotonin] also play critical roles in adult neurogenesis (Liu and Song, 2016).

Allopregnanolone (a prototypic neurosteroid) is an important molecule that can be used as a potential anti-AD molecule to support the concept of endogenous regeneration in the normal as well as in patients with AD (Fig. 8.9) (Brinton, 2013). Allopregnanolone acts as a potent modulator of GABAA receptors. In vitro, allopregnanolone induces neurogenesis in human and rat neural progenitors in a dose-dependent and steroid-specific manner (Wang et al., 2005). In vivo studies with 3xTgAD mouse models have indicated that allopregnanolone significantly increases neurogenesis within the SGZ of dentate gyrus and the SVZ. It also reverses neurogenic deficits in the hippocampus of young and aging 3xTgAD mice. In addition, allopregnanolone reverses the learning and memory deficits in the 3xTgAD mouse and restores both regenerative and cognitive function similar to that of the normal nontransgenic mouse

(Wang et al., 2005). The molecular mechanism of action of allopregnanolone is not fully understood. However, it is suggested that allopregnanolone binds with $GABA_A R$ at sites that differ from GABA-, benzodiazepine-, ethanol-, and barbiturate-binding sites. Stimulation of $GABA_A R$ initiates a cascade of events leading to calcium influx in adult neuroprogenitor cells subsequently inducing the accumulation of a neurogenic transcription factor, NeuroD (Tozuka et al., 2005). It acts as positive or negative modulators of $GABA_A R$ function (Liu and Brinton, 2010). Allopregnanolone is a potent positive allosteric activator of the $GABA_A R$ channels, which at nanomolar concentrations enhance the action of GABA at $GABA_A R$ and at higher concentrations directly activate $GABA_A R$ (Hosie et al., 2006). Chronic administration of allopregnanolone significantly increases the survival of newly generated neurons with simultaneous reduction of $A\beta$ oligomer accumulation and microglia activation in the 3xTgAD mouse (Chen et al., 2011c). Attempts are underway to move allopregnanolone from laboratory to the clinic for human trials in patients with AD (Irwin and Brinton, 2014). However, there are many safety concerns with chronic exposure to allopregnanolone because allopregnanolone possesses sedative hypnotic and/or antiseizure effects (Irwin et al., 2011). Moreover, oral delivery of allopregnanolone presents a big challenge not only because of its low solubility, but also about its metabolism in the digestive tract and liver (Irwin et al., 2011). Attempts to use allopregnanolone to treat cognitive dysfunction in human and animal studies have failed, and chronic treatment with allopregnanolone accelerates the development of AD in $A\beta PP(Swe)PSEN1(\Delta E9)$ mice. Allopregnanolone impairs episodic memory in healthy women, but with a high degree of individual variability (Bengtsson et al., 2012; Kask et al., 2008).

OTHER POTENTIAL THERAPEUTIC TARGETS FOR THE ALZHEIMER'S DISEASE

As stated in Chapter 1, onset of AD is not only accompanied by reduction in brain IR sensitivity (Holscher and Li, 2010) and hypophosphorylation of the IR and IRS-1 (Talbot et al., 2012) but also by attenuation of insulin and IGH receptor expression. In addition, reduction in CSF insulin levels has also been observed in moderate and severe cases of AD (Craft et al., 1998). The IRs occur abundantly in neurons and are highly concentrated in the hippocampus, particularly around the synaptic area (Havrankova et al., 1978; Zhao and Alkon, 2001). Insulin can cross the BBB and can also be synthesized by neurons and released upon membrane depolarization (Clarke et al., 1986). Insulin signaling plays an important role in neuronal survival (Valenciano et al., 2006) and in learning and memory (Dou et al., 2005). Intranasal insulin administration improves memory in young human subjects (Reger et al., 2008) and in patients with cognitive deficits associated with mild-stage AD (Adzovic and Domenici, 2014). Insulin signaling in the brain occurs mainly via the IR pathway

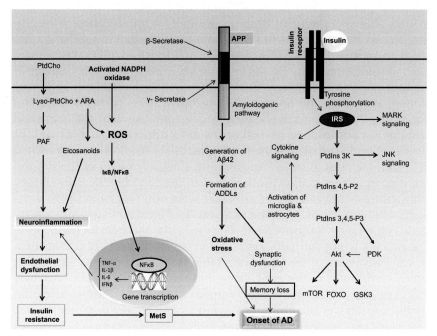

FIGURE 8.10 **Reactions showing crosstalk among insulin signaling, cytokine signaling, and Aβ signaling in Alzheimer's disease.** *ADDL,* soluble Aβ oligomer; *Akt/PKB,* serine/threonine protein kinase; *APP,* amyloid precursor protein; *ARA,* arachidonic acid; *Aβ,* β-amyloid; *FOXOs,* forkhead transcription factors; *GSK 3,* glycogen synthase kinase 3; *IL-1β,* interleukin-1β; *IL-6,* interleukin-6; *IRS,* insulin receptor substrate; *JNK,* c-Jun-N-terminal kinase; *lyso-PtdCho,* lyso-phosphatidylcholine; *mTOR,* mammalian target of rapamycin; *NF-κB,* nuclear factor-κB; *PAF,* platelet-activating factor; *PDK,* phosphoinositide-dependent kinase; *PtdCho,* phosphatidylcholine; *PtdIns 3K,* phosphatidylinositol 3-kinase; *PtdIns(3,4,5)P3,* phosphatidylinositol 3,4,5-trisphosphate; *PtdIns(4,5)P2,* phosphatidylinositol 4,5-bisphosphate; *ROS,* reactive oxygen species; *TNF-α,* tumor necrosis factor-α.

(Fig. 8.10). The binding of insulin with IR results in its activation. This receptor belongs to the family of tyrosine kinase receptors, which require autophosphorylation for its activation. Upon activation, the IR phosphorylates IRS proteins. IRS proteins are also activated upon binding of the IGF-1 ligand to its cognate receptor. PtdIns 3K and the MAP kinases (ERK1/2) are major components of insulin-mediated signaling (Gallagher et al., 2008). PtdIns 3K is the pathway associated with the metabolic effects of insulin, whereas MAP plays a role in growth and mitogenesis (Gallagher et al., 2008). In the PtdIns 3K pathway phosphorylated IRS interacts with the p85 subunit of PtdIns 3K, resulting in the synthesis of phosphatidylinositol 3,4,5-trisphosphate [PtdIns(3,4,5)P3] (Gallagher et al., 2008). Elevated levels of PtdIns(3,4,5)P3 activate phosphoinositide-dependent protein kinase-1 and Akt. Akt represents yet another critical node of interaction with the mammalian target of rapamycin nutrient signaling pathway. Akt targets include GSK3, Akt substrate of 160 kDa (AS160, phosphorylation of which is required for translocation of the glucose transporter (GLUT) 4 to the plasma membrane) and forkhead transcription factors (FOXOs) (Fig. 8.10). Phosphorylation of FOXOs induces their translocation from the nucleus, which causes profound changes in the transcription of key factors implicated in metabolism, cell cycle regulation, apoptosis, and resistance to oxidative stress (Eijkelenboom and Burgering, 2013). Collectively, these studies indicate that the activation of these kinases leads to several of insulin's responses, such as GLUT4 translocation to the membrane (Gallagher et al., 2008). Akt's role related to insulin includes glucose transport and storage, protein synthesis, and stopping lipid degradation (Gallagher et al., 2008). PtdIns 3K is also linked with another critical node of crosstalk with other signaling pathways, such as the JNK stress signaling pathway. Insulin signaling is also associated with memory improvement. The molecular mechanism involved in this process is not fully understood. However, it is reported that insulin can also promote the phosphorylated AMPA receptor at Serine 831, which contributes to the early stage and maintenance of LTP (Adzovic and Domenice, 2014). In the light of these observations, intranasal insulin administration has also been considered as a therapeutic option for AD. This particular route of administration is attractive because it bypasses the BBB. This is very significant because insulin transport to the brain from the periphery depends on active transport mechanisms which are disrupted in patients with AD. In theory, insulin delivery directly to the brain may activate cerebral insulin signaling leading to improvements not only in memory processing but also in promoting neuroprotection (Freiherr et al., 2013; Claxton et al., 2015). An ongoing clinical trial NCT01767909 is evaluating long-term (12 months) efficacy of intranasal insulin (Humulin R U-100) in mild AD. It is predicted that after 12 months of treatment with intranasal insulin, human subjects may show improved performance on a global measure of cognition, on a memory composite, and in daily functional activity. In addition to the examination of CSF biomarkers and hippocampal and entorhinal atrophy, the study aims to examine whether the baseline AD biomarker profile, gender, or APOE-ε4 allele carriage predicts treatment response.

LM11A-31or C31 is a water-soluble amino acid derivative (Fig. 8.9), which is a ligand for p75^{NTR}, a receptor that is abundantly expressed by basal forebrain cholinergic neurons in adulthood (Gibbs et al., 1989). Its signaling positively or negatively modulates the activities of basal forebrain cholinergic neurons depending on the presence of several factors including coreceptors and ligands (Dechant and Barde, 2002; Ibanez and Simi, 2012). LM11A-31 has structural and chemical features similar to the nerve growth factor (NGF) loop 1 domain (Massa et al., 2006). It can cross the BBB and prevents basal forebrain cholinergic neuron atrophy and cognitive deficits in an AD mouse model when administration starts at early pathological stages of the disease, shortly after $A\beta$ plaques appear (Knowles et al., 2013; Nguyen et al., 2014). For example, NGF signaling through p75^{NTR} may promote cell survival or death depending on the presence of its high-affinity receptor TrkA (Dechant and Barde, 2002; Ibanez and Simi, 2012). Interactions between $A\beta40$ and p75^{NTR} also result in neurodegeneration (Yaar et al., 2002). Similarly, the delivery of $A\beta$ oligomer to the brain of wild-type mice, but not p75^{NTR}-deficient mice also results in neurodegeneration (Sotthibundhu et al., 2008), and this neurodegeneration can be retarded by functionally removing the neurotrophin-binding domain of the receptor in an AD mouse model (Knowles et al., 2009) suggesting that p75^{NTR} signaling plays an important role in enabling $A\beta$-induced degeneration, and that this signaling can be used as an AD therapeutic target (Longo and Massa, 2013). Administration of the p75^{NTR} ligand LM11A-31 to APP$^{L/S}$ mice with mid- to late-stage AD pathology slows the progression of cholinergic neurite degeneration in the basal forebrain and cortex. This effect can also be observed in the Tg2576 model. Importantly, LM11A-31 reverses $A\beta$-mediated neurodegeneration when it is administered for a brief duration (1 month) and initiated at an age (12–13 months) that is about 4 times greater than the age of detectable pathology onset (3–4 months old). This suggests that a small molecule like LM11A-31 can penetrate the BBB and modulate a specific target to reduce and/or reverse fundamental AD pathologies (accumulation of $A\beta$) in late-stage AD mice (Simmons et al., 2014).

TAU-DIRECTED ALZHEIMER'S DISEASE THERAPIES

As stated in Chapter 1, Tau is microtubule-associated protein, which is predominantly expressed in the neuronal cell bodies and axons in the brain. However, lower quantities of Tau are also expressed by astrocytes and oligodendrocytes (Yoshiyama et al., 2013). In the brain, human Tau is expressed in six isoforms, which are produced by alternative mRNA splicing from a single gene located on chromosome 17q21, containing 16 exons (Yoshiyama et al., 2013). The size of six isoforms varies from 352 to 441 amino acids. These forms differ from each other by the presence or absence of 29 (Exon 2) or 58 (Exon 2 + Exon 3) amino acids inserts in the amino terminal. Tau protein contains several phosphorylation sites. In the phosphorylated form, Tau binds to microtubules with lesser affinity and has greater propensity to aggregate into paired helical filaments, promoting the formation of NFTs. Based on the importance of Tau phosphorylation in AD, a number of investigators have developed inhibitors for kinases and phosphatases associated with phosphorylation and dephosphorylation of Tau protein (Tell and Hilgeroth, 2013; Anand et al., 2014). As stated in Chapter 1, Tau protein is phosphorylated by several protein kinases including GSK3β, cdk5, p38, JNK, CK, and DYRK1A (Tell and Hilgeroth, 2013). Lithium and valproate are well-known inhibitors of GSK-3β. Treatment of transgenic models with lithium and valproate reduces Tau-related pathology (Engel et al., 2006); however, this effect is not corroborated in small clinical trials for both compounds (Hampel et al., 2009; Tariot and Aisen, 2009). In addition, several inhibitors of GSK-3β have been developed. These inhibitors include SB 216763, CHIR-98014, and SRN-003-556. At present, preclinical studies are performed with these inhibitors (Anand et al., 2014). It is reported that SB 216763 reduces the amount of Tau phosphorylation, but adverse effects are detected in control animals (Tell and Hilgeroth, 2013). Tideglusib is another inhibitor of GSK-3β which is in phase 2b trials for mild to moderate AD. Although, trials on this inhibitor have been completed, the results are not published (clinicaltrials.gov). In a small pilot study, tideglusib produces positive effects in mini-mental status examination, Alzheimer's disease assessment scale-cognitive subscale, geriatric depression scale, and global clinical assessment cognitive scales without statistical significance in the small sample of 30 patients (Del et al., 2013). Rember (methylene blue) is a tau antiaggregation agent. It not only retards Tau aggregation but also promotes Tau assembly disassociation. Preclinical data have indicated that rember reverses learning deficit and can slow down AD progression with a good bioavailability (Deiana et al., 2009; Wischik, 2009). TRx0237 is another methylene blue with improved drug absorption, bioavailability, and tolerability. Since 2008, intensive investigation of this agent has indicated that TRx0237 induces neuroprotection (Wen et al., 2011) not only by inducing Aβ clearance in transgenic mice but also by promoting improvement in spatial learning in rats (Deiana et al., 2009; Riha et al., 2011). Right now three phase 3 studies are ongoing.

Epothilone D (BMS-241,027) is a microtubule stabilizer. On one hand, it inhibits Tau release from microtubule to maintain the transportation function of axon, and on the other hand, it precludes Tau aggregation. This drug restores behavioral and cognitive deficits, inhibits neurodegeneration, and retards the tauopathy in animal models (Hurtado et al., 2010; Zhang et al., 2012). Epothilone can penetrate the BBB and exert a better efficacy at low concentration and now is under a phase 1 clinical trial. Nicotinamide, the precursor of coenzyme NAD^+, reduces phosphorylated Tau and protects microtubule stabilization in mouse model (Green et al., 2008). Nicotinamide has been launched into clinical studies suggesting that it is safe and well tolerated, and a phase 2 clinical trial is ongoing in patients with mild-to-moderate AD. It is tempting to speculate that detailed investigations are needed on these inhibitors of Tau phosphorylation and dephosphorylation in animal models and patients with AD.

CALORIE RESTRICTION AND ALZHEIMER'S DISEASE

These days many investigators are directing their efforts to study effect of calorie restriction on aging and neurodegenerative diseases. Calorie restriction is the only dietary intervention that increases longevity and retards the process of aging in several models in various organisms. In mice and rats, calorie restriction can increase longevity by ~30%, delay physiological aging, and postpone or diminish the morbidity of most age-related diseases (Masoro, 2005). Calorie restriction not only delays brain iron accumulation but also preserves motor performance and longevity in rhesus monkeys (Kastman et al., 2012). However, there is considerable variation in the mortality in both rodents and monkeys. In elderly humans, calorie has been reported to improve memory function (Witte et al., 2009). Calorie restriction has been reported to produce benefits in AD by preventing Aβ deposition in the mouse models of AD. Mechanisms by which calorie restriction mediate beneficial effects in animal models of AD are multifaceted and not fully understood. However, it is proposed that calorie restriction–mediated NAD^+-dependent SIRT1 deacetylase activity may retard AD neuropathology (Qin et al., 2006). SIRT1 directly activates the transcription of the α-secretase through the deacetylation of retinoic acid receptor β (RARβ), which is known to activate α-secretase transcription (Donmez et al., 2010). Moreover, SIRT1 overexpression in mouse neural N2a cells expressing mutant APPs have been reported to elevate ADAM10 and sAPPα levels, and cilostazol (an inhibitor of type III phosphodiesterase) suppresses Aβ production in a SIRT1-RAR-ADAM10-dependent manner (Lee et al., 2014). These observations suggest that neuroprotective effects of SIRT1 is mediated by the stimulation of α-secretase and onset of nonamyloidogenic APP processing. A decline in SIRT1 levels in the aged brain may predispose its neurons to amyloidogenic APP processing and AD.

HEALTHY LIFESTYLE AND ALZHEIMER'S DISEASE

It is well known that healthy nutrition, daily exercise, adequate sleep, mental challenges, and socialization are the foundations for maintaining optimal health. Nutrition modulates brain development and maturation with profound implications of mental health in neurological disorders including AD (Farooqui, 2010, 2014; Jedrziewski et al., 2014). Exercise produces beneficial effects on cognitive function in healthy seniors not only due to increase in mitochondrial biogenesis and upregulation of mitophagy (Bori et al., 2012; Lanza and Sreekumaran Nair, 2012; Guo et al., 2012; Erickson and Kramer, 2009; Erickson et al., 2009; Farooqui, 2014) but also promotes neuroplasticity (Stranahan et al., 2009) and cognitive function (Hillman et al., 2008). Exercise also induces cardiorespiratory and muscular fitness by increasing energy consumption, improving insulin sensitivity, increasing blood flow, strengthening the immune system, reducing neuroinflammation, downregulating oxidative stress, promoting sleep, controlling weight, increasing gray matter volume, and inducing neurogenesis in the dentate gyrus along with the release of BDNF and IGF-1 (Farooqui, 2014). Exercise also reduces the brain's exposure to neurotoxic substances, including Aβ toxicity and excessive glucose levels (Brown et al., 2013; Mehlig et al., 2014). At the same time, exercise has mental stimulatory properties such as those that require eye–hand coordination and visuospatial memory, thus further augmenting their effects on cognitive functioning.

Studies on the effect of exercise in animal models of AD indicate that subjecting transgenic mice (mice expressing the skeletal muscle-specific mutant PSEN2 gene) to treadmill exercise for 3 months not only results in reduction of Aβ-42 deposits but also produces improvement in behavioral function thereby restoring normal concentrations of total cholesterol, HDL cholesterol, LDL cholesterol, and triglyceride (Cho et al., 2003). Furthermore, 16 weeks of exercise on the treadmill in the neuron-specific enolase/Swedish mutation of amyloid precursor protein (APPsw) transgenic mice indicates that exercise not only decreases levels of Aβ-42 peptides and produces antiapoptotic effects but also inhibits the induction of GLUT1 and BDNF (Um et al., 2008). Balance dysfunction, gait disturbances, and falls are common problems in patients with later stages of AD and dementia compared with older people without these conditions (Manckoundia et al., 2006). Although in normal older people, exercise reduces falls and improves their mood, in patients with AD exercise induces marginal effects on fall reduction and mood improvement (Cho et al., 2003). Studies on the effect of aerobic exercise in patients with mild to moderate AD have been controversial. Some studies have indicated that exercise has no effect on global cognition and quality of life except for depression (Yu et al., 2014). Other studies indicate that exercise improves cognitive functions such as tasks mediated by the hippocampus and results in major changes in plasticity in the hippocampus. Interestingly, exercise-induced plasticity is also pronounced in *APOE ε4* carriers who express a risk factor for late-onset AD that may modulate the effect of treatments (Foster et al., 2011). Based on MRI studies, it is suggested that exercise may reduce hypertrophy in the hippocampus and promote the production of BDNF, which enhances neurogenesis and plays a key role in positive cognitive effects (Foster et al., 2011).

CONCLUSION

AD is an age-related progressive neurodegenerative disease, which is pathologically characterized by cerebral atrophy (particularly within the hippocampus and temporal and parietal lobes), senile plaques, NFTs, and neuronal cell death. Biochemically, AD is also characterized by brain insulin deficiency and insulin resistance, loss of Ca^{2+} homeostasis, increase in levels of lipid mediators, elevation in oxidative stress, and onset of neuroinflammation. These observations support the view that AD is a complex multifactorial disorder that may require equally complex approaches to treatment. The success of AD treatment with potential therapeutic drugs that inhibit the generation of β-amyloid has been low. Therefore, due to therapeutic failure in recent years, investigators are looking for alternative hypotheses to explain the causes of the AD and the cognitive loss such as changes related to insulin signaling and energy metabolism. In fact, it is becoming increasingly evident that hypertension, hypercholesterolemia, type 2 diabetes, and MetS are also important risk factors for AD. The term, "type 3 diabetes" has been used to describe AD and there is evidence for altered metabolic changes indicative of "insulin-resistant brain state" in human AD and animal models. Approaches to treat AD should include early detection and control of hypertension, hypercholesterolemia, type 2 diabetes, as well as MetS, combination therapies, and healthy lifestyle choices. A broad range of studies show that inadequate nutrition and lack of physical activity (exercise) can increase the risk of AD development. A healthy diet and regular exercise can not only retard chances of developing hypertension, hypercholesterolemia, type 2 diabetes, and MetS, but also decrease the chances developing AD. However, Mediterranean-type diet, caloric restriction, or antioxidant diet alone can prevent or delay AD. It can be proposed that carefully developed nutrition regimens coupled with combination pharmacotherapies targeting multiple pathways and inclusion of physical activity from childhood to the old age may decrease the onset of AD among seniors.

References

Abramovitz, M., Wong, E., Cox, M.E., Richardson, C.D., Li, C., et al., 1993. 5-lipoxygenase-activating protein stimulates the utilization of arachidonic acid by 5-lipoxygenase. Eur. J. Biochem. 215, 105–111.

Abuznait, A.H., Qosa, H., Busnena, B.A., El Sayed, K.A., Kaddoumi, A., 2013. Olive-oil-derived oleocanthal enhances β-amyloid clearance as a potential neuroprotective mechanism against Alzheimer's disease: in vitro and in vivo studies. ACS Chem. Neurosci. 4, 973–982.

Adzovic, L., Domenici, L., 2014. Insulin induces phosphorylation of the AMPA receptor subunit GluR1, reversed by ZIP, and overexpression of protein kinase M zeta, reversed by amyloid β. J. Neurochem. 31, 582–587.

Aimone, J.B., Deng, W., Gage, F.H., 2010. Adult neurogenesis: integrating theories and separating functions. Trends Cogn. Sci. 14, 325–337.

Alva, G., Cummings, J.L., 2008. Relative tolerability of Alzheimer's disease treatments. Psychiatry (Edgmont) 5, 27–36.

Alvarez-Buylla, A., Kohwi, M., Nguyen, T.M., Merkle, F.T., 2008. The heterogeneity of adult neural stem cells and the emerging complexity of their niche. Cold Spring Harb. Symp. Quant. Biol. 73, 357–365.

Anand, R., Gill, K.D., Mahdi, A.A., 2014. Therapeutics of Alzheimer's disease: past, present and future. Neuropharmacology 76 (Pt. A), 27–50.

Balakumar, P.S., Kathuria, S., 2012. Submaximal PPARγ activation and endothelial dysfunction: new perspectives for the management of cardiovascular disorders. Brit. J. Pharmacol. 166, 1981–1992.

Bateman, R.J., Xiong, C., Benzinger, T.L., Fagan, A.M., Goate, A., et al., 2012. Clinical and biomarker changes in dominantly inherited Alzheimer's disease. N. Engl. J. Med. 367, 795–804.

Bath, K.G., Mandairon, N., Jing, D., Rajagopal, R., Kapoor, R., et al., 2008. Variant brain-derived neurotrophic factor (Val66Met) alters adult olfactory bulb neurogenesis and spontaneous olfactory discrimination. J. Neurosci. 28, 2383–2393.

Beauchamp, G.K., Keast, R.S.J., Morel, D., Lin, J., Pika, J., et al., 2005. Phytochemistry: ibuprofen-like activity in extra-virgin olive oil. Nature 437, 45–46.

Begum, A.N., Jones, M.R., Lim, G.P., Morihara, T., Kim, P., et al., 2008. Curcumin structure-function, bioavailability, and efficacy in models of neuroinflammation and Alzheimer's disease. J. Pharmacol. Exp. Ther. 326, 196–208.

Bengtsson, S.K., Johansson, M., Backstrom, T., Wang, M., 2012. Chronic allopregnanolone treatment accelerates Alzheimer's disease development in AβPP$_{Swe}$ PSEN1Δ$_{E9}$ mice. J. Alzheimers Dis. 31, 71–84.

Bettermann, K., Arnold, A.M., Williamson, J., Rapp, S., Sink, K., et al., 2012. Statins, risk of dementia, and cognitive function: secondary analysis of the ginkgo evaluation of memory study. J. Stroke Cerebrovasc. Dis. 21, 436–444.

Bhalla, K., Hwang, B.J., Choi, J.H., Dewi, R., Ou, L., et al., 2011. N-Acetylfarnesylcysteine is a novel class of peroxisome proliferator-activated receptor γ ligand with partial and full agonist activity *in vitro* and *in vivo*. J. Biol. Chem. 286, 41626–41635.

Birks, J., 2006. Cholinesterase inhibitors for Alzheimer's disease. Cochrane Database Syst. Rev. CD005593.

Blalock, E.M., Phelps, J.T., Pancani, T., Searcy, J.L., Anderson, K.L., et al., 2010. Effects of long-term pioglitazone treatment on peripheral and central markers of aging. PLoS One 5, e10405.

Boccardi, V., Paolisso, G., Mecocci, P., 2016. Nutrition and lifestyle in healthy aging: the telomerase challenge. Aging (Albany NY) 8, 12–15.

Bori, Z., Zhao, Z., Koltai, E., Fatouros, I.G., Jamurtas, A.Z., et al., 2012. The effects of aging, physical training, and a single bout of exercise on mitochondrial protein expression in human skeletal muscle. Exp. Gerontol. 47, 417–424.

Brinton, R.D., 2013. Neurosteroids as regenerative agents in the brain: therapeutic implications. Nat. Rev. Endocrinol. 9, 241–250.

Brookmeyer, R., Gray, S., 2000. Methods for projecting the incidence and prevalence of chronic diseases in aging populations: application to Alzheimer's disease. Stat. Med. 19, 1481–1493.

Burns, A., 2009. Alzheimer's disease: on the verges of treatment and prevention. Lancet Neurol. 8, 4–5.

Brown, B.M., Peiffer, J.J., Taddei, K., Lui, J.K., Laws, S.M., et al., 2013. Physical activity and amyloid-β plasma and brain levels: results from the Australian imaging, biomarkers and lifestyle study of ageing. Mol. Psychiatry 18, 875–881.

Calon, F., Cole, G., 2007. Neuroprotective action of omega-3 polyunsaturated fatty acids against neurodegenerative diseases: evidence from animal studies. Prostaglandis Leukot. Essent. Fat. Acids 77, 287–293.

Castro-Quezada, I., Román-Viñas, B., Serra-Majem, L., 2014. The Mediterranean diet and nutritional adequacy: a review. Nutrients 6, 231–248.

Chandra, V., Pandav, R., Dodge, H.H., Johnston, J.M., Belle, S.H., et al., 2001. Incidence of Alzheimer's disease in a rural community in India: the Indo-US study. Neurology 57, 985–989.

Chauhan, N.B., 2006. Effect of aged garlic extract on APP processing and tau phosphorylation in Alzheimer's transgenic model Tg2576. J. Ethnopharmacol. 108, 385–394.

Chen, Z., Lu, T., Yue, X., Wei, N., Jiang, Y., Chen, M., Ni, G., Liu, X., Xu, G., 2010. Neuroprotective effect of ginsenoside Rb1 on glutamate-induced neurotoxicity: with emphasis on autophagy. Neurosci. Lett. 482, 264–268.

Chen, R., Hu, Z., Wei, L., Ma, Y., Liu, Z., Copeland, J.R., 2011a. Incident dementia in a defined older Chinese population. PLoS One 6, e24817.

Chen, S., Wang, J.M., Irwin, R.W., Yao, J., Liu, L., Brinton, R.D., 2011b. Allopregnanolone promotes regeneration and reduces β-amyloid burden in a preclinical model of Alzheimer's disease. PLoS One 6, e24293.

Chen, S., Wang, J.M., Irwin, R.W., Yao, J., Liu, L., et al., 2011c. Allopregnanolone promotes regeneration and reduces β-amyloid burden in a preclinical model of Alzheimer's disease. PLoS One 6, e24293.

Chen, P.T., Chen, Z.T., Hou, W.C., Yu, L.C., Chen, R.P., 2016. Polyhydroxycurcuminoids but not curcumin upregulate neprilysin and can be applied to the prevention of Alzheimer's disease. Sci. Rep. 6, 29760.

Chiu, C.C., Su, K.P., Cheng, T.C., Liu, H.C., Chang, C.J., et al., 2008. The effects of omega-3 fatty acids monotherapy in Alzheimer's disease and mild cognitive impairment: a preliminary randomized double-blind placebo-controlled study. Prog. Neuropsychopharmacol. Biol. Psychiatry 32, 1538–1544.

Cho, J.Y., Hwang, D.Y., Kang, T.S., Shin, D.H., Hwang, J.H., et al., 2003. Use of NSE/PS2m-transgenic mice in the study of the protective effect of exercise on Alzheimer's disease. J. Sports Sci. 21, 943–951.

Chu, J., Praticò, D., 2011. 5-lipoxygenase as an endogenous modulator of amyloid β formation in vivo. Ann. Neurol. 69, 34–46.

Chu, J., Praticò, D., 2012. Involvement of 5-lipoxygenase activating protein in the amyloidotic phenotype of an Alzheimer's disease mouse model. J. Neuroinflammation 9, 127.

Chu, J., Praticò, D., 2013. 5-Lipoxygenase pharmacological blockade decreases tau phosphorylation in vivo: involvement of the cyclin-dependent kinase-5. Neurobiol. Aging 34, 1549–1554.

Clarke, D.W., Mudd, L., Boyd Jr., F.T., 1986. Insulin is released from rat brain neuronal cells in culture. J. Neurochem. 47, 831–836.

Claxton, A., Baker, L.D., Hanson, A., Trittschuh, E.H., Cholerton, B., et al., 2015. Long-acting intranasal insulin detemir improves cognition for adults with mild cognitive impairment or early-stage Alzheimer's disease dementia. J. Alzheimers Dis. 44, 897–906.

Combs, C.K., Bates, P., Karlo, J.C., Landreth, G.E., 2001. Regulation of β-amyloid stimulated proinflammatory responses by peroxisome proliferator-activated receptor α. Neurochem. Int. 39, 449–457.

Cordle, A., Koenigsknecht-Talboo, J., Wilkinson, B., Limpert, A., Landreth, G., 2005. Mechanisms of statin-mediated inhibition of small G-protein function. J. Biol. Chem. 280, 34202–34209.

Cordle, A., Landreth, G., 2005. 3-Hydroxy-3-methylglutaryl-coenzyme A reductase inhibitors attenuate β-amyloid-induced microglial inflammatory responses. J. Neurosci. 25, 299–307.

Coric, V., van Dyck, C.H., Salloway, S., Andreasen, N., Brody, M., et al., 2012. Safety and tolerability of the γ-secretase inhibitor avagacestat in a phase 2 study of mild to moderate Alzheimer disease. Arch. Neurol. 1–12.

Craft, S., Peskind, E., Schwartz, M.W., Schellenberg, G.D., et al., 1998. Cerebrospinal fluid and plasma insulin levels in Alzheimer's disease: relationship to severity of dementia and apolipoprotein E genotype. Neurology 50, 164–168.

Crump, C.J., Castro, S.V., Wang, F., Pozdnyakov, N., Ballard, T.E., et al., 2012. BMS-708,163 targets presenilin and lacks notch-sparing activity. Biochemistry 51, 7209–7211.

Damar, U., Gersner, R., Johnstone, J.T., Schachter, S., Rotenberg, A., 2016. Huperzine A as a neuroprotective and antiepileptic drug: a review of preclinical research. Expert Rev. Neurother. 16, 671–680.

Dechant, G., Barde, Y.A., 2002. The neurotrophin receptor p75^{NTR}: novel functions and implications for diseases of the nervous system. Nat. Neurosci. 5, 1131–1136.

Deiana, S., Harrington, C.R., Wischik, C.M., Riedel, G., 2009. Methylthioninium chloride reverses cognitive deficits induced by scopolamine: comparison with rivastigmine. Psychopharmacology 202, 53–65.

de la Puerta, R., Ruiz Gutierrez, V., Hoult, J.R., 1999. Inhibition of leukocyte 5-lipoxygenase by phenolics from virgin olive oil. Biochem. Pharmacol. 57, 445–449.

Del, S.T., Steinwachs, K.C., Gertz, H.J., Andres, M.V., et al., 2013. Treatment of Alzheimer's disease with the GSK-3 inhibitor tideglusib: a pilot study. J. Alzheimers Dis. 33, 205–215.

Denis, I., Potier, B., Vancassel, S., Heberden, C., Lavialle, M., 2013. Omega-3 fatty acids and brain resistance to ageing and stress: body of evidence and possible mechanisms. Ageing Res. Rev. 12, 579–594.

Desvergne, B., Michalik, L., Wahli, W., 2006. Transcriptional regulation of metabolism. Physiol. Rev. 86, 465–514.

Donmez, G., Wang, D., Cohen, D.E., Guarente, L., 2010. SIRT1 suppresses β-amyloid production by activating the α-secretase gene ADAM10. Cell 142, 320–332.

Dong, S., Zeng, Q., Mitchell, E.S., Xiu, J., Duan, Y., et al., 2012. Curcumin enhances neurogenesis and cognition in aged rats: implications for transcriptional interactions related to growth and synaptic plasticity. PLoS One 7, e31211.

Doody, R.S., Raman, R., Farlow, M., Iwatsubo, T., Vellas, B., et al., 2013. A phase 3 trial of semagacestat for treatment of Alzheimer's disease. N. Engl. J. Med. 369, 341–350.

Dou, J.T., Chen, M., Dufour, F., Alkon, D.L., Zhao, W.Q., 2005. Insulin receptor signaling in long-term memory consolidation following spatial learning. Learn. Mem. 12, 646–655.

Dysken, M.W., Sano, M., Asthana, S., Vertrees, J.E., Pallaki, M., et al., 2014. Effect of vitamin E and memantine on functional decline in Alzheimer disease: the TEAM-AD VA cooperative randomized trial. JAMA 311, 33–44.

Eggler, A.L., Gay, K.A., Mesecar, A.D., 2008. Molecular mechanisms of natural products in chemoprevention: induction of cytoprotective enzymes by Nrf2. Mol. Nutr. Food Res. 52 (Suppl. 1), S84–S94.

Eijkelenboom, A., Burgering, B.M., 2013. FOXOs: signalling integrators for homeostasis maintenance. Nat. Rev. Mol. Cell Biol. 14, 83–97.

Engel, T., Goni-Oliver, P., Lucas, J.J., Avila, J., Hernandez, F., 2006. Chronic lithium administration to FTDP-17 tau and GSK-3β overexpressing mice prevents tau hyperphosphorylation and neurofibrillary tangle formation, but pre-formed neurofibrillary tangles do not revert. J. Neurochem. 99, 1445–1455.

Endres, M., 2005. Statins and stroke. J. Cereb. Blood Flow Metab. 25, 1093–1110.

Erickson, K.I., Kramer, A.F., 2009. Aerobic exercise effects on cognitive and neural plasticity in older adults. Br. J. Sports Med. 43, 22–24.

Erickson, K.I., Prakash, R.S., Voss, M.W., Chaddock, L., Hu, L., et al., 2009. Aerobic fitness is associated with hippocampal volume in elderly humans. Hippocampus 19, 1030–1039.

Escribano, L., Simón, A.-M., Gimeno, E., Cuadrado-Tejedor, M., López de Maturana, R., et al., 2010. Rosiglitazone rescues memory impairment in Alzheimer's transgenic mice: mechanisms involving a reduced amyloid and tau pathology. Neuropsychopharmacology 35, 1593–1604.

Estruch, R., Ros, E., Salas-Salvadó, J., Covas, M.I., Corella, D., et al., 2013. Primary prevention of cardiovascular disease with a Mediterranean diet. N. Engl. J. Med. 368, 1279–1290.

Fan, X., Sun, D., Tang, X., Cai, Y., Yin, Z.Q., Xu, H., 2014. Stem-cell challenges in the treatment of Alzheimer's disease: a long way from bench to bedside. Med. Res. Rev. 34, 957–978.

Fang, F., Chen, X., Huang, T., Lue, L.F., Luddy, J.S., et al., 2012. Multi-faced neuroprotective effects of Ginsenoside Rg1 in an Alzheimer mouse model. Biochim. Biophys. Acta 1822, 286–292.

Farooqui, A.A., Ong, W.Y., Horrocks, L.A., Chen, P., Farooqui, T., 2007. Comparison of biochemical effects of statins and fish oil in brain: the battle of the titans. Brain Res. Rev. 2007 (56), 443–471.

Farooqui, A.A., 2009. Beneficial Effects of Fish Oil on Human Brain. Springer, New York.

Farooqui, A.A., 2010. Neurochemical Aspects of Neurotraumatic and Neurodegenerative Diseases. Springer, New York.

Farooqui, A.A., 2012. Phytochemicals, Signal Transduction, and Neurological Disorders. Springer, New York.

Farooqui, A.A., 2013. Metabolic Syndrome: an Important Risk Factor for Stroke, Alzheimer Disease, and Depression. Springer, New York.

Farooqui, A.A., 2014. Inflammation and Oxidative Stress in Neurological Disorders: Effect of Lifestyle, Genes, and Age. Springer, New York.

Farooqui, A.A., 2015. High Calorie Diet and the Human Brain: Metabolic Consequences of Long-Term Consumption. Springer, New York.

Farooqui, A.A., 2016. Therapeutic Potentials of Curcumin for Alzheimer Disease. Springer, New York.

Feldman, H.H., Doody, R.S., Kivipelto, M., Sparks, D.L., Waters, D.D., et al., 2010. Randomized controlled trial of atorvastatin in mild to moderate Alzheimer disease: LEADe. Neurology 74, 956–964.

Figueiredo, C.P., Clarke, J.R., Ledo, J.H., Ribeiro, F.C., Costa, C.V., et al., 2013. Memantine rescues transient cognitive impairment caused by high-molecular-weight Aβ oligomers but not the persistent impairment induced by low-molecular-weight oligomers. J. Neurosci. 33, 9626–9634.

Fleisher, A.S., Raman, R., Siemers, E.R., Becerra, L., Clark, C.M., et al., 2008. Phase 2 safety trial targeting amyloid β production with a γ-secretase inhibitor in Alzheimer disease. Arch. Neurol. 65, 1031–1038.

Foster, P.P., Rosenblatt, K.P., Kuljiš, R.O., 2011. Exercise-induced cognitive plasticity, implications for mild cognitive impairment and Alzheimer's disease. Front. Neurol. 2, 28.

Freiherr, J., Hallschmid, M., Frey II, W.H., Brünner, Y.F., Chapman, C.D., et al., 2013. Intranasal insulin as a treatment for Alzheimer's disease: a review of basic research and clinical evidence. CNS Drugs 27, 505–514.

Freir, D.B., Fedriani, R., Scully, D., Smith, I.M., Selkoe, D.J., Walsh, D.M., et al., 2011. Aβ oligomers inhibit synapse remodelling necessary for memory consolidation. Neurobiol. Aging 32, 2211–2218.

Freund-Levi, Y., Eriksdotter-Jonhagen, M., Cederholm, T., Basun, H., Faxen-Irving, G., et al., 2006. Omega-3 fatty acid treatment in 174 patients with mild to moderate Alzheimer disease: OmegAD study: a randomized double-blind trial. Arch. Neurol. 63, 1402–1408.

Gallagher, E.J., LeRoith, D., Karnieli, E., 2008. The metabolic syndrome – from insulin resistance to obesity and diabetes. Endocrinol. Metab. Clin. North Am. 37, 559–579.

Ganguli, M., Chandra, V., Kamboh, M.I., Johnston, J.M., Dodge, H.H., et al., 2000. Apolipoprotein E polymorphism and Alzheimer disease: The Indo-US Cross-National Dementia Study. Arch. Neurol. 57, 824–830.

Ghosh, S., Banerjee, S., Sil, P.C., 2015. The beneficial role of curcumin on inflammation, diabetes and neurodegenerative disease: a recent update. Food Chem. Toxicol. 83, 111–124.

Giannopoulos, P.F., Chu, J., Joshi, Y., Sperow, M., Li, J., et al., 2013. 5-lipoxygenase activating protein reduction ameliorates cognitive deficit, synaptic dysfunction, and neuropathology in a mouse model of Alzheimer's disease. Biol. Psychiatry 74, 348–356.

Giannopoulos, P.F., Chu, J., Joshi, Y.B., Sperow, M., et al., 2014. Gene knockout of 5-lipoxygenase rescues synaptic dysfunction and improves memory in the triple-transgenic model of Alzheimer's disease. Mol. Psychiatry 19, 511–518.

Gibbs, R.B., McCabe, J.T., Buck, C.R., Chao, M.V., Pfaff, D.W., 1989. Expression of NGF receptor in the rat forebrain detected with in situ hybridization and immunohistochemistry. Brain Res. Mol. Brain Res. 6, 275–287.

Ginter, E., Simko, V., 2012. Type 2 diabetes mellitus, pandemic in 21st century. Adv. Exp. Med. Biol. 771, 42–50.

Gold, M., Alderton, C., Zvartau-Hind, M., Egginton, S., Saunders, A.M., et al., 2010. Rosiglitazone monotherapy in mild-to-moderate Alzheimer's disease: results from a randomized, double-blind, placebo-controlled phase III study. Dement. Geriatr. Cogn. Disord. 30, 131–146.

Goldstein, J.L., Brown, M.S., 2009. The LDL receptor. Arterioscler. Thromb. Vasc. Biol. 29, 431–438.

Green, K.N., Steffan, J.S., Martinez-Coria, H., Sun, X., Schreiber, S.S., et al., 2008. Nicotinamide restores cognition in Alzheimer's disease transgenic mice via a mechanism involving sirtuin inhibition and selective reduction of Thr231-phosphotau. J. Neurosci. 28, 11500–11510.

Greig, S.L., 2015. Memantine ER/Donepezil: a review in Alzheimer's disease. CNS Drugs 29, 963–970.

Grossi, C., Rigacci, S., Ambrosini, S., Ed Dami, T., Luccarini, I., 2013. The polyphenol oleuropein aglycone protects TgCRND8 mice against Aß plaque pathology. PLoS One 8, e71702.

Guo, W., Wong, S., Li, M., Liang, W., Liesa, M., et al., 2012. Testosterone plus low-intensity physical training in late life improves functional performance, skeletal muscle mitochondrial biogenesis, and mitochondrial quality control in male mice. PLoS One 7, e51180.

Gupta, S.C., Patchva, S., Koh, W., Aggarwal, B.B., 2012. Discovery of curcumin, a component of golden spice, and its miraculous biological activities. Clin. Exp. Pharmacol. Physiol. 39, 283–299.

Gupta, V.B., Rao, K.S., 2007. Anti-amyloidogenic activity of S-allyl-L-cysteine and its activity to destabilize Alzheimer's β-amyloid fibrils in vitro. Neurosci. Lett. 429, 75–80.

Haag, M.D., Hofman, A., Koudstaal, P.J., Stricker, B.H., Breteler, M.M., 2009. Statins are associated with a reduced risk of Alzheimer disease regardless of lipophilicity. The Rotterdam Study. J. Neurol. Neurosurg. Psychiatry 80, 13–17.

Hamilton, L.K., Aumont, A., Julien, C., Vadnais, A., Calon, F., et al., 2010. Widespread deficits in adult neurogenesis precede plaque and tangle formation in the 3xTg mouse model of Alzheimer's disease. Eur. J. Neurosci. 32, 905–920.

Hampel, H., Ewers, M., Burger, K., Annas, P., Mortberg, A., et al., 2009. Lithium trial in Alzheimer's disease: a randomized, single-blind, placebo-controlled, multicenter 10-week study. J. Clin. Psychiatry 70, 922–931.

Havrankova, J., Roth, J., Brownstein, M., 1978. Insulin receptors are widely distributed in the central nervous system of the rat. Nature 272, 827–829.

Hattiangady, B., Shetty, A.K., 2008. Aging does not alter the number or phenotype of putative stem/progenitor cells in the neurogenic region of the hippocampus. Neurobiol. Aging 29, 129–147.

Hawkes, C.A., Shaw, J.E., Brown, M., Sampson, A.P., McLaurin, J., et al., 2014. MK886 reduces cerebral amyloid angiopathy severity in TgCRND8 mice. Neurodegener. Dis. 13, 17–23.

Heneka, M.T., Sastre, M., Dumitrescu-Ozimek, L., Hanke, A., Dewachter, I., et al., 2005. Acute treatment with the PPARγ agonist pioglitazone and ibuprofen reduces glial inflammation and Aβ1–42 levels in APPV717I transgenic mice. Brain 128, 1442–1453.

Heneka, M.T., Landreth, G.E., Hüll, M., 2007. Drug insight: effects mediated by peroxisome proliferator-activated receptor-γ in CNS disorders. Nat. Clin. Pract. Neurol. 3, 496–504.

Heneka, M.T., Fink, A., Doblhammer, G., 2015. Effect of pioglitazone medication on the incidence of dementia. Ann. Neurol. 78, 284–294.

Higgins, L.S., Depaoli, A.M., 2010. Selective peroxisome proliferator-activated receptor γ (PPARγ) modulation as a strategy for safer therapeutic PPARγ activation. Am. J. Clin. Nut. 91, 72S–267S.

Hillman, C.H., Erickson, K.I., Kramer, A.F., 2008. Be smart, exercise your heart: exercise effects on brain and cognition. Nat. Rev. Neurosci. 9, 58–65.

Ho, S.C., Su, M.S., 2014. Evaluating the anti-neuroinflammatory capacity of raw and steamed garlic as well as five organosulfur compounds. Molecules 19, 17697–17714.

Holmes, C., Boche, D., Wilkinson, D., Yadegarfar, G., Hopkins, V., et al., 2008. Long-term effects of Aβ42 immunisation in Alzheimer's disease: follow-up of a randomised, placebo-controlled phase I trial. Lancet 372, 216–223.

Holscher, C., Li, L., 2010. New roles for insulin-like hormones in neuronal signalling and protection: new hopes for novel treatments of Alzheimer's disease? Neurobiol. Aging 31, 1495–1502.

Holtzman, D.M., Mandelkow, E., Selkoe, D.J., 2012. Alzheimer disease in 2020. Cold Spring Harb. Perspect. Med. 2 pii a011585.

Hosie, A.M., Wilkins, M.E., Da Silva, H.M., Smart, T.G., 2006. Endogenous neurosteroids regulate GABAA receptors through two discrete transmembrane sites. Nature 444, 486–489.

Hu, M., Schurdak, M.E., Puttfarcken, P.S., El Kouhen, R., Gopalakrishnan, M., et al., 2007. High content screen microscopy analysis of Aβ 1-42-induced neurite outgrowth reduction in rat primary cortical neurons: neuroprotective effects of α7 neuronal nicotinic acetylcholine receptor ligands. Brain Res. 1151, 227–235.

Hu, S., Maiti, P., Ma, Q., Zuo, X., Jones, M.R., et al., 2015. Clinical development of curcumin in neurodegenerative disease. Expert Rev. Neurother. 15, 629–637.

Huang, X.T., Qian, Z.M., He, X., Gong, Q., Wu, K.C., et al., 2014. Reducing iron in the brain: a novel pharmacologic mechanism of huperzine A in the treatment of Alzheimer's disease. Neurobiol. Aging 35, 1045–1054.

Hurtado, D.E., Molina-Porcel, L., Iba, M., et al., 2010. Aβ accelerates the spatiotemporal progression of tau pathology and augments tau amyloidosis in an Alzheimer mouse model. Am. J. Pathol. 177, 1977–1988.

Hussein, M.A., 2011. A convenient mechanism for the free radical scavenging activity of resveratrol. Int. J. Phytomed. 3, 459–469.

Hwang, S.H., Shin, E.J., Shin, T.J., Lee, B.H., Choi, S.H., et al., 2012. Gintonin, a ginseng-derived lysophosphatidic acid receptor ligand, attenuates Alzheimer's disease-related neuropathies: involvement of non-amyloidogenic processing. J. Alzheimers Dis. 31, 207–223.

Iadecola, C., 2013. The pathobiology of vascular dementia. Neuron 80, 844–866.

Ibanez, C.F., Simi, A., 2012. p75 neurotrophin receptor signaling in nervous system injury and degeneration: paradox and opportunity. Trends Neurosci. 35, 431–440.

Imbimbo, B.P., Giardina, G.A., 2011. γ-secretase inhibitors and modulators for the treatment of Alzheimer's disease: disappointments and hopes. Curr. Top. Med. Chem. 11, 1555–1570.

Imfeld, P., Bodmer, M., Jick, S.S., Meier, C.R., 2012. Metformin, other antidiabetic drugs, and risk of Alzheimer's disease: a population-based case-control study. J. Am. Geriatr. Soc. 60, 916–921.

Irwin, R.W., Brinton, R.D., 2014. Allopregnanolone as regenerative therapeutic for Alzheimer's disease: translational development and clinical promise. Prog. Neurobiol. 113, 40–55.

Irwin, R.W., Wang, J.M., Chen, S., Brinton, R.D., 2011. Neuroregenerative mechanisms of allopregnanolone in Alzheimer's disease. Front. Endocrinol. (Lausanne) 2, 117.

Jedrziewski, M.K., Ewbank, D.C., Wang, H., Trojanowski, J.Q., 2014. The impact of exercise, cognitive activities, and socialization on cognitive function: results from the national long-term care survey. Am. J. Alzheimer's Dis. Other Demen. 29, 372–378.

Jia, N., Han, K., Kong, J.J., Zhang, X.M., Sha, S., Ren, G.R., et al., 2013. (-)-Epigallocatechin-3-gallate alleviates spatial memory impairment in APP/PS1 mice by restoring IRS-1 signaling defects in the hippocampus. Mol. Cell. Biochem. 380, 211–218.

Jiao, Y., Wilkinson, J.T., Christine Pietsch, E., Buss, J.L., Wang, W., et al., 2006. Iron chelation in the biological activity of curcumin. Free Radic. Biol. Med. 40, 1152–1160.

Kask, K., Backstrom, T., Nilsson, L.G., Sundstrom-Poromaa, I., 2008. Allopregnanolone impairs episodic memory in healthy women. Psychopharmacol. Berl. 199, 161–168.

Kastman, E.K., Willette, A.A., Coe, C.L., Bendlin, B.B., Kosmatka, K.J., et al., 2012. A calorie-restricted diet decreases brain iron accumulation and preserves motor performance in old rhesus monkeys. J. Neurosci. 32, 11897–11904.

Katsouri, L., Parr, C., Bogdanovic, N., Willem, M., Sastre, M., 2011. PPARγ co-activator-1α (PGC-1α) reduces amyloid-β generation through a PPARγ-dependent mechanism. J. Alzheimers Dis. 25, 151–162.

Kim, G.Y., Kim, K.H., Lee, S.H., Yoon, M.S., Lee, H.J., et al., 2005. Curcumin inhibits immunostimulatory function of dendritic cells: MAPKs and translocation of NF-κB as potential targets. J. Immunol. 174, 8116–8124.

Kim, J., Kim, S.H., Lee, D.S., Lee, D.J., Kim, S.H., et al., 2013. Effects of fermented ginseng on memory impairment and β-amyloid reduction in Alzheimer's disease experimental models. J. Ginseng Res. 37, 100–107.

Kim, Y., Kim, C., Jang, H.Y., Mook-Jung, I., Kim, B., 2016. Inhibition of cholesterol biosynthesis reduces γ-secretase activity and amyloid-β generation. J. Alzheimers Dis. 51, 1057–1068.

Knowles, J.K., Rajadas, J., Nguyen, T.V., Yang, T., LeMieux, M.C., et al., 2009. The p75 neurotrophin receptor promotes amyloid-β(1-42)-induced neuritic dystrophy in vitro and in vivo. J. Neurosci. 29, 10627–10637.

Knowles, J.K., Simmons, D.A., Nguyen, T.V., Vander Griend, L., Xie, Y., et al., 2013. Small molecule p75NTR ligand prevents cognitive deficits and neurite degeneration in an Alzheimer's mouse model. Neurobiol. Aging 34, 2052–2063.

Kojro, E., Gimpl, G., Lammich, S., März, W., Fahrenholz, F., 2001. Low cholesterol stimulates the nonamyloidogenic pathway by its effect on the α-secretase ADAM 10. Proc. Natl. Acad. Sci. U. S. A. 98, 5815–5820.

Kosuge, Y., Koen, Y., Ishige, K., Minami, K., Urasawa, H., Saito, H., Ito, Y., 2003. S-allyl-L-cysteine selectively protects cultured rat hippocampal neurons from amyloid β-protein-and tunicamycin-induced neuronal death. Neuroscience 122, 885–895.

Kounnas, M.Z., Danks, A.M., Cheng, S., Tyree, C., Ackerman, E., et al., 2010. Modulation of γ-secretase reduces β-amyloid deposition in a transgenic mouse model of Alzheimer's disease. Neuron 67, 769–780.

Kummer, M.P., Heneka, M.T., 2008. PPARs in Alzheimer's disease. PPAR Res. 403896.

Lanza, I.R., Sreekumaran Nair, K., 2010. Regulation of skeletal muscle mitochondrial function: genes to proteins. Acta Physiol. (Oxf.) 199, 529–547.

Lazarov, O., Mattson, M.P., Peterson, D.A., Pimplikar, S.W., van Praag, H., 2010. When neurogenesis encounters aging and disease. Trends Neurosci. 10, 1016.

Lee, H.R., Shin, H.K., Park, S.Y., Kim, H.Y., Lee, W.S., et al., 2014. Cilostazol suppresses β-amyloid production by activating a disintegrin and metalloproteinase 10 via the upregulation of SIRT1-coupled retinoic acid receptor-β. J. Neurosci. Res. 92, 1581–1590.

Li, G., Larson, E.B., Sonnen, J.A., Shofer, J.B., Petrie, E.C., et al., 2007. Statin therapy is associated with reduced neuropathologic changes of Alzheimer disease. Neurology. 69, 878–885.

Li, W., Sperry, J.B., Crowe, A., Trojanowski, J.Q., Smith 3rd, A.B., 2009. Inhibition of tau fibrillization by oleocanthal via reaction with the amino groups of tau. J. Neurochem. 110, 1339–1351.

Li, G., Shofer, J.B., Rhew, I.C., Kukull, W.A., Peskind, E.R., et al., 2010. Age-varying association between statin use and incident Alzheimer's disease. J. Am. Geriatr. Soc. 58, 1311–1317.

Lim, G.P., Calon, F., Morihara, T., Yang, F., Teter, B., et al., 2005. A diet enriched with the omega-3 fatty acid docosahexaenoic acid reduces amyloid burden in an aged Alzheimer mouse model. J. Neurosci. 25, 3032–3040.

Liu, H., Yang, J., Du, F., Gao, X., Ma, X., et al., 2009. Absorption and disposition of ginsenosides after oral administration of Panax notoginseng extract to rats. Drug Metab. Dispos. 37, 2290–2298.

Liu, Z.J., Zhao, M., Zhang, Y., Xue, J.F., Chen, N.H., 2010. Ginsenoside Rg1 promotes glutamate release via a calcium/calmodulin-dependent protein kinase II-dependent signaling pathway. Brain Res. 1333, 1–8.

Liu, L., Brinton, R.D., 2010. Gonadal hormones, neurosteroids and clinical progestins as neurogenic regenerative agents: therapeutic implications. In: Gravanis, A.G., Mellon, S.H. (Eds.), Hormones in Neurodegeneration, Neuroprotection, and Neurogenesis. Wiley-Blackwell, Hoboken, NJ, pp. 281–303.

Liu, W., Xu, Z., Deng, Y., Xu, B., Wei, Y., et al., 2013. Protective effects of memantine against methylmercury-induced glutamate dyshomeostasis and oxidative stress in rat cerebral cortex. Neurotox. Res. 24, 320–337.

Liu, H., Song, N., 2016. Molecular mechanism of adult neurogenesis and its association with human brain diseases. J. Cent. Nerv. Syst. Dis. 8, 5–11.

Longo, F.M., Massa, S.M., 2013. Small-molecule modulation of neurotrophin receptors: a strategy for the treatment of neurological disease. Nat. Rev. Drug Discov. 12, 507–525.

Lopez-Miranda, J., Delgado-Lista, J., Perez-Martinez, P., Jimenez-Gómez, Y., Fuentes, F., et al., 2007. Olive oil and the haemostatic system. Mol. Nutr. Food Res. 51, 1249–1259.

López-Miranda, V., Soto-Montenegro, M.L., Vera, G., Herradón, E., Desco, M., et al., 2012. Resveratrol: a neuroprotective polyphenol in the Mediterranean diet. Rev. Neurol. 54, 349–356.

Lowinus, T., Bose, T., Busse, S., Busse, M., Reinhold, D., Schraven, B., Bommhardt, U.H., July 22, 2016. Immunomodulation by memantine in therapy of Alzheimer's disease is mediated through inhibition of Kv1.3 channels and T cell responsiveness. Oncotarget (Epub ahead of print).

Lu, C., Guo, Y., Li, J., Yao, M., Liao, Q., et al., 2012. Design, synthesis, and evaluation of resveratrol derivatives as $A\beta_{1-42}$ aggregation inhibitors, antioxidants, and neuroprotective agents. Bioorg. Med. Chem. Lett. 22, 7683–7687.

Lu, C., Guo, Y., Yan, J., Luo, Z., Luo, H.-B., et al., 2013. Design, synthesis, and evaluation of multitarget-directed resveratrol derivatives for the treatment of Alzheimer's disease. J. Med. Chem. 56, 5843–5859.

Luchtman, D.W., Song, C., 2013. Cognitive enhancement by omega-3 fatty acids from childhood to old age: findings from animal and clinical studies. Neuropharmacology 64, 550–565.

Ma, Q.L., Yang, F., Rosario, E.R., Ubeda, Q.J., Beech, W., et al., 2009a. β-amyloid oligomers induce phosphorylation of tau and inactivation of insulin receptor substrate via c-Jun N-terminal kinase signaling: suppression by omega-3 fatty acids and curcumin. J. Neurosci. 29, 9078–9089.

Ma, T., Zhao, Y., Kwak, Y.D., Yang, Z., Thompson, R., et al., 2009b. Statin's excitoprotection is mediated by sAPP and the subsequent attenuation of calpain-induced truncation events, likely via rho-ROCK signaling. J. Neurosci. 29, 11226–11236.

Ma, L., Shao, Z., Wang, R., Zhao, Z., Dong, W., et al., 2015. Rosiglitazone improves learning and memory ability in rats with type 2 diabetes through the insulin signaling pathway. Am. J. Med. Sci. 350, 121–128.

MacMillan, K.S., Naidoo, J., Liang, J., et al., 2011. Development of proneurogenic, neuroprotective small molecules. J. Am. Chem. Soc. 133, 1428–1437.

Maiti, P., Manna, J., Veleri, S., Frautschy, S., 2014. Molecular chaperone dysfunction in neuro-degenerative diseases and effects of curcumin. Biomed. Res. Int. 495091.

Mans, R.A., Chowdhury, N., Cao, D., McMahon, L.L., Li, L., 2010. Simvastatin enhances hip-pocampal long-term potentiation in C57BL/6 mice. Neuroscience 166, 435–444.

Mans, R.A., McMahon, L.L., Li, L., 2012. Simvastatin-mediated enhancement of long-term potentiation is driven by farnesyl-pyrophosphate depletion and inhibition of farnesyl-ation. Neuroscience 202, 1–9.

Mancini, J.A., Abramovitz, M., Cox, M.E., Wong, E., Charleson, S., et al., 1993. 5-lipoxygen-ase-activating protein is an arachidonate binding protein. FEBS Lett. 318, 277–281.

Mandel, S.A., Amit, T., Kalfon, L., Reznichenko, L., Youdim, M.B., 2008. Targeting multi-ple neurodegenerative diseases etiologies with multimodal-acting green tea catechins. J. Nutr. 138, 1578S–1583S.

Manckoundia, P., Pfitzenmeyer, P., d'Athis, P., Dubost, V., Mourey, F., 2006. Impact of cogni-tive task on the posture of elderly subjects with Alzheimer's disease compared to healthy elderly subjects. Mov. Disord. 21, 236–241.

Mandrekar-Colucci, S., Karlo, J.C., Landreth, G.E., 2012. Mechanisms underlying the rapid peroxisome proliferator-activated receptor-γ-mediated amyloid clearance and reversal of cognitive deficits in a murine model of Alzheimer's disease. J. Neurosci. 32, 10117–10128.

Marambaud, P., Zhao, H., Davies, P., 2005. Resveratrol promotes clearance of Alzheimer's disease amyloid-β peptides. J. Biol. Chem. 280, 37377–37382.

Maruszak, A., Pilarski, A., Murphy, T., Branch, N., Thuret, S., 2014. Hippocampal neurogen-esis in Alzheimer's disease: is there a role for dietary modulation? J. Alzheimers Dis. 38, 11–38.

Masoro, E.J., 2005. Overview of caloric restriction and ageing. Mech. Ageing Dev. 126, 913–922.

Massa, S.M., Xie, Y., Yang, T., Harrington, A.W., Kim, M.L., et al., 2006. Small, nonpeptide p75NTR ligands induce survival signaling and inhibit proNGF-induced death. J. Neurosci. 26, 5288–5300.

McShane, R., Areosa Sastre, A., Minakaran, N., 2006. Memantine for dementia. Cochrane Database Syst. Rev. CD003154.

Mehlig, K., Skoog, I., Waern, M., Jonasson, J.M., Lapidus, L., et al., 2014. Physical activity, weight status, diabetes and dementia: a 34-year follow-up of the population study of women in Gothenburg. Neuroepidemiology 42, 252–259.

Mitani, Y., Yarimizu, J., Saita, K., Uchino, H., Akashiba, H., et al., 2012. Differential effects between γ-secretase inhibitors and modulators on cognitive function in amyloid precur-sor protein-transgenic and nontransgenic mice. J. Neurosci. 32, 2037–2050.

Mondragon-Rodriguez, S., Basurto-Islas, G., Lee, H.G., Perry, G., Zhu, X., et al., 2010. Causes versus effects: the increasing complexities of Alzheimer's disease pathogenesis. Expert Rev. Neurother. 10, 683–691.

Morris, M.C., Evans, D.A., Bienias, J.L., Tangney, C.C., Bennett, D.A., et al., 2003. Consumption of fish and n-3 fatty acids and risk of incident Alzheimer disease. Arch. Neurol. 60, 940–946.

Mu, Y., Gage, F.H., 2011. Adult hippocampal neurogenesis and its role in Alzheimer's dis-ease. Mol. Neurodegener. 6, 85.

Muñoz-Torrero, D., 2008. Acetylcholinesterase inhibitors as disease-modifying therapies for Alzheimer's disease. Curr. Med. Chem. 15, 2433–2455.

Narlawar, R., Baumann, K., Czech, C., Schmidt, B., 2007. Conversion of the LXR-agonist TO-901317–from inverse to normal modulation of γ-secretase by addition of a carboxylic acid and a lipophilic anchor. Bioorg. Med. Chem. Lett. 17, 5428–5431.

Narlawar, R., Pickhardt, M., Leuchtenberger, S., Baumann, K., Krause, S., et al., 2008. Curcumin-derived pyrazoles and isoxazoles: Swiss army knives or blunt tools for Alzheimer's disease? ChemMedChem 3, 165–172.

Ng, T.P., Chiam, P.C., Lee, T., Chua, H.C., Lim, L., et al., 2006. Curry consumption and cognitive function in the elderly. Am. J. Epidemiol. 164, 898–906.

Nguyen, T.V., Shen, L., Vander Griend, L., Quach, L.N., Belichenko, N.P., et al., 2014. Small molecule p75^{NTR} ligands reduce pathological phosphorylation and misfolding of tau, inflammatory changes, cholinergic degeneration, and cognitive deficits in AβPP$^{L/S}$ transgenic mice. J. Alzheimer's Dis. 42, 459–483.

Nordberg, A., 1999. PET studies and cholinergic therapy in Alzheimer's disease. Rev. Neurol. Paris 155 (Suppl. 4), S53–S63.

Oehlrich, D., Berthelot, D.J., Gijsen, H.J., 2011. γ-Secretase modulators as potential disease modifying anti-Alzheimer's drugs. J. Med. Chem. 54, 669–698.

Ong, W.Y., Farooqui, T., Kokotos, G., Farooqui, A.A., 2015. Synthetic and natural inhibitors of phospholipases A$_2$: their importance for understanding and treatment of neurological disorders. ACS Chem. Neurosci. 6, 814–831.

Palop, J.J., Mucke, L., 2010. Amyloid-β-induced neuronal dysfunction in Alzheimer's disease: from synapses toward neural networks. Nat. Neurosci. 13, 812–818.

Papadopoulos, P., Rosa-Neto, P., Rochford, J., Hamel, E., 2013. Pioglitazone improves reversal learning and exerts mixed cerebrovascular effects in a mouse model of Alzheimer's disease with combined amyloid-β and cerebrovascular pathology. PLoS One 8, e68612.

Park, S.Y., Kim, H.S., Cho, E.K., Kwon, B.Y., Phark, S., et al., 2008. Curcumin protected PC12 cells against β-amyloid-induced toxicity through the inhibition of oxidative damage and tau hyperphosphorylation. Food Chem. Toxicol. 46, 2881–2887.

Parsons, C.G., Danysz, W., Bartmann, A., Spielmanns, P., Frankiewicz, T., et al., 1999. Aminoalkyl-cyclohexanes are novel uncompetitive NMDA receptor antagonists with strong voltage-dependency and fast blocking kinetics: *in vitro* and *in vivo* characterization. Neuropharmacology 38, 85–108.

Pasinetti, G.M., 2012. Novel role of red wine-derived polyphenols in the prevention of Alzheimer's disease dementia and brain pathology: experimental approaches and clinical implications. Planta Med. 78, 1614–1619.

Pedrini, S., Carter, T.L., Prendergast, G., Petanceska, S., Ehrlich, M.E., et al., 2005. Modulation of statin-activated shedding of Alzheimer APP ectodomain by ROCK. PLoS Med. 2, e18.

Petroni, A., Blasevich, M., Salami, M., Papini, N., Montedoro, G.F., et al., 1995. Inhibition of platelet aggregation and eicosanoid production by phenolic components of olive oil. Thromb. Res. 78, 151–160.

Picada, J., Flores, E., Cappelari, S., Pereira, P., 2011. Effects of memantine, a non-competitive N-methyl-D-aspartate receptor antagonist, on genomic stability. Basic Clin. Pharmacol. Toxicol. 109, 413–417.

Pitt, J., Thorner, M., Brautigan, D., Larner, J., Klein, W.L., 2013. Protection against the synaptic targeting and toxicity of Alzheimer's-associated Aβ oligomers by insulin mimetic chiro-inositols. FASEB J. 27, 199–207.

Priyadarsini, K.I., 2013. Chemical and structural features influencing the biological activity of curcumin. Curr. Pharm. Des. 19, 2093–2100.

Qin, W., Chachich, M., Lane, M., Roth, G., Bryant, M., de Cabo, R., Ottinger, M.A., Mattison, J., Ingram, D., Gandy, S., Pasinetti, G.M., 2006. Calorie restriction attenuates Alzheimer's disease type brain amyloidosis in Squirrel monkeys (*Saimiri sciureus*). J. Alzheimer's Dis. 10, 417–422.

Qin, W., Haroutunian, V., Katsel, P., Cardozo, C.P., Ho, L., et al., 2009. PGC-1α expression decreases in the Alzheimer disease brain as a function of dementia. Arch. Neurol. 66, 352–361.

Qosa, H., Batarseh, Y.S., Mohyeldin, M.M., El Sayed, K.A., Keller, J.N., et al., 2015. Oleocanthal enhances amyloid-β clearance from the brains of TgSwDI mice and in vitro across a human blood–brain barrier model. ACS Chem. Neurosci. 6, 1849–1859.

Quintero-Monzon, O., Martin, M.M., Fernandez, M.A., Cappello, C.A., Krzysiak, A.J., et al., 2011. Dissociation between the processivity and total activity of γ-secretase: implications for the mechanism of Alzheimer's disease-causing presenilin mutations. Biochemistry 50, 9023–9035.

Radmark, O., Werz, O., Steinhilber, D., Samuelsson, B., 2007. 5-Lipoxygenase: regulation of expression and enzyme activity. Trends Biochem. Sci. 32, 332–341.

Rafii, M.S., Walsh, S., Little, J.T., Behan, K., Reynolds, B., et al., 2011. A phase II trial of huperzine A in mild to moderate Alzheimer disease. Neurology 76, 1389–1394.

Raina, P., Santaguida, P., Ismaila, A., Patterson, C., Cowan, D., et al., 2008. Effectiveness of cholinesterase inhibitors and memantine for treating dementia: evidence review for a clinical practice guideline. Ann. Intern. Med. 148, 379–397.

Ray, B., Chauhan, N.B., Lahiri, D.K., 2011. Oxidative insults to neurons and synapse are prevented by aged garlic extract and S-allyl-L-cysteine treatment in the neuronal culture and APP-Tg mouse model. J. Neurochem. 117, 388–402.

Reisberg, B., Doody, R., Stöffler, A., Schmitt, F., Ferris, S., et al., 2003. Memantine in moderate-to-severe Alzheimer's disease. N. Engl. J. Med. 348, 1333–1341.

Reger, M.A., Watson, G.S., Green, P.S., Wilkinson, C.W., Baker, L.D., Cholerton, B., et al., 2008. Intranasal insulin improves cognition and modulates β-amyloid in early AD. Neurology 70, 440–448.

Ribeiro Do Couto, B., Aguilar, M.A., Manzanedo, C., Rodriguez-Arias, M., Minarro, J., 2004. Effects of NMDA receptor antagonists (MK-801 and memantine) on the acquisition of morphine-induced conditioned place preference in mice. Prog. Neuropsychopharmacol. Biol. Psychiatry 28, 1035–1043.

Riha, P.D., Rojas, J.C., Gonzalez-Lima, F., 2011. Beneficial network effects of methylene blue in an amnestic model. NeuroImage 54, 2623–2634.

Riviere, C., Richard, T., Quentin, L., Krisa, S., Mérillon, J.M., et al., 2007. Inhibitory activity of stilbenes on Alzheimer's β-amyloid fibrils in vitro. Bioorg. Med. Chem. 15, 1160–1167.

Rizvi, S.M., Shaikh, S., Waseem, S.M., Shakil, S., Abuzenadah, A.M., et al., 2015. Role of anti-diabetic drugs as therapeutic agents in Alzheimer's disease. EXCLI J. 14, 684–696.

Rockwood, K., Kirkland, S., Hogan, D.B., MacKnight, C., Merry, H., et al., 2002. Use of lipid-lowering agents, indication bias, and the risk of dementia in community-dwelling elderly people. Arch. Neurol. 59, 223–227.

Rodriguez, J.J., Jones, V.C., Tabuchi, M., Allan, S.M., Knight, E.M., et al., 2008. Impaired adult neurogenesis in the dentate gyrus of a triple transgenic mouse model of Alzheimer's disease. PLoS One 3, e2935.

Rodriguez, J.J., Jones, V.C., Verkhratsky, A., 2009. Impaired cell proliferation in the subventricular zone in an Alzheimer's disease model. Neuroreport 20, 907–912.

Roy, A., Jana, M., Corbett, G.T., Ramaswamy, S., Kordower, J.H., et al., 2013. Regulation of cyclic AMP response element binding and hippocampal plasticity-related genes by peroxisome proliferator-activated receptor α. Cell Rep. 4, 724–737.

Rybak, M.E., Calvey, E.M., Harnly, J.M., 2004. Quantitative determination of allicin in garlic: supercritical fluid extraction and standard addition of alliin. J. Agric. Food Chem. 52, 682–687.

Sahiner, M., Erken, G., Kursunluoglu, R., Genc, O., Sahiner, T., 2011. Memantine improves learning in kindled rats. J. Neurol. Sci. 28, 322–329.

Sano, M., Bell, K.L., Galasko, D., Galvin, J.E., Thomas, R.G., et al., 2011. A randomized, double-blind, placebo-controlled trial of simvastatin to treat Alzheimer disease. Neurology 77, 556–563.

Scapagnini, G., Colombrita, C., Amadio, M., D'Agata, V., Arcelli, E., et al., 2006. Curcumin activates defensive genes and protects neurons against oxidative stress. Antioxid. Redox Signal. 8, 395–403.

Schaefer, E.J., Bongard, V., Beiser, A.S., Lamon-Fava, S., Robins, S.J., et al., 2006. Plasma phosphatidylcholine docosahexaenoic acid content and risk of dementia and Alzheimer disease: the Framingham Heart Study. Arch. Neurol. 63, 1545–1550.

Schupp, M., Clemenz, M., Gineste, R., Witt, H., Janke, J., et al., 2005. Molecular character-ization of new selective peroxisome proliferator-activated receptor γ modulators with angiotensin receptor blocking activity. Diabetes 54, 3442–3452.

Selkoe, D., Mandelkow, E., Holtzman, D., 2012. Deciphering Alzheimer disease. Cold Spring Harb. Perspect. Med. 2, a011460.

Sfeir, A.J., Chai, W., Shay, J.W., Wright, W.E., 2005. Telomere end processing: the terminal nucleotides of human chromosomes. Mol. Cell 18, 131–138.

Shankar, G.M., Bloodgood, B.L., Townsend, M., Walsh, D.M., Selkoe, D.J., et al., 2007. Natural oligomers of the Alzheimer amyloid-beta protein induce reversible synapse loss by mod-ulating an NMDA-type glutamate receptor-dependent signaling pathway. J. Neurosci. 27, 2866–2875.

Shinohara, M., Sato, N., Shimamura, M., Kurinami, H., Hamasaki, T., et al., 2014. Possible modification of Alzheimer's disease by statins in midlife: interactions with genetic and non-genetic risk factors. Front. Aging Neurosci. 6, 71.

Shytle, R.D., Tan, J., Bickford, P.C., Rezai-Zadeh, K., Hou, L., et al., 2012. Optimized turmeric extract reduces β-Amyloid and phosphorylated tau protein burden in Alzheimer's trans-genic mice. Curr. Alzheimer Res. 9, 500–506.

Simmons, D.A., Knowles, J.K., Belichenko, N.P., Banerjee, G., Finkle, C., et al., 2014. A small molecule p75^{NTR} ligand, LM11A-31, reverses cholinergic neurite dystrophy in Alzheimer's disease mouse models with mid- to late-stage disease progression. PLoS One 9, e102136.

Smith, A.B., Han, Q., Breslin, P.A.S., Beauchamp, G.K., 2005. Synthesis and assignment of absolute configuration of (-)-oleocanthal: a potent, naturally occurring non-steroidal anti-inflammatory and anti-oxidant agent derived from extra virgin olive oils. Org. Lett. 7, 5075–5078.

Smith, A., Giunta, B., Bickford, P.C., Fountain, M., Tan, J., Shytle, R.D., 2010. Nanolipidic particles improve the bioavailability and α-secretase inducing ability of epigallocatechin-3-gallate (EGCG) for the treatment of Alzheimer's disease. Int. J. Pharm. 389, 207–212.

Sofi, F., Cesari, F., Abbate, R., Gensini, G.F., Casini, A., 2008. Adherence to Mediterranean diet and health status: meta-analysis. BMJ 337, a1344.

Sofi, F., Macchi, C., Abbate, R., Gensini, G.F., Casini, A., 2010. Effectiveness of the Mediterranean diet: can it help delay or prevent Alzheimer's disease? J. Alzheimer's Dis. 20, 795–801.

Sofi, F., Macchi, C., Abbate, R., Gensini, G.F., Casini, A., 2013. Mediterranean diet and health. Biofactors 39, 335–342.

Sotthibundhu, A., Sykes, A.M., Fox, B., Underwood, C.K., Thangnipon, W., et al., 2008. β-Amyloid$_{1-42}$ induces neuronal death through the p75 neurotrophin receptor. J. Neurosci. 28, 3941–3946.

Song, M.S., Rauw, G., Baker, G.B., Kar, S., 2008. Memantine protects rat cortical cultured neurons against β-amyloid-induced toxicity by attenuating tau phosphorylation. Eur. J. Neurosci. 28, 1989–2002.

Sparks, D.L., Kryscio, R.J., Sabbagh, M.N., Connor, D.J., Sparks, L.M., et al., 2008. Reduced risk of incident AD with elective statin use in a clinical trial cohort. Curr. Alzheimer Res. 5, 416–421.

Sperling, R.A., Aisen, P.S., Beckett, L.A., Bennett, D.A., Craft, S., et al., 2011. Toward defin-ing the preclinical stages of Alzheimer's disease: recommendations from the National Institute on Aging-Alzheimer's Association workgroups on diagnostic guidelines for Alzheimer's disease. Alzheimers Dement. 7, 280–292.

St-Laurent-Thibault, C., Arseneault, M., Longpre, F., Ramassamy, C., 2011. Tyrosol and hydroxytyrosol, two main components of olive oil, protect N2a cells against amyloid-β-induced toxicity. Involvement of the NF-κB signaling. Curr. Alzheimer Res. 8, 543–551.

Stranahan, A.M., Zhou, Y., Martin, B., Maudsley, S., 2009. Pharmacomimetics of exercise: novel approaches for hippocampally-targeted neuroprotective agents. Curr. Med. Chem. 16, 4668–4678.

Suzuki, T., Yamamoto, M., 2015. Molecular basis of the Keap1-Nrf2 system. Free Radic. Biol. Med. 88 (Pt. B), 93–100.

Suzuki, T., Motohashi, H., Yamamoto, M., 2013. Toward clinical application of the Keap1-Nrf2 pathway. Trends Pharmacol. Sci. 34, 340–346.

Talbot, K., Wang, H.Y., Kazi, H., Han, L.Y., Bakshi, K.P., et al., 2012. Demonstrated brain insulin resistance in Alzheimer's disease patients is associated with IGF-1 resistance, IRS-1 dysregulation, and cognitive decline. J. Clin. Invest. 122, 1316–1338.

Tamayev, R., D'Adamio, L., 2012. Inhibition of γ-secretase worsens memory deficits in a genetically congruous mouse model of Danish dementia. Mol. Neurodegener. 7, 19.

Tariot, P.N., Aisen, P.S., 2009. Can lithium or valproate untie tangles in Alzheimer's disease? J. Clin. Psychiatry 70, 919–921.

Taupin, P., 2009. Adult neurogenesis, neural stem cells and Alzheimer's disease: developments, limitations, problems and promises. Curr. Alzheimer Res. 6, 461–470.

Tell, V., Hilgeroth, A., 2013. Recent developments of protein kinase inhibitors as potential AD therapeutics. Front. Cell. Neurosci. 7, 189.

Thies, W., Bleiler, L., 2012. Alzheimer's association: Alzheimer's disease facts and figures. Alzheimer's Dement. 8, 131–168.

Thies, W., Bleiler, L., 2013. Alzheimer's disease: facts and figures. Alzheimer Dement. 9, 208–215.

Tozuka, Y., Fukuda, S., Namba, T., Seki, T., Hisatsune, T., 2005. GABAergic excitation promotes neuronal differentiation in adult hippocampal progenitor cells. Neuron 47, 803–815.

Tsai, S.J., Chiu, C.P., Yang, H.T., Yin, M.C., 2011. S-allyl cysteine, S-ethyl cysteine, and S-propyl cysteine alleviate β-amyloid, glycative, and oxidative injury in brain of mice treated by D-galactose. J. Agric. Food Chem. 59, 6319–6326.

Tu, L.H., Ma, J., Liu, H.P., Wang, R.R., Luo, J., 2009. The neuroprotective effects of ginsenosides on calcineurin activity and tau phosphorylation in SY5Y cells. Cell. Mol. Neurobiol. 29, 1257–1264.

Um, H.S., Kang, E.B., Leem, Y.H., Cho, I.H., Yang, C.H., et al., 2008. Exercise training acts as a therapeutic strategy for reduction of the pathogenic phenotypes for Alzheimer's disease in an NSE/APPsw-transgenic model. J. Mol. Med. 22, 529–539.

Valenciano, A.I., Corrochano, S., de Pablo, F., de la Villa, P., de la Rosa, E.J., 2006. Proinsulin/insulin is synthesized locally and prevents caspase- and cathepsin-mediated cell death in the embryonic mouse retina. J. Neurochem. 99, 524–536.

van de Rest, O., Berendsen, A.A., Haveman-Nies, A., de Groot, L.C., 2015. Dietary patterns, cognitive decline, and dementia: a systematic review. Adv. Nutr. 6, 154–168.

Velayutham, P., Babu, A., Liu, D., 2008. Green tea catechins and cardiovascular health: an update. Curr. Med. Chem. 15, 1840–1850.

Vingtdeux, V., Marambaud, P., 2012. Identification and biology of α-secretase. J. Neurochem. 120, 34–45.

Villemagne, V.L., Burnham, S., Bourgeat, P., Brown, B., Ellis, K.A., et al., 2013. Amyloid β deposition, neurodegeneration, and cognitive decline in sporadic Alzheimer's disease: a prospective cohort study. Lancet Neurol. 12, 357–367.

Wagner, S.L., Zhang, C., Cheng, S., Nguyen, P., Zhang, X., et al., 2014. Soluble γ-secretase modulators selectively inhibit the production of the 42-amino acid amyloid β peptide variant and augment the production of multiple carboxy-truncated amyloid β species. Biochemistry 53, 702–713.

Wang, R., Tang, X.C., 2005. Neuroprotective effects of huperzine A. A natural cholinesterase inhibitor for the treatment of Alzheimer's disease. Neurosignals 14, 71–82.

Wang, J., Ho, L., Zhao, Z., Seror, I., Humala, N., Dickstein, D.L., et al., 2006. Moderate consumption of Cabernet Sauvignon attenuates Aβ neuropathology in a mouse model of Alzheimer's disease. FASEB J. 20, 2313–2320.

Wang, B.S., Wang, H., Wei, Z.H., Song, Y.Y., Zhang, L., et al., 2009. Efficacy and safety of natural acetylcholinesterase inhibitor huperzine A in the treatment of Alzheimer's disease: an updated meta-analysis. J. Neural Transm. 116, 457–465.

Wang, J.M., Johnston, P.B., Ball, B.G., Brinton, R.D., 2005. The neurosteroid allopregnanolone promotes proliferation of rodent and human neural progenitor cells and regulates cell-cycle gene and protein expression. J. Neurosci. 25, 4706–4718.

Wang, J.M., Singh, C., Liu, L., Irwin, R.W., Chen, S., et al., 2010. Allopregnanolone reverses neurogenic and cognitive deficits in mouse model of Alzheimer's disease. Proc. Natl. Acad. Sci. U. S. A. 107, 6498–6503.

Wang, Y., Tang, X.C., Zhang, H.Y., 2012. Huperzine A alleviates synaptic deficits and modulates amyloidogenic and nonamyloidogenic pathways in APPswe/PS1dE9 transgenic mice. J. Neurosci. Res. 90, 508–517.

Wang, X.Y., Tang, S.S., Hu, M., Long, Y., Li, Y.Q., et al., 2013. Leukotriene D4 induces amyloid-β generation via $CysLT_1R$-mediated NF-κB pathways in primary neurons. Neurochem. Int. 62, 340–347.

Wen, Y., Li, W., Poteet, E.C., Xie, L., Tan, C., et al., 2011. Alternative mitochondrial electron transfer as a novel strategy for neuroprotection. J. Biol. Chem. 286, 16504–16515.

Willett, W.C., Sacks, F., Trichopoulou, A., Drescher, G., Ferro-Luzzi, A., et al., 1995. Mediterranean diet pyramid: a cultural model for healthy eating. Am. J. Clin. Nutr. 61 (Suppl.), 1402S–1406S.

Willard, L.B., Hauss-Wegrzyniak, B., Danysz, W., Wenk, G.L., 2000. The cytotoxicity of chronic neuroinflammation upon basal forebrain cholinergic neurons of rats can be attenuated by glutamatergic antagonism or cyclooxygenase-2 inhibition. Exp. Brain Res. 134, 58–65.

Williams, R.J., Spencer, J.P., 2012. Flavonoids, cognition, and dementia: actions, mechanisms, and potential therapeutic utility for Alzheimer disease. Free Radic. Biol. Med. 52, 35–45.

Wischik, C., 2009. Rember: issues in design of a phase 3 disease modifying clinical trial of tau aggregation inhibitor therapy in Alzheimer's disease. Alzheimer's Dement. J. Alzheimer's Assoc. 5, P74.

Witte, A.V., Fobker, M., Gellner, R., Knecht, S., Flöel, A., 2009. Caloric restriction improves memory in elderly humans. Proc. Natl. Acad. Sci. U. S. A. 106, 1255–1260.

Wolfe, M.S., Citron, M., Diehl, T.S., Xia, W., Donkor, I.O., 1998. A substrate-based difluoro ketone selectively inhibits Alzheimer's γ-secretase activity. J. Med. Chem. 41, 6–9.

Wolfe, M.S., Xia, W., Moore, C.L., Leatherwood, D.D., Ostaszewski, B., 1999. Peptidomimetic probes and molecular modeling suggest Alzheimer's γ-secretases are intramembrane-cleaving aspartyl proteases. Biochemistry 38, 4720–4727.

Wolfe, M.S., 2012. γ-Secretase inhibitors and modulators for Alzheimer's disease. J. Neurochem. 120 (Suppl. 1), 89–98.

Wong, W.B., Lin, V.W., Boudreau, D., Devine, E.B., 2013. Statins in the prevention of dementia and Alzheimer's disease: a meta-analysis of observational studies and an assessment of confounding. Pharmacoepidemiol. Drug Saf. 22, 345–358.

Xiao, J., Ching, Y.P., Liong, E.C., Nanji, A.A., Fung, M.L., Tipoe, G.L., 2013. Garlic-derived S-allylmercaptocysteine is a hepato-protective agent in non-alcoholic fatty liver disease in vivo animal model. Eur. J. Nutr. 52, 179–191.

Xu, S., Guan, Q., Wang, C., Wei, X., Chen, X., et al., 2014. Rosiglitazone prevents the memory deficits induced by amyloid-β oligomers via inhibition of inflammatory responses. Neurosci. Lett. 578, 7–11.

Yang, F., Lim, G.P., Begum, A.N., Ubeda, O.J., Simmons, M.R., et al., 2005. Curcumin inhibits formation of amyloid β oligomers and fibrils, binds plaques, and reduces amyloid *in vivo*. J. Biol. Chem. 280, 5892–5901.

Yao, E.C., Xue, L., 2014. Therapeutic effects of curcumin on Alzheimer's disease. Adv. Alzheimer Dis. 3, 145–159.

Yaar, M., Zhai, S., Fine, R.E., Eisenhauer, P.B., Arble, B.L., et al., 2002. Amyloid β binds trimers as well as monomers of the 75-kDa neurotrophin receptor and activates receptor signaling. J. Biol. Chem. 277, 7720–7725.

Ye, J., Zhang, Y., 2012. Curcumin protects against intracellular amyloid toxicity in rat primary neurons. Int. J. Clin. Exp. Med. 5, 44–49.

Yin, W., Zhang, X., Li, Y., 2012. Protective effects of curcumin in APPswe transfected SH-SY5Y cells. Neural Regen. Res. 7, 405–412.

Yoshiyama, Y., Lee, V.M., Trojanowski, J.Q., 2013. Therapeutic strategies for tau mediated neurodegeneration. J. Neurol. Neurosurg. Psychiatry 84, 784–795.

Yu, F., Bronas, U.G., Konety, S., Nelson, N.W., Dysken, M., et al., 2014. Effects of aerobic exercise on cognition and hippocampal volume in Alzheimer's disease: study protocol of a randomized controlled trial (The FIT-AD trial). Trials 15, 394.

Yuan, H.Y., Kuang, S.Y., Zheng, X., Ling, H.Y., Yang, Y.B., et al., 2008. Curcumin inhibits cellular cholesterol accumulation by regulating SREBP-1/caveolin-1 signaling pathway in vascular smooth muscle cells. Acta Pharmacol. Sin. 29, 555–563.

Zhang, Z., Wang, X., Chen, Q., Shu, L., Wang, J., et al., 2002. Clinical efficacy and safety of huperzine α in treatment of mild to moderate Alzheimer disease, a placebo-controlled, double-blind, randomized trial. Zhonghua Yi Xue Za Zhi 82, 941–944.

Zhang, H.Y., Zheng, C.Y., Yan, H., Wang, Z.F., Tang, L.L., et al., 2008. Potential therapeutic targets of huperzine A for Alzheimer's disease and vascular dementia. Chem. Biol. Interact. 175, 396–402.

Zhang, B., Carroll, J., Trojanowski, J.Q., Yao, Y., Iba, M., et al., 2012. The microtubule-stabilizing agent, epothilone D, reduces axonal dysfunction, neurotoxicity, cognitive deficits, and Alzheimer-like pathology in an interventional study with aged tau transgenic mice. J. Neurosci. 32, 3601–3611.

Zhao, W.Q., Alkon, D.L., 2001. Role of insulin and insulin receptor in learning and memory. Mol. Cell. Endocrinol. 177, 125–134.

Zimmermann, M., Gardoni, F., Di Luca, M., 2005. Molecular rationale for the pharmacological treatment of Alzheimer's disease. Drugs Aging 22 (Suppl. 1), 27–37.

Immunotherapy for the Treatment of Alzheimer's Disease

INTRODUCTION

Alzheimer's disease (AD) is the most prevalent form of age-related and fatal neurodegenerative disease, which is characterized by the accumulation of β-amyloid (Aβ) plaques consisting primarily of aggregated Aβ proteins and neurofibrillary tangles (NFTs) formed from hyperphosphorylated Tau protein. Both Aβ and hyperphosphorylated Tau produce toxic effects in in vivo and in vitro studies. AD is also characterized by progressive loss of neurons in the hippocampus, nucleolus basalis, and cortex resulting in the shrinkage of brain. The loss of neurons results in decline in cognitive and behavioral functions like memory, thinking, and language skills (O'Brien and Wong, 2011). AD affects ~50 million people worldwide with estimated 115 million cases of AD by 2050, providing an unsustainable health-care challenge due to a lack of effective treatments.

Neurochemical Aspects of Alzheimer's Disease
http://dx.doi.org/10.1016/B978-0-12-809937-7.00009-4

In addition, to plaques and tangles, AD brain is characterized by gliosis, neuroinflammation, neuritic dystrophy, neuron loss, and changes in neurotransmitter levels (Hardy and Selkoe, 2002). AD is classified into familial AD and sporadic AD; the two kinds of AD share similar clinical and pathological features, including progressive cognitive impairment, plaque deposits in the brain, axonal transport defects, loss of synapse, and selective neurodegeneration in the nucleus basalis and hippocampal regions. Only 5%–7% of the AD cases are of familial type, whereas 93%–95% of AD cases are sporadic (Thal and Fandrich, 2015). Another classification system for AD based on age is also emerging. It is observed that 1%–6% of the AD cases emerge in 30- to 60-year-old people, which is known as early-onset AD, whereas late-onset AD occurs in people older than 60 years with 90% prevalence (Mullane and Williams, 2013).

In the brain, Aβ is synthesized and degraded continuously. As stated in Chapter 1, oligomerization and aggregation of Aβ and subsequent plaque deposition in AD are concentration-dependent processes. Excessive accumulation of both soluble and insoluble Aβ may occur not only as a result of aberrant amyloid precursor protein (APP) processing by β- and γ-secretase enzymes (Fig. 9.1), but also may be caused by inefficient removal of newly generated Aβ. Reduction in activities of Aβ-degrading enzymes, such

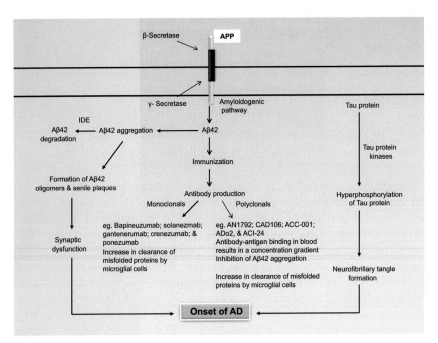

FIGURE 9.1 **Production of polyclonal and monoclonal antibodies against Aβ.** *AD,* Alzheimer's disease; *APP,* amyloid precursor protein; *Aβ,* Aβ-amyloid.

as neprilysin, insulin-degrading enzyme, and angiotensin-converting enzyme 1, may also provoke an imbalance between the Aβ generation and clearance (Nalivaeva et al., 2008; Baranello et al., 2015; Bates et al., 2009). So for the treatment of AD, the removal of existing soluble and insoluble Aβ assemblies is essential. This process may not only stabilize brain function but also slow down the cognitive decline. Based on the "amyloid cascade hypothesis," antiamyloid therapies have been developed by many investigators to treat AD (Robinson et al., 2004). Therapeutic strategies aimed at lowering cerebral Aβ accumulation are currently under extensive development. One strategy is to immunize patients with AD with Aβ peptides so that they will develop antibodies that bind to Aβ protein and enhance its clearance. In vivo and in vitro studies have indicated that Aβ antibodies may act through the prevention of Aβ aggregation and Fc-receptor-mediated phagocytosis by microglial cells (Solomon et al., 1997; Bard et al., 2000). A third mechanism of action of antibodies may involve a shift in the gradient of Aβ transport across the blood–brain barrier (BBB) resulting in an increase in efflux from the brain to blood (Lemere et al., 2003), and a fourth mechanism proposes that antibodies bind to Aβ oligomers may neutralize the toxic effects of Aβ species on synapses (Klyubin et al., 2005). These mechanisms are not mutually exclusive for each form of Aβ and may overlap with each other under certain conditions. Furthermore, the mechanism of action may depend on the disease state. For example, a prevention vaccine may not require that the antibodies cross the BBB to induce Aβ phagocytosis, whereas a therapeutic vaccine (once plaque deposition is well underway) would likely benefit from the transport of Aβ antibodies into the brain.

ACTIVE IMMUNIZATION FOR ALZHEIMER'S DISEASE

Active immunization engages the cellular and humoral immune system, including T cells and B cells, to promote the production of anti-antigen antibodies. Active immunization involves an antigen and an immune-boosting adjuvant to ensure high antibody titers. On the one hand, active immunotherapy is an attractive process because it results in long-term antibody production in a large population, while being cost-effective. However, active immunization–mediated antibody production is accompanied by a T-cell response that can increase the risk of a deleterious immune response (i.e., release of proinflammatory cytokines), especially, if the T cell recognizes the antigen as a self-protein. T-cell response takes long time to "shut off" an active immunization immune response. An active immunization leads to a polyclonal antibody response, which means that it generates antibodies recognizing multiple, sometimes overlapping epitopes on the target protein. This may be important for broad coverage or, it may be less useful if the goal is to lower a specific form of a protein but not all forms. Immunotherapy for AD has been designed and developed to promote removal of Aβ and reducing the Aβ load in patients with AD. Active immunization with either Aβ42 (predominant form found in senile plaques) in transgenic (Tg) mouse models of AD has indicated some success in animal models (Gilman et al., 2005; Gandy and Sano, 2015). Anti-Aβ antibodies activate microglia, and the microglial activation is important for the clearance of compact amyloid deposits. These antibodies can not only catalytically disaggregate amyloid deposits in vitro but also promote the efflux of Aβ from the brain via a mechanism called the peripheral sink (DeMattos et al., 2001). In Tg mouse models of AD, preclinical studies have indicated not only a dramatic reduction of cerebral Aβ burden but also a reduction in cognitive decline (Schenk et al., 1999; Fu et al., 2010).

These results in Tg mice did not translate well in humans. Clinical trials conducted by Elan/Wyeth in 2001, using Aβ peptide delivered in QS-21 adjuvant (AN1792), resulted in a meningoencephalitis reaction in 6% of the treated patients. The trial then was immediately stopped. In these studies, assays have been performed on the stimulation of B cells and T cells, and immune responses occur through activation of the phagocytic capacity of microglia. Immunotherapy not only reduces cerebral Aβ levels but also improves cognitive deficits in mouse models of AD (Delrieu et al., 2012; Morgan et al., 2000). Similar results on lowering of senile plaque load are obtained in nonhuman primates (Lemere et al., 2004). The development of a safe and effective vaccine requires therapeutic levels of anti-Aβ antibodies without causing adverse autoimmune responses. This is particularly challenging in the elderly due to immunosenescence. T-cell mediated inappropriate immune responses can be an adverse reaction from the adjuvant (Wilcock and Colton, 2008). Several phase 2 clinical trials of active immunization using full-length human Aβ42 peptide (AN1792) and a strong Th1-biased adjuvant, QS-21, have been stopped prematurely in 2002 because of the onset of meningoencephalitis and neuroinflammation in approximately 6%–8% of the patients with AD enrolled in these studies (Gilman et al., 2005). Neuroimaging studies have indicated the absence of white matter lesions with or without the presence of brain edema. Consistent with data from animal models, postmortem analyses in a subgroup of trial patients revealed dramatic clearance of plaques in the brain parenchyma, thus validating the efficacy of this approach for amyloid clearance in humans (Nicoll et al., 2006; Bombois et al., 2007). These observations suggest the onset of immune response as a double-edged sword and raises concerns about the beneficial effects of the vaccine, which are counterbalanced by deleterious T-cell responses. Histopathological studies have revealed that there are broad stretches of cerebral cortex devoid of plaques, interspersed with areas that contain residual plaques. These persistent plaques have a "moth–eaten" appearance or seemed to have a "naked" dense core. Additionally, these plaques have been seen along with microglia, which are immunoreactive for Aβ, supporting the view that amyloid clearance may be associated with the process of phagocytosis without cognitive impairment (Foster et al., 2009). Interestingly, Tau aggregates in neuropil threads and dystrophic neurites, often associated with plaques, are reduced by AN1792 vaccination, but no changes were observed in Tau accumulation within neuronal cell bodies (Boche et al., 2010). The exact cause of the meningoencephalitis in the AN1792 trial is unknown; however, possible causes may include recognition of the antigen (full-length Aβ peptide) by Aβ-specific T cells, the strong Th1-biased adjuvant, or possibly, the reformulation of the vaccine with polysorbate 80. At present, a number of next-generation Aβ-active immunization trials in either phase 1 or 2 (www.clinicaltrials.gov) are under way (Lemere et al., 2004; Lemere, 2013). Novartis Pharmaceuticals has launched a phase 1 result with an active vaccine called CAD106 (Winblad et al., 2012; Wisniewski, 2012). This vaccine is conjugated to a carrier (Qβ bacteriophage) and targets against a small peptide fragment of Aβ (Aβ1-6). CAD106 is designed to not only ensure repetitive antigen (Aβ1-6) presentation and strong B-cell response but also stimulation of Aβ-specific antibody generation while avoiding initiation of Aβ-specific T-cell responses (Wiessner et al., 2011). Mild to moderate probable subjects with AD with Mini-Mental State Examination (MMSE) scores ranging from 16 to 26 were enrolled in CAD106 trial. They were randomized to two cohorts, the first cohort received three injections of 50 g CAD106 (24 test and 7 placebo subjects, cohort 1) and a second cohort received 150 g CAD106 (22 test and 5 placebo subjects, cohort 2). With a treatment time span of 1 year and a 2-year follow-up period, the study indicated that 75% and 100% patients developed anti-Aβ immunoglobulin M (IgM) titers in cohorts 1 and 2, respectively, while 67% and 82% developed anti-Aβ IgG titers, respectively, in cohorts 1 and 2. Nine patients reported serious adverse reactions, but none have been thought to be secondary to the immunogen. None of the subjects developed meningitis, meningoencephalitis, or vasogenic edema either clinically or by neuroimaging during the initial trial or 2-year follow-up period. No significant change in cerebrospinal fluid (CSF) biomarkers is observed in the CAD106 subjects. However, some differences are observed in cohort 2 in treated subjects compared with controls for free plasma Aβ1-40. An important limitation of this trial is that it was not sufficiently powered to demonstrate a significant clinical difference between the treatment and control arms. This trial was completed in February 2013, but results have not been published.

Another phase 2 immunization trial being performed by Janssen and Pfizer is ACC-001. It uses the Aβ1-6 fragment coupled to a carrier protein, and surface-active saponin adjuvant QS21 (Ryan and Grundman, 2009).

Additionally, Affiris AG in collaboration with GlaxoSmithKline (GSK) has used AFFITOME technology to produce synthetic antigenic peptides called mimotopes to target the unmodified Aβ N-terminus in their AD02 trials (Schneeberger et al., 2009). Affiris AG has also started another phase 1 trial with AFFTOME to target a pyroglutamic-3-modified Aβ N-terminus. This posttranslational modification for Aβ is prone to more aggregation and occurs mainly in senile plaques or vascular amyloid (Saido et al., 1995; Frost et al., 2013). It should be noted that pyroglutamic-3-modified Aβ is present only in plaques and vascular amyloid deposits but is not detectable in the CSF or plasma. Pyroglutamic-3-modified Aβ is only found in CSF and plasma during therapeutic interventions where deposited Aβ has been mobilized (DeMattos et al., 2012). Another company (AC Immune) has initiated phase 1/2 trials by utilizing their product, ACI-24, which works by inducing a humoral immune response to Aβ in a primarily β-sheet conformation. The studies are based on earlier studies by this group in an AD Tg model, where a tetrapalmitoylated amyloid 1–15 peptide that exists chiefly in a β-sheet conformation has been used as an immunogen (Hickman et al., 2011). The results of this trial have not been published. It should be noted that there are many shortcomings to current Tg models of AD pathology. These include the fact that Tg amyloid deposits typically lack the extensive posttranslational modifications of AD amyloid and thus are much more soluble, presumably allowing them to be cleared more easily (Roher and Kokjohn, 2002). The rodent immune system is quite different from the human immune system, leading to significant differences in the toxic responses to amyloid deposition (Tenner and Fonseca, 2006).

At present, phase 2 trials are currently underway to test ACC-001-QS-21, a conjugated N-terminal peptide attached to an immunostimulatory carrier protein (Ryan and Grundman, 2009). Converging evidence suggests that the aforementioned trials with second-generation antibodies specifically target Aβ with more specificity and less cross-reactivity leading to less chances of autoimmune toxicity. Since immunogens are still derived from the Aβ sequence, some element of cross-reactivity to normal Aβ peptides may also exist with the plausible risk of inflammatory toxicity (Sabharwal and Wisniewski, 2014). To prevent the side effects induced by active immunization, passive immunization has been utilized. Two other vaccines, UB 311 (Aβ 1–14) and V950 (Aβ N-terminal conjugated to ISCO-MATRIX), both containing B-cell epitopes, have also finished phase 1 trial. Results on these studies have not been published.

Human intravenous immunoglobulins (hIVIG) have been used as a safe and effective treatment for certain neurological conditions such as Guillain-Barré syndrome (van Doorn et al., 2010). Intravenous immunoglobulins (IVIG) can cross the BBB and bind to Aβ deposits in brain parenchyma (Dodel et al., 2004). In neuronal cell cultures and Tg mouse models of AD, IVIG treatment reduces hallmark AD pathology, including amyloid plaques, NFTs, neuroinflammation, glial activation, and oxidative stress (Counts et al., 2014; Sudduth et al., 2013; Bailey et al., 2012; Lahiri and Ray, 2013). IVIG administration has also shown beneficial effects on inflammation and synaptic function in AD Tg mice (Counts et al., 2014; Sudduth et al., 2013). In tissue culture system, IVIG treatment increases the levels of synaptic proteins in primary rat hippocampal neurons (Bailey et al., 2012; Lahiri and Ray, 2013). Furthermore, the IVIG treatment also protects neurons against oxidative stress–mediated insults supporting the view that IVIG can be used for the treatment of AD. A small fraction of antibodies in hIVIG has been reported to react with Aβ protein (O'Nuallain et al., 2008). This observation suggests that hIVIG can be used for the treatment of AD. A small trial of hIVIG in patients with AD has indicated that hIVIG not only produces promising effects on Aβ levels in the CSF but also shows some positive effects on cognitive status of patients with AD (Dodel et al., 2004; Relkin et al., 2009). It is suggested that hIVIG promotes Aβ clearance by virtue of its constituent anti-Aβ antibodies, which can account for its beneficial effect in patients with AD. In addition, due to its antiinflammatory properties, hIVIG may modulate the neuroinflammation around amyloid plaques leading to neuroprotective effects in patients with AD. Thus there is an urgent need to test the mechanisms of hIVIG action not only in an animal model of AD but also in patients with AD. A preparation of covalently stabilized Aβ (1–40)Cys26 dimers, which is free of Aβ monomer or fibrils has been used as an immunogen to screen hybridomas for their ability to produce antibodies that discriminate between reduced non-cross-linked monomer and covalently linked dimers. Two murine IgM monoclonal antibodies (mAbs), referred to as 3C6 and 4B5, preferentially bind covalent Aβ dimer assemblies, but not Aβ monomer or fibrils formed by other amyloidogenic proteins. Notably, mAb 3C6, but not an IgM isotype-matched control antibody, ameliorates the synaptic plasticity–disrupting effect of aqueous extracts of AD brain Aβ on rodent long-term potentiation (O'Nuallain et al., 2011). It is proposed that these preparations can be used as immunogens and recommended for further investigation of the therapeutic and diagnostic utility of mAbs raised to such assemblies.

PASSIVE IMMUNIZATION FOR ALZHEIMER'S DISEASE

Passive transfer of anti-Aβ antibodies is an alternative strategy, which is as effective as active immunization in the mouse model of AD. In this therapy, the risk of meningoencephalitis can be minimized because the antibodies are administered without adjuvant. Passive immunotherapy involves the direct injection of mAbs (or fragments thereof). There are several benefits of passive immunotherapy. Passive immunotherapy can be stopped immediately if there are any adverse reactions, and one can target specific epitopes or pathogenic conformations without disturbing the other forms of the protein of interest. In addition, passive immunotherapy not only requires the production of expensive humanized mAbs but also requires repeated injections of mAbs (Asuni et al., 2006). Over time, this process may lead to the formation of anti-antibodies, which may either produce a neutralizing effect or lead to side effects such as glomerulonephritis and vasculitis. In AD, passive immunization studies have been performed by the administration of mAbs directed against Aβ peptide. This therapy avoids immune responses caused by the direct administration of antibodies and consists of the intravenous administration of anti-Aβ antibodies to the patient with AD. An advantage of passive immunization over active immunization is that the proinflammatory T-cell-mediated immune reactions do not occur. Passive immunization not only has similar potency to remove amyloid plaques and rescue neuritic and glial pathology (Bard et al., 2000) but also reverses abnormal hippocampus synaptic plasticity (Klyubin et al., 2005). In Tg mice, passive immunization reduces cerebral amyloid load and improves cognition, even when the amyloid plaque numbers are not significantly reduced. These results can be attributed to the neutralization of soluble amyloid oligomers, which are increasingly recognized to play a fundamental role in the pathogenesis of AD.

Bapineuzumab (AAB-001) and solanezumab (LY2062430) are two mAbs that have reached advanced stages of clinical development. Bapineuzumab is a humanized, IgG1 mAb derived from the murine antibody, 3D6. Bapineuzumab binds with high affinity to monomeric Aβ (Basi and Jacobsen, 2006), soluble Aβ oligomers (Zago et al., 2012), and fibrillar Aβ (Schenk et al., 1999), but does not recognize APP or the product of β-secretase cleavage of APP (Johnson-Wood et al., 1997). Babineuzumab also binds with vascular amyloid and Aβ plaques in hAPP Tg mice brains and human AD brains and clears vascular amyloid and Aβ plaques in hAPP Tg mice brains (Racke et al., 2005; Basi et al., 2002). This antibody is designed to reduce senile plaque formation and promote clearance of Aβ (Tayeb et al., 2013; Salloway et al., 2014). In phase 2 studies in patients with mild to moderate AD, bapineuzumab reduces not only phosphorylated Tau (p-Tau) protein in CSF but also ^{11}C-Pittsburgh compound B (^{11}C-PiB) average uptake visualized by positron emission tomography (PET) (Blennow et al., 2012; Rinne et al., 2010). Bapineuzumab contributes not only to the clearance of Aβ by passing the BBB and subsequent engulfing microglial cells but also to the creation of peripheral sink. The findings provide a rationale for conducting separate trials in apolipoprotein E (APOE) ε4 allele carriers and noncarriers and for limiting the bapineuzumab dose in carriers to minimize risk of amyloid-related imaging abnormalities with edema or effusion (ARIA-E; previously termed *vasogenic edema*)

(Salloway et al., 2009). In phase 3 clinical trials, the infusion of bapineuzumab 0.5 or 1.0 mg/kg every 13 weeks for up to 3.5 years indicates that this antibody is well tolerated. Phase 3 clinical studies of bapineuzumab in patients with mild to moderate AD failed to show a significant difference in clinical endpoints of Alzheimer's Disease Assessment Scale Cognitive subscale ADAS-Cog 11 and disability assessment for dementia scores or other clinical endpoints (Salloway et al., 2014). Among APOE ε4 carriers (but not noncarriers), bapineuzumab infusion results in reduction of CSF p-Tau concentrations, a marker of neurodegeneration. Carriers also show a decrease in amyloid accumulation based on PiB-PET findings. In Tg mouse, bapineuzumab has been shown to bind plaques in the brain and elicit Fc-receptor-mediated microglial phagocytosis of Aβ plaques. Bapineuzumab produces significant decrease in brain amyloid plaques and p-Tau in CSF. However, the treatment failed to produce significant improvements of cognitive function. Collective evidence suggests that bapineuzumab has failed to show overall clinical improvement or any clear disease-modifying results (Salloway et al., 2014; Doody et al., 2014; Tayeb et al., 2013). AAB-003/PF-05236812 is a humanized 3D6 (i.e., bapineuzumab) with mutations in the Fc domain (Black et al., 2009) to reduce effector functions and thereby ARIAs. Therefore an improved clinical safety profile information on AAB-003 compared to bapineuzumab will be important to have for future studies. Currently, two clinical phase I trials are ongoing to evaluate the safety and the tolerability of AAB-003.

Solanezumab (LY2062430) is a humanized mAb that binds to a large Aβ epitope (960 Å^2 buried interface over residues 16–26) that forms extensive contacts and hydrogen bonds to the antibody, largely via main-chain Aβ atoms and a deeply buried Phe19-Phe20 dipeptide core (Crespi et al., 2015). This antibody is unique among Aβ antibodies because it does not bind fibrillar Aβ. Solanezumab has high-affinity soluble Aβ, although the exact nature of the soluble Aβ that it binds has not reported. It has been suggested that 266 is a conformation-specific antibody that solely recognizes soluble Aβ and readily binds monomeric Aβ (Racke et al., 2005), without binding APP or the C-terminal APP cleavage product of α-secretase (Seubert et al., 1992). It has been reported that 266 selectively sequesters Aβ monomer and dimer in the periphery of 3-month-old APP Tg mice (DeMattos et al., 2001), and definitive evidence for 266 binding of monomeric Aβ has also been reported (Yamada et al., 2009). The infusions of 400 mg of solanezumab or placebo once a month for 80 weeks in patients with mild to moderate AD result in some improvement in cognitive function in mild AD; however, they are not statistically significant (Tayeb et al., 2013) and no changes are observed in biomarkers such as Tau, p-Tau, hippocampal volume, whole brain volume, or amyloid accumulation (http://www.ctad.fr/07-download/Congres2012/PressRelease/Sola-Release_29Oct2012.pdf). The sponsors are apparently not discouraged by the data and continue the clinical development of solanezumab in an open-label extension study. It is reported that repeated treatments with solanezumab do not show a significant benefit in data obtained from patients with mild or moderate AD dementia, but a slowing of cognitive decline has been observed in approximately 34% of patients with mild AD, diagnosed as ADAS-Cog 14 (Farlow et al., 2012; Goure et al., 2014), supporting the view that amyloid-targeted therapy can be more effective when applied at earlier stages of AD or before visible symptoms appear (Wisniewski and Goni, 2014; Carrillo et al., 2013). Specific immunization of the neurotoxic Aβ oligomer may be beneficial to circumvent inhibitory damage to the protective physiological benefits of Aβ. Further ongoing studies may reveal the efficacy of these antibodies in the treatment of patients with AD. Aβ immunotherapies are being currently used in clinical trials (Goure et al., 2014). At present, solanezumab is in phase 3 trials not only in patients with AD (NCT01127633 and NCT01900665) but also in older individuals who have normal thinking and memory function but who may be at risk of developing AD in the future (NCT02008357) (Panza et al., 2014). Eli Lilly & Co is trying to develop an intravenous formulation of solanezumab to treatment of mild to moderate AD. Acute and subchronic treatment with solanezumab of Tg mice attenuates or reverses memory deficits with no effects on incidence or severity of cerebral amyloid angiopathy-associated microhemorrhages, a severe side effect associated with bapineuzumab, another mAb (Imbimbo et al., 2012). Phase 2 studies in patients with AD have indicated a good safety profile with encouraging indications on CSF and plasma biomarkers.

Gantenerumab (RO4909832 or R1450) is another fully human anti-Aβ mAb directed to both N-terminal and central regions of Aβ. Gantenerumab also binds soluble Aβ oligomers with high affinity and, to a lesser extent, Aβ monomer. The reported binding constants (Kds) for gantenerumab binding to fibrillar Aβ, soluble Aβ oligomers, and Aβ monomer are 0.6, 1.2, and 17 nM, respectively (Bohrmann et al., 2012). The dissociation constants (Kds) for gantenerumab–Aβ complexes are 2.8×10^{-4}, 4.9×10^{-4}, and 1.2×10^{-2}, respectively, for fibrillar Aβ, Aβ oligomers, and Aβ monomer, suggesting a more rapid exchange of antibody–monomer complex compared with antibody–fibril or antibody–oligomer complexes (Bohrmann et al., 2012). A 6-month PET study in 16 patients with AD has indicated that gantenerumab treatment reduces brain Aβ deposition in a dose-dependent manner, possibly stimulating microglial-mediated phagocytosis. Furthermore, gantenerumab is being tested in human subjects who are at risk of developing presenile AD due to genetic mutations. An NCT01760005 trial is still recruiting participants and will determine the efficacy of both gantenerumab and solanezumab in the prodromal disease stages (Novakovic et al., 2013; Bohrmann et al., 2012; Jacobsen et al., 2014). In addition, two more phase 3 trials of gantenerumab in patients with mild AD (NCT02051608) and prodromal AD (NCT01224106) are ongoing. Gantenerumab is a fully human IgG1 antibody designed to interact with high affinity to a conformational epitope on the Aβ fibers. Microglia recruitment and ensuing phagocytosis will presumably result in amyloid plaque degradation. GSK933776 is a humanized mAb directed against the N-terminus of Aβ, believed as linear epitope (http://www.gskclinical-studyregister.com/result_comp_list.jsp?phase=All&studyType=All&population=All&marketing=No&compound=GSK933776). The Fc domain of GSK933776 is mutated to reduce the risk for vasogenic edema. Its further testing for retarding AD has been discontinued after phase 1 in 2011.

Crenezumab (MABT5102A) is another humanized mAb obtained by utilizing IgG4 backbone (Jindal et al., 2014). This antibody is engineered to reduce Fcγ receptor–mediated microglia activation and minimize adverse effects due to vasogenic edema and cerebral microhemorrhage (Pfeifer et al., 2008; Adolfsson et al., 2012). Crenezumab targets multiple conformational protofibrillar epitopes of Aβ, including oligomeric forms in hAPP Tg mice and human AD brain tissues, while inhibiting aggregation and promoting disaggregation of Aβ (Adolfsson et al., 2012). A phase 2 clinical trial to assess the safety and efficacy of this antibody in patients with mild to moderate AD (NCT01343966) has been completed in April 2014, but results have not yet been published. Another phase 2 trial aiming to evaluate the safety and efficacy of crenezumab in asymptomatic carriers of E280A autosomal-dominant mutation of PSEN1 has also been completed in November 2013 (NCT01998841). However, results have not been announced. Comparative studies on the immunohistochemical staining profiles of solanezumab, crenezumab, and bapineuzumab in human formalin-fixed, paraffin-embedded tissue and human fresh frozen tissue have indicated that these antibodies show similar staining patterns (Bouter et al., 2015). All three antibodies detect plaques, cerebral amyloid angiopathy, and intraneuronal Aβ in a similar fashion. Remarkably, solanezumab shows a strong binding affinity to plaques. Bapineuzumab does not recognize N-truncated or modified Aβ. However, solanezumab and crenezumab can detect N-terminally modified Aβ peptides Aβ4–42 and pyroglutamate Aβ3–42. With the use of solanezumab, crenezumab, and bapineuzumab similar, staining patterns are obtained in different mouse models (5XFAD, Tg4-42, TBA42, APP/PS1KI, 3xTg) (Bouter et al., 2015).

Aducanumab is a human mAb that selectively targets aggregated Aβ with high affinity. In a Tg mouse model of AD, aducanumab has been reported to enter the brain, bind with parenchymal Aβ, and reduce soluble and insoluble Aβ in a dose-dependent manner. In patients with mild AD, 1 year of monthly intravenous infusions of aducanumab reduces brain Aβ in a dose- and time-dependent manner. This is accompanied by a slowing of clinical decline measured by Clinical Dementia Rating-Sum of Boxes and MMSE scores (Sevigny et al., 2016). Furthermore, patients with AD can tolerate aducanumab without side effects. These results justify further development of aducanumab for the treatment of AD. At present, aducanumab is undergoing phase 3 clinical trials (Sevigny et al., 2016) (Table 9.1).

BIIB037/BART is a novel fully human IgG1, which is generated using a reverse translational medicine approach screening endogenous anti-Aβ antibodies from a patient with AD with an unusual stable clinical course. BART not only shows a high affinity/avidity for insoluble fibrillar Aβ, but a 100-fold less affinity for Aβ monomers. In vitro BiiB037 interacts with fibrillar Aβ42 with high affinity but does not bind soluble Aβ40 (Dunstan et al., 2011a). Although details of the binding characteristics of BiiB037 have not been published, it is reported that BiiB037 is very similar, or identical, to antibody NI-101.11, which binds with high affinity to Aβ plaques

TABLE 9.1   Antibodies Used for the Treatment of Alzheimer's Disease in Human Patients

Name	Company	Trial Outcome	Mechanism	References
AN1792 (polyclonal)	Janssen/Pfizer	Failed and stopped	Active Aβ42	Gilman et al. (2005) and Boche et al. (2010)
ACC-001 (polyclonal)	Janssen/Pfizer	Failed and stopped	Active N-terminal Aβ	Gilman et al. (2005)
CAD106	Novartis Pharmaceuticals	Failed and stopped	Active Aβ 1–6	Winblad et al. (2012) and Wisniewski (2012)
Bapineuzumab (monoclonal)	J & J/Elan/ Pfizer	Failed and stopped	Passive (N-terminal Aβ epitope	Salloway et al. (2014), Doody et al. (2014), and Tayeb et al. (2013)
Solanezumab (monoclonal)	Eli Lilly	Failed and stopped	Passive (central domain epitope)	Tayeb et al. (2013) and Novakovic et al. (2013)
Gantenerumab (monoclonal)	Hoffmann La Roche	Failed and stopped	Passive (N-terminal plus central domain epitope of Aβ)	Ostrowitzki et al. (2012)
Crenezumab/ MABT5102A (monoclonal)	Genentech	Failed and stopped	Passive oligomeric fibrillary and soluble Aβ	Ostrowitzki et al. (2012)

Aβ, β-amyloid.

in human AD brain tissue samples (Nitsch et al., 2008). Such binding is not blocked by monomeric Aβ16 or Aβ42, showing selective affinity for fibrillar Aβ versus Aβ monomers (Nitsch et al., 2008). In APP Tg mice BART reduces amyloid burden while Aβ plasma increase is not observed. Microglia appears to play a pivotal role in clearing plaques (Dunstan et al., 2011b). A Clinical Phase I trial is currently ongoing.

There are many other mAbs that have been developed against Aβ. These antibodies include PF-04360365 (ponezumab), which targets the free carboxy-terminal amino acids 33–40 of the Aβ peptide; MABT5102A, which binds to Aβ monomers, oligomers, and fibrils with equally high affinity; and GSK933776A, which is similar to bapineuzumab and interacts with the N-terminal Aβ(1-5). In addition, other passive immunotherapies mostly in phase 1 clinical development include NI-101, SAR-228810, and BAN-2401(Panza et al., 2014; Tayeb et al., 2013; Jacobsen et al., 2014; Jindal et al., 2014). Gammagard is a preparation of antibody from human plasma. Its safety for human use had been previously demonstrated in certain autoimmune conditions. Gammagard effects were evaluated in a small number of patients with AD (NCT00818662). It is believed that this mixture contains a small fraction of polyclonal antibodies against the Aβ peptide. In addition, this preparation may possess immunomodulatory properties that could potentially enhance microglial phagocytosis (Dodel et al., 2013; Relkin et al., 2009; Szabo et al., 2010). Collectively, the aforementioned studies on active and passive immunization suggest that active immunization maintains the body with a constant high concentration of immunoglobulin, so this strategy calls for fewer follow-up injections with a reasonable expense. However, to tackle with the T-cell-mediated neuroinflammation is an important issue. In contrast, passive immunization is a more effective method for elderly human patients considering their weaker responsiveness to vaccines (McElhaney and Effros, 2009). Selection of safe epitopes and better antibody titer can be readily done in human subjects. However, the risk of development of vasogenic edema and cerebral amyloid angiopathy remains significantly high. Despite the availability of many mAbs and some polyclonal antibodies, there are still no effective therapies that modify the pathogenesis of AD.

Collective evidence suggests that almost all clinical trials using mAbs (solanezumab, crenezumab, and bapineuzumab) raised against different epitopes within the Aβ1–42 molecule have failed to produce beneficial effects and improve cognitive decline in patients with AD. It is not known whether or not the mAb treatment sufficiently eliminates Aβ deposits, including senile plaques and Aβ-derived diffusible ligands from neural cells (Rosenblum, 2014). At least one study has reported that treatment of a patient with AD with bapineuzumab neither significantly reduces the number of Aβ plaques nor produces any improvement in mild to moderate dementia (Roher et al., 2011). In addition, PET imaging studies with

[11]C-PiB of bapineuzumab-treated patients show very little reduction in the positive signal (Rinne et al., 2010) suggesting that passive immunotherapy produces weak Aβ reduction effects compared with active immunotherapy. Thus it is difficult to evaluate the efficacy of mAbs and significance of their treatment in patients with AD. It is not known whether unsatisfactory results are obtained from poor treatment efficacy (such as insufficient administration, timing of treatment, or mAb inefficiency, including its epitope specificity) or from limitations of the treatment itself (e.g., failure despite successful reduction of targeted Aβ species; the strategic limitation) remains undetermined (Kohyama and Matsumoto, 2015).

TAU ANTIBODIES FOR THE TREATMENT OF ALZHEIMER'S DISEASE

As stated in Chapter 1, Tau is a cytoplasmic neuronal protein, which is involved in microtubule dynamics. It binds to tubulin and stabilizes microtubules, which are responsible for the neuronal transport of vesicles and cargo to the synapse. In addition, Tau also plays a key role in maintaining cellular signaling and axonal transport (Fig. 9.2). The dynamic relationship that exists between monomeric Tau and microtubule proteins is driven by the phosphorylation state of Tau, which is under the control of a variety of kinases and phosphatases (Blennow et al., 2010). In AD, the hyperphosphorylated form of Tau undergoes aggregation and misfolding. Oligomeric hyperphosphorylated forms of Tau induce neurotoxicity (Jack and Holtzman, 2013; Tepper et al., 2014). Hyperphosphorylated Tau no longer binds to microtubule proteins (Iqbal et al., 2013). This leads to higher cytosolic concentrations of unbound Tau. Hyperphosphorylated Tau is susceptible to aggregation, protein trapping, and misfolding (Ballatore et al., 2007). The molecular mechanisms associated with these processes are not fully understood. However, two mechanisms have been proposed. The first mechanism involves Tau-mediated destabilization of microtubule that arrests anterograde axonal transportation contributing to neuronal damage and loss of neuronal function along with neurodegeneration (Fig. 9.2) (LaPointe et al., 2009). The second mechanism involves the association of Tau with ribosomes in AD brains, and this association impairs protein synthesis (Meier et al., 2015, 2016). Overall, these interactions result in a stark reduction of nascent proteins, including those that participate in synaptic plasticity, a process that is crucial for learning and memory. These data mechanistically link a common pathologic sign, such as the appearance of pathological Tau inside brain cells, with cognitive impairments evident in virtually all tauopathies including AD (Meier et al., 2016). In addition, AD-brain-derived Tau oligomers disrupt memory and propagate the abnormal conformation of endogenous Tau in wild-type mice,

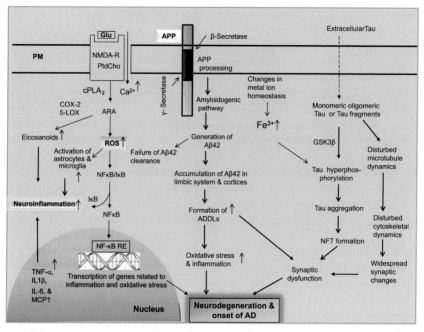

FIGURE 9.2 **Processes associated with the pathogenesis of AD in the brain.** *5-LOX*, 5-lipoxygenase; *AD*, Alzheimer's disease; *ADE*, amyloid-degrading enzyme; *APP*, amyloid precursor protein; *ARA*, arachidonic acid; *COX-2*, cyclooxygenase-2; *cPLA₂*, cytosolic phospholipase A_2; *Glu*, glutamate; *IL-1β*, interleukin-1β; *IL-6*, interleukin-6; *IκB*, inhibitory subunit of NF-κB; *lyso-PtdCho*, lyso-phosphatidylcholine; *MCP-1*, monocyte chemoattractant protein-1; *NEP*, neprilysin; *NF-κB*, nuclear factor-κB; *NF-κB-RE*, nuclear factor-κB-response element; *NMDA-R*, N-methyl-D-aspartate receptor; *PtdCho*, phosphatidylcholine; *ROS*, reactive oxygen species; *TNF-α*, tumor necrosis factor-α.

supporting the involvement of Tau oligomers in the spread of pathology but not NFTs (Lasagna-Reeves et al., 2012a,b). Furthermore, widespread Tau oligomers accumulate at the synapses (Henkins et al., 2012), impairing the ubiquitin–proteasome system (Tai et al., 2012) and contributing to synaptic dysfunction and loss. Furthermore, animal model studies have also revealed that Tau may also promote excitotoxicity via the sensitization of N-methyl-D-aspartate receptors and makes neurons more vulnerable to Aβ toxicity (Rapoport et al., 2002). Stereologic studies on human AD also indicate that neurodegeneration actually precedes NFT formation (Terry, 2000). Granular Tau oligomers have been detected and biochemically isolated at very early stages of the disease, when clinical symptoms of AD and NFTs are absent (Maeda et al., 2007), and Tau-positive fine granules have been found in postmortem tissue from the parkinsonism–dementia complex of Guam tauopathy (Yamazaki et al., 2005). Furthermore, Tau oligomers have been detected in platelets from patients with AD, demonstrating that this species of Tau protein may serve as a new biological marker for AD (Neumann et al., 2011).

It is known that NFTs can be labeled in situ with antibodies against a variety of neuronal proteins, including vimentin, actin, ubiquitin, MAP2, and Aβ. In crude preparations, paired helical filaments (PHFs) can be labeled with antibodies against MAP2, neurofilament, ubiquitin, and Tau (Santa-Maria et al., 2012). In animal models, treatment with a p-Tau peptide (containing the phosphorylated PHF-1 epitopes Ser 396 and Ser 404) injected prior to the onset of pathology produces retardation of Tau aggregates in the Tg P301L mouse Tau model (Asuni et al., 2007). Phosphorylation at these specific epitopes has been reported to increase the fibrillogenic character of Tau and enhances PHF formation (Fath et al., 2002). This model has been reported to develop NFTs in several brain regions and the spinal cord. In an experiment, two groups were immunized from 2 to 5 months and from 2 to 8 months. Immunohistochemical analysis using PHF-1 and MC1 antibodies show a significant reduction in Tau-related pathology, compared to controls. In addition, amelioration in the vaccinated groups is seen on a number of sensorimotor tasks. Antibodies generated by this vaccination can not only cross the BBB but also bind to p-Tau and reduce pathology without significant adverse effect, thus providing strong support in favor of the idea that it is possible to reduce Tau-related pathology with active immunization (Asuni et al., 2007). These results have also been confirmed in a similar study done in an htau/PSEN1 Tau pathology model (Boutajangout et al., 2010). As the Tg mice used in these studies have severe locomotor deficits, a major limitation of this work is that cognition cannot be assessed as a therapeutic endpoint. Thus, some progress has been made on immunotherapy of AD in animal models. However, almost all studies have failed to give beneficial effects in patients with AD.

It is also reported that targeting oligomeric Tau species through immunotherapy may be superior to targeting NFTs across the spectrum of neurodegenerative tauopathies, and therefore this approach has outstanding potential as a disease-modifying intervention for AD (Lasagna-Reeves et al., 2011a,b; Ubhi and Masliah, 2011; Castillo-Carranza et al., 2013a,b). Tau is an endogenous protein with normal cellular functions; treatments with native Tau carry the risk of inducing autoimmunity and/or other related complications (Rosenmann et al., 2006). To protect against negative side effects, a novel anti-Tau oligomer-specific mouse mAb (anti-TOMA) has been developed (Castillo-Carranza et al., 2014). This antibody recognizes Tau oligomers specifically and does not recognize monomeric functional Tau or mature metastable NFTs. This antibody has been used to study Tau oligomers in a mouse model of tauopathy, JNPL3. JNPL3 is expressed in the mutant human Tau protein (P301L), which is responsible for frontotemporal dementia (Lewis et al., 2000).

TOMAs have ideal characteristics for immunotherapy. TOMAs are not only unique for their specificity for Tau oligomers but also have high affinity and the ability to sequester Tau oligomer toxicity in vitro. Furthermore, other IgG antibodies have been shown to enter the brain in P301L mice, supporting the ability of TOMA to cross the BBB if administered intravenously (Asuni et al., 2007). Treatment of male homozygous P301L mice with TOMA indicates that TOMA has ability to modulate the pathological effects of Tau oligomers in vivo. TOMA dose used in the aforementioned studies was 5- to 10-fold lower than those previously used to nonselectively target Tau aggregates by passive immunotherapy (Boutajangout et al., 2010, 2011; Chai et al., 2011). This may contribute to increased safety of antibodies while maintaining efficacy because oligomeric Tau represents a small percentage of total Tau in the brain. Moreover, these experiments have provided evidence that intravenous delivery of IgG antibodies can be detected within the P301L mouse brain, suggesting that the BBB may be impaired in this mouse model (Asuni et al., 2007), as in APP Tg mice (Bard et al., 2000; Yamada et al., 2009) and synuclein mice (Masliah et al., 2011). Because the BBB has also been shown to be impaired in patients with AD patients (Bowman et al., 2007), immunization with IgG antibody may represent a good therapeutic approach in humans.

Large double-blind trials on TOMA have not been performed in patients with AD. So nothing is known about the ability of TOMA for crossing the BBB as well as affinity and specificity of TOMA in normal human subjects and patients with AD.

An active vaccine based on the cysteinated Tau peptide C-[294]KDNI-KHVPGGGS[305], which contains the structural determinant [299]HVP-GGG[304] (Kontsekova et al., 2014a,b), has been prepared. This peptide is conjugated via its N-terminus to a carrier (KLH) and formulated so that a single dose contained 100 µg of the conjugated peptide. Immunization with this Tau peptide in Tg mice has indicated that this immunization improves various neurobehavioral parameters in treated animals. Tau vaccination not only increases the time in which mice are able to stay on the accelerating rotarod but also reduces the number of foot slips in the traverse beam task (Asuni et al., 2007) and improve short-term memory in the Y maze (Troquier et al., 2012). Studies on the efficacy of the Tau peptide vaccine using the NeuroScale—a battery of behavioral tests featuring a novel scoring system for the phenotyping of the Tg rat model of tauopathy (Korenova et al., 2009)—show that vaccination improves the sensorimotor impairment of Tg rats (Kontsekova et al., 2014a). Collectively, this study indicate that Tau vaccine elicits a robust and specific antibody response. Applied vaccine induces the production of high-affinity antibodies targeting pathological Tau in immunized animals. Moreover, the immune response to the active vaccine has shifted toward the Th2 phenotype, underlining its safety. This antibody has been used in a clinical trial in patients with AD (NCT01850238 and NCT02031198), and it is reported that this antibody recognizes Tau lesions, including NFTs and neuropil threads in brains from patients with AD. Therefore, it is reasonable to assume that this antibody will exert the same pattern of therapeutic activity in patients with AD as it displayed in Tg rats (Kontsekova et al., 2014b).

Other approaches for targeting Tau in AD have also been used. These include reducing Tau level itself, preventing Tau hyperphosphorylation, and inhibiting the aggregation (Grill and Cummings, 2010). Many investigators have focused on retarding hyperphosphorylation of Tau, an earlier event that causes detachment of Tau protein from microtubule. As stated in Chapter 1, Tau is phosphorylated by several protein kinases (glycogen synthase kinase-3β, cyclin-dependent kinase-5, and microtubule affinity–regulating kinase) and dephosphorylated by a phosphatase (protein phosphatase 2A). These enzymes can be used as possible therapeutic targets (Schneider and Mandelkow, 2008). AZD 1080 (AstraZeneca) and NP-12/tideglusib (Noscria) are the most promising glycogen synthase kinase-3β inhibitors; however, AZD 1080 has been withdrawn from AD therapeutic market due to the nephrotoxic side effect in phase 1 clinical trials (Domínguez et al., 2012; Eldar-Finkelman and Martinez, 2011). Since then, NP-12/tideglusib has been extensively used as an effective glycogen synthase kinase -3β inhibitor. A pilot clinical study has been completed using this inhibitor, and results show a positive trend in MMSE, ADAS-Cog, Global Deterioration Scale, and Global cortical atrophy (Del Ser et al., 2013). A phase 2 clinical trial (ClinicalTrials.gov identifier: NCT01350362) is also ongoing.

NONIMMUNOGENIC ALZHEIMER'S DISEASE THERAPIES RELATED TO TAU PROTEIN

The aggregation of Tau occurs in a stepwise manner: initially Tau molecules bind to each other through disulfide binding of their Cys residues (Sahara et al., 2007) to form soluble Tau oligomers (Sahara et al., 2007; Maeda et al., 2007); in a second step, these oligomers, composed of ~40 Tau molecules, grow and precipitate as granular Tau oligomers with β-sheet structure; last, the granular Tau oligomers bind to each other and form Tau fibrils (Maeda et al., 2007). Granular Tau oligomers can be detected in the prefrontal cortex at Braak stage I, whereas NFTs appear much later (Braak stage V) (Maeda et al., 2006), indicating that their formation represents a crucial early pathogenic event. Some investigators are trying to develop inhibitors of Tau aggregation (Soeda et al., 2015). These inhibitors bind to Cys residues in Tau that prevent Tau oligomerization and formation of insoluble Tau aggregates. Isoproterenol, an adrenergic receptor agonist, is such an inhibitor. This compound has 1,2-dihydroxybenzene. It can not only penetrate the brain but also reduces the levels of insoluble Tau and neuronal loss and reverses NFT-associated brain dysfunction in a dose-dependent manner (Soeda et al., 2015). It is suggested that isoproterenol can reverse emotional disturbances associated with the expression of P301L Tau and stimulate neural activity in the prelimbic frontal cortex; the latter is accompanied by contemporaneous reductions in the levels of sarkosyl-insoluble Tau in the prelimbic cortex supporting the view that isoproterenol targets the cysteine residues of Tau slowing or stopping the progression of AD and other tauopathies (Soeda et al., 2015). In addition, several compounds are able to inhibit formation of Tau oligomers and fibrils. These compounds have been tested in different animal models (Wischik et al., 2015; Paranjape et al., 2015).

In a phase 2 clinical trial, treatment with methylene blue for 50 weeks shows improvement in cognitive function (as determined by ADAS-Cog) in both subjects with mild and moderate AD (Mori et al., 2014; Wischik et al., 2015). Positive effects have been observed at 138 mg/day, while a higher dose (228 mg/day) is not effective. The failure of dose response has been hypothesized to be due to differences in redox processing and absorption (Baddeley et al., 2015). A methylene blue derivative is currently being tested in phase 3 clinical trials for AD and frontotemporal dementia (Seripa et al., 2016; Panza et al., 2016; Al Mansouri et al., 2012). Both in vivo and in vitro studies have indicated that methylthioninium chloride (MTC) or TRx0237 reduces Tau protein aggregation in AD through proteasomal (O'Leary et al., 2010) and macroautophagic (Congdon et al., 2012; Xie et al., 2013) degradation of the protein. Other potential effects of MTC not only include the oxidation of cysteine sulfhydryl groups in the Tau repeat domain retarding the formation of disulfide bridges to preserve Tau monomers (Akoury et al., 2013) but also inhibition of many enzymes and receptors and signaling systems (Table 9.2). In another study, MTC treatment decreases detergent-insoluble p-Tau levels in the JNPL3 (P301L) Tau Tg mice. Treatment of 3-month-old rTg4510 mice for 12 weeks with oral MTC retards behavioral deficits and reduces soluble Tau levels in the brain (O'Leary et al., 2010). JNPL3 (P301L) mice treated with MTC for 2 weeks show reduction in soluble Tau levels without affecting insoluble Tau levels (Congdon et al., 2012). Collectively, these studies indicate that MTC treatment reduces soluble Tau levels and prevents cognitive decline when treatment begins at a time point before NFTs are present in the brain (O'Leary et al., 2010). Curiously, methylene blue has already been used by Cajal to stain dendritic spines (DeFelipe, 2015), a structure that contains Tau protein (Avale et al., 2013). These studies suggest that a more comprehensive understanding of the molecular mechanisms regulating oligomerization will provide us with a

TABLE 9.2   Effect of Methylthioninium Chloride on Various Enzymes and Neurotransmitter Systems

Enzyme	Effect	References
Acetylcholinesterase	Inhibition	Pfaffendorf et al. (1997)
Nitric oxide synthase	Inhibition	Mayer et al. (1993)
Monoamine oxidase B	Inhibition	Ramsay et al. (2007)
Guanylate cyclase	Inhibition	Mayer et al. (1993)
β-Secretase activity	Inhibition	
Noradrenaline uptake inhibition	Inhibition	Chies et al. (2003)
Glutamatergic inhibition	Inhibition	Vutskits et al. (2008)
α7-Nicotinic acetylcholine receptors	Inhibition	
Aggregation of Aβ peptides	Inhibition	Necula et al. (2007)
Mitochondrial amyloid–binding alcohol dehydrogenase	Improved	Zakaria et al. (2016)
Mitochondrial antioxidant properties	Improved	Wen et al. (2011)
Nrf2/antioxidant response element	Improved	Stack et al. (2014)

better outline of Tau cellular networking and, hopefully, offer new clues for designing more efficient approaches to tackle AD and tauopathies in the near future.

Converging evidence from the aforementioned studies indicate that investigators have made some progress on immunotherapy of AD in animal models. However, immunotherapy in patients with AD has failed. To this end, investigators have to overcome many experimental and clinical problems such as long-term effects, clinical safety, and efficacy of mono- and polyclonal antibodies in patients with AD and age-matched normal human subjects. mAbs or polyclonal antibodies with high titers are required for effectively crossing the BBB and entering the brain tissue. The use of chaperone proteins or bispecific antibodies has also been proposed (Lemere, 2013). Since the removal or clearance of Aβ after neurodegeneration has been ineffective in animal model and patients with AD, the administration of antibodies should be started earlier and tested for longer time periods. Furthermore, a better understanding of the clearance of Aβ/anti-Aβ immune complexes is needed to avoid clogging of the clearance pathway during long-term treatment. Particular attention is required on the immune effects of immunotherapy in the elderly patients with AD. Detailed investigations should be performed on potential autoimmunity of antibody preparation as well as on the use of strong proinflammatory adjuvants. Finally, improving the sensitivity of biomarkers, including imaging of preamyloid diffuse plaques, and cognitive/functional tests to detect the earliest changes in AD will allow for better patient selection for clinical trials and more sensitive outcome measures (Lemere, 2013).

CONCLUSION

According to Aβ hypothesis, the accumulation of highly toxic oligomeric Aβ peptides is the primary cause of neurodegeneration and pathogenic driver of not only downstream Tau hyperphosphorylation but also NFT formation and synaptic toxicity in AD. In addition, a progressive neuroinflammatory reaction has been observed surrounding Aβ plaques, resulting in astrogliosis and microglial activation. This inflammatory process may independently lead to neural dysfunction and cell death, thus establishing a self-perpetuating vicious cycle, which further contributes to neurodegeneration and enhances the pathological hallmarks of the disease.

Immunotherapy is an important way to reduce the pathological lesions of AD and slowing or reversing cognitive decline. Active immunization can be performed by injecting of fragments of Aβ peptide conjugated to an adjuvant that stimulates the host immune response. The goal of active immunization is to produce antibodies against Aβ that can promote the removal of Aβ plaques from the brain of patients with AD. A safe and effective vaccine requires therapeutic levels of anti-Aβ antibodies, which do not cause adverse autoimmune responses. This is particularly challenging in the elderly human patients due to immunosenescence. T-cell-mediated inappropriate immune responses can produce adverse reaction from the adjuvant. Several polyclonal antibodies have been produced (AN1792, ACC-001, and CAD106) and used for the therapy of patients with AD. Immunotherapy with polyclonal antibodies has worked in animal models of AD but has failed in human patients, and 6%–8% patients with AD developed meningoencephalitis. Passive immunization involves direct administration of anti-Aβ mAbs (bapineuzumab, solanezumab, gantenerumab, crenezumab). The advantages of passive immunization over active immunization are greater dose control and easy withdrawal of treatment if adverse effects arise. The disadvantages of passive immunization are the need for repeated injections and the fact that this therapy can never be preventative since the development of clinical symptoms proceeds treatment. Despite the availability of many mAbs, there are still no effective therapies that modify the pathogenesis of AD in human patients and presently available mAbs have failed to show overall clinical improvement or any clear disease.

References

Adolfsson, O., Pihlgren, M., Toni, N., Varisco, Y., Buccarello, A.L., et al., 2012. An effector-reduced anti-β-amyloid (Aβ) antibody with unique Aβ binding properties promotes neuroprotection and glial engulfment of Aβ. J. Neurosci. 32, 9677–9689.

Akoury, E., Pickhardt, M., Gajda, M., Biernat, J., Mandelkow, E., et al., 2013. Mechanistic basis of phenothiazine-driven inhibition of Tau aggregation. Angew. Chem.—Int. Ed. 52, 3511–3515.

Al Mansouri, A.S., Lorke, D.E., Nurulain, S.M., Ashoor, A., Keun-Hang, S., et al., 2012. Methylene blue inhibits the function of α7-nicotinic acetylcholine receptors. Cent. Nerv. Syst. Neurol. Disord. Drug Targets 11, 791–800.

Asuni, A., Boutajangout, A., Scholtzova, H., Knudsen, E., Li, Y., et al., 2006. Aβ derivative vaccination in alum adjuvant prevents amyloid deposition and does not cause brain microhemorrhages in Alzheimer's model mice. Eur. J. Neurosci. 24, 2530–2542.

Asuni, A.A., Boutajangout, A., Quartermain, D., Sigurdsson, E.M., 2007. Immunotherapy targeting pathological tau conformers in a tangle mouse model reduces brain pathology with associated functional improvements. J. Neurosci. 27, 9115–9129.

Avale, M.E., Rodriguez-Martin, T., Gallo, J.M., 2013. Trans-splicing correction of tau isoform imbalance in a mouse model of tau mis-splicing. Hum. Mol. Genet. 22, 2603–2611.

Baddeley, T.C., McCaffrey, J., Storey, J.M., Cheung, J.K., Melis, V., et al., 2015. Complex disposition of methylthioninium redox forms determines efficacy in tau aggregation inhibitor therapy for Alzheimer's disease. J. Pharmacol. Exp. Ther. 352, 110–118.

Bailey, J.A., Ray, B., Lahiri, D.K., 2012. Intravenous immunoglobulin (IVIG) protects neuronal viability and synaptic markers in cultured degenerating primary hippocampal neurons. Soc. Neurosci. Abstr. 852, 04.

Ballatore, C., Lee, V.M., Trojanowski, J.Q., 2007. Tau-mediated neurodegeneration in Alzheimer's disease and related disorders. Nat. Rev. Neurosci. 8, 663–672.

Baranello, R.J., Bharani, K.L., Padmaraju, V., Chopra, N., Lahiri, D.K., et al., 2015. Amyloid-beta protein clearance and degradation (ABCD) pathways and their role in Alzheimer's disease. Curr. Alzheimer Res. 12, 32–46.

Bard, F., Cannon, C., Barbour, R., Burke, R.L., Games, D., et al., 2000. Peripherally administered antibodies against amyloid β-peptide enter the central nervous system and reduce pathology in a mouse model of Alzheimer disease. Nat. Med. 6, 916–919.

Basi, G., Jacobsen, J.S., 2006. Humanized Abeta Antibodies for Use in Improving Cognition (United States Patent Application Publication Number US 2006/0198851 A1).

Basi, G., Saldanha, J., Yednock, T., 2002. Humanized Antibodies that Recognize Beta Amyloid Peptide (World Intellectual Property Organization International Publication Number WO 02/46237 A2).

Bates, K.A., Verdile, G., Li, Q.X., Ames, D., Hudson, P., et al., 2009. Clearance mechanisms of Alzheimer's amyloid-beta peptide: implications for therapeutic design and diagnostic tests. Mol. Psychiatry 14, 469–486.

Black, R., Ekman, L., Lieberburg, I., Grundman, M., Callaway, J., et al., 2009. Immunotherapy Regimes Dependent on ApoE Status (US Patent).

Blennow, K., Hampel, H., Weiner, M., Zetterberg, H., 2010. Cerebrospinal fluid and plasma biomarkers in Alzheimer disease. Nat. Rev. Neurol. 6, 131–144.

Blennow, K., Zetterberg, H., Rinne, J.O., Salloway, S., Wei, J., Black, R., et al., 2012. Effect of immunotherapy with bapineuzumab on cerebrospinal fluid biomarker levels in patients with mild to moderate Alzheimer disease. Arch. Neurol. 69, 1002–1010.

Boche, D., Donald, J., Love, S., Harris, S., Neal, J.W., et al., 2010. Reduction of aggregated Tau in neuronal processes but not in the cell bodies after Aβ42 immunisation in Alzheimer's disease. Acta Neuropathol. 8, 13–20.

Bohrmann, B., Baumann, K., Benz, J., Gerber, F., Huber, W., et al., 2012. Gantenerumab: a novel human anti-Aβ antibody demonstrates sustained cerebral amyloid-β binding and elicits cell-mediated removal of human amyloid-β. J. Alzheimer's Dis. 28, 49–69.

Bombois, S., Maurage, C.A., Gompel, M., Deramecourt, V., kowiak-Cordoliani, M.A., et al., 2007. Absence of beta-amyloid deposits after immunization in Alzheimer disease with Lewy body dementia. Arch. Neurol. 64, 583–587.

Boutajangout, A., Quartermain, D., Sigurdsson, E.M., 2010. Immunotherapy targeting pathological tau prevents cognitive decline in a new tangle mouse model. J. Neurosci. 30, 16559–16566.

Boutajangout, A., Ingadottir, J., Davies, P., Sigurdsson, E.M., 2011. Passive immunization targeting pathological phospho-tau protein in a mouse model reduces functional decline and clears tau aggregates from the brain. J. Neurochem. 118, 658–667.

Bouter, Y., Lopez Noguerola, J.S., Tucholla, P., Crespi, G.A., Parker, M.W., et al., 2015. Abeta targets of the biosimilar antibodies of Bapineuzumab, Crenezumab, Solanezumab in comparison to an antibody against N-truncated Abeta in sporadic Alzheimer disease cases and mouse models. Acta Neuropathol. 130, 713–729.

Bowman, G.L., Kaye, J.A., Moore, M., Waichunas, D., Carlson, N.E., et al., 2007. Blood-brain barrier impairment in Alzheimer disease: stability and functional significance. Neurology 68, 1809–1814.

Carrillo, M.C., Brashear, H.R., Logovinsky, V., Ryan, J.M., Feldman, H.H., et al., 2013. Can we prevent Alzheimer's disease? Secondary "prevention" trials in Alzheimer's disease. Alzheimer's Dement. 9, 123–131.

Castillo-Carranza, D.L., Lasagna-Reeves, C.A., Kayed, R., 2013a. Tau aggregates as immunotherapeutic targets. Front. Biosci. (Schol Ed.) 5, 426–438.

Castillo-Carranza, D.L., Guerrero-Muñoz, M.J., Kayed, R., 2013b. Immunotherapy for the treatment of Alzheimer's disease: amyloid-β or tau, which is the right target? ImmunoTargets Ther. 2014, 19–28.

Castillo-Carranza, D.L., Sengupta, U., Guerrero-Muñoz, M.J., Lasagna-Reeves, C.A., Gerson, J.E., et al., 2014. Passive immunization with Tau oligomer monoclonal antibody reverses tauopathy phenotypes without affecting hyperphosphorylated neurofibrillary tangles. J. Neurosci. 34, 4260–4272.

Chai, X., Wu, S., Murray, T.K., Kinley, R., Cella, C.V., et al., 2011. Passive immunization with anti-Tau antibodies in two transgenic models: reduction of Tau pathology and delay of disease progression. J. Biol. Chem. 286, 34457–34467.

Chies, A.B., Custódio, R.C., de Souza, G., Corrêa, F.M.A., Pereira, O.C.M., 2003. Pharmacological evidence that methylene blue inhibits noradrenaline neuronal uptake in the rat vas deferens. Pol. J. Pharmacol. 55, 573–579.

Congdon, E.E., Wu, J.W., Myeku, N., Figueroa, Y.H., Herman, M., et al., 2012. Methylthioninium chloride (methylene blue) induces autophagy and attenuates tauopathy in vitro and in vivo. Autophagy 8, 609–622.

Counts, S., Perez, S., He, B., Mufson, E., 2014. Intravenous immunoglobulin reduces tau pathology and preserves neuroplastic gene expression in the 3xTg mouse model of Alzheimer's disease. Curr. Alzheimer Res. 11, 655–663.

Crespi, G.A., Hermans, S.J., Parker, M.W., Miles, L.A., 2015. Molecular basis for mid-region amyloid-β capture by leading Alzheimer's disease immunotherapies. Sci. Rep. 5, 9649.

DeFelipe, J., 2015. The dendritic spine story: an intriguing process of discovery. Front. Neuroanat. 9, 14.

Delrieu, J., Ousset, P.J., Caillaud, C., Vellas, B., 2012. 'Clinical trials in Alzheimer's disease': immunotherapy approaches. J. Neurochem. 120 (Suppl. 1), 186–193.

Del Ser, T., Steinwachs, K.C., Gertz, H.J., Andrés, M.V., Gómez-Carrillo, B., et al., 2013. Treatment of Alzheimer's disease with the GSK-3 inhibitor tideglusib: a pilot study. J. Alzheimer's Dis. 33, 205–215.

DeMattos, R.B., Bales, K.R., Cummins, D.J., Dodart, J.C., Paul, S.M., et al., 2001. Peripheral anti-A beta antibody alters CNS and plasma A beta clearance and decreases brain A beta burden in a mouse model of Alzheimer's disease. Proc. Natl. Acad. Sci. U.S.A. 98, 8850–8855.

DeMattos, R.B., Lu, J., Tang, Y., Racke, M.M., Delong, C.A., et al., 2012. A plaque – specific antibody clears existing beta – amyloid plaques in Alzheimer's disease mice. Neuron 76, 908–920.

Dodel, R.C., Du, Y., Depboylu, C., Hampel, H., Frolich, L., et al., 2004. Intravenous immunoglobulins containing antibodies against beta-amyloid for the treatment of Alzheimer's disease. J. Neurol. Neurosurg. Psychiatry 75, 1472–1474.

Dodel, R., Rominger, A., Bartenstein, P., Barkhof, F., Blennow, K., et al., 2013. Intravenous immunoglobulin for treatment of mild-to-moderate Alzheimer's disease: a phase 2, randomised, double-blind, placebo-controlled, dose-finding trial. Lancet Neurol. 12, 233–243.

Domínguez, J.M., Fuertes, A., Orozco, L., del Monte-Millán, M., Delgado, E., et al., 2012. Evidence for irreversible inhibition of glycogen synthase kinase-3β by tideglusib. J. Biol. Chem. 287, 893–904.

Doody, R.S., Thomas, R.G., Farlow, M., Iwatsubo, T., Vellas, B., et al., 2014. Alzheimer's Disease Cooperative Study Steering Committee; Solanezumab Study Group. Phase 3 trials of solanezumab for mild-to-moderate Alzheimer's disease. N. Engl. J. Med. 370, 311–321.

Dunstan, R., Bussiere, T., Rhodes, K., Engber, T., Maier, M., Weinreb, P., Grimm, J., Nitsch, R., Arustu, M., Qian, F., Li, M., 2011a. Molecular characterization and preclinical efficacy. Alzheimer's Dement. 7, S457.

Dunstan, R., Bussiere, T., Engber, T., Weinreb, P., Maier, M., et al., 2011b. The role of brain macrophages on the clearance of amyloid plaques following the treatment of Tc2576 with BIIB037. [abstr.] Alzheimer's Dement. 7, S700.

Eldar-Finkelman, H., Martinez, A., 2011. GSK-3 inhibitors: preclinical and clinical focus on CNS. Front. Mol. Neurosci. 4, 32.

Farlow, M., Arnold, S.E., van Dyck, C.H., Aisen, P.S., Snider, B.J., et al., 2012. Safety and biomarker effects of solanezumab in patients with Alzheimer's disease. Alzheimer's Dement. 8, 261–271.

Fath, T., Eidenmuller, J., Brandt, R., 2002. Tau-mediated cytotoxicity in a pseudohyperphosphorylation model of Alzheimer's disease. J. Neurosci. 22, 9733–9741.

Foster, J.K., Verdile, G., Bates, K.A., Martins, R.N., 2009. Immunization in Alzheimer's disease: naive hope or realistic clinical potential? Mol. Psychiatry 14, 239–251.

Frost, J.L., Le, K.X., Cynis, H., Ekpo, E., Kleinschmidt, M., et al., 2013. Pyroglutamate – 3 amyloid – beta deposition in the brains of humans, non-human primates, canines, and Alzheimer disease-like transgenic mouse models. Am. J. Pathol. 183, 369–381.

Fu, H.J., Liu, B., Frost, J.L., Lemere, C.A., 2010. Amyloid-β immunotherapy for Alzheimer's disease. CNS Neurol. Disord. Drug Targets 9, 197–206.

Gandy, S., Sano, M., 2015. Alzheimer disease: Solanezumab—prospects for meaningful interventions in AD? Nat. Rev. Neurol. 11, 669–670.

Gilman, S., Koller, M., Black, R.S., et al., 2005. Clinical effects of Aβ immunization (AN1792) in patients with AD in an interrupted trial. Neurology 64, 1553–1562.

Goure, W.F., Krafft, G.A., Jerecic, J., Hefti, F., 2014. Targeting the proper amyloid-β neuronal toxins: a path forward for Alzheimer's disease immunotherapeutics. Alzheimer's Res. Ther. 6, 42.

Grill, J.D., Cummings, J.L., 2010. Current therapeutic targets for the treatment of Alzheimer's disease. Expert Rev. Neurother. 10, 711–728.

Hardy, J., Selkoe, D.J., 2002. The amyloid hypothesis of Alzheimer's disease: progress and problems on the road to therapeutics. Science 297, 353–356.

Henkins, K.M., Sokolow, S., Miller, C.A., Vinters, H.V., Poon, W.W., et al., 2012. Extensive p-tau pathology and SDS-stable p-tau oligomers in Alzheimer's cortical synapses. Brain Pathol. 22, 826–833.

Hickman, D.T., López-Deber, M.P., Ndao, D.M., Silva, A.B., Nand, D., Pihlgren, M., Giriens, V., Madani, R., St-Pierre, A., Karastaneva, H., Nagel-Steger, L., Willbold, D., Riesner, D., et al., 2011. Sequence – independent control of peptide conformation in liposomal vaccines for targeting protein misfolding diseases. J. Biol. Chem. 286, 13966–13976.

Imbimbo, B.P., Ottonello, S., Frisardi, V., Solfrizzi, V., Greco, A., et al., 2012. Solanezumab for the treatment of mild-to-moderate Alzheimer's disease. Expert Rev. Clin. Immunol. 8, 135–149.

Iqbal, K., Gong, C.X., Liu, F., 2013. Hyperphosphorylation-induced tau oligomers. Front. Neurol. 4, 112.

Jack Jr., C.R., Holtzman, D.M., 2013. Biomarker modeling of Alzheimer's disease. Neuron 80, 1347–1358.

Jacobsen, H., Ozmen, L., Caruso, A., Narquizian, R., Hilpert, H., et al., 2014. Combined treatment with a BACE inhibitor and anti-Aβ antibody gantenerumab enhances amyloid reduction in APPLondon mice. J. Neurosci. 34, 11621–11630.

Jindal, H., Bhatt, B., Sk, S., Malik, J.S., 2014. Alzheimer disease immunotherapeutics: then and now. Hum. Vaccin. Immunother. 10, 2741–2743.

Johnson-Wood, K., Lee, M., Motter, R., Hu, K., Gordon, G., et al., 1997. Amyloid precursor protein processing and Aβ$_{42}$ deposition in a transgenic mouse model of Alzheimer disease. Proc. Natl. Acad. Sci. U.S.A. 94, 1550–1555.

Klyubin, I., Walsh, D.M., Lemere, C.A., Cullen, W.K., Shankar, G.M., et al., 2005. Amyloid β protein immunotherapy neutralizes Aβ oligomers that disrupt synaptic plasticity in vivo. Nat. Med. 11, 556–561.

Kohyama, K., Matsumoto, Y., 2015. Alzheimer's disease and immunotherapy: what is wrong with clinical trials? ImmunoTargets Ther. 4, 27–34.

Kontsekova, E., Zilka, N., Kovacech, B., Skrabana, R., Novak, M., 2014a. Identification of structural determinants on tau protein essential for its pathological function: novel therapeutic target for tau immunotherapy in Alzheimer's disease. Alzheimer's Res. Ther. 6, 45.

Kontsekova, E., Zilka, N., Kovacech, B., Novak, P., Novak, M., 2014b. First-in-man tau vaccine targeting structural determinants essential for pathological tau-tau interaction reduces tau oligomerisation and neurofibrillary degeneration in an Alzheimer's disease model. Alzheimer's Res. Ther. 6, 44.

Korenova, M., Zilka, N., Stozicka, Z., Bugos, O., Vanicky, I., et al., 2009. NeuroScale, the battery of behavioral tests with novel scoring system for phenotyping of transgenic rat model of tauopathy. J. Neurosci. Methods 177, 108–114.

Lahiri, D.K., Ray, B., 2013. Effect of IVIG in preserving human primary neurons and protecting them against oxidative stress. Alzheimer's Dement. 9 (Suppl.), P800.

LaPointe, N.E., Morfini, G., Pigino, G., Gaisina, I.N., Kozikowski, A.P., et al., 2009. The amino terminus of tau inhibits kinesin-dependent axonal transport: implications for filament toxicity. J. Neurosci. Res. 87, 440–451.

Lasagna-Reeves, C.A., Castillo-Carranza, D.L., Sengupta, U., Clos, A.L., Jackson, G.R., et al., 2011a. Tau oligomers impair memory and induce synaptic and mitochondrial dysfunction in wild-type mice. Mol. Neurodegener. 6, 39.

Lasagna-Reeves, C.A., Castillo-Carranza, D.L., Jackson, G.R., Kayed, R., 2011b. Tau oligomers as potential targets for immunotherapy for Alzheimer's disease and tauopathies. Curr. Alzheimer Res. 8, 659–665.

Lasagna-Reeves, C.A., Castillo-Carranza, D.L., Sengupta, U., Sarmiento, J., Troncoso, J., et al., 2012a. Identification of oligomers at early stages of tau aggregation in Alzheimer's disease. FASEB J. 26, 1946–1959.

Lasagna-Reeves, C.A., Castillo-Carranza, D.L., Sengupta, U., Guerrero-Munoz, M.J., Kiritoshi, T., et al., 2012b. Alzheimer brain-derived tau oligomers propagate pathology from endogenous tau. Sci. Rep. 2, 700.

Lemere, C.A., Spooner, E.T., LaFrancois, J., Malester, B., Mori, C., et al., 2003. Evidence for peripheral clearance of cerebral Abeta protein following chronic, active Abeta immunization in PSAPP mice. Neurobiol. Dis. 14, 10–18.

Lemere, C.A., Beierschmitt, A., Iglesias, M., Spooner, E.T., Bloom, J.K., et al., 2004. Alzheimer's disease abeta vaccine reduces central nervous system abeta levels in a non-human primate, the Caribbean vervet. Am. J. Pathol. 165, 283–297.

Lemere, C.A., 2013. Immunotherapy for Alzheimer's disease: hoops and hurdles. Mol. Neurodegener. 8, 36.

Lewis, J., McGowan, E., Rockwood, J., Melrose, H., Nacharaju, P., et al., 2000. Neurofibrillary tangles, amyotrophy and progressive motor disturbance in mice expressing mutant (P301L) tau protein. Nat. Genet. 25, 402–405.

Maeda, S., Sahara, N., Saito, Y., Murayama, M., Yoshiike, Y., 2007. Granular tau oligomers as intermediates of tau filaments. Biochemistry 46, 3856–3861.

Maeda, S., Sahara, N., Saito, Y., Murayama, S., Ikai, A., Takashima, A., et al., 2006. Increased levels of granular tau oligomers: an early sign of brain aging and Alzheimer's disease. Neurosci. Res. 54, 197–201.

Masliah, E., Rockenstein, E., Mante, M., Crews, L., Spencer, B., et al., 2011. Passive immunization reduces behavioral and neuropathological deficits in an alpha-synuclein transgenic model of Lewy body disease. PLoS One 6, e19338.

Mayer, B., Brunner, F., Schmidt, K., 1993. Inhibition of nitric oxide synthesis by methylene blue. Biochem. Pharmacol. 45, 367–374.

McElhaney, J.E., Effros, R.B., 2009. Immunosenescence: what does it mean to health outcomes in older adults? Curr. Opin. Immunol. 21, 418–424.

Meier, S., Bell, M., Lyons, D.N., Ingram, A., Chen, J., et al., 2015. Identification of novel tau interactions with endoplasmic reticulum proteins in Alzheimer's disease brain. J. Alzheimer's Dis. 48, 687–702.

Meier, S., Bell, M., Lyons, D.N., Rodriguez-Rivera, J., Ingram, A., et al., 2016. Pathological tau promotes neuronal damage by impairing ribosomal function and decreasing protein synthesis. J. Neurosci. 36, 1001–1007.

Morgan, D., Diamond, D.M., Gottschall, P.E., Ugen, K.E., Dickey, C., et al., 2000. A beta peptide vaccination prevents memory loss in an animal model of Alzheimer's disease. Nature 408, 982–985.

Mori, T., Koyama, N., Segawa, T., Maeda, M., Maruyama, N., et al., 2014. Methylene blue modulates β-secretase, reverses cerebral amyloidosis, and improves cognition in transgenic mice. J. Biol. Chem. 289, 30303–30317.

Mullane, K., Williams, M., 2013. Alzheimer's therapeutics: continued clinical failures question the validity of the amyloid hypothesis-but what lies beyond? Biochem. Pharmacol. 85, 289–305.

Nalivaeva, N.N., Fisk, L.R., Belyaev, N.D., Turner, A.J., 2008. Amyloid-degrading enzymes as therapeutic targets in Alzheimer's disease. Curr. Alzheimer Res. 5, 212–224.

Necula, M., Breydo, L., Milton, S., Kayed, R., van der Veer, W.E., et al., 2007. Methylene blue inhibits amyloid Aβ oligomerization by promoting fibrillization. Biochemistry 46, 8850–8860.

Neumann, K., Farias, G., Slachevsky, A., Perez, P., Maccioni, R.B., 2011. Human platelets tau: a potential peripheral marker for Alzheimer's disease. J. Alzheimer's Dis. 25, 103–109.

Nicoll, J.A., Barton, E., Boche, D., Neal, J.W., Ferrer, I., et al., 2006. Abeta species removal after abeta42 immunization. J. Neuropathol. Exp. Neurol. 65, 1040–1048.

Nitsch, R., Hock, C., Esslinger, C., Knobloch, M., Tissot, K., 2008. Method of Providing Disease-Specific Binding Molecules and Targets (World Intellectual Property Organization International Publication Number WO 2008/081008 A1).

Novakovic, D., Feligioni, M., Scaccianoce, S., Caruso, A., Piccinin, S., et al., 2013. Profile of gantenerumab and its potential in the treatment of Alzheimer's disease. Drug Des. Dev. Ther. 7, 1359–1364.

O'Brien, R.J., Wong, P.C., 2011. Amyloid precursor protein processing and Alzheimer's disease. Annu. Rev. Neurosci. 34, 185–204.

O'Leary, J.C., Li, Q., Marinec, P., Blair, L.J., Congdon, E.E., et al., 2010. Phenothiazine-mediated rescue of cognition in tau transgenic mice requires neuroprotection and reduced soluble tau burden. Mol. Neurodegener. 5, 45.

O'Nuallain, B., Acero, L., Williams, A.D., Koeppen, H.P., Weber, A., et al., 2008. Human plasma contains cross-reactive Abeta conformer-specific IgG antibodies. Biochemistry 47, 12254–12256.

O'Nuallain, B., Klyubin, I., Mc Donald, J.M., Foster, J.S., Welzel, A., et al., 2011. A monoclonal antibody against synthetic Aβ dimer assemblies neutralizes brain-derived synaptic plasticity-disrupting Aβ. J. Neurochem. 119, 189–201.

Ostrowitzki, S., Deptula, D., Thurfjell, L., Barkhof, F., Bohrmann, B., Brooks, D.J., et al., 2012. Mechanism of amyloid removal in patients with Alzheimer disease treated with gantenerumab. Arch. Neurol. 69, 198–207.

Panza, F., Solfrizzi, V., Imbimbo, B.P., Tortelli, R., Santamato, A., et al., 2014. Amyloid-based immunotherapy for Alzheimer's disease in the time of prevention trials: the way forward. Expert Rev. Clin. Immunol. 10, 405–419.

Panza, F., Solfrizzi, V., Seripa, D., Imbimbo, B.P., Lozupone, M., et al., 2016. Tau-centric targets and drugs in clinical development for the treatment of Alzheimer's disease. Biomed. Res. Int. 2016, 3245935.

Paranjape, S.R., Riley, A.P., Somoza, A.D., Oakley, C.E., Wang, C.C., et al., 2015. Azaphilones inhibit tau aggregation and dissolve tau aggregates in vitro. ACS Chem. Neurosci. 6, 751–760.

Pfaffendorf, M., Bruning, T.A., Batink, H.D., Van Zwieten, P.A., 1997. The interaction between methylene blue and the cholinergic system. Br. J. Pharmacol. 122, 95–98.

Pfeifer, A., Pihlgren, M., Muhs, A., Watts, R., 2008. Humanized Antibodies to Amyloid Beta (World Intellectual Property Organization International Publication Number WO 2008/156622 A1).

Racke, M.M., Boone, L.I., Hepburn, D.L., Parsadainian, M., Bryan, M.T., et al., 2005. Exacerbation of cerebral amyloid angiopathy-associated microhemorrhage in amyloid precursor protein transgenic mice by immunotherapy is dependent on antibody recognition of deposited forms of amyloid β. J. Neurosci. 25, 629–636.

Ramsay, R.R., Dunford, C., Gillman, P.K., 2007. Methylene blue and serotonin toxicity: inhibition of monoamine oxidase A (MAO A) confirms a theoretical prediction. Br. J. Pharmacol. 152, 946–951.

Rapoport, M., Dawson, H.N., Binder, L.I., Vitek, M.P., Ferreira, A., 2002. Tau is essential to beta-amyloid-induced neurotoxicity. Proc. Nat. Acad. Sci. U.S.A. 99, 6364–6369.

Relkin, N.R., Szabo, P., Adamiak, B., Burgut, T., Monthe, C., et al., 2009. 18-month study of intravenous immunoglobulin for treatment of mild Alzheimer disease. Neurobiol. Aging 30, 1728–1736.

Rinne, J.O., Brooks, D.J., Rossor, M.N., Fox, N.C., Bullock, R., et al., 2010. ^{11}C-PiB PET assessment of change in fibrillar amyloid-β load in patients with Alzheimer's disease treated with bapineuzumab: a phase 2, double-blind, placebo-controlled, ascending-dose study. Lancet Neurol. 9, 363–372.

Robinson, S.R., Bishop, G.M., Lee, H.G., Münch, G., 2004. Lessons from the AN 1792 Alzheimer vaccine: lest we forget. Neurobiol. Aging 25, 609–615.

Roher, A.E., Kokjohn, T.A., 2002. Of mice and men: the relevance of transgenic mice Abeta immunizations to Alzheimer's disease. J. Alzheimers Dis. 4, 431–434.

Roher, A.E., Maarouf, C.L., Daugs, I.D., et al., 2011. Neuropathology and amyloid-beta spectrum in a bapineuzumab immunotherapy recipient. J. Alzheimers Dis. 24, 315–325.

Rosenblum, W.I., 2014. Why Alzheimer trials fail: removing soluble oligomeric beta amyloid is essential, inconsistent, and difficult. Neurobiol. Aging 35, 969–974.

Rosenmann, H., Grigoriadis, N., Karussis, D., Boimel, M., Touloumi, O., et al., 2006. Tauopathy-like abnormalities and neurologic deficits in mice immunized with neuronal tau protein. Arch. Neurol. 63, 1459–1467.

Ryan, J.M., Grundman, M., 2009. Anti-amyloid-beta immunotherapy in Alzheimer's disease: ACC-001 clinical trials are ongoing. J. Alzheimers Dis. 17, 243.

Sabharwal, P., Wisniewski, T., 2014. Novel immunological approaches for the treatment of Alzheimer's disease. Zhongguo Xian Dai Shen Jing Ji Bing Za Zhi 14, 139–151.

Sahara, N., Maeda, S., Murayama, M., Suzuki, T., Dohmae, N., et al., 2007. Assembly of two distinct dimers and higher-order oligomers from full-length tau. Eur. J. Neurosci. 25, 3020–3029.

Saido, T.C., Iwatsubo, T., Mann, D.M., Shimada, H., Ihara, Y., et al., 1995. Dominant and differential deposition of distinct beta-amyloid peptide species, Abeta N3(pE), in senile plaques. Neuron 14, 457–466.

Salloway, S., Sperling, R., Gilman, S., Fox, N.C., Blennow, K., et al., 2009. A phase 2 multiple ascending dose trial of bapineuzumab in mild to moderate Alzheimer disease. Neurology 73, 2061–2070.

Salloway, S., Sperling, R., Fox, N.C., Blennow, K., Klunk, W., et al., 2014. Bapineuzumab 301 and 302 Clinical Trial Investigators. Two phase 3 trials of bapineuzumab in mild-to-moderate Alzheimer's disease. N. Engl. J. Med. 370, 322–333.

Santa-Maria, I., Varghese, M., Ksiezak-Reding, H., Dzhun, A., Wang, J., et al., 2012. Paired helical filaments from Alzheimer disease brain induce intracellular accumulation of tau protein in aggresomes. J. Biol. Chem. 287, 20522–20533.

Schenk, D., Barbour, R., Dunn, W., et al., 1999. Immunization with amyloid-β attenuates Alzheimer disease-like pathology in the PDAPP mouse. Nature 400, 173–177.

Schneeberger, A., Mandler, M., Otawa, O., Zauner, W., Mattner, F., et al., 2009. Development of AFFITOPE vaccines for Alzheimer's disease (AD): from concept to clinical testing. J. Nutr. Health Aging 13, 264–267.

Schneider, A., Mandelkow, E., 2008. Tau-based treatment strategies in neurodegenerative diseases. Neurotherapeutics 5, 443–457.

Seripa, D., Solfrizzi, V., Imbimbo, B.P., Daniele, A., Santamato, A., et al., 2016. Tau-directed approaches for the treatment of Alzheimer's disease: focus on leuco-methylthioninium. Expert Rev. Neurother. 16, 259–277.

Seubert, P., Vigo-Pelfrey, C., Esch, F., Lee, M., Dovey, H., et al., 1992. Isolation and quantification of soluble Alzheimer's β-peptide from biological fluids. Nature 359, 325–327.

Sevigny, J., Chiao, P., Bussière, T., Weinreb, P.H., Williams, L., et al., 2016. The antibody aducanumab reduces Aβ plaques in Alzheimer's disease. Nature 537, 50–56.

Soeda, Y., Yoshikawa, M., Almeida, O.F., Sumioka, A., et al., 2015. Oxic tau oligomer formation blocked by capping of cysteine residues with 1,2-dihydroxybenzene groups. Nat. Commun. 6, 10216.

Solomon, B., Koppel, R., Frenkel, D., Hanan-Aharon, E., 1997. Disaggregation of Alzheimer β-amyloid by site-directed mAb. Proc. Natl. Acad. Sci. U.S.A. 94, 4109–4112.

Stack, C., Jainuddin, S., Elipenahli, C., Gerges, M., Starkova, N., et al., 2014. Methylene blue upregulates Nrf2/ARE genes and prevents tau-related neurotoxicity. Hum. Mol. Genet. 23 (14), 3716–3732.

Sudduth, T.L., Greenstein, A., Wilcock, D.M., 2013. Intracranial injection of Gammagard, a human IVIg, modulates the inflammatory response of the brain and lowers Aβ in APP/PS1 mice along a different time course than anti-Aβ antibodies. J. Neurosci. 33, 9684–9692.

Szabo, P., Mujalli, D.M., Rotondi, M.L., Sharma, R., Weber, A., et al., 2010. Measurement of anti-beta amyloid antibodies in human blood. J. Neuroimmunol. 227, 167–174.

Tai, H.C., Serrano-Pozo, A., Hashimoto, T., Frosch, M.P., Spires-Jones, T.L., et al., 2012. The synaptic accumulation of hyperphosphorylated tau oligomers in Alzheimer disease is associated with dysfunction of the ubiquitin-proteasome system. Am. J. Pathol. 181, 1426–1435.

Thal, D.R., Fandrich, M., 2015. Protein aggregation in Alzheimer's disease: a beta and tau and their potential roles in the pathogenesis of AD. Acta Neuropathol. 129, 163–165.

Tayeb, H.O., Murray, E.D., Price, B.H., Tarazi, F.I., 2013. Bapineuzumab and solanezumab for Alzheimer's disease: is the 'amyloid cascade hypothesis' still alive? Expert Opin. Biol. Ther. 13, 1075–1084.

Tenner, A.J., Fonseca, M.I., 2006. The double-edged flower: roles of complement protein C1q in neurodegenerative diseases. Adv. Exp. Med. Biol. 586, 153–176.

Tepper, K., Biernat, J., Kumar, S., Wegmann, S., Timm, T., et al., 2014. Oligomer formation of tau protein hyperphosphorylated in cells. J. Biol. Chem. 289, 34389–34407.

Terry, R.D., 2000. Do neuronal inclusions kill the cell? J. Neural Transm. Suppl. 59, 91–93.

Troquier, L., Caillierez, R., Burnouf, S., Fernandez-Gomez, F.J., Grosjean, M.E., et al., 2012. Targeting phospho-Ser422 by active Tau immunotherapy in the THYTau22 mouse model: a suitable therapeutic approach. Curr. Alzheimer Res. 9, 397–405.

Ubhi, K., Masliah, E., 2011. Recent advances in the development of immunotherapies for tauopathies. Exp. Neurol. 230, 157–161.

van Doorn, P.A., Kuitwaard, K., Walgaard, C., van Koningsveld, R., Ruts, L., et al., 2010. IVIG treatment and prognosis in Guillain-Barre syndrome. J. Clin. Immunol. 1, S74–S78.

Vutskits, L., Briner, A., Klauser, P., Gascon, E., Dayer, A.G., et al., 2008. Adverse effects of methylene blue on the central nervous system. Anesthesiology 108, 684–692.

Wen, Y., Li, W., Poteet, E.C., Xie, L., Tan, C., et al., 2011. Alternative mitochondrial electron transfer as a novel strategy for neuroprotection. J. Biol. Chem. 286, 16504–16515.

Wilcock, D.M., Colton, C.A., 2008. Anti-amyloid-beta immunotherapy in Alzheimer's disease: relevance of transgenic mouse studies to clinical trials. J. Alzheimers Dis. 15, 555–569.

Winblad, B., Andreasen, N., Minthon, L., Floesser, A., Imbert, G., et al., 2012. Safety, tolerability, and antibody response of active Abeta immunotherapy with CAD106 in patients with Alzheimer's disease: randomised, double-blind, placebo-controlled, first-in-human study. Lancet Neurol. 11, 597–604.

Wiessner, C., Wiederhold, K.H., Tissot, A.C., Frey, P., Danner, S., et al., 2011. The second-generation active Aβ immunotherapy CAD106 reduces amyloid accumulation in APP transgenic mice while minimizing potential side effects. J. Neurosci. 31, 9323–9331.

Wischik, C.M., Staff, R.T., Wischik, D.J., Bentham, P., Murray, A.D., et al., 2015. Tau aggregation inhibitor therapy: an exploratory phase 2 study in mild or moderate Alzheimer's disease. J. Alzheimers Dis. 44, 705–720.

Wisniewski, T., 2012. Active immunotherapy for Alzheimer's disease. Lancet Neurol. 11, 571–572.

Wisniewski, T., Goni, F., 2014. Immunotherapy for Alzheimer's disease. Biochem. Pharmacol. 88, 499–507.

Xie, L., Li, W., Winters, A., Yuan, F., Jin, K., Yang, S.-H., 2013. Methylene blue induces macroautophagy through 5′ adenosine monophosphate-activated protein kinase pathway to protect neurons from serum deprivation. Front. Cell. Neurosci. 7, 56.

Yamada, K., Yabuki, C., Seubert, P., Schenk, D., Hori, Y., et al., 2009. Abeta immunotherapy: intracerebral sequestration of Abeta by an anti-Abeta monoclonal antibody 266 with high affinity to soluble Abeta. J. Neurosci. 29, 11393–11398.

Yamazaki, M., Hasegawa, M., Mori, O., Murayama, S., Tsuchiya, K., et al., 2005. Tau-positive fine granules in the cerebral white matter: a novel finding among the tauopathies exclusive to parkinsonism-dementia complex of Guam. J. Neuropathol. Exp. Neurol. 64, 839–846.

Zago, W., Buttini, M., Comery, T.A., Nishioka, C., Gardai, S.J., et al., 2012. Neutralization of soluble, synaptotoxic amyloid β species by antibodies is epitope specific. J. Neurosci. 32, 2696–2702.

Zakaria, A., Hamdi, N., Abdel-Kader, R.M., 2016. Methylene blue improves brain mitochondrial ABAD functions and decreases Aβ in a neuroinflammatory Alzheimer's disease mouse model. Mol. Neurobiol. 53, 1220–1228.

CHAPTER

10

Perspective, Summary, and Directions for Future Research on Alzheimer's Disease

INTRODUCTION

Alzheimer's disease (AD) is the most common form of dementia. The risk of AD includes age, sedentary lifestyle, genes, environmental factors, and epigenetic factors. AD is characterized by two of main pathological features, namely, amyloid plaques, which are composed of β-amyloid (Aβ), and neurofibrillary tangles (NFTs), which are formed from hyperphosphorylated Tau protein. The progression of AD pathology is also accompanied by a substantial neurodegeneration and loss of synapses. Early events in AD include alterations in insulin and insulinlike growth factor signaling in the brain, mitochondrial dysfunction, and Tau hyperphosphorylation. These processes contribute to the loss of synapse in brains of patients with AD, an important pathological feature that best correlates with cognitive impairment (Farooqui, 2010, 2016; Pedrós et al., 2014).

Neurochemical Aspects of Alzheimer's Disease
http://dx.doi.org/10.1016/B978-0-12-809937-7.00010-0

361

Indeed, AD pathology includes a substantial decrease in the number of dendritic spines (Penzes et al., 2011; Knobloch and Mansuy, 2008), which are thought to be a structural correlate of learning and memory (Alvarez and Sabatini, 2007). Late events in the pathophysiology of AD include Aβ plaque formation and impaired glucose and insulin tolerance (Farooqui, 2010, 2016; Pedrós et al., 2014). These processes may also contribute to cognitive impairment and amyloid plaque load in patients with AD (Nelson et al., 2012). Collectively, these studies suggest that the pathogenesis of AD is not only multifactorial and progressive but also complex and irreversible. It is driven by metabolic (type 2 diabetes, hypertension, and neuroinflammation), environmental (exposure to neurotoxins), and genetic factors [APOE, presenilin (PSEN) 1, and PSEN2] as well as by traumatic brain injury (TBI) (Shinohara et al., 2014). Although most of the AD cases (93%) are sporadic with an obscure cause, some forms of AD are inherited (7%) and mutations in genes coding for amyloid precursor protein (APP; chromosome 21), PSEN1 (chromosome 14), and PSEN2 (chromosome 1) have been implicated in familial forms of the disease (Bateman et al., 2012). AD produces progressive memory loss with difficulty in performing daily activities, lack of coordination, social withdrawal, vision problems, and poor judgment. In case of sporadic AD (SAD), a significant number of AD cases (approximately 25%) are carriers of the e4 allele of the APOE gene (apolipoprotein E; chromosome 19). This protein transports lipids transport protein. The exact mechanisms through which APOE contributes to increase in senile plaques and tangles is currently unknown (Haass et al., 2012; Mucke and Selkoe, 2012). Accumulation of plaques and tangles alone does not explain the pathogenesis of AD. About 30% people older than 70 years have elevated Aβ levels and are cognitively normal (Dickson et al., 1992; van Duinen et al., 1987). The National Institute on Aging and the Alzheimer's Association have revised the criteria and guidelines that will better identify the disease progression from its preclinical stage to the clinical stage (Sperling et al., 2011). It is stated that neuronal loss in AD starts much earlier (7–10 years) than the manifestation of the symptoms and clinical diagnosis. As the baby boomer generation ages, the cost of caring for patients with AD is expected to increase markedly in the near future. So, novel therapeutic regimens that delay the onset of AD are highly desirable. As stated in Chapter 2, increase in age, Western diet, sedentary lifestyle, family history (genes), TBI, and epigenetics are important risk factors for AD (Fig. 10.1). Additional risk factors include heart disease, vascular abnormalities, diabetes, stroke, hypertension, and high cholesterol (Farooqui, 2010, 2016). The incidence of AD increases exponentially from about 1% in 65-year-old people to 30% in all 85-year-old people. With the current global increase in average life span, approximately 115 million people are estimated to be suffering from AD in 2050. Besides the tremendous suffering in affected

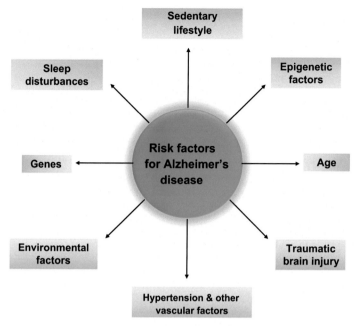

FIGURE 10.1   Risk factors for Alzheimer's disease.

individuals and their close relatives, the cost for the society is estimated to be more than US $200 billion in 2013 and US $1200 billion by 2050 in the United States alone, a number that will put the health-care system under enormous strain (World Health Organization, 2012; Alzheimer's Association, 2013; World Population Review, 2013; World Alzheimer Report, 2013, 2014). It has also been stated in earlier chapters that there are multiple interconnected mechanisms that contribute to the pathogenesis of AD, such as increase in oxidative stress and neuroinflammation, decrease in autophagy, and accumulation of Aβ peptide and hyperphosphorylated Tau protein, along with impaired neuronal metabolism and mitochondrial dysfunction (Farooqui, 2010, 2016).

β-AMYLOID-INDUCED NEUROTOXICITY IN ALZHEIMER'S DISEASE

As stated in Chapter 1, Aβ is derived from APP, which is cleaved by β-secretase (BACE1) at the N-terminal of an Aβ sequence to form a 99-amino-acid fragment C99. This peptide is subsequently cleaved by γ-secretase producing an Aβ fragment and APP intracellular domain (Hardy and Higgins, 1992; Hardy and Selkoe, 2002; De Strooper, 2010). This process produces Aβ consisting of 36–43 amino acids; Aβ40 is the most abundant species (90% of the total Aβ peptide) in normal and AD brains followed by Aβ42 (Santos et al., 2011). Under physiological conditions, Aβ is involved in a variety of important functions in healthy subjects (Lahiri and Maloney, 2010). These functions include activation of kinases, regulation of cholesterol transport, mediation of synaptic plasticity, and proinflammatory activity (Fig. 10.2) (Tabaton et al., 2010; Baruch-Suchodolsky and Fischer, 2009; Igbavboa et al., 2009; Soscia et al., 2010). In addition to its deposition in and around neural elements of the brain, Aβ promotes the maintenance of the integrity of cerebral vascular membranes. Any disruptions in this function may underlie cerebral amyloidal angiopathy (Yamada and Naiki, 2012). In AD, three distinct pools of Aβ species have been reported to occur under in vivo and in vitro conditions (Aβ monomers, soluble Aβ oligomers, and insoluble fibrillar Aβ). Each of these pool is composed of an array of individual species. Thus monomeric Aβ peptides are composed of various isoforms, including Aβ40, Aβ42, and Aβ43, as well as numerous N-terminal truncated isoforms (Tekirian et al., 1999). Insoluble fibrillar Aβ aggregates are also known to be heterogeneous in structure and composed of various Aβ isoforms, both full-length as well as N- and C-terminal truncated isoforms (Roher et al., 1993; Wang et al., 1999). Aβ also undergoes oligomerization and forms Aβ-derived diffusible ligands (ADDLs). The size of ADDLs ranges from tetramers to dodecamers. ADDL is considered as an initiator of AD not only due to inducing synaptic loss and progressive cognitive decline but also by mediating the development of Tau pathology and synaptic dysfunction (Tu et al., 2014; Klein et al., 2001; Viola and Klein, 2015; Selkoe, 2008; Bloom, 2014). In axons, ADDLs impair transport of cargoes such as mitochondria and

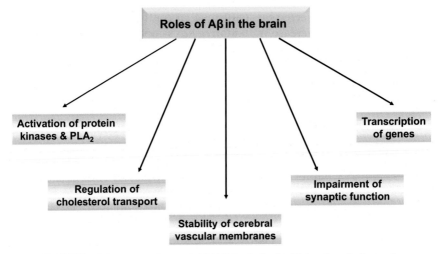

FIGURE 10.2   Roles of β-amyloid (Aβ) in the brain. *PLA₂*, phospholipase A₂.

vesicles containing brain-derived neurotrophic factor (BDNF) (Decker et al., 2010), which are required for the maintenance neuronal functions. Mitochondria are needed at presynaptic boutons to maintain neurotransmission by producing ATP and buffering synaptic calcium (Ca^{2+}) (Sheng, 2014). Once secreted from axon terminals, BDNF increases spine density and the proportion of mature spines by interacting with postsynaptic TrkB receptors at the target cell membrane (Luine and Frankfurt, 2012). Thus impairment in transport of mitochondria and BDNF may contribute to synaptic dysfunction in AD (Scharfman and Chao, 2013). In addition, at very low concentrations (nanomolar), ADDL not only inhibits long-term potentiation in brain slices and produces dendritic spine retraction from pyramidal cells but also impairs rodent spatial memory (Fig. 10.3) (Lacor et al., 2004; Selkoe, 2008). The buildup of high amounts of ADDLs induces neuroinflammation, causes oxidative damage, and negatively influences multiple signal transduction pathways including the activation of glycogen synthase kinase-3β (GSK-3β) (Muyllaert et al., 2008), a pivotal kinase that not only plays an important role in AD but also is associated with memory consolidation. For instance, activation of GSK-3β inhibits adult hippocampal neurogenesis and promotes neuroinflammation and apoptosis (Llorens-Martín et al., 2014). Furthermore, its upregulation leads to phosphorylation of the APP and Tau; accumulation of both these proteins is associated with the pathogenesis of AD (Avila et al., 2012). In addition, ADDLs have the ability to cause neurodegeneration in hippocampal neurons at nanomolar concentrations. Furthermore, increased exposure of ADDLs accelerates neural cell dysfunction and death, not only through binding of ADDLs to the cellular prion protein (PrP^C), which activates

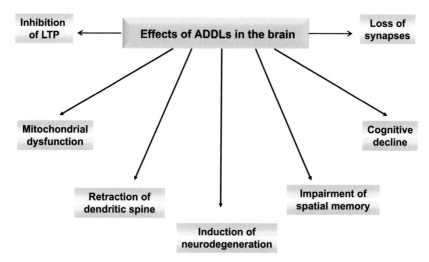

FIGURE 10.3 Effects of β-amyloid (Aβ)-derived diffusible ligands (ADDLs) in the brain.

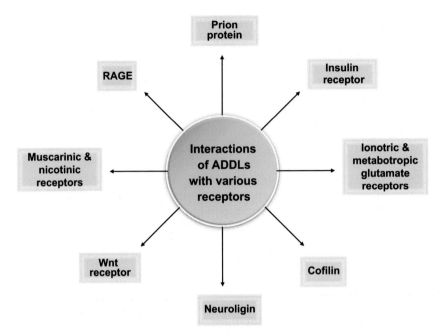

FIGURE 10.4   The binding of Aβ-derived diffusible ligands (ADDLs) with various receptors and proteins.

the protein tyrosine kinase Fyn, but also through the synaptic dysfunction induced by ADDLs. This process causes perturbations in insulin signaling (Zhao et al., 2008). In addition, ADDLs interact with a variety of receptors (Fig. 10.4). It is proposed that ADDL binding with its receptors triggers a redistribution of critical synaptic proteins and induces hyperactivity in metabotropic and ionotropic glutamate receptors. This results in Ca^{2+} influx and overload and investigates major facets of AD neuropathology, including Tau hyperphosphorylation, insulin resistance, oxidative stress, neuroinflammation, and synapse loss (Ong et al., 2013; Viola and Klein, 2015). A careful analysis of oligomeric toxicity of ADDL is required because much of the aforementioned information has been obtained in different laboratories using sodium dodecyl sulfate (SDS)-polyacrylamide gel electrophoresis under varying conditions. This procedure is not a reliable method for analyzing amyloid oligomers. SDS has been reported to artificially induce Aβ aggregation and conformational changes (Bitan et al., 2005). Memory loss at the early stage of AD may be partly due to the synaptic dysfunction induced by amyloid oligomers, which cause perturbations in insulin signaling (Zhao et al., 2008; Zhao and Alkon, 2001).

As stated in Chapter 1, interactions of ADDLs with the PrP^C activates Src family of kinases (Fyn), resulting in the disruption of synaptic plasticity (Barry et al., 2011; Um et al., 2012). ADDLs that are isolated from

AD brain also interact with frizzled receptors (Fz receptors) leading to inhibition of Wnt signaling, which in turn causes Tau phosphorylation and promotes the formation of NFTs (Kayed and Lasagna-Reeves, 2013). ADDLs not only induce oxidative stress and endoplasmic reticulum (ER) stress but also promote calcium influx and Tau phosphorylation leading to the sensitization of neurons to excitotoxicity (Pei and Hugon, 2008; Chaudhari et al., 2014). Although these findings underpin the amyloid cascade hypothesis, it only accounts for less than 2%–5% of AD cases. Importantly, most information supporting the amyloid cascade hypothesis comes mainly from studies using animal models of autosomal-dominant Alzheimer's disease (ADAD). There are important limitations to Aβ cascade hypothesis. First, a direct correlation between ADDL and dementia severity has not been clearly established, because some patients with AD without amyloid deposition display severe memory deficits, whereas other patients with cortical Aβ deposits have no dementia symptoms (Aizenstein et al., 2008). These observations along with the failure of anti-Aβ therapies (see Chapter 9) to preserve or rescue cognitive function suggests that Aβ may not be universally neurotoxic (Extance, 2010; Robakis, 2011; Tayeb et al., 2013), but other mechanisms directly or indirectly related to ADDL may contribute to the pathogenesis of AD (Bourgeat et al., 2010; Farooqui, 2016). These mechanisms include activation of kinases (Tabaton et al., 2010) and increase in levels of glycerophospholipid-derived lipid mediators and increased levels of cholesterol (Farooqui, 2011). Second important point against Aβ cascade hypothesis is that it is very difficult to prepare and study aggregation and molarity of each ADDLs in AD pathophysiological experiments under in vitro and in vivo conditions (Puzzo and Arancio, 2013). This is due to the conformational changes that occur in ADDLs, when they are dissolved in water or buffer. Moreover, ADDLs stick to the test tubes, altering the final concentration of the Aβ solution. Furthermore, behavioral abnormalities observed in animal models overexpressing APP on which Aβ cascade hypothesis is based cannot unambiguously be assigned to ADDLs because APP is metabolized to a large number of APP-derived fragments some of which are neurotoxic and others are not. ADDLs have been isolated not only from extracts of postmortem AD brain tissue but also from transgenic AD animal models (Gong et al., 2003). Importantly, studies suggest a correlation between cerebrospinal fluid (CSF) levels of ADDLs and cognitive deficits in human patients with AD (Savage et al., 2014). These findings support the view that ADDLs interfere acutely with normal synaptic functions and contribute significantly to the memory loss and cognitive dysfunction characteristic of AD. Neuroimaging studies indicate that a sizable proportion of patients clinically diagnosed with AD do not display the accumulation of Aβ, even though the process of neurodegeneration is in progress (Jack et al., 2013; Hyman et al., 2012). Remarkably, rather than concluding that

Aβ status is not a reliable marker for the early stages of clinical AD, a consensus has been reached in which patients with clinically diagnosed AD without Aβ are classified as not suffering from AD. These ideas are not scientifically warranted for younger researchers, as there is no evidence to assume that clinical AD cases with and without Aβ accumulation are etiologically different. Furthermore, in the absence of correlations between Aβ deposits and cognition, detection in normal individuals of Aβ loads similar to AD and animal models with behavioral abnormalities independent of Aβ are inconsistent with this theory (Robakis, 2011). Thus at present, evidence supporting the view that AD is caused by ADDLs is rather elusive (Giliberto et al., 2010). So more studies are required on the validity of Aβ cascade hypothesis for the pathogenesis of SAD. Studies have also indicated that Aβ42 oligomer can enter into the nucleus through channels like Aβ pores. The nuclear localization of internalized Aβ42 peptides has not only been confirmed by both confocal and transmission electron microscopy but also by chromatin immunoprecipitation assay (Barucker et al., 2014). In the nucleus, Aβ42 specifically interacts with the *LRP1* and *KAI1* promoter. At higher concentrations, Aβ42 acts as a repressor of transcription of genes *LRP1* and *KAI1*. These results have been confirmed by quantification of the messenger RNA (mRNA) levels in the examined candidate genes by quantitative reverse transcription polymerase chain reaction. Furthermore, it is also reported that Aβ42 increases the transcription of its own precursor gene *APP* suggesting that in the nucleus, Aβ can modulate gene transcription (Barucker et al., 2014).

Neuroinflammation (elevation in proinflammatory lipid mediators, cytokines, and chemokines) is an important component of AD. Aβ oligomers have been implicated in initiating the inflammatory processes (Minter et al., 2016). Aβ oligomers promote the maturation of infiltrating peripheral monocytes into phagocytic microglia, whereas monomeric Aβ or fibrillar Aβ produce no effect (Crouse et al., 2009). Aβ oligomers not only interact with many membrane proteins at the synapse (Dinamarca et al., 2012) but also induce changes in astrocytes and microglial cell metabolism (Walker et al., 2015). Many of the synaptic receptors increase the Ca^{2+} concentration in the postsynapse leading to neuroinflammation and cell death. These receptors include N-methyl-D-aspartate (NMDA) and mGluR receptors (Dinamarca et al., 2012). In addition to aberrant Ca^{2+} concentrations, Aβ oligomers have been reported to bind with receptors for advanced glycation end products (RAGEs) which are situated on microglial cells. These interactions lead to the activation of proinflammatory pathways and are found to be elevated in AD (Walker et al., 2015). RAGE is also implicated in hypertension, cardiovascular diseases, diabetes, stroke, and TBI, which are all known risk factors for the development of AD. These risk factors are all linked to RAGE and Aβ production through its many ligands (Matrone et al., 2015). In addition, there is mounting evidence supporting the view that components of the blood–brain barrier (BBB) can respond and potentiate the neuroinflammatory cycle. The BBB is a unique anatomical entity, which is essential for maintaining homeostasis in the human brain parenchymal microenvironment (Abbott et al., 2010). Cerebral endothelial cells and astrocytes are among the key players in the human brain inflammatory responses, initiated by inflammatory events in the brain's environment. As stated in Chapter 6, astrocytes are complex, highly differentiated cells that are present throughout the entire central nervous system (CNS) and may be activated in response to a wide variety of inflammatory stimuli changing their morphology and molecular expression accordingly (Abbott et al., 2010; Sofroniew, 2009). The cells of the BBB are highly responsive to the neuroinflammatory processes and can be modulated by neuroinflammatory mediators (proinflammatory eicosanoids, cytokines, and chemokines) of both the systemic nervous system and CNS. A major consequence of this neuroinflammation is the loss of barrier integrity. In AD, many proinflammatory mediators such as tumor necrosis factor (TNF)-α or interleukin (IL)-1β cause loss of "tightness" that increases BBB permeability (Abbott, 2000; Farooqui, 2014). This increase in BBB permeability allows immune cells to enter the brain parenchyma and worsen pathology. As stated earlier, there is a link between Aβ, cytokine release, the BBB, and AD progression. For example, Aβ not only activates the inflammatory Toll-like receptors (Reed-Geaghan et al., 2009) and RAGE (Ramasamy et al., 2009; Farooqui, 2014) but also binds with the complement factor C1 activating the classical complement pathway of cytotoxicity (Rogers et al., 1992; Farooqui, 2014) and leading to cytokine secretion.

TAU-INDUCED NEUROTOXICITY IN ALZHEIMER'S DISEASE

Tau is a major protein in the brain and paired helical filament-Tau is phosphorylated at several serine and threonine residues. The presence of abnormally phosphorylated Tau has been reported in selected subcortical areas long before its presence in the cerebral cortex indicating that diseaselike phosphorylation occurs already in early, preclinical states of AD (Attems et al., 2012). In addition, Tau also undergoes several posttranslational modifications including O-glycosylation, ubiquitination, methylation, and acetylation (Morris et al., 2015). Acetylation of Tau at Lys174 has been identified as an early modification in brains from patients with AD. This acetylation not only slows down Tau turnover in a mouse model but also promotes Tau aggregation (Min et al., 2015). The proteolytic cleavage of Tau is relevant to the development and progression of AD because it may generate neurotoxic fragments (Lang et al., 2014; Park and Ferreira, 2005). The primary role of Tau protein in neurons is the regulation and stabilization of microtubule dynamics (Binder et al., 2005; Lee et al., 2001). As stated in Chapters 1 and 9, there is a dynamic relationship between Tau and microtubule proteins. This relationship is driven by the phosphorylation state of Tau, which is regulated by a variety of kinases and phosphatases (Blennow et al., 2010). The onset of AD is accompanied by increase in hyperphosphorylation of Tau (Jack and Holtzman, 2013; Tepper et al., 2014). Hyperphorsphorylated Tau does not bind with microtubule proteins (Iqbal et al., 2013) leading to higher cytosolic concentrations of unbound Tau. At high concentrations, hyperphosphorylated Tau has tendency to aggregate, binding with proteins, and undergo misfolding (Ballatore et al., 2007). Based on these properties, it is proposed that oligomeric hyperphosphorylated Tau produces neurodegeneration in patients with AD (Fig. 10.5) (Lasagna-Reeves et al., 2011, 2012a). Moreover, injections of Tau oligomers (isolated from AD brains) into wild-type mice disrupt memory and propagate

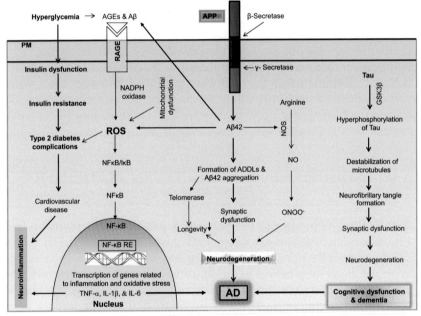

FIGURE 10.5   Molecular mechanisms contributing to neurodegeneration in Alzheimer's disease. *Aβ*, β-amyloid; *AD*, Alzheimer's disease; *AGEs*, advanced glycation end products; *APP*, amyloid precursor protein; *GSK-3β*, glycogen synthase kinase 3β; *IL-1β*, interleukin-1β; *IL-6*, interleukin-6; *NADP oxidase*, nicotinamide adenine dinucleotide phosphate oxidase; *NF-κB*, nuclear factor-κB; *NF-κB-RE*, nuclear factor-κB response element; *NFT*, neurofibrillary tangle; *NO*, nitric oxide; *NOS*, nitric oxide synthase; *ONOO⁻*, peroxynitrite; *PM*, plasma membrane; *RAGE*, receptor for advanced glycation end products; *ROS*, reactive oxygen species; *TNF-α*, tumor necrosis factor-alpha.

abnormal Tau conformation of endogenous Tau after prolonged incubation (Lasagna-Reeves et al., 2012b). The molecular mechanism associated with oligomeric hyperphosphorylated Tau-mediated neurodegeneration is not fully understood. However, it is demonstrated that the accumulation of Tau oligomers at the synapse may be critical for neurodegeneration. Using recombinant Tau oligomers, it is shown that Tau oligomers display amnesic effects and synaptic dysfunction when administered intracranially to wild-type mice (Lasagna-Reeves et al., 2011; 2012b). It is suggested that the redistribution of pathological Tau from the axon to the cell body and dendrites may contribute to spine loss observed in AD (Zempel et al., 2010). At the molecular level, Aβ either binds preferentially to neuronal dendrites promoting Tau missorting (Zempel and Mandelkow, 2012) or induces Tau pathology by directly interacting in a prionlike manner. In vitro assays have shown that Aβ oligomers can seed Tau oligomerization, providing evidence that this phenomenon may occur in vivo (Lasagna-Reeves et al., 2010). The induction of Tau misfolding in transgenic mice overexpressing APP (Castillo-Carranza et al., 2015) and mice infused with brain extract from aged APP23 transgenic mice (Bolmont et al., 2007) suggest that Aβ can seed Tau oligomerization in vivo as well. Collectively, these studies support the view that a direct interaction between Aβ and oligomeric Tau may be involved in the induction of synaptic dysfunction as Tau and Aβ coexist within synaptic compartments in AD brain (Zempel et al., 2010; Tai et al., 2012; Miller et al., 2014). However, the complexity and diversity of Aβ aggregates have made it difficult to elucidate the chemical nature of interactions between the two proteins. The molecular mechanisms associated with aforementioned processes are not fully understood. However, it is proposed that Tau-mediated destabilization of microtubule arrests anterograde axonal transportation contributing to the neuronal damage and loss of neuronal function along with neurodegeneration (LaPointe et al., 2009). It still remains unclear whether the synaptotoxic and neurotoxic actions of Aβ are separate mechanistic processes or if their actions follow a common molecular mechanism (Cappai and Barnham, 2008). As stated in Chapter 1, Aβ monomers have very low toxicity and Aβ toxicity rises substantially after ADDL formation (Jin et al., 2011). However, it is still challenging to establish a relationship between the degree of increasing toxicity and formation of oligomers not only because of the lack characterization of ADDLs in vivo in human brain but also due to the decrease in Aβ neurotoxicity at higher Aβ oligomers.

In some animal models, which overexpress Tau, the onset of neurodegeneration occurs in the absence of overt NFT pathology (Andorfer et al., 2005). Memory function and neuronal loss can be restored in a tauopathy mouse model despite the ongoing accumulation of NFTs (Santacruz et al., 2005). Moreover, persistent occurrence of NFTs in neurons for 20–30 years makes them unlikely candidates for inducing immediate neurotoxicity (Kordower et al., 2001; Guillozet-Bongaarts et al., 2006). In fact, a large immunohistochemical study on cholinergic basal forebrain neurons in the nucleus basalis using an early Tau marker showing pretangled neurons and neutrophil thread staining correlates extremely well with cognitive decline, which occurs prior to the emergence of significant NFT pathology (Vana et al., 2011). Finally, the loss of synapses correlates better with cognitive decline than NFTs, again supporting the possibility of a different mechanism for Tau neurotoxicity (Crews and Masliah, 2010). These observations suggest that more studies are needed on the role of NFTs in the pathogenesis of AD. In Tg2576 mouse model, the removal of Tau oligomers by immunotherapy shifts the Aβ aggregation pathway to amyloid plaques, while improving cognition in mice (Castillo-Carranza et al., 2015). These observations may explain the presence of amyloid plaques in individuals without clinical symptoms of AD, thus termed high-pathology controls or NDAN subjects (Bjorklund et al., 2012), and in unsuccessful clinical trials even after removing amyloid plaques (Hardy, 2009).

CHALLENGES TO BASIC RESEARCH IN ALZHEIMER'S DISEASE

Other important challenges that are faced by investigators working on the pathogenesis and treatment of AD are the availability of ideal animal models that can replicate the pathogenesis and symptoms of AD (Farooqui, 2016). Lack of specific biomarkers to detect the pace AD progression makes it difficult to monitor the effects of therapeutic agents. Current animal models of AD have been mainly generated from ADAD genes that facilitate the onset and pathogenesis of AD process so that the pathological changes and memory deficits typical of AD can be observed at a younger age. However, age is an important risk factor for AD, especially in SAD or late-onset AD, which is much more prevalent among patients with AD than in those with early-onset AD (Farooqui, 2016). Discovery of specific biomarker for the detection and progression of AD in the CSF is an important area of research. Multimodal biomarkers are needed for the detection and progression of AD. The prediction of AD progression should not only be closely related to the clinical utility, which varies across the different stages of the disease, but also with the cognitive dysfunction in patients with AD. To this end, investigators are focusing their research on combination of genetic, biochemical, and imaging biomarkers [amyloid positron emission tomography (PET), fluorodeoxyglucose PET, and magnetic resonance imaging] as well as measures of cognitive function that may provide for improved diagnostic accuracy at different stages of the disease. Although a number of genetic and CSF biomarkers have been discovered, only few studies have been performed on biomarkers in the blood and biopsy and autopsy samples of AD patients (Schreitmüller et al., 2012; Hampel et al., 2014). At present, we are empowered with technological advances in lipidomics, proteomics, and genomics. Using these techniques, researchers are performing analysis of hundreds to thousands of mRNAs, proteins, and genes simultaneously; dissecting signaling pathways; determining levels of lipid mediators; characterizing enzyme proteins; and cloning genes from cells in biological fluids, such as CSF and brain extracellular fluid (ECF), mostly obtained by cerebral microdialysis (Maurer, 2010). Both materials have high diagnostic value in clinical neurology (Jaeger et al., 2016). CSF and ECF samples from patients with AD can be analyzed for lipid biomarkers, protein variants, and genes associated with them. This information can be used for mapping the disease-specific molecular networks related to AD. Lipidomics can detect minute amounts of lipid mediators in biological fluids (German et al., 2007). Proteomics can be used to characterize various proteins in biopsy, autopsy, and CSF samples not only from animal models of AD but also in samples from patients with AD. Combining lipidomics and proteomics will not only enhance existing knowledge of disease pathology but also increase the likelihood of discovering specific markers associated with AD. Studies on AD biomarker profiling in the blood have indicated alterations in 18 plasma proteins in patients with AD compared to healthy subjects with high specificity. However, other studies have not confirmed these results (Patel et al., 2011; Ray et al., 2007). Thus more studies are required on this important topic in blood of patients with AD.

Healthy brain astrocytes do not express β-secretase. Therefore astrocytes are not capable of producing Aβ. In contrast, exposure of astrocytes to chronic stress results in expression of β-secretase leading to the production of Aβ (Rossner et al., 2005). This process induces glial activation. Astroglial expression of β-secretase has been detected in various mice models of AD (Rossner et al., 2001; Heneka et al., 2005) suggesting that astrocytes can contribute to the pathogenesis of AD. Changes in astrocytes in aging and AD are highly heterogeneous and region specific. In animal models of AD, astrocytes undergo neurodegeneration and atrophy at the early stages of pathological progression. These processes may alter the homeostatic reserve of the brain and promote early cognitive deficits. At later stages of AD, reactive astrocytes are associated with neurite plaques, a feature commonly found in animal models of AD and in patients with AD. In animal models of AD, the onset of reactive astrogliosis occurs in the hippocampus, but not in other brain regions, supporting the view that hippocampus may be closely associated with the pathogenesis of AD. It should be noted that in entorhinal and prefrontal cortices, astrocytes do not mount gliotic response to emerging Aβ deposits. Thus, 3xTG-AD mice respond to Aβ-mediated astrocytic changes in the hippocampus, but not in the prefrontal or entorhinal cortices (Yeh et al., 2011; Kulijewicz-Nawrot et al., 2012). This deficit of astrocytic response is correlated with the absence of Aβ-mediated remodeling of Ca^{2+} signaling in the entorhinal astrocytes. At the same time exposure of hippocampal astrocytes to Aβ substantially upregulates the expression of main components of Ca^{2+} signaling tool kit [plasmalemmal glutamate metabotropic receptors and inositol 1,4,5-trisphosphate receptors of the ER (Grolla et al., 2013)], which further supports the role of hippocampus in the pathogenesis of AD. These studies also suggest that astroglial morphology and function can be regulated through environmental stimulation and/or medication. Thus astrocytes can be regarded as a target for therapies aimed at the prevention and cure of AD.

Another important challenge in AD research today is which Aβ oligomer species trigger the Aβ cascade. In addition to amyloid plaques, which contain precipitates of fibrillar Aβ, several types of soluble Aβ assemblies ("oligomers") have been described in the brains of patients with AD and transgenic mouse models of AD (Verkhratsky et al., 2016). Thus determining the nature of oligomer is important in understanding the pathogenesis of AD. Studies have described the isolation of Aβ dimers from the cortex of patients with AD. At subnanomolar concentration, these Aβ dimers increase the hyperphosphorylation of Tau, followed by microtubule cytoskeleton disruption and neuritic degeneration (Ahn et al., 2010; Vossel et al., 2010). Thus the development of reliable assays for Aβ (both monomers and oligomers) will be important for the early differential diagnosis of dementia, predicting the progression of AD, as well as monitoring the effectiveness of novel anti-Aβ drugs for AD (Zhou et al., 2016). Is inhibition of ADDL formation sufficient to delay or stop the progression of AD or do researchers have to investigate factors that modulate the progression of AD along with multifactorial treatments? Immunotherapy targeting Aβ oligomers is specifically effective in rodents but not in humans. The multifactorial nature of AD pathogenesis suggests that treatment with multiple therapeutic agents will be needed to stop AD progression. Additionally, information is needed on which process appears first, induction of neuroinflammation or formation of Aβ oligomers. Answering this question may prove to be difficult due to the potential for positive feedback in the neuroinflammatory pathway but may have a profound impact on how one views the pathogenesis of AD. Understanding the sequence of events of these processes may provide insight into the mechanisms by which the pathology of AD spreads. For example, does chronic inflammation, which may appear in patients with AD due to long-term consumption of Western diet, lead to the initial formation of Aβ oligomer seeds spreading the AD pathology? Or perhaps as the Aβ oligomers spread and seed aggregation in new regions, they may induce inflammation and accelerate degeneration. These questions are not only relevant for Aβ oligomers but also for other amyloids, especially Tau. Tau oligomers are known to spread locally and via synaptic connections (Boluda et al., 2015; Ahmed et al., 2014; Asai et al., 2015; Farooqui, 2014).

It is also reported that Aβ is present in neurons in infant brains. It increases up to 8 years of age, a period of intense myelination and high brain plasticity. In this period, about half of the neurons are Aβ-immunopositive (Puzzo et al., 2015). In adulthood, Aβ is present in the major part of the neurons, whereas in aged people, there is a 20% reduction. What role if any Aβ plays in the infant brain is not known. Furthermore, the deposition of Aβ in the brain only requires Aβ seeding and the age of animal is not important (Caroli et al., 2015). However, if the age of the animal is not important, then what are other factors that contribute to Aβ seed formation? Only the answers to these critical questions will open a pathway for the treatment of AD. Now it is becoming more and more clear that a rise of Aβ concentrations is neither inevitable nor a natural consequence of aging. Rather other factors such as exercise, diet, and sleep may play important roles.

Supporters of Aβ cascade hypothesis still believe that Aβ has a pathogenic role in AD. However, the mechanisms contributing to the pathogenesis of AD remain poorly understood. In addition, Aβ aggregates activate microglial cells and astrocytes. This activation induces the production of nitric oxide (NO), reactive oxygen species, proinflammatory cytokines (TNF-α, IL-1β, IL-6), chemokines (e.g., IL-18), and prostaglandins (prostaglandin E_2). These metabolites promote neuronal death, along with more oxidative stress and neuroinflammation. Neuroinflammation is essential for the recovery from a neuronal injury, but it may also play detrimental roles in neurodegenerative processes (Farooqui, 2014). Another system that is linked with the production of Aβ is the RAGE. Pro-oxidant environments, such as an inflammatory milieu, the activation of RAGE by Aβ peptide, as well as Aβ oligomers on the surface of microglia, astrocytes, vascular endothelial cells, and neurons results in activation of microglia and astrocytes. Blocking the interaction of Aβ with RAGE impairs the activation of microglia and reduces the production of proinflammatory mediators (Ramasamy et al., 2009). RAGE is also suggested to play an important role in the clearance of Aβ and to be involved in APOE-mediated cellular processing and signaling (Bu, 2009). It is becoming increasingly evident that a well-regulated neuroinflammatory process is essential for tissue homeostasis and proper cellular function. However, uncontrolled and excessive neuroinflammation can be the result of either direct immunologic insults (e.g., bacteria, viruses, or their products) or a secondary reaction to neural trauma, genetic predisposition, or environmental toxins. Because CNS has a restricted cell renewal and regenerative capacity, brain is extremely vulnerable to uncontrolled autodestructive immune and neuroinflammatory processes. To date, a vast literature indicates that neuroinflammation contributes to the neuronal loss in most neurodegenerative diseases but whether or how neuroinflammation decisively affects the chronic progression of AD is largely unknown. Thus more studies are needed on the molecular mechanism of progression of neuroinflammation in AD.

As stated in Chapter 3, brain levels of cholesterol are significantly reduced in the hippocampus and cerebral cortex in patients with AD (Mulder et al., 1998). These brain areas not only contain senile plaques and tangles, the pathological hallmarks of AD, but also are enriched in synaptic membrane cholesterol in the following quantitative order: hippocampus > cortex > cerebellum (Leduc et al., 2010; Chochina et al., 2001). However, the levels of cholesterol in the brain of patients with AD vary considerably (Wood et al., 2005). It is also reported that there is a relationship among APP processing, Tau hyperphosphorylation, and neural membrane cholesterol levels. Thus, the posttranscriptional processing of APP occurs in cholesterol-rich membrane domains (Shobab et al., 2005). Furthermore, there is an indirect relationship between cholesterol and Tau hyperphosphorylation (Rahman et al., 2005). The strongest support on the involvement of cholesterol in the pathogenesis of AD comes from studies in animal models of AD and in vitro studies using primary and immortalized cells. A majority of the animal and cell culture studies indicate that increasing cholesterol levels increases $A\beta$ abundance, whereas the opposite effects are noted when cholesterol levels are reduced (Posse de Chaves, 2012; Maulik et al., 2013; Ong et al., 2013). However, the role of cholesterol as a causative factor in the progression of AD is still debated and controversial. For example, elevated serum and plasma cholesterol levels do not appear to be risk factors for AD (Wood et al., 2005), and the brain cholesterol level that is associated with AD remains unclear (i.e., reduced cholesterol levels, increased cholesterol levels, and no changes in cholesterol levels have been observed in patients AD compared with controls) (Wood et al., 2014). However, other findings have indicated that a set of 10 lipids from the peripheral blood can predict phenoconversion to amnestic mild cognitive impairment (which is associated with a high risk for AD) or AD within a 2- to 3-year time frame with over 90% accuracy (Mapstone et al., 2014). Moreover, progressive deterioration of cholesterol homeostasis as a central component of AD pathophysiology, and use of statins and maintenance cholesterol homeostasis may retard the onset of AD (Mapstone et al., 2014; Leduc et al., 2010). Inclusion of cholesterol metabolism is an important addition to the amyloid cascade hypothesis and Tau phosphorylation studies. Solid data are needed to fully understand the role of cholesterol in the pathogenesis of AD.

As stated in Chapters 8 and 9 that potential treatments of AD are focused on several aspects of AD. These treatments are aimed to modify the known processes that contribute to the pathogenesis of AD. Current potential treatments fall into categories that target:

1. The metabolic changes in brain lipids, proteins, and DNA along with insulin resistance
2. The neuroinflammation associated with certain regions of the brain that are affected in AD
3. The production of pathological forms of Tau protein (inhibition of Tau phosphorylation) and prevention of Tau aggregation
4. The production, $A\beta$ plaque formation, and clearance of $A\beta$
5. Intranasal insulin treatment
6. Antiaging phytochemicals (resveratrol) and
7. Gene therapy

The best characterized of these approaches, and the agents that have shown the greatest promise, have been those that modify $A\beta$ levels. Anti-$A\beta$ strategies include pharmaceutical compounds with distinct mechanisms of action, namely, drugs that (1) facilitate the clearance, (2) inhibit the production, or (3) prevent the aggregation of $A\beta$ (Salomone et al., 2012). As mentioned in Chapter 8, many pharmacological compounds have been developed to tackle the "amyloid cascade," with the prospect of reducing the $A\beta$ burden in the brain of mild to moderately demented patients with AD (Jack et al., 2009). However, most of these pharmacological compounds have failed to produce beneficial effects and there is evidence that these interventions must be 5–7 years before the onset of AD. Furthermore, as stated in Chapter 9, both active and passive immunization target the reduction of intracerebral $A\beta$ load by eliciting humoral response against the $A\beta$ peptide, facilitating its clearance from the brain by immune-mediated mechanisms (Panza et al., 2011). Highly encouraging results are obtained in preclinical studies in transgenic mice. It is reported that these strategies not only prove effective in reducing the amount of $A\beta$ in the mouse brain but also promote the improvements in behavior and cognition (Morgan et al., 2000). However, immune therapy in patients with AD have failed and all human trials are halted due to the occurrence of severe adverse events in a substantial proportion of subjects (6%) who received this intervention, which was associated with the induction of a cytotoxic T-cell reaction in brain vessels leading to acute meningoencephalitis (Nicoll et al., 2003). Reason for the failure of monoclonal and polyclonal antibodies treatment in human patients is not fully understood. However, the failure may be due to the detection of the wrong conformation of $A\beta$ oligomer, namely, monomers or fibrils in patients with AD. The treatment of patients with AD with second generation of active anti-$A\beta$ immunotherapeutic agents is designed to minimize the risk of eliciting such secondary inflammatory responses or vasogenic edema, by stimulating soluble Aβ derivative immunogens. These vaccines produce the immune response to raise antibodies against Aβ monomers and oligomers. Studies with the vaccine CAD106 in phase 1 show that it can reduce Aβ accumulation in cortical and subcortical brain regions by binding to Aβ aggregates and blocking cellular toxicity, with no evidence of microhemorrhage, vasogenic edema, or inflammatory reactions subsequent to activation of T cells (Wiessner et al., 2011).

CONCLUSION

AD, the most common neurodegenerative disease in elderly population, is complex, slowly progressive, and irreversible. Neuropathologically, AD is characterized by the aberrant accumulation and deposition of misfolded proteins, in particular, extracellular neuritic plaques, which are made up of aggregated forms of neurotoxic Aβ and intracellular deposits of abnormally phosphorylated Tau protein in NFTs. AD is also characterized by an early and progressive memory loss and other cognitive and behavioral disturbances. The accumulation of Aβ and NFTs lead to a deadly cascade of impairment of axonal transport, synaptic alterations, microglial and astrocytic activation, progressive neuronal loss associated with multiple neurotransmitter deficiencies, and cognitive failure. Molecular mechanisms of neurochemical processes mentioned earlier are interrelated supporting the view that neurodegenerative processes in AD are multifactorial, which is complicated by the fact that degenerating neurons also mount prosurvival responses to protect themselves from the aforementioned neurochemical and neuropathological disease-related changes. This results in a very complex situation in which degenerating neurons go through a struggle between prodeath factors and prosurvival responses causing the disruption of normal neuronal function but eventually leading to neural cell death. Increasing evidence supports the view that cognitive impairment in SAD can occur independent of amyloid deposition, which may in fact represent a downstream result and not the cause of the disease. Although Tau pathology seems to correlate better with cognitive decline suggesting the potential value of Tau-targeted therapy, there have been no promising candidates so far.

For some years, the treatment of AD has relied on the symptomatic effects of cholinesterase inhibitors (tacrine, donepezil, rivastigmine, metrifonate, and galantamine) and NMDA receptor antagonist (memantine). Many promising compounds have been validated by experimental models as candidate disease-modifying drugs for AD; however, only a few of these appear in the pipeline of drug development or have been clinically tested by randomized controlled trials. Overall results from these trials have so far been negative. Most phase 2 and 3 trials with candidate drugs (peroxisome proliferator-activated receptors agonists, β- and γ-secretase inhibitors, n-3 fatty acids, immunization with Aβ and Tau antibodies) for AD in the past decade have failed to provide unequivocal clinical benefits, or have been suspended due to severe adverse events. So, the treatment of AD still remains a great challenge, and the development of new strategies is an active area of research. To this end, current efforts are directed toward a broad range of alternative mechanisms of AD including mitochondrial dysfunction, metabolic stress, altered insulin signaling, and cerebrovascular dysfunction. Accumulating evidence in fact supports the notion that cerebrovascular or neurovascular dysfunction may represent a primary initiator of a cascade of pathogenic events leading to neurodegeneration in AD.

References

Abbott, N.J., 2000. Inflammatory mediators and modulation of blood–brain barrier permeability. Cell Mol. Neurobiol. 20, 131–147.

Abbott, N.J., Patabendige, A.A., Dolman, D.E., Yusof, S.R., Begley, D.J., 2010. Structure and function of the blood–brain barrier. Neurobiol. Dis. 37, 13–25.

Ahn, H.J., Zamolodchikov, D., Cortes-Canteli, M., Norris, E.H., Glickman, J.F., 2010. Alzheimer's disease peptide beta-amyloid interacts with fibrinogen and induces its oligomerization. Proc. Natl. Acad. Sci. USA 107, 21812–21817.

Ahmed, Z., Cooper, J., Murray, T., Garn, K., Mcnaughton, E., et al., 2014. A novel in vivo model of tau propagation with rapid and progressive neurofibrillary tangle pathology: the pattern of spread is determined by connectivity, not proximity. Acta Neuropathol. 127, 667–683.

Aizenstein, H.J., Nebes, R.D., Saxton, J.A., Price, J.C., Mathis, C.A., Tsopelas, N.D., et al., 2008. Frequent amyloid deposition without significant cognitive impairment among the elderly. Arch. Neurol. 65, 1509–1517.

Alvarez, V.A., Sabatini, B.L., 2007. Anatomical and physiological plasticity of dendritic spines. Annu. Rev. Neurosci. 30, 79–97.

Alzheimer's Association, 2013. Alzheimer's Disease Facts and Figures. Alzheimer's Association, Chicago, IL. Available at: http://www.alz.org/alzheimers_disease_facts_and_figures.asp.

Andorfer, C., Acker, C.M., Kress, Y., Hof, P.R., Duff, K., et al., 2005. Cell-cycle reentry and cell death in transgenic mice expressing nonmutant human tau isoforms. J. Neurosci. 25, 5446–5454.

Asai, H., Ikezu, S., Tsunoda, S., Medalla, M., Luebke, J., et al., 2015. Depletion of microglia and inhibition of exosome synthesis halt tau propagation. Nat. Neurosci. 18, 1584–1593.

Attems, J., Thomas, A., Jellinger, K., 2012. Correlations between cortical and subcortical tau pathology. Neuropathol. Appl. Neurobiol. 38, 582–590.

Avila, J., León-Espinosa, G., García, E., García-Escudero, V., Hernández, F., et al., 2012. Tau phosphorylation by GSK3 in different conditions. Int. J. Alzheimers Dis. 2012, 578373.

Ballatore, C., Lee, V.M., Trojanowski, J.Q., 2007. Tau-mediated neurodegeneration in Alzheimer's disease and related disorders. Nat. Rev. Neurosci. 8, 663–672.

Barry, A.E., Klyubin, I., McDonald, J.M., Mably, A.J., Farrell, M.A., et al., 2011. Alzheimer's disease brain-derived amyloid-β-mediated inhibition of LTP in vivo is prevented by immunotargeting cellular prion protein. J. Neurosci. 31, 7259–7263.

Barucker, C., Harmeier, A., Weiske, J., Fauler, B., Albring, K.F., et al., 2014. Nuclear translocation uncovers the amyloid peptide Aβ42 as a regulator of gene transcription. J. Biol. Chem. 289, 20182–20191.

Baruch-Suchodolsky, R., Fischer, B., 2009. Abeta40, either soluble or aggregated, is a remarkably potent antioxidant in cell-free oxidative systems. Biochemistry 48, 4354–4370.

Bateman, R.J., Xiong, C., Benzinger, T.L., Fagan, A.M., Goate, A., et al., 2012. Dominantly inherited Alzheimer network. N. Engl. J. Med. 367, 795–804.

Binder, L.I., Guillozet-Bongaarts, A.L., Garcia-Sierra, F., Berry, R.W., 2005. Tau, tangles, and Alzheimer's disease. Biochim. Biophys. Acta 1739, 216–223.

Bitan, G., Fradinger, E.A., Spring, S.M., Teplow, D.B., 2005. Neurotoxic protein oligomers—what you see is not always what you get. Amyloid 12, 88–95.

Bjorklund, N.L., Sadagoparamanujam, R.L., Ghirardi, V.-M., Woltjer, V., Taglialatela, G., 2012. Absence of amyloid ββ oligomers at the postsynapse and regulated synaptic Zn^{2++} in cognitively intact aged individuals with Alzheimer's disease neuropathology. Mol. Neurodegener. 7, 23.

Blennow, K., Hampel, H., Weiner, M., Zetterberg, H., 2010. Cerebrospinal fluid and plasma biomarkers in Alzheimer disease. Nat. Rev. Neurol. 6, 131–144.

Bloom, G.S., 2014. Amyloid-β and Tau: the trigger and bullet in Alzheimer disease pathogenesis. JAMA Neurol. 71, 505–508.

Bolmont, T., Clavaguera, F., Meyer-Luehmann, M., Herzig, M.C., Radde, R., et al., 2007. Induction of tau pathology by intracerebral infusion of amyloid-beta -containing brain extract and by amyloid-beta deposition in APP x Tau transgenic mice. Am. J. Pathol. 171, 2012–2020.

Boluda, S., Iba, M., Zhang, B., Raible, K., Lee, V.-Y., et al., 2015. Differential induction and spread of tau pathology in young PS19 tau transgenic mice following intracerebral injections of pathological tau from Alzheimer's disease or corticobasal degeneration brains. Acta Neuropathol. 129, 221–237.

Bourgeat, P., Chételat, G., Villemagne, V.L., Fripp, J., Raniga, P., et al., 2010. Beta-amyloid burden in the temporal neocortex is related to hippocampal atrophy in elderly subjects without dementia. Neurology 74, 121–127.

Bu, G., 2009. Apolipoprotein E and its receptors in Alzheimer's disease: pathways, pathogenesis and therapy. Nat. Rev. Neurosci. 10, 333–344.

Cappai, R., Barnham, K.J., 2008. Delineating the mechanism of Alzheimer's disease Aβ peptide neurotoxicity. Neurochem. Res. 33, 526–532.

Caroli, A., Prestia, A., Galluzzi, S., Ferrari, C., van der Flier, W.M., et al., 2015. Mild cognitive impairment with suspected nonamyloid pathology (SNAP): prediction of progression. Neurology 84, 508–515.

Castillo-Carranza, D.L., Guerrero-Munoz, M.J., Sengupta, U., Hernandez, C., Barret, A., et al., 2015. Tau immunotherapy modulates both pathological tau and upstream amyloid pathology in an AD mouse model. J. Neurosci. 35, 4857–4868.

Chaudhari, N., Talwar, P., Parimisetty, A., Lefebvre d'Hellencourt, C., Ravanan, P., 2014. A molecular web: endoplasmic reticulum stress, inflammation, and oxidative stress. Front. Cell. Neurosci. 8, 213.

Chochina, N. V., Avdulov, N. A., Igbavboa, U., Cleary, J.P., O'Hare, E.O.et al., 2001. Amyloid beta-peptide1–40 increases neuronal membrane fluidity: role of cholesterol and brain region. J. Lipid Res 42, 1292–1297.

Crews, L., Masliah, E., 2010. Molecular mechanisms of neurodegeneration in Alzheimer's disease. Hum. Mol. Genet. 19 (R1), R12–R20.

Crouse, N.R., Ajit, D., Udan, M.L.D., Nichols, M.R., 2009. Oligomeric amyloid-β(1–42) induces THP-1 human monocyte adhesion and maturation. Brain Res. 1254, 109–119.

Decker, H., Lo, K.Y., Unger, S.M., Ferreira, S.T., Silverman, M.A., 2010. Amyloid-beta peptide oligomers disrupt axonal transport through an NMDA receptor-dependent mechanism that is mediated by glycogen synthase kinase 3beta in primary cultured hippocampal neurons. J. Neurosci. 30, 9166–9171.

De Strooper, B., 2010. Proteases and proteolysis in Alzheimer disease: a multifactorial view on the disease process. Physiol. Rev. 90, 465–494.

Dickson, D.W., Crystal, H.A., Mattiace, L.A., Masur, D.M., Blau, A.D., et al., 1992. Identification of normal and pathological aging in prospectively studied nondemented elderly humans. Neurobiol. Aging 13, 179–189.

Dinamarca, M.C., Rios, J.A., Inestrosa, N.C., 2012. Postsynaptic receptors for amyloid-beta oligomers as mediators of neuronal damage in Alzheimer's disease. Front. Physiol. 3, 464.

Extance, A., 2010. Alzheimer's failure raises questions about disease-modifying strategies. Nat. Rev. Drug Discov. 9, 749–751.

Farooqui, A.A., 2010. Neurochemical Aspects of Neurotraumatic and Neurodegenerative Diseases. Springer, New York.

Farooqui, A.A., 2014. Inflammation and Oxidative Stress in Neurological Disorders. Springer, New York.

Farooqui, A.A., 2011. Lipid Mediators and Their Metabolism in the Brain. Springer, New York.

Farooqui, A.A., 2016. Therapeutic Potentials of Curcumin for Alzheimer Disease. Springer, New York.

German, J.B., Gillies, L.A., Smilowitz, J.T., Zivkovic, A.M., Watkins, S.M., 2007. Lipidomics and lipid profiling in metabolomics. Curr. Opin. Lipidol. 18, 66–71.

Giliberto, L., d'Abramo, C., Acker, C.M., Davies, P., D'Adamio, L., 2010. Transgenic expression of the amyloid-β precursor protein-intracellular domain does not induce Alzheimer's disease–like traits in vivo. PLoS One 5, e11609.

Gong, Y., Chang, L., Viola, K.L., Lacor, P.N., Lambert, M.P., et al., 2003. Alzheimer's disease-affected brain: presence of oligomeric Aβ ligands (ADDLs) suggests a molecular basis for reversible memory loss. Proc. Natl. Acad. Sci. USA 100, 10417–10422.

Grolla, A.A., Sim, J.A., Lim, D., Rodriguez, J.J., Genazzani, A.A., et al., 2013. Amyloid-β and Alzheimer's disease type pathology differentially affects the calcium signalling toolkit in astrocytes from different brain regions. Cell Death Dis. 4, e623.

Guillozet-Bongaarts, A.L., Cahill, M.E., Cryns, V.L., Reynolds, M.R., Berry, R.W., Binder, L.I., 2006. Pseudophosphorylation of tau at serine422 inhibits caspase cleavage: in vitro evidence and implications for tangle formation in vivo. J. Neurochem. 97, 1005–1014.

Hampel, H., Lista, S., Teipel, S.J., et al., 2014. Perspective on future role of biological markers in clinical therapy trials of Alzheimer's disease: a long-range point of view beyond 2020. Biochem. Pharmacol. 88, 426–449.

Hardy, J.A., Higgins, G.A., 1992. Alzheimer's disease: the amyloid cascade hypothesis. Sci. Hypothesis 256, 184–185.

Hardy, J., Selkoe, D.J., 2002. The amyloid hypothesis of Alzheimer's disease: progress and problems on the road to therapeutics. Science 297, 353–356.

Hardy, J., 2009. The amyloid hypothesis for Alzheimer's disease: a critical reappraisal. J. Neurochem. 110, 1129–1134.

Haass, C., Kaether, C., Thinakaran, G., Sisodia, S., 2012. Trafficking and proteolytic processing of APP. Cold Spring Harb. Perspect. Med. 2, a006270.

Heneka, M.T., Sastre, M., Dumitrescu-Ozimek, L., Dewachter, I., Walter, J., et al., 2005. Focal glial activation coincides with increased BACE1 activation and precedes amyloid plaque deposition in APPV717I transgenic mice. J. Neuroinflam. 2, 22.

Hyman, B.T., Phelps, C.H., Beach, T.G., Bigio, E.H., Cairns, N.J., et al., 2012. National Institute on aging-Alzheimer's Association guidelines for the neuropathologic assessment of Alzheimer's disease. Alzheimers Dement. J. Alzheimers Assoc. 8, 1–13.

Igbavboa, U., Sun, G.Y., Weisman, G.A., He, Y., Wood, W.G., 2009. Amyloid beta-protein stimulates trafficking of cholesterol and caveolin-1 from the plasma membrane to the golgi complex in mouse primary astrocytes. Neuroscience 162, 328–338.

Iqbal, K., Gong, C.X., Liu, F., 2013. Hyperphosphorylation-induced tau oligomers. Front. Neurol. 4, 112.

Jack Jr., C.R., Lowe, V.J., Weigand, S.D., Wiste, H.J., Senjem, M.L., et al., 2009. Serial PIB and MRI in normal, mild cognitive impairment and Alzheimer's disease: implications for sequence of pathological events in Alzheimer's disease. Brain 132, 1355–1365.

Jack Jr., C.R., Holtzman, D.M., 2013. Biomarker modeling of Alzheimer's disease. Neuron 80, 1347–1358.

Jack, C.R., Wiste, H.J., Weigand, S.D., Knopman, D.S., Lowe, V., et al., 2013. Amyloid-first and neurodegeneration-first profiles characterize incident amyloid PET positivity. Neurology 81, 1732–1740.

Jaeger, P.A., Lucin, K.M., Britschgi, M., Vardarajan, B., Huang, R.P., et al., 2016. Network-driven plasma proteomics expose molecular changes in the Alzheimer's brain. Mol. Neurodegener. 11, 31.

Jin, M., Shepardson, N., Yang, T., Chen, G., Walsh, D., et al., 2011. Soluble amyloid β-protein dimers isolated from Alzheimer cortex directly induce Tau hyperphosphorylation and neuritic degeneration. Proc. Natl. Acad. Sci. USA 108, 5819–5824.

Kayed, R., Lasagna-Reeves, C.A., 2013. Molecular mechanisms of amyloid oligomers toxicity. J. Alzheimers Dis. 33, S67–S78.

Klein, W.L., Krafft, G.A., Finch, C.E., 2001. Targeting small Abeta oligomers: the solution to an Alzheimer's disease conundrum? Trends Neurosci. 24, 219–224.

Knobloch, M., Mansuy, I.M., 2008. Dendritic spine loss and synaptic alterations in Alzheimer's disease. Mol. Neurobiol. 37, 73–82.

Kordower, J.H., Chu, Y., Stebbins, G.T., DeKosky, S.T., Cochran, E.J., et al., 2001. Loss and atrophy of layer II entorhinal cortex neurons in elderly people with mild cognitive impairment. Ann. Neurol. 49, 202–213.

Kulijewicz-Nawrot, M., Verkhratsky, A., Chvatal, A., Sykova, E., Rodriguez, J.J., 2012. Astrocytic cytoskeletal atrophy in the medial prefrontal cortex of a triple transgenic mouse model of Alzheimer's disease. J. Anat. 221, 252–262.

Lacor, P.N., Buniel, M.C., Chang, L., Fernandez, S.J., Gong, Y., Viola, K.L., Lambert, M.P., et al., 2004. Synaptic targeting by Alzheimer's-related amyloid beta oligomers. J. Neurosci. 24, 10191–10200.

Lahiri, D.K., Maloney, B., 2010. Beyond the signaling effect role of amyloid-ß42 on the processing of APP, and its clinical implications. Exp. Neurol. 225, 51–54.

Lang, A.E., Riherd Methner, D.N., Ferreira, A., 2014. Neuronal degeneration, synaptic defects, and behavioral abnormalities in tau$_{45-230}$ transgenic mice. Neuroscience 275, 322–339.

LaPointe, N.E., Morfini, G., Pigino, G., Gaisina, I.N., Kozikowski, A.P., et al., 2009. The amino terminus of tau inhibits kinesin-dependent axonal transport: implications for filament toxicity. J. Neurosci. Res. 87, 440–451.

Lasagna-Reeves, C.A., Castillo-Carranza, D.L., Guerrero-Muoz, M.J., Jackson, G.R., Kayed, R., 2010. Preparation and characterization of neurotoxic tau oligomers. Biochemistry 49, 10039–10041.

Lasagna-Reeves, C.A., Castillo-Carranza, D.L., Sengupta, U., Clos, A.L., Jackson, G.R., et al., 2011. Tau oligomers impair memory and induce synaptic and mitochondrial dysfunction in wild-type mice. Mol. Neurodegener. 6, 39.

Lasagna-Reeves, C.A., Castillo-Carranza, D.L., Sengupta, U., Sarmiento, J., Troncoso, J., et al., 2012a. Identification of oligomers at early stages of tau aggregation in Alzheimer's disease. FASEB J. 26, 1946–1959.

Lasagna-Reeves, C.A., Castillo-Carranza, D.L., Sengupta, U., Guerrero-Munoz, M.J., Kiritoshi, T., et al., 2012b. Alzheimer brain-derived tau oligomers propagate pathology from endogenous tau. Sci. Rep. 2, 700.

Leduc, V., Jasmin-Bélanger, S., Poirier, J., 2010. APOE and cholesterol homeostasis in Alzheimer's disease. Trends Mol. Med. 16, 469–477.

Lee, V.M., Goedert, M., Trojanowski, J.Q., 2001. Neurodegenerative tauopathies. Annu. Rev. Neurosci. 24, 1121–1159.

Llorens-Martín, M., Jurado, J., Hernández, F., Avila, J., 2014. GSK-3β, a pivotal kinase in Alzheimer disease. Front. Mol. Neurosci. 7, 46.

Luine, V., Frankfurt, M., 2012. Interactions between estradiol, BDNF and dendritic spines in promoting memory. Neuroscience 239, 34–45.

Mapstone, M., Cheema, A.K., Fiandaca, M.S., Zhong, X., Mhyre, T.R., et al., 2014. Plasma phospholipids identify antecedent memory impairment in older adults. Nat. Med. 20, 415–418.

Matrone, C., Djelloul, M., Taglialatela, G., Perrone, L., 2015. Inflammatory risk factors and pathologies promoting Alzheimer's disease progression: is RAGE the key? Histol. Histopathol. 30, 125–139.

Maulik, M., Westaway, D., Jhamandas, J.H., Kar, S., 2013. Role of cholesterol in APP metabolism and its significance in Alzheimer's disease pathogenesis. Mol. Neurobiol. 47, 37–63.

Maurer, M.H., 2010. Proteomics of brain extracellular fluid (ECF) and cerebrospinal fluid (CSF). Mass Spectrom. Rev. 29, 17–28.

Miller, E.C., Teravskis, P.J., Dummer, B.W., Zhao, X., Huganir, R.L., et al., 2014. Tau phosphorylation and tau mislocalization mediate soluble Abeta oligomer-induced AMPA glutamate receptor signaling deficits. Eur. J. Neurosci. 39, 1214–1224.

Min, S.W., Chen, X., Tracy, T.E., Li, Y., Zhou, Y., et al., 2015. Critical role of acetylation in tau-mediated neurodegeneration and cognitive deficits. Nat. Med. 21, 1154–1162.

Minter, M.R., Taylor, J.M., Crack, P.J., 2016. The contribution of neuro-inflammation to amyloid toxicity in Alzheimer's disease. J. Neurochem. 136, 457–474.

Morgan, D., Diamond, D.M., Gottschall, P.E., Ugen, K.E., Dickey, C., et al., 2000. A beta peptide vaccination prevents memory loss in an animal model of Alzheimer's disease. Nature 408, 982–985.

Morris, M., Knudsen, G.M., Maeda, S., Trinidad, J.C., Ioanoviciu, A., et al., 2015. Tau post-translational modifications in wild-type and human amyloid precursor protein transgenic mice. Nat. Neurosci. 18, 1183–1189.

Mucke, L., Selkoe, D.J., 2012. Neurotoxicity of amyloid β-protein: synaptic and network dysfunction. Cold Spring Harb. Perspect. Med. 2, a006338.

Mulder, M., Ravid, R., Swaab, D.F., de Kloet, E.R., Haasdijk, E.D., et al., 1998. Reduced levels of cholesterol, phospholipids, and fatty acids in cerebrospinal fluid of Alzheimer disease patients are not related to apolipoprotein E4. Alzheimer Dis. Assoc. Disord. 12, 198–203.

Muyllaert, D., Kremer, A., Jaworski, T., Borghgraef, P., Devijver, H., et al., 2008. Glycogen synthase kinase-3beta, or a link between amyloid and tau pathology? Genes Brain Behav. 7 (Suppl. 1), 57–66.

Nelson, P.T., Alafuzoff, I., Bigio, E.H., Bouras, C., Braak, H., et al., 2012. Correlation of Alzheimer disease neuropathologic changes with cognitive status: a review of the literature. J. Neuropathol. Exp. Neurol. 71, 362–381.

Nicoll, J.A., Wilkinson, D., Holmes, C., Steart, P., Markham, H., et al., 2003. Neuropathology of human Alzheimer disease after immunization with amyloid-beta peptide: a case report. Nat. Med. 9, 448–452.

Ong, W.Y., Tanaka, K., Dawe, G.S., Ittner, L.M., Farooqui, A.A., 2013. Slow excitotoxicity in Alzheimer's disease. J. Alzheimers Dis. 35, 643–668.

Park, S.Y., Ferreira, A., 2005. The generation of a 17 kDa neurotoxic fragment: an alternative mechanism by which tau mediates beta-amyloid-induced neurodegeneration. J. Neurosci. 25, 5365–5375.

Pedrós, I., Petrov, D., Allgaier, M., Sureda, F., Barroso, E., et al., 2014. Early alterations in energy metabolism in the hippocampus of APPswe/PS1dE9 mouse model of Alzheimer's disease. Biochim. Biophys. Acta 1842, 1556–1566.

Pei, J.J., Hugon, J., 2008. mTOR-dependent signalling in Alzheimer's disease. J. Cell. Mol. Med. 12, 2525–2532.

Panza, F., Frisardi, V., Imbimbo, B.P., Seripa, D., Paris, F., et al., 2011. Anti-β-amyloid immunotherapy for Alzheimer's disease: focus on bapineuzumab. Curr. Alzheimer Res. 8, 808–817.

Patel, S., Shah, R.J., Coleman, P., Sabbagh, M., 2011. Potential peripheral biomarkers for the diagnosis of Alzheimer's disease. Int. J. Alzheimer's Dis. 2011, 9.

Penzes, P., Cahill, M.E., Jones, K.A., VanLeeuwen, J.-E., Woolfrey, K.M., 2011. Dendritic spine pathology in neuropsychiatric disorders. Nat. Neurosci. 14, 285–293.

Posse de Chaves, E.I., 2012. Reciprocal regulation of cholesterol and beta amyloid at the subcellular level in Alzheimer's disease. Can. J. Physiol. Pharmacol. 90, 753–764.

Puzzo, D., Arancio, O., 2013. Amyloid-β peptide: Dr. Jekyll or Mr. Hyde? J. Alzheimers Dis. 33 (Suppl. 1), S111–S120.

Puzzo, D., Gulisano, W., Arancio, O., Palmeri, A., 2015. The keystone of Alzheimer pathogenesis might be sought in Aβ physiology. Neuroscience 307, 26–36.

Rahman, A., Akterin, S., Flores-Morales, A., Crisby, M., Kivipelto, M., et al., 2005. High cholesterol diet induces tau hyperphosphorylation in apolipoprotein E deficient mice. FEBS Lett. 579, 6411–6416.

Ramasamy, R., Yan, S.F., Schmidt, A.M., 2009. RAGE: therapeutic target and biomarker of the inflammatory response-the evidence mounts. J. Leukoc. Biol. 86, 505–512.

Ray, S., Britschgi, M., Herbert, C., et al., 2007. Classification and prediction of clinical Alzheimer's diagnosis based on plasma signaling proteins. Nat. Med. 13, 1359–1362.

Reed-Geaghan, E.G., Savage, J.C., Hise, A.G., Landreth, G.E., 2009. CD14 and toll-like receptors 2 and 4 are required for fibrillar Abeta-stimulated microglial activation. J. Neurosci. 29, 11982–11992.

Robakis, N.K., 2011. Mechanisms of AD neurodegeneration may be independent of Abeta and its derivatives. Neurobiol. Aging 32, 372–379.

Rogers, J., Cooper, N.R., Webster, S., Schultz, J., McGeer, P.L., et al., 1992. Complement activation by beta-amyloid in Alzheimer disease. Proc. Natl. Acad. Sci. USA 89, 10016–10020.

Roher, A.E., Lowenson, J.D., Clarke, S., Woods, A.S., Cotter, R.J., Gowing, E., Ball, M.L., 1993. β-Amyloid-(1–42) is a major component of cerebrovascular amyloid deposits: implications for the pathology of Alzheimer disease. Proc. Natl. Acad. Sci. USA 90, 10836–10840.

Rossner, S., Apelt, J., Schliebs, R., Perez-Polo, J.R., Bigl, V., 2001. Neuronal and glial β-secretase (BACE) protein expression in transgenic Tg2576 mice with amyloid plaque pathology. J. Neurosci. Res. 64, 437–446.

Rossner, S., Lange-Dohna, C., Zeitschel, U., Perez-Polo, J.R., 2005. Alzheimer's disease β-secretase BACE1 is not a neuron-specific enzyme. J. Neurochem. 92, 226–234.

Salomone, S., Caraci, F., Leggio, G.M., Fedotova, J., et al., 2012. New pharmacological strategies for treatment of Alzheimer's disease: focus on disease modifying drugs. Br. J. Clin. Pharmacol. 73, 504–517.

Santacruz, K., Lewis, J., Spires, T., Paulson, J., Kotilinek, L., et al., 2005. Tau suppression in a neurodegenerative mouse model improves memory function. Science 309, 476–481.

Santos, C.R.A., Cardoso, I., Goncalves, I., 2011. Key enzymes and proteins in amyloid-β production and clearance. In: de la Monte, S. (Ed.), Alzheimer's Disease Pathogenesis—Core Concepts, Shifting Paradigms and Therapeutic Targets. In Tech, Shanghai, China, pp. 53–86.

Savage, M.J., Kalinina, J., Wolfe, A., Tugusheva, K., Korn, R., et al., 2014. A sensitive Aβ oligomer assay discriminates Alzheimer's and aged control cerebrospinal fluid. J. Neurosci. 34, 2884–2897.

Scharfman, H.E., Chao, M.V., 2013. The entorhinal cortex and neurotrophin signaling in Alzheimer's disease and other disorders. Cogn. Neurosci. 4, 123–135.

Schreitmüller, B., Leyhe, T., Stransky, E., Köhler, N., Laske, C., 2012. Elevated angiopoietin-1 serum levels in patients with Alzheimer's disease. Int. J. Alzheimer's Dis. 2012, 5.

Selkoe, D.J., 2008. Soluble oligomers of the amyloid β-protein impair synaptic plasticity and behavior. Behav. Brain Res. 192, 106–113.

Sheng, Z.H., 2014. Mitochondrial trafficking and anchoring in neurons: new insight and implications. J. Cell Biol. 204, 1087–1098.

Shinohara, M., Sato, N., Shimamura, M., Kurinami, H., Hamasaki, T., et al., 2014. Possible modification of Alzheimer's disease by statins in midlife: interactions with genetic and non-genetic risk factors. Front. Aging Neurosci. 6, 71.

Shobab, L.A., Hsiung, G.Y., Feldman, H.H., 2005. Cholesterol in Alzheimer's disease. Lancet Neurol. 4, 841–852.

Sofroniew, M.V., 2009. Molecular dissection of reactive astrogliosis and glial scar formation. Trends Neurosci. 32, 638–647.

Soscia, S.J., Kirby, J.E., Washicosky, K.J., Tucker, S.M., Ingelsson, M., et al., 2010. The Alzheimer's disease-associated amyloid beta-protein is an antimicrobial peptide. PLoS One 5, e9505.

Sperling, R.A., Aisen, P.S., Beckett, L.A., Bennett, D.A., Craft, S., et al., 2011. Toward defining the preclinical stages of Alzheimer's disease: recommendations from the National Institute on Aging-Alzheimer's Association workgroups on diagnostic guidelines for Alzheimer's disease. Alzheimers Dement. 7, 280–292.

Tabaton, M., Zhu, X., Perry, G., Smith, M.A., Giliberto, L., 2010. Signaling effect of amyloid-beta(42) on the processing of AbetaPP. Exp. Neurol. 221, 18–25.

Tai, H.C., Serrano-Pozo, A., Hashimoto, T., Frosch, M.P., Spires-Jones, T.L., et al., 2012. The synaptic accumulation of hyperphosphorylated tau oligomers in Alzheimer disease is associated with dysfunction of the ubiquitin-proteasome system. Am. J. Pathol. 181, 1426–1435.

Tayeb, H.O., Murray, E.D., Price, B.H., Tarazi, F.I., 2013. Bapineuzumab and solanezumab for Alzheimer's disease: is the 'amyloid cascade hypothesis' still alive? Expert Opin. Biol. Ther. 13, 1075–1084.

Tekirian, T.L., Yang, A.Y., Glabe, C., Geddes, J.W., 1999. Toxicity of pyroglutaminated amyloid β-peptides 3(pE)-40 and -42 is similar to that of Aβ1-40 and -42. J. Neurochem. 73, 1584–1589.

Tepper, K., Biernat, J., Kumar, S., Wegmann, S., Timm, T., et al., 2014. Oligomer formation of tau protein hyperphosphorylated in cells. J. Biol. Chem. 289, 34389–34407.

Tu, S., Okamoto, S., Lipton, S.A., Xu, H., 2014. Oligomeric Aβ-induced synaptic dysfunction in Alzheimer's disease. Mol. Neurodegener. 9, 48.

Um, J.W., Nygaard, H.B., Heiss, J.K., Kostylev, M.A., Stagi, M., et al., 2012. Alzheimer amyloid-β oligomer bound to postsynaptic prion protein activates Fyn to impair neurons. Nat. Neurosci. 15, 1227–1235.

Vana, L., Kanaan, N.M., Ugwu, I.C., Wuu, J., Mufson, E.J., et al., 2011. Progression of tau pathology in cholinergic basal forebrain neurons in mild cognitive impairment and Alzheimer's disease. Am. J. Pathol. 179, 2533–2550.

van Duinen, S.G., Castano, E.M., Prelli, F., Bots, G.T., Luyendijk, W., et al., 1987. Hereditary cerebral hemorrhage with amyloidosis in patients of Dutch origin is related to Alzheimer disease. Proc. Natl. Acad. Sci. USA 84, 5991–5994.

Verkhratsky, A., Zorec, R., Rodriguez, J.J., Parpura, V., 2016. Pathobiology of neurodegeneration: the role of astroglia. Opera Med. Physiol. 1, 13–22.

Viola, K.L., Klein, W.L., 2015. Amyloid β oligomers in Alzheimer's disease pathogenesis, treatment, and diagnosis. Acta Neuropathol. 129, 183–206.

Vossel, K.A., Zhang, K., Brodbeck, J., Daub, A.C., Sharma, P., et al., 2010. Tau reduction prevents Aβ-induced defects in axonal transport. Science 330, 198.

Walker, D., Lue, L.F., Paul, G., Patel, A., Sabbagh, M.N., 2015. Receptor for advanced glycation endproduct modulators: a new therapeutic target in Alzheimer's disease. Expert Opin. Investig. Drugs 24, 393–399.

Wang, J., Dickson, D.W., Trojanowski, J.Q., Lee, V.M., 1999. The levels of soluble versus insoluble brain Abeta distinguish Alzheimer's disease from normal and pathologic aging. Exp. Neurol. 158, 328–337.

Wiessner, C., Wiederhold, K.H., Tissot, A.C., Frey, P., Danner, S., et al., 2011. The second-generation active Aβ immunotherapy CAD106 reduces amyloid accumulation in APP transgenic mice while minimizing potential side effects. J. Neurosci. 31, 9323–9331.

Wood, W.G., Li, L., Müller, W.E., Eckert, G.P., 2014. Cholesterol as a causative factor in Alzheimer's disease: a debatable hypothesis. J. Neurochem. 129, 559–572.

Wood, W.G., Igbavboa, U., Eckert, G.P., Johnson-Anuna, L.N., Müller, W.E., 2005. Is hypercholesterolemia a risk factor for Alzheimer's disease? Mol. Neurobiol. 31, 185–192.

World Alzheimer Report, September 2013. An analysis of long-term care for dementia. Alzheimer's Disease International (ADI), London.

World Alzheimer Report, September 2014. An analysis of long-term care for dementia. Alzheimer's Disease International (ADI), London.

World Health Organization, 2012. Dementia Fact Sheet No. 362 April 2012. Available at: http://www.who.int/mediacentre/factsheets/fs362/en/.

World Population Review, 2013. Available at: http://worldpopulationreview.com/continents/world-population/.

Yamada, M., Naiki, H., 2012. Cerebral amyloid angiopathy. Prog. Mol. Biol. Transl. Sci. 107, 41–78.

Yeh, C.Y., Vadhwana, B., Verkhratsky, A., Rodriguez, J.J., 2011. Early astrocytic atrophy in the entorhinal cortex of a triple transgenic animal model of Alzheimer's disease. ASN Neuro 3, 271–279.

Zhao, W.Q., de Felice, F.G., Fernandez, S., Chen, H., et al., 2008. Amyloid β oligomers induce impairment of neuronal insulin receptors. FASEB J. 22, 246–260.

Zempel, H., Mandelkow, E.M., 2012. Linking amyloid-beta and tau: amyloid-beta induced synaptic dysfunction via local wreckage of the neuronal cytoskeleton. Neurodegener. Dis. 10, 64–72.

Zempel, H., Thies, E., Mandelkow, E., Mandelkow, E.M., 2010. Abeta oligomers cause localized Ca^{2+} elevation, missorting of endogenous Tau into dendrites, Tau phosphorylation, and destruction of microtubules and spines. J. Neurosci. 30, 11938–11950.

Zhao, W.Q., Alkon, D.L., 2001. Role of insulin and insulin receptor in learning and memory. Mol. Cell. Endocrinol. 177, 125–134.

Zhou, Y., Liu, L., Hao, Y., Xu, M., 2016. Detection of Aβ monomers and oligomers: early diagnosis of Alzheimer's disease. Chem. Asian J. 11, 805–817.

Index

Printed in the United States
By Bookmasters